Graduate Texts in Mathematics
45

M. Loève

Probability Theory I

4th Edition

Springer-Verlag

New York   Heidelberg   Berlin

M. Loève
Departments of Mathematics and Statistics
University of California at Berkeley
Berkeley, California 94720

Editorial Board

AMS Subject Classifications
28–01, 60A05, 60Bxx, 60E05, 60Fxx

Library of Congress Cataloging in Publication Data

Loève, Michel, 1907–
    Probability theory.

    (Graduate texts in mathematics; 45)
    Bibliography p.
    Includes index.
    1. Probabilities. I. Title. II. Series.
QA273.L63 1977   519.2   76–28332

Originally published in the University Series in Higher Mathematics
(D. Van Nostrand Company); edited by M. H. Stone, L. Nirenberg, and
S. S. Chern.

Printed in the United States of America.

ISBN 0–387–90210–4   Springer-Verlag New York

ISBN 3–540–90210–4   Springer-Verlag Berlin Heidelberg

To LINE

and

To the students and teachers
of the School in
the Camp de Drancy

PREFACE TO THE FOURTH EDITION

This fourth edition contains several additions. The main ones concern three closely related topics: Brownian motion, functional limit distributions, and random walks. Besides the power and ingenuity of their methods and the depth and beauty of their results, their importance is fast growing in Analysis as well as in theoretical and applied Probability.

These additions increased the book to an unwieldy size and it had to be split into two volumes.

About half of the first volume is devoted to an elementary introduction, then to mathematical foundations and basic probability concepts and tools. The second half is devoted to a detailed study of Independence which played and continues to play a central role both by itself and as a catalyst.

The main additions consist of a section on convergence of probabilities on metric spaces and a chapter whose first section on domains of attraction completes the study of the Central limit problem, while the second one is devoted to random walks.

About a third of the second volume is devoted to conditioning and properties of sequences of various types of dependence. The other two thirds are devoted to random functions; the last Part on Elements of random analysis is more sophisticated.

The main addition consists of a chapter on Brownian motion and limit distributions.

It is strongly recommended that the reader begin with less involved portions. In particular, the starred ones ought to be left out until they are needed or unless the reader is especially interested in them.

I take this opportunity to thank Mrs. Rubalcava for her beautiful typing of all the editions since the inception of the book. I also wish to thank the editors of Springer-Verlag, New York, for their patience and care.

M. L.

January, 1977
Berkeley, California

PREFACE TO THE THIRD EDITION

This book is intended as a text for graduate students and as a reference for workers in Probability and Statistics. The prerequisite is honest calculus. The material covered in Parts Two to Five inclusive requires about three to four semesters of graduate study. The introductory part may serve as a text for an undergraduate course in elementary probability theory.

The Foundations are presented in:

the Introductory Part on the background of the concepts and problems, treated without advanced mathematical tools;

Part One on the Notions of Measure Theory that every probabilist and statistician requires;

Part Two on General Concepts and Tools of Probability Theory.

Random sequences whose general properties are given in the Foundations are studied in:

Part Three on Independence devoted essentially to sums of independent random variables and their limit properties;

Part Four on Dependence devoted to the operation of conditioning and limit properties of sums of dependent random variables. The last section introduces random functions of second order.

Random functions and processes are discussed in:

Part Five on Elements of random analysis devoted to the basic concepts of random analysis and to the martingale, decomposable, and Markov types of random functions.

Since the primary purpose of the book is didactic, methods are emphasized and the book is subdivided into:

unstarred portions, independent of the remainder; starred portions, which are more involved or more abstract;

complements and details, including illustrations and applications of the material in the text, which consist of propositions with frequent hints; most of these propositions can be found in the articles and books referred to in the Bibliography.

Also, for teaching and reference purposes, it has proved useful to name most of the results.

Numerous historical remarks about results, methods, and the evolution of various fields are an intrinsic part of the text. The purpose is purely didactic: to attract attention to the basic contributions while introducing the ideas explored. Books and memoirs of authors whose contributions are referred to and discussed are cited in the Bibliography, which parallels the text in that it is organized by parts and, within parts, by chapters. Thus the interested student can pursue his study in the original literature.

This work owes much to the reactions of the students on whom it has been tried year after year. However, the book is definitely more concise than the lectures, and the reader will have to be armed permanently with patience, pen, and calculus. Besides, in mathematics, as in any form of poetry, the reader has to be a poet *in posse*.

This third edition differs from the second (1960) in a number of places. Modifications vary all the way from a prefix ("sub" martingale in lieu of "semi"-martingale) to an entire subsection (§36.2). To preserve pagination, some additions to the text proper (especially 9, p. 656) had to be put in the Complements and Details. It is hoped that moreover most of the errors have been eliminated and that readers will be kind enough to inform the author of those which remain.

I take this opportunity to thank those whose comments and criticisms led to corrections and improvements: for the first edition, E. Barankin, S. Bochner, E. Parzen, and H. Robbins; for the second edition, Y. S. Chow, R. Cogburn, J. L. Doob, J. Feldman, B. Jamison, J. Karush, P. A. Meyer, J. W. Pratt, B. A. Sevastianov, J. W. Woll; for the third edition, S. Dharmadhikari, J. Fabius, D. Freedman, A. Maitra, U. V. Prokhorov. My warm thanks go to Cogburn, whose constant help throughout the preparation of the second edition has been invaluable. This edition has been prepared with the partial support of the Office of Naval Research and of the National Science Foundation.

M. L.

April, 1962
Berkeley, California

CONTENTS OF VOLUME I

GRADUATE TEXTS IN MATHEMATICS VOL. 45

PART THREE: INDEPENDENCE

CHAPTER V: SUMS OF INDEPENDENT RANDOM VARIABLES

CHAPTER VI: CENTRAL LIMIT PROBLEM

CONTENTS OF VOLUME II

GRADUATE TEXTS IN MATHEMATICS VOL. 46

Introductory Part

ELEMENTARY PROBABILITY THEORY

Probability theory is concerned with the mathematical analysis of the intuitive notion of "chance" or "randomness," which, like all notions, is born of experience. The quantitative idea of randomness first took form at the gaming tables, and probability theory began, with Pascal and Fermat (1654), as a theory of games of chance. Since then, the notion of chance has found its way into almost all branches of knowledge. In particular, the discovery that physical "observables," even those which describe the behavior of elementary particles, were to be considered as subject to laws of chance made an investigation of the notion of chance basic to the whole problem of rational interpretation of nature.

A theory becomes mathematical when it sets up a mathematical model of the phenomena with which it is concerned, that is, when, to describe the phenomena, it uses a collection of well-defined symbols and operations on the symbols. As the number of phenomena, together with their known properties, increases, the mathematical model evolves from the early crude notions upon which our intuition was built in the direction of higher generality and abstractness.

In this manner, the inner consistency of the model of random phenomena became doubtful, and this forced a rebuilding of the whole structure in the second quarter of this century, starting with a formulation in terms of axioms and definitions. Thus, there appeared a branch of pure mathematics—probability theory—concerned with the construction and investigation *per se* of the mathematical model of randomness.

The purpose of the Introductory Part (of which the other parts of this book are independent) is to give "intuitive meaning" to the concepts and problems of probability theory. First, by analyzing briefly

1

some ideas derived from everyday experience—especially from games of chance—we shall arrive at an elementary axiomatic setup; we leave the illustrations with coins, dice, cards, darts, etc., to the reader. Then, we shall apply this axiomatic setup to describe in a precise manner and to investigate in a rigorous fashion a few of the "intuitive notions" relative to randomness. No special tools will be needed, whereas in the nonelementary setup measure-theoretic concepts and Fourier-Stieltjes transforms play a prominent role.

I. INTUITIVE BACKGROUND

1. Events. The primary notion in the understanding of nature is that of *event*—the occurrence or nonoccurrence of a phenomenon. The abstract concept of event pertains only to its occurrence or nonoccurrence and not to its nature. This is the concept we intend to analyze. We shall denote events by A, B, C, $\cdots$ with or without affixes.

To every event A there corresponds a contrary event "not A," to be denoted by A^c; A^c occurs if, and only if, A does not occur. An event may imply another event: A *implies* B if, when A occurs, then B necessarily occurs; we write $A \subset B$. If A implies B and also B implies A, then we say that A and B are *equivalent;* we write $A = B$. The nature of two equivalent events may be different, but as long as we are concerned only with occurrence or nonoccurrence, they can and will be identified. Events are combined into new events by means of operations expressed by the terms "and," "or" and "not."

A "*and*" B is an event which occurs if, and only if, both the event A and the event B occur; we denote it by $A \cap B$ or, simply, by AB. If AB cannot occur (that is, if A occurs, then B does not occur, and if B occurs, then A does not occur), we say that the event A and the event B are *disjoint* (exclude one another, are mutually exclusive, are incompatible).

A "*or*" B is an event which occurs if, and only if, at least one of the events A, B occurs; we denote it by $A \cup B$. If, and only if, A and B are disjoint, we replace "or" by $+$. Similarly, more than two events can be combined by means of "and," "or"; we write

$$A_1 \cap A_2 \cap \cdots \cap A_n \quad \text{or} \quad A_1 A_2 \cdots A_n \quad \text{or} \quad \bigcap_{k=1}^{n} A_k,$$

$$A_1 \cup A_2 \cup \cdots \cup A_n \quad \text{or} \quad \bigcup_{k=1}^{n} A_k, \quad A_1 + A_2 + \cdots A_n \quad \text{or} \quad \sum_{k=1}^{n} A_k.$$

There are two combinations of events which can be considered as "boundary events"; they are the first and the last events—in terms of

3

implication. Events of the form $A + A^c$ can be said to represent an "always occurrence," for they can only occur. Since, whatever be the event A, the events $A + A^c$ and the events they imply are equivalent, all such events are to be identified and will be called the *sure event*, to be denoted by Ω. Similarly, events of the form AA^c and the events which imply them, which can be said to represent a "never occurrence" for they cannot occur, are to be identified, and will be called the *impossible event*, to be denoted by $\emptyset$; thus, the definition of disjoint events A and B can be written $AB = \emptyset$. The impossible and the sure events are "first" and "last" events, for, whatever be the event A, we have $\emptyset \subset A \subset \Omega$.

The interpretation of symbols $\subset$, $=$, $\cap$, $\cup$, in terms of occurrence and nonoccurrence, shows at once that

if $A \subset B$, then $B^c \subset A^c$, and conversely;

$$AB = BA, \quad A \cup B = B \cup A;$$

$$(AB)C = A(BC), \quad (A \cup B) \cup C = A \cup (B \cup C);$$

$$A(B \cup C) = AB \cup AC, \quad A \cup BC = (A \cup B)(A \cup C);$$

$$(AB)^c = A^c \cup B^c, \quad (A \cup B)^c = A^c B^c, \quad A \cup B = A + A^c B;$$

more generally

$$(\bigcap_{k=1}^{n} A_k)^c = \bigcup_{k=1}^{n} A_k{}^c, \quad (\bigcup_{k=1}^{n} A_k)^c = \bigcap_{k=1}^{n} A_k{}^c,$$

and so on.

We recognize here the rules of operations on sets. In terms of sets, Ω is the space in which lie the sets A, B, C, $\cdots$, $\emptyset$ is the *empty* set, A^c is the set *complementary* to the set A; AB is the *intersection*, $A \cup B$ is the *union* of the sets A and B, and $A \subset B$ means that A is contained in B.

In science, or, more precisely, in the investigation of "laws of nature," events are classified into conditions and outcomes of an experiment. *Conditions* of an experiment are events which are known or are made to occur. *Outcomes* of an experiment are events which *may* occur when the experiment is performed, that is, when its conditions occur. All (finite) combinations of outcomes by means of "not," "and," "or," are outcomes; in the terminology of sets, the outcomes of an experiment form a *field* (or an "algebra" of sets). The conditions of an experiment, together with its field of outcomes, constitute a *trial*. Any (finite) number of trials can be combined by "conditioning," as follows:

The collective outcomes are combinations by means of "not," "and," "or," of the outcomes of the constituent trials. The conditions are conditions of the first constituent trial together with conditions of the second to which are added the observed outcomes of the first, and so on. Thus, given the observed outcomes of the preceding trials, every constituent trial is performed under supplementary conditions: it is conditioned by the observed outcomes. When, for every constituent trial, any outcome occurs if, and only if, it occurs without such conditioning, we say that the trials are *completely independent*. If, moreover, the trials are identical, that is, have the same conditions and the same field of outcomes, we speak of *repeated trials* or, equivalently, *identical and completely independent trials*. The possibility of repeated trials is a basic assumption in science, and in games of chance: *every trial can be performed again and again, the knowledge of past and present outcomes having no influence upon future ones.*

2. Random events and trials. Science is essentially concerned with permanencies in repeated trials. For a long time *Homo sapiens* investigated *deterministic trials* only, where the conditions (causes) determine completely the outcomes (effects). Although another type of permanency has been observed in games of chance, it is only recently that *Homo sapiens* was led to think of a rational interpretation of nature in terms of these permanencies: nature plays the greatest of all games of chance with the observer. This type of permanency can be described as follows:

Let the *frequency* of an outcome A in n repeated trials be the ratio n_A/n of the number n_A of occurrences of A to the total number n of trials. If, in repeating a trial a large number of times, the observed frequencies of any one of its outcomes A cluster about some number, the trial is then said to be *random*. For example, in a game of dice (two homogeneous ones) "double-six" occurs about once in 36 times, that is, its observed frequencies cluster about 1/36. The number 1/36 is a permanent numerical property of "double-six" under the conditions of the game, and the observed frequencies are to be thought of as measurements of the property. This is analogous to stating that, say, a bar at a fixed temperature has a permanent numerical property called its "length" about which the measurements cluster.

The outcomes of a random trial are called *random* (chance) *events*. The number measured by the observed frequencies of a random event A is called the *probability* of A and is denoted by PA. Clearly, $P\emptyset = 0$,

$P\Omega = 1$ and, for every A, $0 \leqq PA \leqq 1$. Since the frequency of a sum $A_1 + A_2 + \cdots + A_n$ of disjoint random events is the sum of their frequencies, we are led to assume that

$$P(A_1 + A_2 + \cdots + A_n) = PA_1 + PA_2 + \cdots + PA_n.$$

Furthermore, let n_A, n_B, n_{AB} be the respective numbers of occurrences of outcomes A, B, AB in n repeated random trials. The frequency of outcome B in the n_A trials in which A occurs is

$$\frac{n_{AB}}{n_A} = \frac{n_{AB}}{n} : \frac{n_A}{n}$$

and measures the ratio PAB/PA, to be called probability of B *given A* (given that A occurs); we denote it by $P_A B$ and have

$$PAB = PA \cdot P_A B.$$

Thus, when to the original conditions of the trial is added the fact that A occurs, then the probability PB of B is transformed into the probability $P_A B$ of B given A. This leads to defining B as being *stochastically independent* of A if $P_A B = PB$ or

$$PAB = PA \cdot PB.$$

Then it follows that A is stochastically independent of B, for

$$P_B A = \frac{PAB}{PB} = PA,$$

and it suffices to say that *A and B are stochastically independent.* (We assumed in the foregoing ratios that the denominators were not null.)

Similarly, if a collective trial is such that the probability of any outcome of any constituent random trial is independent of the observed outcomes of preceding constituents, we say that the constituent random trials are *stochastically independent.* Clearly, complete independence defined in terms of occurrences implies stochastic independence defined in terms of probability alone. Thus, as long as we are concerned with stochastic independence only, the concept of repeated trials reduces to that of *identical and stochastically independent trials.*

3. Random variables. For a physicist, the outcomes are, in general, values of an observable. From the gambler's point of view, what counts is not the observed outcome of a random trial but the corresponding gain or loss. In either case, when there is only a finite number of possible outcomes, the sure outcome Ω is partitioned into a number of disjoint outcomes $A_1, A_2, \cdots, A_m$. The *random variable X*, say, the chance gain of the gambler, is stated by assigning to these outcomes numbers $x_{A_1}, x_{A_2}, \cdots, x_{A_m}$, which may be positive, null, or negative. The "average gain" in n repeated random trials is

$$x_{A_1} \frac{n_{A_1}}{n} + x_{A_2} \frac{n_{A_2}}{n} + \cdots + x_{A_m} \frac{n_{A_m}}{m}.$$

Since the trial is random, this average clusters about $x_{A_1} P A_1 + x_{A_2} P A_2 + \cdots + x_{A_m} P A_m$ which is defined as the *expectation EX* of the random variable X. It is easily seen that the averages of a sum of two random variables X and Y cluster about the sum of their averages, that is,

$$E(X + Y) = EX + EY.$$

The concept of random variable is more general than that of a random event. In fact, we can assign to every random event A a random variable—its indicator $I_A = 1$ or 0 according as A occurs or does not occur. Then, the observed value of I_A tells us whether or not A occurred, and conversely. Furthermore, we have $EI_A = 1 \cdot PA + 0 \cdot PA^c = PA$.

A physical observable may have an infinite number of possible values, and then the foregoing simple definitions do not apply. The evolution of probability theory is due precisely to the consideration of more and more complicated observables.

II. AXIOMS; INDEPENDENCE AND THE BERNOULLI CASE

We give now a consistent model for the intuitive concepts which appeared in the foregoing brief analysis; we shall later see that this model has to be extended.

1. Axioms of the finite case. Let Ω or the *sure event* be a space of points ω; the empty set (set containing no points ω) or the *impossible event* will be denoted by $\emptyset$. Let $\mathcal{C}$ be a nonempty class of sets in Ω, to be called *random events* or, simply, *events*, since no other type of events will be considered. Events will be denoted by capitals A, B, $\cdots$ with or without affixes. Let P or *probability* be a numerical function defined on $\mathcal{C}$; the value of P for an event A will be called the *probability of A* and will be denoted by PA. The pair $(\mathcal{C}, P)$ is called a *probability field* and the triplet $(\Omega, \mathcal{C}, P)$ is called a *probability space*.

Axiom I. $\mathcal{C}$ *is a field:* complements A^c, finite intersections $\bigcap\limits_{k=1}^{n} A_k$, and finite unions $\bigcup\limits_{k=1}^{n} A_k$ of events are events.

Axiom II. *P on $\mathcal{C}$ is normed, nonnegative, and finitely additive:*

$$P\Omega = 1, \quad PA \geqq 0, \quad P\sum_{k=1}^{n} A_k = \sum_{k=1}^{n} PA_k.$$

It suffices to assume additivity for two arbitrary disjoint events, since the general case follows by induction.

Since $\emptyset$ is disjoint from any event A and $A + \emptyset = A$, we have

$$PA = P(A + \emptyset) = PA + P\emptyset,$$

so that $P\emptyset = 0$. Furthermore, it is immediate that, if $A \subset B$, then $PA \leqq PB$, and also that

$$P\bigcup_{k=1}^{n} A_k = PA_1 + PA_1{}^c A_2 + \cdots + PA_1{}^c A_2{}^c \cdots A_{n-1}{}^c A_n \leqq \sum_{k=1}^{n} PA_k.$$

The axioms are consistent.

To see this, it suffices to construct an example in which the axioms are both verified: take as the field $\mathcal{C}$ of events Ω and $\emptyset$ only, and set $P\Omega = 1$, $P\emptyset = 0$. A less trivial example is that of a *simple probability field:* 1° The events, except $\emptyset$, are formed by all sums of disjoint events A_1, $A_2, \cdots, A_n$ which form a finite partition of the sure event: $A_1 + A_2 + \cdots + A_n = \Omega$; 2° to every event A_k of the partition is assigned a probability $p_k = PA_k$ such that every $p_k \geq 0$ and $\sum_{k=1}^{n} p_k = 1$—this is always possible. Then P is defined on $\mathcal{C}$, consistently with axiom II, by assigning to every event A as its probability the sum of probabilities of those A_k whose sum is A.

2. Simple random variables. Let the probability field ($\mathcal{C}$, P) be fixed. In order to introduce the concept of random variables, it will be convenient to begin with very special ones, which permit operations on events to be transformed into ordinary algebraic operations.

To every event A we assign a function I_A on Ω with values $I_A(\omega)$, such that $I_A(\omega) = 1$ or 0 according as ω belongs or does not belong to A; I_A will be called the *indicator* of A (in terms of occurrences, $I_A = 1$ or 0, according as A occurs or does not occur). Thus, $I_A{}^2 = I_A$ and the boundary cases are those of $I_\emptyset = 0$ and $I_\Omega = 1$ (if, in a relation containing functions of an argument, the argument does not figure, then the relation holds for all values of the argument unless otherwise stated).

The following properties are immediate:

$$\text{if} \quad A \subset B, \quad \text{then} \quad I_A \leq I_B, \quad \text{and conversely;}$$

$$\text{if} \quad A = B, \quad \text{then} \quad I_A = I_B, \quad \text{and conversely;}$$

$$I_{A^c} = 1 - I_A, \quad I_{AB} = I_A I_B, \quad I_{A+B} = I_A + I_B,$$

$$I_{A \cup B} = I_{A + A^c B} = I_A + I_B - I_{AB}$$

and, more generally,

$$I_{\bigcap_{k=1}^{n} A_k} = \prod_{k=1}^{n} I_{A_k}, \quad I_{\sum_{k=1}^{n} A_k} = \sum_{k=1}^{n} I_{A_k}$$

$$I_{\bigcup_{k=1}^{n} A_k} = I_{A_1} + (1 - I_{A_1})I_{A_2} + \cdots + (1 - I_{A_1}) \cdots (1 - I_{A_{n-1}})I_{A_n}.$$

Linear combinations $X = \sum_{j=1}^{m} x_j I_{A_j}$ of indicators of events A_j of a finite partition of Ω, where the x_j are (finite) numbers, are called *simple random variables*, to be denoted by capitals X, Y, $\cdots$, with or without affixes. By convention, every written linear combination of indicators will be that of indicators of disjoint events whose sum is the sure event; however, when $x_j = 0$, we may drop the corresponding null term $x_j I_{A_j} = 0$ from the linear combination. The set of values $P A_j$ which correspond to the values x_j of X, assumed all distinct, is called the *probability distribution* and the A_j form the *partition* of X. The *expectation EX* of a simple random variable $X = \sum\limits_{j=1}^{m} x_j I_{A_j}$ is defined by

$$EX = \sum_{j=1}^{m} x_j P A_j.$$

Clearly, any constant c is a simple random variable, and the sum or the product of two simple random variables is a simple random variable; $E(c) = c$, $EcX = cEX$; if $X \geq 0$, that is, all its values $x_j \geq 0$, then $EX \geq 0$; if $X \leq Y$, then $EX \leq EY$. Furthermore, expectations possess the following basic property.

ADDITION PROPERTY. *The expectation of a sum of (a finite number of) simple random variables is the sum of their expectations.*

It suffices to prove the assertion for a sum of two simple random variables

$$X = \sum_{j=1}^{m} x_j I_{A_j}, \quad Y = \sum_{k=1}^{n} y_k I_{B_k},$$

since the general case follows by induction. Because of the properties of probabilities and indicators given above,

$$EX + EY = \sum_{j=1}^{m} x_j P A_j + \sum_{k=1}^{n} y_k P B_k = \sum_{j=1}^{m} \sum_{k=1}^{n} (x_j + y_k) P A_j B_k$$

while

$$E(X + Y) = E \sum_{j=1}^{m} \sum_{k=1}^{n} (x_j + y_k) I_{A_j B_k}$$

$$= \sum_{j=1}^{m} \sum_{k=1}^{n} (x_j + y_k) P A_j B_k.$$

and the conclusion is reached.

Application to probabilities of combinations of events. To begin with, we observe that

$$EI_A = 1 \cdot P A + 0 \cdot P A^c = P A.$$

Therefore, from

$$I_{A \cup B} = I_A + I_B - I_{AB}$$

it follows, upon taking expectations of both sides, that

$$P(A \cup B) = PA + PB - PAB.$$

Similarly, from

$$I_{A \cup B \cup C} = I_A + (1 - I_A)I_B + (1 - I_A)(1 - I_B)I_C$$

it follows, upon expanding the right-hand side and taking expectations, that

$$P(A \cup B \cup C) = PA + PB + PC - PAB - PBC - PCA + PABC,$$

and so on.

The foregoing properties of expectations lead to the celebrated

TCHEBICHEV INEQUALITY. *If X is a simple random variable, then, for every $\epsilon > 0$,*

$$P[|X| \geq \epsilon] \leq \frac{1}{\epsilon^2} EX^2.$$

$[|X| \geq \epsilon]$ is to be read: the union of all those events for which the values of $|X|$ are $\geq \epsilon$.

The inequality follows from

$$EX^2 = E(X^2 I_{[|X| \geq \epsilon]}) + E(X^2 I_{[|X| < \epsilon]}) \geq E(X^2 I_{[|X| \geq \epsilon]}) \geq \epsilon^2 E I_{[|X| \geq \epsilon]}$$
$$= \epsilon^2 P[|X| \geq \epsilon].$$

3. Independence. Two events A_1, A_2 are said to be *stochastically independent* or, simply, *independent* (no other type of independence of events will be considered) if

$$PA_1 A_2 = PA_1 PA_2.$$

More generally, events A_k, $k = 1, 2, \cdots, n$ are *independent*, if, for every $m \leq n$ and for arbitrary distinct integers $k_1, k_2, \cdots, k_m \leq n$,

$$PA_{k_1} A_{k_2} \cdots A_{k_m} = PA_{k_1} PA_{k_2} \cdots PA_{k_m}.$$

If this property holds for all events A_k selected arbitrarily each within a different class $\mathcal{Q}_k$, we say that these classes are *independent*. Simple random variables X_k, $k = 1, 2, \cdots, n$, are said to be *independent* if the partitions on which they are defined are independent. A basic property of independent simple random variables is the following

MULTIPLICATION PROPERTY. *The expectation of a product (of a finite number) of independent simple random variables is the product of their expectations.*

It suffices to give the proof for two independent simple random variables,

$$X = \sum_{j=1}^{m} x_j I_{A_j}, \quad Y = \sum_{k=1}^{n} y_k I_{B_k}, \text{ all } x_j(y_k) \text{ distinct,}$$

since the general case follows by induction. Because of independence,

$$EXY = E \sum_{j=1}^{m} \sum_{k=1}^{n} x_j y_k I_{A_j B_k} = \sum_{j=1}^{m} \sum_{k=1}^{n} x_j y_k PA_j PB_k$$

$$= (\sum_{j=1}^{m} x_j PA_j)(\sum_{k=1}^{n} y_k PB_k) = EXEY,$$

and the conclusion is reached.

The expectation $E(X - EX)^2$, called the *variance* of X, is denoted by $\sigma^2 X$. By the additive property,

$$\sigma^2 X = E(X^2 - 2XEX + E^2 X) = EX^2 - E^2 X.$$

The celebrated Bienaymé equality follows from the additive and multiplicative properties.

BIENAYMÉ EQUALITY. *If X_k, $k = 1, 2, \cdots, n$, are independent, then*

$$\sigma^2 \sum_{k=1}^{n} X_k = \sum_{k=1}^{n} \sigma^2 X_k.$$

Since

$$E(X_k - EX_k) = EX_k - EX_k = 0$$

and independence of the X_k implies independence of the $X_k - EX_k$, it follows that

$$\sigma^2 \sum_{k=1}^{n} X_k = E (\sum_{k=1}^{n} X_k - \sum_{k=1}^{n} EX_k)^2 = E\{ \sum_{k=1}^{n} (X_k - EX_k)\}^2$$

$$= \sum_{k=1}^{n} E(X_k - EX_k)^2 + \sum_{j \neq k=1}^{n} E(X_j - EX_j)(X_k - EX_k)$$

$$= \sum_{k=1}^{n} \sigma^2 X_k + \sum_{j \neq k=1}^{n} E(X_j - EX_j)E(X_k - EX_k) = \sum_{k=1}^{n} \sigma^2 X_k.$$

Observe that we used independence of the X_k considered *two by two* only.

4. Bernoulli case. A simple case of independence has played a central role in the evolution of probability theory. This is the *Bernoulli* *case* of events A_k, $k = 1, 2, \cdots$, which are independent whatever be their total number n under consideration and such that their probabilities PA_k have the same value p.

We observe that independence of the A_k, $k = 1, 2, \cdots, n$ implies independence of the $\overline{A}_k = A_k$ or $A_k{}^c$, and, more generally, of the n fields $\mathfrak{a}_k = \{\emptyset, A_k, A_k{}^c, \Omega\}$. For example,

$$PA_{k_1}{}^c A_{k_2} \cdots A_{k_m} = PA_{k_2} A_{k_3} \cdots A_{k_m} - PA_{k_1} A_{k_2} \cdots A_{k_m}$$

$$= PA_{k_2} PA_{k_3} \cdots PA_{k_m} - PA_{k_1} PA_{k_2} \cdots PA_{k_m}$$

$$= (1 - PA_{k_1}) PA_{k_2} \cdots PA_{k_m} = PA_{k_1}{}^c PA_{k_2} \cdots PA_{k_m},$$

where the subscripts are all distinct and $\leq n$. These fields correspond to repeated random trials where an outcome A at the kth trial is represented by A_k.

The number of occurrences of outcome A in n repeated trials is represented by a simple random variable $S_n = \sum_{k=1}^{n} I_{A_k}$. To write S_n in the usual form, that is, with values assigned to events of a partition of the sure event, we observe that

$$I_{A_k} = I_{A_k} \prod_{\substack{j=1 \\ j \neq k}}^{n} (I_{A_j} + I_{A_j{}^c}).$$

It follows, upon substituting in S_n and expanding, that

$$S_n = \sum_{j=0}^{n} j I_{B_j},$$

where

$$I_{B_j} = \sum I_{A_{k_1}} \cdots I_{A_{k_j}} I_{A_{k_{j+1}}{}^c} \cdots I_{A_{k_n}{}^c}.$$

The summation is over all permutations of subscripts $k = 1, 2, \cdots, n$, classified into two groups, one having j terms and the other having $n - j$ terms.

On account of the independence, the expectations of the terms under the summation sign are

$$PA_{k_1} PA_{k_2} \cdots PA_{k_j} PA_{k_{j+1}}{}^c \cdots PA_{k_n}{}^c = p^j q^{n-j}, \quad q = 1 - p,$$

and, therefore, the probability of j occurrences in n trials is given by

$$P[S_n = j] = PB_j = \frac{n!}{i!(n-j)!} p^j q^{n-j}, \quad j = 0, 1, \cdots, n.$$

yields nothing. A first and obvious way to pass from the finite to the infinite is to extrapolate, that is, to postulate that properties of the finite case continue to hold in the infinite case. Yet these extrapolations have to be meaningful and consistent.

In set theory, intersections $\bigcap\limits_{n=1}^{\infty} A_n$ and unions $\bigcup\limits_{n=1}^{\infty} A_n$ of sets A_n, where n runs over the denumerable set of integers, continue to be defined as the sets of points which belong to every A_n and to at least one A_n, respectively. We still have that

$$(\bigcap_{n=1}^{\infty} A_n)^c = \bigcup_{n=1}^{\infty} A_n{}^c, \quad (\bigcup_{n=1}^{\infty} A_n)^c = \bigcap_{n=1}^{\infty} A_n{}^c,$$

$$\bigcup_{n=1}^{\infty} A_n = A_1 + A_1{}^c A_2 + A_1{}^c A_2{}^c A_3 + \cdots \quad \text{ad infinitum}$$

and, correspondingly,

$$I_{\bigcap\limits_{n=1}^{\infty} A_n} = \prod_{n=1}^{\infty} I_{A_n}, \quad I_{\sum\limits_{n=1}^{\infty} A_n} = \sum_{n=1}^{\infty} I_{A_n}$$

$$I_{\bigcup\limits_{n=1}^{\infty} A_n} = I_{A_1} + I_{A_1{}^c} I_{A_2} + I_{A_1{}^c} I_{A_2{}^c} I_{A_3} + \cdots.$$

If we want all *countable* (finite or denumerable) combinations of events by means of "not," "and," "or," to be events and their probabilities to be defined, then axioms I and II become

AxIOM I'. *Events form a σ-field* α: Complements A^c, countable intersections $\bigcap_{j} A_j$, and countable unions $\bigcup_{j} A_j$ of events are events.

AxIOM II'. *Probability P on* α *is normed, nonnegative, and σ-additive:*

$$P\Omega = 1, \quad PA \geqq 0, \quad P\sum_{j} A_j = \sum_{j} PA_j.$$

It follows that

CovERING RULE: $P\bigcup_{j} A_j = PA_1 + PA_1{}^c A_2 + PA_1{}^c A_2{}^c A_3 + \cdots$

$\leqq \sum_{j} PA_j.$

These axioms are consistent, since the examples constructed for the finite case continue to apply trivially. A nontrivial example in the infinite case is that of nonsimple *elementary probability fields*: 1° The events, except $\emptyset$, are formed by all countable sums of events A_n which form a denumerable partition of the sure event: $\sum_{n=1}^{\infty} A_n = \Omega$; 2° to every event A_n of the partition is assigned probability $p_n = PA_n$ such that every $p_n \geq 0$ and $\sum_{n=1}^{\infty} p_n = 1$—this is always possible. Then P is defined on $\mathcal{C}$, consistently with axiom II', by assigning to every event A as its probability the sum (finite sum or convergent series) of probabilities of those A_n whose sum is A.

6. Elementary random variables. A linear combination $X = \sum_j x_j I_{A_j}$ of a countable number of indicators of disjoint events A_j is an *elementary random variable* X; if j varies over a finite set, then X reduces to a simple random variable. Clearly a sum or a product of two elementary random variables is an elementary random variable. We may still try to define the expectation EX by

$$EX = \sum_j x_j PA_j.$$

But, if the sum is a divergent series, it has no content or is infinite. Furthermore, even if it is a convergent series, it may not be absolutely convergent, so that by changing the order of terms we can change its value, and the expectation is no longer well defined if no ordering is specified; this is undesirable according to the very meaning of an expectation. We are therefore led to define EX by the foregoing expression *only when the right-hand side is absolutely convergent*, so that

if EX exists and is finite, then $E|X|$ exists and is finite; and conversely.

(We recognize here an integrable elementary function in the sense of Lebesgue with respect to the measure P.)

The argument used to prove the addition property of simple random variables continues to apply to finite sums of elementary random variables whose expectations exist and are finite, provided σ-additivity of P is used. We obtain:

If the expectations of a finite number of elementary random variables exist and are finite, then the expectation of their sum exists and is finite and is the sum of their expectations.

Also, Tchebichev's inequality remains valid, provided its right-hand side exists and is finite.

Independence of a countable number of events A_j, or σ-fields $\mathcal{C}_j$ contained in $\mathcal{C}$, is defined to be independence of every finite number of these events, or σ-fields. Independence of a countable number of elementary random variables $X_k = \sum_j x_{j_k} I_{A_{jk}}$ is defined to be independence of every finite number of events $A_{j_k,k}$ as k varies. The argument used to prove the multiplication property yields:

If the expectations of a finite number of independent elementary random variables exist and are finite, then the expectation of their product exists and is the finite product of their expectations.

Also, Bienaymé's equality remains valid, provided its right-hand side exists and is finite.

In the Bernoulli law of large numbers only simple random variables figure and only finite additivity of the probability P is used, so that nothing is to be changed. However, now we can introduce probabilities of denumerable combinations of events and use the supplementary requirement that the additive property of P remains valid for denumerable sums. Therefore, in the present setup we can expect a more precise interpretation of the "clustering of frequencies." This is the celebrated Borel strong law of large numbers derived below.

Let $X_1, X_2, \cdots$ be a sequence of elementary random variables. We investigate the convergence to 0 of the sequence; the limits are taken as $n \to \infty$. It will be more convenient to consider the contrary case—X_n does not converge to 0 or, equivalently, there exists at least one integer m such that to every integer n there corresponds at least one integer ν for which $|X_{n+\nu}| \geq \dfrac{1}{m}$. Since "at least one" corresponds to "$\bigcup$" while "every" corresponds to "$\bigcap$," we can write

$$[X_n \nrightarrow 0] = \bigcup_{m=1}^{\infty} \bigcap_{n=1}^{\infty} \bigcup_{\nu=1}^{\infty} \left[|X_{n+\nu}| \geq \frac{1}{m} \right];$$

the right-hand side is an event. Thus, the condition $X_n \nrightarrow 0$ determines the event $[X_n \nrightarrow 0]$, the contrary condition $X_n \to 0$ determines the complementary event $[X_n \to 0]$, and the probabilities of these two events add up to 1.

We are interested in $X_n \to 0$ with probability 1 or, equivalently, $X_n \nrightarrow 0$ with probability 0, and require the following proposition.

If, for every integer m, $\displaystyle\sum_{n=1}^{\infty} P\left[|X_n| \geq \frac{1}{m} \right] < \infty$, then $P[X_n \nrightarrow 0] = 0$.

We set $A_{nm} = \displaystyle\bigcup_{\nu=1}^{\infty} \left[|X_{n+\nu}| \geq \frac{1}{m} \right]$ and $A_m = \displaystyle\bigcap_{n=1}^{\infty} A_{nm}$ and observe that, by the covering rule and the hypothesis, for every m,

$$PA_{nm} = P \bigcup_{k=n+1}^{\infty} \left[|X_k| \geq \frac{1}{m} \right]$$

$$\leq \sum_{k=n+1}^{\infty} P\left[|X_k| \geq \frac{1}{m} \right] \to 0 \quad \text{as} \quad n \to \infty.$$

Since whatever be n',

$$PA_m = P \bigcap_{n=1}^{\infty} A_{nm} \leq PA_{n'm}.$$

it follows upon letting $n' \to \infty$ that $PA_m = 0$. Therefore, by the covering rule

$$P[X_n \nrightarrow 0] = P \bigcup_{m=1}^{\infty} A_m \leq \sum_{m=1}^{\infty} PA_m = 0$$

and the proposition is proved.

We can now pass to

BOREL'S STRONG LAW OF LARGE NUMBERS (1909). *In the Bernoulli case*

$$P\left[\frac{S_n}{n} \to p \right] = 1.$$

We recall that in the Bernoulli case

$$X_n = \frac{S_n}{n} = \frac{1}{n} \sum_{j=1}^{n} I_{A_j}$$

where the A_j are independent events of common probability p whatever be n, and $EX_n = p$, $\sigma^2 X_n = pq/n$ (observe that only independence two by two is used). Since for every m

$$\sum_{k=1}^{\infty} P\left[|X_{k^2} - p| \geq \frac{1}{m} \right] \leq m^2 pq \sum_{k=1}^{\infty} \frac{1}{k^2} < \infty,$$

it follows by the foregoing proposition that $X_{k^2} \to p$ with probability 1 as $k \to \infty$. But to every n there corresponds an integer $k = k(n)$ with $k^2 \leq n < (k+1)^2$; hence $0 \leq n - k^2 \leq 2k$ and $n \to \infty$ implies $k \to \infty$.

Since

$$| X_n - X_{k^2} | = \left| \left(\frac{1}{n} - \frac{1}{k^2}\right) \sum_{j=1}^{k^2} I_{A_j} + \frac{1}{n} \sum_{j=k^2+1}^{n} I_{A_j} \right|$$

$$\leqq \frac{(n - k^2)k^2}{nk^2} + \frac{n - k^2}{n} \leqq \frac{4}{k}$$

so that

$$| X_n - p | \leqq | X_n - X_{k^2} | + | X_{k^2} - p | \leqq \frac{4}{k} + | X_{k^2} - p |,$$

it follows that $X_n \to p$ with probability 1 as $n \to \infty$, and Borel's result is proved.

Application. Let X be an elementary random variable. We set $F(x - 0) = F(x) = P[X < x]$, $F(x + 0) = P[X \leqq x]$ so that $P[X = x] = F(x + 0) - F(x)$. The function F so defined determines the probability distribution of X, that is, the probabilities of all values of X; it is called the *distribution function* of X. We organize repeated independent trials where we observe the values of X; in other words, we consider independent random variables X_1, X_2, $\cdots$ with the same probability distribution as X.

If k is the number of values observed in n of those trials and which are less than x or, equivalently, if k is the number of independent events $[X_1 < x]$, $[X_2 < x]$, $\cdots$ $[X_n < x]$ (with common probability $p = F(x)$) which occur, we set $F_n(x - 0) = F_n(x) = k/n$. Thus, $F_n(x)$ is a random variable with

$$P\left[F_n(x) = \frac{k}{n} \right] = \frac{n!}{k!(n - k)!} \{F(x)\}^k \{1 - F(x)\}^{n-k}.$$

The function F_n is called *empirical distribution function* of X in n trials. According to Borel's strong law of large numbers, this frequency $F_n(x)$ of occurrences of the outcome $[X < x]$ converges to $F(x)$ with probability 1. In other words, the observations permit us to find with probability 1 every value $F(x)$ of the distribution function of X. In fact, Borel's result yields more (Glivenko-Cantelli):

CENTRAL STATISTICAL THEOREM. *If F is the distribution function of a random variable X and F_n is the empirical distribution function of X in n independent and identical trials, then*

$$P[\sup_{-\infty < x < +\infty} | F_n(x) - F(x) | \to 0] = 1.$$

In other words, with probability 1, $F_n(x) \to F(x)$ uniformly in x.

Let x_{jk} be the smallest value x such that

$$F(x) \leq \frac{j}{k} \leq F(x + 0).$$

Since the frequency of the event $[X < x_{jk}]$ is $F_n(x_{jk})$ and its probability is $F(x_{jk})$, it follows by Borel's result that $PA'_{jk} = 1$ where $A'_{jk} = [F_n(x_{jk}) \rightarrow F(x_{jk})]$. Similarly, $PA''_{jk} = 1$ where $A''_{jk} = [F_n(x_{jk} + 0) \rightarrow F(x_{jk} + 0)]$. Let $A_{jk} = A'_{jk}A''_{jk}$ and let $\theta = \pm 0$

$$A_k = \bigcap_{j=1}^{k} A_{jk} = \lceil \sup_{1 \leq j \leq k} | F_n(x_{jk} + \theta) - F(x_{jk} + \theta) | \rightarrow 0].$$

By the covering rule and by what precedes

$$PA_k{}^c = P \bigcup_{i=1}^{k} A_{jk}{}^c \leq \sum_{j=1}^{k} PA_{jk}{}^c = 0$$

and, hence, $PA_k = 1$. Upon setting $A = \bigcap_{k=1}^{\infty} A_k$, it follows similarly that $PA = 1$.

On the other hand, for every x between x_{jk} and $x_{j+1,k}$

$$F(x_{jk} + 0) \leq F(x) \leq F(x_{j+1,k}), \quad F_n(x_{jk} + 0) \leq F_n(x) \leq F_n(x_{j+1,k})$$

while for every x_{jk}

$$0 \leq F(x_{j+1,k}) - F(x_{jk} + 0) \leq \frac{1}{k}.$$

Therefore,

$$F_n(x) - F(x) \leq F_n(x_{j+1,k}) - F(x_{jk} + 0) \leq F_n(x_{j+1,k}) - F(x_{j+1,k}) + \frac{1}{k}$$

and

$$F_n(x) - F(x) \geq F_n(x_{jk} + 0) - F(x_{j+1,k})$$

$$\geq F_n(x_{jk} + 0) - F(x_{jk} + 0) - \frac{1}{k}.$$

It follows that, whatever be x and k,

$$| F_n(x) - F(x) | \leq \sup_{1 \leq j \leq k} | F_n(x_{jk} + \theta) - F(x_{jk} + \theta) | + \frac{1}{k}$$

or

$$\Delta_n = \sup_{-\infty < x < +\infty} | F_n(x) - F(x) | \leq \sup_{1 \leq j \leq k} | F_n(x_{jk} + \theta) - F(x_{jk} + \theta) | + \frac{1}{k}.$$

Hence $P[\Delta_n \rightarrow 0] \geq PA = 1$, and the theorem is proved.

*REMARK. The foregoing proof and hence the theorem remain valid when the random variable X is not elementary.

7. Need for nonelementary random variables. The sophisticated mathematician prefers to work with "closed" models—such that the operations defined for the entities within the model yield only entities within the model. While elementary random variables can be obtained as limits of sequences of simple random variables, *all* limits of sequences of simple and, more generally, of elementary random variables are not necessarily elementary—families of elementary random variables are not necessarily closed under passages to the limit. If this closure is required, then the concept of a random variable has to be extended so as to include "measurable functions." This will be done in the following parts. In fact, the need for further expansion of the model in order to include random variables with a noncountable set of values appeared quite early in the development of probability theory, once more in connection with the Bernoulli case. This is the celebrated (as the reader observes, all results obtained in or used for the Bernoulli case are "celebrated")

DE MOIVRE-LAPLACE THEOREM. *In the Bernoulli case with $p > 0$, $q = 1 - p > 0$, as $n \to \infty$,*

de Moivre (1732):

$$P_n(x) = P[S_n = j] \sim \frac{1}{\sqrt{2\pi npq}} e^{-x^2/2}, \quad x = \frac{j - np}{\sqrt{npq}},$$

uniformly on every finite interval $[a, b]$ of values of x;

Laplace (1801):

$$P\left[a \leqq \frac{S_n - np}{\sqrt{npq}} \leqq b \right] \to \frac{1}{\sqrt{2\pi}} \int_a^b e^{-x^2/2} \, dx.$$

The relation $a_n \sim b_n$ means that $a_n/b_n \to 1$. The integer j varies with n, so that $x = x(n)$ remains within a fixed finite interval $[a, b]$ and

$$j = np + x\sqrt{npq} \to \infty, \quad k = n - j = nq - x\sqrt{npq} \to \infty.$$

We apply Stirling's formula

$$m! = \sqrt{2\pi m} \cdot m^m e^{-m} e^{\theta_m}, \quad 0 < \theta_m < \frac{1}{12m}$$

to the binomial probabilities $P_n(x) = \frac{n!}{i!k!} p^j q^k$. Thus

$$P_n(x) = \frac{\sqrt{2\pi n} \cdot n^n e^{-n}}{\sqrt{2\pi j} \cdot j^j e^{-j} \sqrt{2\pi k} \cdot k^k e^{-k}} \, p^j q^k e^{\theta_n - \theta_j - \theta_k}$$

$$= \frac{1}{\sqrt{2\pi}} \sqrt{\frac{n}{jk}} \left(\frac{np}{j}\right)^j \left(\frac{nq}{k}\right)^k e^{\theta}$$

where, uniformly on $[a, b]$,

$$|\theta| < \frac{1}{12}\left(\frac{1}{n} + \frac{1}{j} + \frac{1}{k}\right)$$

and

$$\frac{jk}{n} = n\left(p + x\sqrt{\frac{pq}{n}}\right)\left(q - x\sqrt{\frac{pq}{n}}\right) \sim npq.$$

Therefore, uniformly on $[a, b]$,

$$P_n(x) \sim \frac{1}{\sqrt{2\pi npq}} \left(\frac{np}{j}\right)^j \left(\frac{nq}{k}\right)^k$$

and

$$\log\left(\frac{np}{j}\right)^j \left(\frac{nq}{k}\right)^k = -(np + x\sqrt{npq})\left[x\sqrt{\frac{q}{np}} - \frac{1}{2}\frac{qx^2}{np} + O\left(\frac{1}{n^{3/2}}\right)\right]$$

$$- (nq - x\sqrt{npq})\left[-x\sqrt{\frac{p}{nq}} - \frac{1}{2}\frac{px^2}{nq} + O\left(\frac{1}{n^{3/2}}\right)\right]$$

$$= -\frac{x^2}{2} + O\left(\frac{1}{\sqrt{n}}\right).$$

The first assertion follows.

Let x_{nj} be those numbers of the form $\dfrac{j - np}{\sqrt{npq}}$ which belong to the interval $[a, b]$; consecutive x_{nj}'s differ by $1/\sqrt{npq}$. On account of the first assertion, uniformly in j,

$$P_n(x_{nj}) \sim \frac{1}{\sqrt{2\pi npq}} e^{-x_{nj}^2/2}$$

and

$$P\left[a \leqq \frac{S_n - np}{\sqrt{npq}} \leqq b\right] = \sum_j P_n(x_{nj}) \sim \frac{1}{\sqrt{2\pi}} \cdot \frac{1}{\sqrt{npq}} \sum_j e^{-x_{nj}^2/2}.$$

Since the last expression is a Riemann sum approximating the integral $\dfrac{1}{\sqrt{2\pi}} \displaystyle\int_a^b e^{-x^2/2}\, dx$, the second assertion follows.

III. DEPENDENCE AND CHAINS

1. Conditional probabilities. Let A be an event with $PA > 0$. The ratio PAB/PA is called the *conditional probability of B given A* or, simply, *probability of B given A* and is denoted by $P_A B$, so that

$$PAB = PA P_A B.$$

By induction we obtain the *multiplication rule:*

$$P(AB \cdots KL) = PA P_A B \cdots P_{AB \cdots K} L.$$

Furthermore, if $\sum_j A_j = \Omega$, then, from

$$PB = P\Omega B = \sum_j PA_j B,$$

follows the *total probability rule:*

$$PB = \sum_j PA_j P_{A_j} B.$$

Bayes' theorem,

$$P_B A_k = \frac{PA_k P_{A_k} B}{\sum_j PA_j P_{A_j} B},$$

follows upon replacing PB by the foregoing expression in the relation

$$PA_k B = PA_k P_{A_k} B = PB P_B A_k.$$

All events which figure as subscripts are supposed to be of positive probability. However, *if, say, $P_A B$ is given, then every given PA, whether zero or not, determines correctly PAB by $PAB = PA P_A B$, since $PA = 0$ implies $PAB = 0$.*

The set of all probabilities of events given a fixed A with $PA > 0$ defines a function P_A on $\mathcal{C}$, to be called the *conditional probability given A* or, simply, the *probability given A*. It follows at once from the definition that P_A obeys axiom II′: it is normed, nonnegative, and σ-additive on $\mathcal{C}$. Therefore, the pair $(\mathcal{C}, P_A)$ is a probability field *given A* for which all definitions and general properties of probability fields remain valid.   In particular, if $X = \sum_j x_j I_{A_j}$ is an elementary random variable, the expectation of X with respect to P_A or the *conditional expectation of X given A* or, simply, the *expectation of X given A* is defined by

$$E_A X = \sum_j x_j P_A A_j = \frac{1}{PA} \sum_j x_j P A A_j;$$

clearly, if EX exists and is finite, then $E_A X$ exists and is finite.   In terms of trials, the probability field given A represents the original trial with the occurrence of outcome A added to the original conditions.

It is easily verified that the events A_j of a countable set are independent if, and only if, for every finite subset $j_1, j_2, \cdots, j_k$ of indices

$$P_{A_{i_1} A_{i_2} \cdots A_{i_{k-1}}} (A_{j_k}) = P A_{j_k},$$

provided the "given" events have positive probability.

2. Asymptotically Bernoullian case.   Let A_n, $n = 1, 2, \cdots$, be an arbitrary sequence of events, and let $X_n = \frac{1}{n} \sum_{k=1}^{n} I_{A_k}$ be the random frequency of occurrence of the n first ones.   We set

$$p_1(n) = \frac{1}{n} \sum_{k=1}^{n} P A_k, \quad p_2(n) = \frac{2}{n(n-1)} \sum_{1 \leq j < k \leq n} P A_j A_k$$

so that $p_1(n)$ and $p_2(n)$ are bounded by 0 and 1.   It follows, by elementary computations, that

$$EX_n = p_1(n), \quad \sigma^2 X_n = p_2(n) - p_1^2(n) + \frac{p_1(n) - p_2(n)}{n}.$$

In the Bernoulli case

$$d_n = p_2(n) - p_1^2(n) = p^2 - p^2 = 0,$$

and we can consider the quantity d_n as some sort of measure of "deviation" from the Bernoulli case.   To make this precise, let us first prove a

KOLMOGOROV INEQUALITY.   *If X is an elementary random variable bounded by 1 (in absolute value), then, for every $\epsilon > 0$,*

$$P[|X| \geq \epsilon] \geq EX^2 - \epsilon^2.$$

We proceed as for the proof of Tchebichev's inequality: the inequality follows from

$$EX^2 = E(X^2 I_{[|X| \geq \epsilon]}) + E(X^2 I_{[|X| < \epsilon]}) \leq E I_{[|X| \geq \epsilon]} + \epsilon^2$$
$$= P[|X| \geq \epsilon] + \epsilon^2.$$

EXTENDED BERNOULLI LAW OF LARGE NUMBERS. *Bernoulli's result, that for every $\epsilon > 0$*

$$P[|\,X_n - EX_n\,| \geq \epsilon] \to 0,$$

remains valid for the sequence of events A_n, independent or not, if, and only if,

$$d_n = p_2(n) - p_1{}^2(n) \to 0.$$

Since $|\,X_n\,| \leq 1$, we can apply Kolmogorov's inequality as well as Tchebichev's, so that

$$\sigma^2 X_n - \epsilon^2 \leq P[|\,X_n - EX_n\,| \geq \epsilon] \leq \sigma^2 X_n / \epsilon^2.$$

Therefore, the asserted property holds if, and only if, $\sigma^2 X_n \to 0$. But

$$\left|\,\sigma^2 X_n - d_n\,\right| = \frac{|\,p_1(n) - p_2(n)\,|}{n} \leq \frac{1}{n} \to 0,$$

and the extension follows.

If $d_n \to 0$ at least as fast as $1/n$, then (asymptotically) we are even "closer" to the Bernoulli case. In fact,

EXTENDED BOREL STRONG LAW OF LARGE NUMBERS. *If $d_n = O(1/n)$, then Borel's result remains valid:*

$$P[X_n - EX_n \to 0] = 1.$$

The hypothesis means that there exists a fixed finite number c such that $|\,nd_n\,| \leq c$. Upon referring to the proof of Borel's result, we observe that it suffices to show that $\sum_{k=1}^{\infty} \sigma^2 X_{k^2} < \infty$. Since

$$n\sigma^2 X_n \leq |\,nd_n\,| + |\,p_1(n) - p_2(n)\,| \leq c + 1,$$

it follows by setting $n = k^2$ that

$$\sum_{k=1}^{\infty} \sigma^2 X_{k^2} \leq (c + 1) \sum_{k=1}^{\infty} \frac{1}{k^2} < \infty,$$

and the extension follows.

It is easily shown that both extensions apply to the events A_n which are independent but otherwise arbitrary.

3. Recurrence. The decomposition

$$\sigma^2 X_n = p_2(n) - p_1{}^2(n) + \frac{p_1(n) - p_2(n)}{n}$$

yields at once a proposition which leads very simply to the celebrated

Poincaré's recurrence theorem and its known refinements. Since $\sigma^2 X_n \geqq 0$ and $p_1(n)$, $p_2(n)$ are bounded by 0 and 1, it follows that, for any fixed $\epsilon > 0$, if $n \geqq 1/\epsilon$, then

$$p_2(n) = p_1^2(n) + \frac{p_2(n) - p_1(n)}{n} + \sigma^2 X_n \geqq p_1^2(n) - \frac{1}{n} \geqq p_1^2(n) - \epsilon.$$

But $p_2(n)$ is the arithmetic mean of PA_jA_k for $1 \leqq j < k \leqq n$. Therefore,

Whatever be the events A_n, if $n \geqq 1/\epsilon$, then there exist at least two events A_j, A_k, $1 \leqq j < k \leqq n$, such that $PA_jA_k \geqq p_1^2(n) - \epsilon$.

In particular, if $PA_n \geqq p > 0$ whatever be n, then every subsequence of these events contains at least two events A_j, A_k such that $PA_jA_k \geqq p^2 - \epsilon$; if this inequality holds, we say that A_j "ϵ-intersects" A_k. In fact, there exists then a subsequence whose first term ϵ-intersects every other term. For, if there is no such subsequence, then there exist integers m_n such that no event A_n ϵ-intersects events $A_{n'}$ with $n' \geqq n + m_n$, no two events of the subsequence $A_{n_1}, A_{n_2}, A_{n_3}, \cdots$ with $n_1 = 1$, $n_2 = n_1 + m_{n_1}$, $n_3 = n_2 + m_{n_2}, \cdots$, ϵ-intersect, and this contradicts the particular case of the foregoing proposition. Thus, let $A_{11}, A_{21}, A_{31}, \cdots$, be a subsequence such that the first term ϵ-intersects every other term. Let $A_{12}, A_{22}, A_{32}, \cdots$, be a subsequence of $A_{21}, A_{31}, \cdots$, with same property, and so on indefinitely. The sequence $A_{11}, A_{12}, \cdots$, is such that every one of its terms ϵ-intersects every other term. Hence

RECURRENCE THEOREM. *If $PA_n \geqq p > 0$ whatever be n, then for every $\epsilon > 0$ there exists a subsequence of events A_n such that $PA_jA_k \geqq p^2 - \epsilon$ whatever be the terms A_j, A_k of this subsequence.*

We observe that, if $PA_n = p$, then $PA_jA_k \geqq p^2 - \epsilon$ while, if the A_n are two by two independent, then $PA_jA_k = p^2$. Thus, however small be $\epsilon > 0$, for every sequence A_n of events, independent or not, there exists a subsequence which behaves as if its terms were two by two semi-independent *up to* ϵ ("semi" only since we do not have necessarily $PA_jA_k \leqq p^2 + \epsilon$).

A phenomenological interpretation of the foregoing theorem is as follows. Consider integer values of time and an incompressible fluid in motion filling a container of unit volume. Any portion of the fluid which at time 0 occupies a position A of volume $PA = p > 0$ occupies at time m a position A_m of same volume $PA_m = p$. The theorem says that, for every $\epsilon > 0$, the portion occupies in its motion an infinity of positions such that the volume of the intersection of any two of these positions is $\geq p^2 - \epsilon$. In particular, if the motion is "second order stationary," that is, $PA_jA_{j+k} = PAA_k$, then it intersects infinitely often its initial position—this is Poincaré's recurrence theorem (he assumes "stationarity")—and the intersections may be selected to be of volume $\geq p^2 - \epsilon$—this is Khintchine's refinement.

4. Chain dependence. There is a type of dependence, studied by Markov and frequently called Markov dependence, which is of considerable phenomenological interest. It represents the chance (random, stochastic) analogue of nonhereditary systems, mechanical, optical, $\cdots$, whose known properties constitute the bulk of the present knowledge of laws of nature.

A system is subject to laws which govern its evolution. For example, a particle in a given field of forces is subject to Newton's laws of motion, and its positions and velocities at times 1, 2, $\cdots$, describe the "states" (events) that we observe; crudely described, a very small particle in a given liquid is subject to Brownian laws of motion, and its positions (or positions and velocities) at times $t = 1, 2, \cdots$, are the "states" (events) that we observe. While Newton's laws of motion are deterministic in the sense that, given the present state of the particle, the future states are uniquely determined (are sure outcomes), Brownian laws of motion are stochastic in the sense that only the *probabilities* of future states are determined. Yet both systems are "nonhereditary" in the sense that the future (described by the sure outcomes or probabilities of outcomes, respectively) is determined by the last observed state only—the "present." It is sometimes said that nonhereditary systems obey the "Huygens principle." The mathematical concept of nonheredity in a stochastic context is that of Markov or chain dependence, and appears as a "natural" generalization of that of independence.

Events A_j, where j runs over an ordered countable set, are said to be *chained* if the probability of every A_j given any finite set of the preceding ones depends only upon the last given one; in symbols, for every finite subset of indices $j_1 < j_2 < \cdots j_k$, we have

$$P_{A_{j_1}A_{j_2}\cdots A_{j_{k-1}}}(A_{j_k}) = P_{A_{j_{k-1}}}(A_{j_k}).$$

Classes $\mathcal{C}_j = \{A_{j1}, A_{j2}, \cdots\}$ of events are said to be *chained* if events A_{jk} selected arbitrarily—one in each $\mathcal{C}_j$—are chained.

An elementary chain is a sequence of chained elementary partitions $\sum_k A_{nk} = \Omega$, $n = 1, 2, \cdots$; in particular, if $X_n = \sum_k x_{nk}I_{A_{nk}}$ with distinct x_{n1}, x_{n2}, $\cdots$ are elementary random variables, then the X_n are said to be *chained*, or to form a *chain*, when the corresponding partitions are chained.

It will be convenient to use a phenomenological terminology. Events of the nth partition will be called *states at time n*, or *at the nth* step, of the system described by the chain. The totality of all states of the system is countable; we shall denote them by the letters j, k, h, $\cdots$, and summations over, say, states k will be over the set of all states, unless otherwise stated.

The *evolution* of the system is described by the probabilities of its states given the last known one. The probabilities $P_{jk}^{m,n}$ of passage from a state j at time m to a state k at time $m + n$ (in n steps) form a matrix $P^{m,n}$. Since "probability given j" is a probability, and the probability given j at time m to pass to some state in n steps is one, we have

$$P_{jk}^{m,n} \geqq 0, \quad \sum_k P_{jk}^{m,n} = 1.$$

Furthermore, by the definition of chain dependence, the probability given j at time m to pass to state k in $n + n'$ steps equals the probability given j at time m to pass to some state in n steps and then to pass to k in n' steps, we have

$$P_{jk}^{m,n+n'} = \sum_h P_{jh}^{m,n} P_{hk}^{m+n,n'}$$

or, in matrix notation,

$$P^{m,n+n'} = P^{m,n} P^{m+n,n'}$$

An elementary chain is said to be *constant* if $P_{jk}^{m,n}$ is independent of m whatever be j, k, and n. Then we denote this probability by P_{jk}^n, and call it *transition probability from j to k in n steps*. The corresponding matrix P^n is called *transition matrix in n steps*; if $n = 1$ we drop it. The foregoing relations become the *basic constant chain relations:*

$$P_{jk}^n \geqq 0, \quad \sum_k P_{jk}^n = 1, \quad P_{jk}^{n+n'} = \sum_h P_{jh}^n P_{hk}^{n'}.$$

The last one can also be written as a matrix product $P^{n+n'} = P^n P^{n'}$. Hence P^n is the nth power of the transition matrix $P = P^1$, so that P determines all transition probabilities. In fact, for an elementary chain to be constant it suffices that the matrix $P^{m,1}$ be independent of m: $P^{m,1} = P$, since then

$$P^{m,2} = P^{m,1} P^{m+1,1} = P^2, \quad P^{m,3} = P^{m,2} P^{m+2,1} = P^3, \quad \cdots.$$

We observe that in P_{jk}^n and in every symbol to be introduced below, superscripts are not power indices, unless so stated.

We investigate the evolution of a system subject to constant chain laws described by a transition matrix P. In particular, we want to find its asymptotic behavior according to the state from which it starts. In phenomenological terms the system is a nonhereditary one subject to constant laws (independent of the time) and we ask what happens to the system in the long run. The "direct" method we use—requiring no special tools and which has a definite appeal to the intuition—has been developed by Kolmogorov (1936) and by Döblin (1936, 1937) after Hadamard (1928) introduced it. But the concept of chain and the basic pioneering work are due to Markov (1907).

*5. Types of states and asymptotic behavior. According to the total probability rule and the definition of chain dependence, the probability Q_{jk}^n of passage from j to k *in exactly n* steps, that is, without passing through k before the nth step, is given by

$$Q_{jk}^n = \sum_{h_1 \neq k, h_2 \neq k, \cdots, h_{n-1} \neq k} P_{jh_1} P_{h_1 h_2} \cdots P_{h_{n-1}k}.$$

The *central relation* in our investigation is

(1) $$P_{jk}^n = \sum_{m=1}^{n} Q_{jk}^m P_{kk}^{n-m}, \quad n = 1, 2, \cdots,$$

the expressions $P_{kk}^0 = 1$ (obtained for $m = n$) are the diagonal elements of the unit matrix P^0.

The proof is immediate upon applying the total probability rule. The system passes from j to k in n steps if, and only if, it passes from j to k for the first time in exactly m steps, $m = 1, 2, \cdots, n$, and then passes from k to k in the remaining $n - m$ steps. These "paths" are disjoint events, and their probabilities are given by $Q_{jk}^m P_{kk}^{n-m}$.

Summing over $n = 1, 2, \cdots, N$, the central relation yields

$$\sum_{n=1}^{N} P_{jk}^n = \sum_{n=1}^{N} \sum_{m=1}^{n} Q_{jk}^m P_{kk}^{n-m} = \sum_{m=1}^{N} \left(Q_{jk}^m \sum_{n=m}^{N} P_{kk}^{n-m} \right)$$

and, therefore,

$$\left(1 + \sum_{n=1}^{N} P_{kk}^n\right) \sum_{m=1}^{N} Q_{jk}^m \geqq \sum_{n=1}^{N} P_{jk}^n \geqq \left(1 + \sum_{n=1}^{N-N'} P_{kk}^n\right) \sum_{m=1}^{N'} Q_{jk}^m, \quad N' < N.$$

It follows. upon dividing by $1 + \sum_{n=1}^{N} P_{kk}^n$ and letting first $N \to \infty$ and then $N' \to \infty$, that

(2)
$$\sum_{m=1}^{\infty} Q_{jk}^m = \lim_{N \to \infty} \frac{\sum_{n=1}^{N} P_{jk}^n}{1 + \sum_{n=1}^{N} P_{kk}^n} ;$$

in particular,

(3)
$$1 - \sum_{m=1}^{\infty} Q_{jj}^m = \lim_{N \to \infty} \frac{1}{1 + \sum_{n=1}^{N} P_{jj}^n} .$$

The sum

$$q_{jk} = \sum_{m=1}^{\infty} Q_{jk}^m$$

is the probability, starting at j, of passing through k *at least once*; for $k = j$ it is the probability of *returning to j at least once*. More generally, the probability q_{jk}^n, starting at j, of passing through k *at least n times* is given by

$$q_{jk}^n = \left(\sum_{m=1}^{\infty} Q_{jk}^m \right) q_{kk}^{n-1} = q_{jk} q_{kk}^{n-1} .$$

In particular, the probability q_{jj}^n of *returning to j at least n times* is given by

$$q_{jj}^n = q_{jj} q_{jj}^{n-1} = (q_{jj})^2 q_{jj}^{n-2} = \cdots = (q_{jj})^n .$$

Its limit,

(4) $\quad r_{jj} = \lim_{n \to \infty} (q_{jj})^n = 0 \quad$ or $\quad 1, \quad$ according as $\quad q_{jj} < 1 \quad$ or $\quad q_{jj} = 1,$

is the probability of *returning to j infinitely often*. It follows that the probability, starting at j, of passing through k *infinitely often* is

$$r_{jk} = \lim_{n \to \infty} q_{jk}^n = q_{jk} \lim_{n \to \infty} q_{kk}^{n-1} = q_{jk} r_{kk},$$

so that

(5) $\qquad r_{jk} = 0 \quad$ or $\quad q_{jk}, \quad$ according as $\quad q_{kk} < 1 \quad$ or $\quad q_{kk} = 1.$

Upon singling out the states j such that $q_{jj} = 0$ (noreturn) and $q_{jj} = 1$ (return with probability 1), we are led to two dichotomies of states:

j is a *return state* or a *noreturn state* according as $q_{jj} > 0$ or $q_{jj} = 0$; j is a *recurrent state* or a *nonrecurrent state* according as $q_{jj} = 1$ or $q_{jj} < 1$ or, on account of (4), according as $r_{jj} = 1$ or $r_{jj} = 0$.

Clearly, noreturn states are boundary cases of nonrecurrent states and recurrent states are boundary cases of return states. In terms of transition probabilities, we have the following criteria.

RETURN CRITERION. *A state j is a return or a noreturn state according as $P_{jj}^n > 0$ for at least one n or $P_{jj}^n = 0$ for all n.*

This follows at once from the fact that

$$\sup_{n \geq 1} P_{jk}^n \leq q_{jk} \leq \sum_{n=1}^{\infty} P_{jk}^n.$$

RECURRENCE CRITERION. *A state j is a recurrent or a nonrecurrent state according as the series $\sum_{n=1}^{\infty} P_{jj}^n$ is divergent or convergent.*
This follows from (3).

Less obvious *types* of states are described in terms of "mean frequency of returns," as follows:

Let ν_{jk} be the *passage time*, from j to k, taking values $m = 1, 2, \cdots$, with probabilities Q_{jk}^m. If $q_{jk} = 1$, then ν_{jk} are elementary random variables. If $q_{jk} < 1$, then, to avoid exceptions, we say that $\nu_{jk} = \infty$ with probability $1 - q_{jk}$. The symbol ∞ is subject to the rules

$$\frac{1}{\infty} = 0, \infty + c = \infty, \text{ and } \infty \times c = \infty \text{ or } 0 \text{ according as } c > 0 \text{ or } c = 0.$$

We define the expected passage time τ_{jk} from j to k by

$$\tau_{jk} = \sum_{m=1}^{\infty} m Q_{jk}^m + \infty(1 - q_{jk});$$

we call τ_{jj} the *expected return time to j and the mean frequency of returns to j* is $\frac{1}{\tau_{jj}}$.

We can now define the following dichotomy of states. A state j is *null* or *positive* according as $\frac{1}{\tau_{jj}} = 0$ or $\frac{1}{\tau_{jj}} > 0$. Clearly, a noreturn and, more generally, a nonrecurrent state is null while a positive state is recurrent.

We shall now establish a criterion for this new dichotomy of states in terms of transition probabilities. To make it precise, we have to introduce the concept of period of a state.

Let j be a return state; then let d_j be the period of the Q_{jj}^m, that is, the greatest integer such that a return to j can occur with positive probability only after multiples of d_j steps: $Q_{jj}^n = 0$ for all $n \neq 0$ (modulo d_j), and $Q_{jj}^{nd_j} > 0$ for some n. Let d_j' be the period of the P_{jj}^n defined similarly. We prove that $d_j = d_j'$ and call it the *period* (of return) *of j*.

The proof is immediate. If $Q_{jj}^{nd_j} > 0$, then $P_{jj}^{nd_j} \geqq Q_{jj}^{nd_j} > 0$ so that $d_j' \leqq d_j$. Thus, if $d_j = 1$, then $d_j' = 1$. If $d_j > 1$ and $r = 1, \cdots, d_j - 1$, then the central relation yields

$$P_{jj}^r = 0, \quad P_{jj}^{d_j+r} = Q_{jj}^{d_j} P_{jj}^r = 0,$$

$$P_{jj}^{2d_j+r} = Q_{jj}^{d_j} P_{jj}^{d_j+r} + Q_{jj}^{2d_j} P_{jj}^r = 0, \text{ etc. } \cdots,$$

so that $d_j \leqq d_j'$ and, hence, $d_j = d_j'$.

If j is a noreturn state, then we say that its period is infinite.

POSITIVITY CRITERION. *A state j is null or positive according as*
$\limsup\limits_{n \to \infty} P_{jj}^n = 0$ *or* > 0.

More precisely, if j is a null state, then $P_{jj}^n \to 0$, and if j is a positive state, then $P_{jj}^{nd_j} \to \dfrac{d_j}{\tau_{jj}} > 0$, while $P_{jj}^n = 0$ for all $n \neq 0$ (modulo d_j).

Since the proof is involved, we give it in several steps.

1° If j is nonrecurrent, then it is null and, by the recurrence criterion, the series $\sum\limits_{n=1}^{\infty} P_{jj}^n$ converges so that $P_{jj}^n \to 0$.

If j is recurrent, then, by definition of its period d_j, $P_{jj}^n = 0$ for all $n \neq 0$ (modulo d_j). Therefore, it suffices to prove that, if j is recurrent, then $P_{jj}^{nd_j} \to \dfrac{d_j}{\tau_{jj}}$; for, if j is null, then $\dfrac{1}{\tau_{jj}} = 0$ implies $\dfrac{d_j}{\tau_{jj}} = 0$, and if j is positive, then $\dfrac{d_j}{\tau_{jj}} > 0$.

Assume, for the moment, that, if the period d_j of the positive recurrent state j is 1, then $P_{jj}^n \to \dfrac{1}{\tau_{jj}}$. In the general case, take d_j for the unit step and set $P' = P^{d_j}$, so that $P_{jj}^{'n} = P_{jj}^{nd_j}$; hence $Q_{jj}^{'n} = Q_{jj}^{nd_j}$. Then, since $\tau_{jj}' = \sum\limits_{n=1}^{\infty} n Q_{jj}^{'n} = \dfrac{\tau_{jj}}{d_j}$, the assertion follows by

$$P_{jj}^{nd_j} = P_{jj}^{'n} \to \frac{1}{\tau_{jj}'} = \frac{d_j}{\tau_{jj}}.$$

Thus, it suffices to prove that, if j is recurrent with period $d_j = 1$, then

$$P_{jj}^n \to \frac{1}{\tau_{jj}}.$$

2° Let j be recurrent with $d_j = 1$. To simplify the writing, we drop the subscripts j and, to avoid confusion with matrices, we write superscripts as subscripts. We follow now Erdös, Feller, and Pollard.

Let $\alpha = \lim \sup P_n$ so that there is a subsequence n' of integers such that $P_{n'} \to \alpha$ as $n' \to \infty$. Since $q = \sum\limits_{m=1}^{\infty} Q_m = 1$, it follows that, given $\epsilon > 0$, there exists n_ϵ such that, for $n \geq n_\epsilon$, $\sum\limits_{m=n+1}^{\infty} Q_m < \epsilon$. Therefore, for $n' \geq n \geq n_\epsilon$ and every $p \leq n'$ with $Q_p > 0$, the central relation yields

$$P_{n'} \leq Q_p P_{n'-p} + \sum_{m \leq n, m \neq p} Q_m P_{n'-m} + \epsilon.$$

Since for n' sufficiently large, $P_{n'} > \alpha - \epsilon$ and $P_{n'-m} < \alpha + \epsilon$ for $m \leq n$, it follows that

$$\alpha - \epsilon \leq Q_p P_{n'-p} + (1 - Q_p)(\alpha + \epsilon) + \epsilon$$

hence

$$\alpha + \epsilon - \frac{3\epsilon}{Q_p} < P_{n'-p} < \alpha + \epsilon.$$

Therefore, letting $n' \to \infty$ and then $\epsilon \to 0$, we obtain $P_{n'-p} \to \alpha$, and, repeating the argument, we have, for every fixed integer m,

$$P_{n'-mp} \to \alpha \quad \text{as} \quad n' \to \infty.$$

3° Let us assume, for the moment, that $Q_1 > 0$ so that $P_{n'-m} \to \alpha$ for every fixed m. We introduce the expected return time τ and use the fact that j is recurrent, so that, setting $q_n = \sum\limits_{m=n+1}^{\infty} Q_m$, we have $q_0 = 1$. The expected return time τ can be written

$$\tau = \sum_{m=1}^{\infty} m Q_m = \sum_{m=1}^{\infty} m(q_{m-1} - q_m) = \sum_{m=0}^{\infty} q_m,$$

and the central relation can be written

$$P_n = \sum_{m=1}^{n} Q_m P_{n-m} = \sum_{m=1}^{n} (q_{m-1} - q_m) P_{n-m},$$

so that

$$\sum_{m=0}^{n} q_m P_{n-m} = \sum_{m=0}^{n-1} q_m P_{n-1-m} = \cdots = q_0 P_0 = 1.$$

Therefore, for $n < n'$,

$$\sum_{m=0}^{n} q_m P_{n'-m}' \leq 1$$

and, letting $n' \to \infty$ and then $n \to \infty$, we obtain $\alpha \leq 1/\tau$. If $\tau = \infty$, then $\alpha = 0$; hence $P_n \to 1/\tau = 0$. Thus let $\tau < \infty$. The same argument for $\beta = \liminf_{n \to \infty} P_n$ shows that, for a subsequence n'' such that $P_{n''} \to \beta$ as $n'' \to \infty$, we have $P_{n''-m} \to \beta$ for every fixed m and, from

$$\sum_{m=0}^{n} q_m P_{n''-m} + \epsilon \geq 1 \quad \text{for} \quad n_\epsilon \leq n < n'',$$

it follows as above that $\beta \geq \dfrac{1}{\tau}$. Therefore, $P_n \to \dfrac{1}{\tau}$, and the assertion is proved under the assumption that $Q_1 > 0$.

4° To get rid of the last assumption, we appeal to elementary number theory. Consider the set of all those p for which $Q_p > 0$. It contains a finite subset $\{p_i\}$ whose greatest common divisor is the period $d(=1)$. As above, if $P_{n'} \to \alpha$, then $P_{n'-m_i p_i} \to \alpha$ for every fixed m_i and p_i, and it follows that $P_{n'-m} \to \alpha$ for every fixed linear combination $m = \sum_i m_i p_i$. But every multiple of the period $md = m \geq \prod_i p_i$ can be written in this form, so that, starting with n' sufficiently large, $P_{n'-m} \to \alpha$ for every fixed m, and the assertion follows as above. This concludes the proof.

Since, for a state j with period d_j there exists a finite number of integers p_i such that $P_{jj}^{p_i} > 0$ and, for m sufficiently large $md_j = \sum m_i p_i$, it follows, by $P_{jj}^{md_j} \geq \prod_i P_{jj}^{m_i p_i} > 0$, that

If d_j is the period of j, then $P_{jj}^{md_j} > 0$ for all sufficiently large values of m.

In other words, after some time elapses the system returns to j with positive probability after *every* interval of time d_j.

We can now describe the asymptotic behavior of the system. If k is a return state of period d_k, set

$$q_{jk}(r) = \sum_{m=0}^{\infty} Q_{jk}^{md_k+r}, \quad r = 1, 2, \cdots, d_k,$$

so that $q_{jk}(r)$ is the probability of passage from j to k in $n = r$ (modulo d_k) steps and

$$\sum_{r=1}^{d_k} q_{jk}(r) = q_{jk}.$$

ASYMPTOTIC PASSAGE THEOREM. *For every state j*

if k is a null state, then $P_{jk}^n \to 0$;

if k is a positive state, then $P_{jk}^{nd_k+r} \to q_{jk}(r) \dfrac{d_k}{\tau_{kk}}$;

and, whatever be the state k,

$$\frac{1}{n} \sum_{m=1}^{n} P_{jk}^m \to \bar{P}_{jk} = \frac{q_{jk}}{\tau_{kk}}.$$

The theorem results from the positivity criterion and the central relation, as follows:

If k is a null state, then $P_{kk}^n \to 0$. Therefore,

$$P_{jk}^n \leqq \sum_{m=1}^{n'} Q_{jk}^m P_{kk}^{n-m} + \sum_{m=n'+1}^{n} Q_{jk}^m,$$

and it follows, upon letting $n \to \infty$ and then $n' \to \infty$, that $P_{jk}^n \to 0$.

If k is a positive state, then $P_{kk}^{nd_k+r} = 0$ for $r < d_k$ and $P_{kk}^{nd_k} \to d_k/\tau_{kk}$. Therefore, from

$$0 \leqq P_{jk}^{nd_k+r} - \sum_{m=1}^{n'} Q_{jk}^{md_k+r} P_{kk}^{(n-m)d_k} \leqq \sum_{m=n'+1}^{n} Q_{jk}^{md_k+r}$$

it follows, upon letting $n \to \infty$ and then $n' \to \infty$, that $P_{jk}^{nd_k+r} \to q_{jk}(r)d_k/\tau_{kk}$.

The last assertion follows from the first two assertions.

*6. Motion of the system. To investigate the motion of the system we have to consider the probabilities of passage from one state to another. But, first, let us introduce a convenient terminology.

A state j is an *everreturn* state if, for every state k such that $q_{jk} > 0$, we have $q_{kj} > 0$. Two states j and k are *equivalent* and we write $j \sim k$ if $q_{jk} > 0$ and $q_{kj} > 0$; they are *similar* if they have the same period and are of the same type. A class of similar states will be qualified according to the common type of its states.

A class of states is *indecomposable* if any two of its states are equivalent, and it is *closed* if the probability of staying within the class is one. For example, the class of all states is closed but not necessarily indecomposable.

The motion of the system is described by the foregoing asymptotic behavior of the probabilities of passage from a given state to another given state, and also by the following theorem.

DECOMPOSITION THEOREM. *The class of all return states splits into equivalence classes which are indecomposable classes of similar states.*

A not everreturn equivalence class is not closed. An everreturn equivalence class is closed; if its period d > 1, then it splits into d cyclic subclasses C(1), C(2), $\cdots$, C(d) such that the system passes from a state in C(r) to a state in C(r + 1) (C(d + 1) = C(1)) with probability 1.

The proof is simple but somewhat long. To begin with, we observe that, if j and k are two equivalent states, distinct or not, then there exist two integers, say m and p, such that $P_{jk}^m > 0$, $P_{kj}^p > 0$.

1° The set of all states which are equivalent to some state coincides with the set of all return states. For, on the one hand, every return state is equivalent to itself and, on the other hand, if $j \sim k$, then $q_{jj} \geq P_{jj}^{m+p} \geq P_{jk}^m P_{kj}^p > 0$. Thus, the relation $j \sim k$, symmetric by definition, is reflexive: $j \sim j$. It is also transitive, for $j \sim k$ implies $P_{jk}^m > 0$, $k \sim h$ implies $P_{kh}^n > 0$ for some integer n and, hence, $q_{jh} \geq P_{jh}^{m+n} \geq P_{jk}^m P_{kh}^n > 0$; similarly for q_{hj}. Therefore, the relation $j \sim k$ has the usual properties of an equivalence relation and the set of all return states splits into indecomposable equivalence classes.

We prove now that, if $j \sim k$, then they are similar. We know already that they are both return states; let d_j and d_k be their respective periods. There exists an integer n such that $P_{kk}^n > 0$; hence $P_{kk}^{2n} \geq P_{kk}^n P_{kk}^n > 0$ and $P_{jj}^{m+n+p} \geq P_{jk}^m P_{kk}^n P_{kj}^p > 0$; similarly, $P_{jj}^{m+2n+p} > 0$. Therefore, d_j, being a divisor of $m + n + p$ and of $m + 2n + p$, is a divisor of every such n and hence of d_k. By interchanging j and k, it follows that j and k have the same period.

If j is an everreturn state and $P_{kh}^q > 0$, then, from $P_{jh}^{m+q} \geq P_{jk}^m P_{kh}^q > 0$, it follows that there exists an integer r such that $P_{hj}^r > 0$; hence $P_{hk}^{r+p} \geq P_{hj}^r P_{jk}^p > 0$, and k is an everreturn state. By interchanging j and k, it follows that they are both either everreturn or not everreturn states.

If k is recurrent, then, by the recurrence criterion,

$$\sum_{n=1}^{\infty} P_{jj}^n \geq \sum_{n=1}^{\infty} P_{jj}^{m+n+p} \geq P_{jk}^m \left(\sum_{n=1}^{\infty} P_{kk}^n \right) P_{kj}^p = \infty$$

and j is recurrent. By interchanging j and k, it follows that they are both either recurrent or nonrecurrent.

If d is the common period of the two equivalent states j and k, then, from

$$P_{jj}^{m+nd+p} \geq P_{jk}^m P_{kk}^{nd} P_{kj}^p,$$

it follows that d is a divisor of $m + p$ and $\lim_{n \to \infty} P_{kk}^{nd} > 0$ implies $\lim_{n \to \infty} P_{jj}^{nd}$

> 0. Hence, upon applying the positivity criterion and then interchanging j and k, both are either positive or null. This completes the proof of the first assertion.

2° If j is a return but not an everreturn state, then there exists a state h such that $q_{jh} > 0$ while $q_{hj} = 0$, so that h is not equivalent to j and there is a positive probability of leaving the equivalence class of j.

If j is an everreturn state, then $q_{jk} > 0$ entails $q_{kj} > 0$ so that k belongs to the equivalence class of j. Therefore, the probability of passage from j to a state which does not belong to the equivalence class of j is zero and, the class of all states being countable, the probability of leaving this class is zero.

Finally, we split an everreturn equivalence class C of period $d > 1$ as follows: Let j and k belong to C. Since $P_{jj}^{m+p} \geqq P_{jk}^{m} P_{kj}^{p} > 0$, d is a divisor of $m + p$ and, if m_1 and m_2 are two values of m, then $m_1 = m_2$ (modulo d). Thus, fixing j, to every k belonging to C there corresponds a unique integer $r = 1$ or 2, $\cdots$, or d such that, if $P_{jk}^{m} > 0$, then $m = r$ (modulo d). The states belonging to C with the same value of r form a subclass $C(r)$ and C splits into subclasses $C(1)$, $C(2)$, $\cdots$ $C(d)$. It follows that, if k and k' belong respectively to $C(r)$ and $C(r')$, then $P_{kk'}^{n}$ can be positive only for $n = |r - r'|$ (modulo d). Moreover, according to the proposition which follows the positivity criterion, $P_{kk'}^{n} > 0$ for all such n sufficiently large. Thus no subclass $C(r)$ is empty and the system moves cyclically from $C(r)$ to $C(r + 1) \cdots$ with $C(d + 1) = C(1)$. This proves the second assertion.

COROLLARY 1. *The states of an everreturn equivalence class C are linked in a constant chain whose transition matrix is obtained from the initial transition matrix P by deleting all those P_{jk} for which j or k or both do not belong to C.*

COROLLARY 2. *The states of a cyclic subclass $C(r)$ of an everreturn equivalence class with period d are linked in a constant chain whose transition matrix P' is obtained from P^d by deleting all those P_{jk}^{d} for which j or k or both do not belong to $C(r)$.*

COROLLARY 3. *An everreturn null equivalence class C is either empty or infinite. In particular, a finite chain has no everreturn null states.*

Let C be finite nonempty. By the asymptotic passage theorem, $P_{jk}^{n} \to 0$ for $k \in C$. But C is closed, so that $1 = \sum_{k \in C} P_{jk}^{n} \to 0$ for $j \in C$, and we reach a contradiction.

COROLLARY 4. *If j and k are nonequivalent everreturn states, then* $P_{jk}^n = 0$.

If j and k are equivalent positive states, with period d, then

$$P_{jk}^{nd+r} \to d/\tau_{kk} \quad \text{for some} \quad r = r(j, k)$$

$$P_{jk}^{nd+r'} = 0 \quad \text{for} \quad r' \not\equiv r \,(\text{modulo } d).$$

This follows by the asymptotic passage theorem.

*7. Stationary chains. The evolution of a system is determined by the laws which govern the system. In the case of constant elementary chains these laws are represented by the transition matrix P with elements P_{jk}. While P determines probabilities of *passage* from one state to another, it does not determine the probability that at a given time the system *be* in a given state. To obtain such probabilities we have to know the initial conditions. In the deterministic case this is the state at time 0. In our case it is the probability distribution at time 0, that is, the set of probabilities P_j for the system to be in the state j at time 0. Then, according to the total probability rule, the probability P_k^n that the system be in the state k at time $n = 1, 2, \cdots$, is

$$P_k^n = \sum_j P_j P_{jk}^n.$$

The notion of statistical equilibrium corresponds to the concept of stationarity in time. In our case of a constant elementary chain with transition matrix P, it is *stationary* if $P_k^n = P_k$ for every state k and every $n = 1, 2, \cdots$.

Given the laws of evolution represented by a transition matrix, the problem arises whether or not there exist initial conditions represented by the initial probability distribution such that the chain is stationary; in other words, whether or not there exists a probability distribution $\{\bar{P}_j\}$ which remains *invariant* under transitions. In general, one expects that if, under given laws of evolution, an equilibrium is possible, then it is attained in the long run. To this somewhat vague idea corresponds the following

INVARIANCE THEOREM. *For states j belonging to a cyclic subclass of a positive equivalence class with period d, the set of values $\bar{P}_j = \dfrac{d}{\tau_{jj}}$ is an invariant and the only invariant distribution under the transition matrix of the subclass.*

According to Corollary 2 of the decomposition theorem, it suffices to consider the chain formed by the subclass, that is, by one cyclic positive class with some transition matrix P of period one. According to the asymptotic passage theorem,

$$P_{jk}^n \rightarrow \frac{1}{\tau_{kk}} = \bar{P}_k > 0.$$

Since

$$\sum_k P_{jk}^n = 1 \quad \text{and} \quad P_{jk}^{n+m} = \sum_h P_{jh}^n P_{hk}^m,$$

it follows, upon taking arbitrary but finite sets of states and letting $n \rightarrow \infty$, that

$$\sum_k \bar{P}_k \leqq 1, \quad \bar{P}_k \geqq \sum_h \bar{P}_h P_{hk}^m.$$

But if, for some k, the second inequality is strict, then summing over all states k, we obtain

$$1 \geqq \sum_k \bar{P}_k > \sum_h \bar{P}_h$$

so that, *ab contrario*,

$$\bar{P}_k = \sum_h \bar{P}_h P_{hk}^m.$$

Since $\sum_h \bar{P}_h$ is finite, we can pass to the limit under the summation sign, so that, by letting $m \rightarrow \infty$, we obtain

$$\bar{P}_k = (\sum_h \bar{P}_h)\bar{P}_k$$

and, $\bar{P}_k$ being positive, it follows that $\sum_h \bar{P}_h = 1$. Thus, the set of values $\bar{P}_k$ is a probability distribution invariant under P.

It remains for us to prove that, if a set of values P_k has the same properties, then $P_k = \bar{P}_k$. But from

$$P_k = \sum_h P_h P_{hk}^m$$

it follows, as before, that $P_k = (\sum_h P_h)\bar{P}_k = \bar{P}_k$, and the conclusion is reached.

COROLLARY. *If C is a positive equivalence class, then*

$$\sum_{j \in C} \frac{1}{\tau_{jj}} = 1.$$

This follows from

$$d \sum_{j \in C} \frac{1}{\tau_{jj}} = \sum_{r=1}^d \sum_{j \in C(r)} \frac{d}{\tau_{jj}} = d.$$

STATIONARITY THEOREM. *A constant elementary chain with transition matrix P is stationary with initial probability distribution* $\{\bar{P}_k\}$ *if, and only if,* $\bar{P}_k = 0$ *for all null states k, and* $\bar{P}_k = \dfrac{p_t}{\tau_{kk}}$ *for all states belonging to positive equivalence classes* C_t, *with* $\sum_t p_t = 1$.

Let the probability distribution $\{\bar{P}_k\}$ be invariant under the transition matrix P so that

$$\bar{P}_k = \sum_j \bar{P}_j P_{jk}^n.$$

If k is a null state, then, by the asymptotic passage theorem, $P_{jk}^n \to 0$. $\sum_j \bar{P}_j$ being finite, we can pass to the limit under the summation sign. It follows, upon letting $n \to \infty$, that $\bar{P}_k = 0$. Hence, by summing over positive states only, $\sum' \bar{P}_k = 1$.

If k belongs to a positive equivalence class C_t, then, by the asymptotic passage theorem, we have that $P_{jk}^n = 0$ for every j which does not belong to C_t and $\dfrac{1}{n}\sum_{m=1}^{n} P_{jk}^m \to \dfrac{1}{\tau_{kk}}$ for every j belonging to C_t. It follows that

$$\bar{P}_k = \sum_{j \in C_t} \bar{P}_j P_{jk}^n = \sum_{j \in C_t} \bar{P}_j \left(\frac{1}{n}\sum_{m=1}^{n} P_{jk}^m\right) \to \frac{p_t}{\tau_{kk}}$$

where

$$p_t = \sum_{j \in C_t} \bar{P}_j \quad \text{and} \quad \sum_t p_t = \sum' \bar{P}_k = 1.$$

This proves the "only if" assertion.

Conversely, let the conditions on the $\bar{P}_k$ hold and use

$$P_k^m = \sum_j' \bar{P}_j P_{jk}^m$$

where the summation is over positive states j only, since $\bar{P}_j = 0$ for j null.

Therefore, if k is null, then $P_{jk}^m = 0$ and $P_k^m = 0$ for every m. If k belongs to a positive equivalence class C_t, then, since C_t is closed, $P_{jk}^m = 0$ for all states j which do not belong to C_t, and, C'_t being a finite subclass of C_t such that $\sum \bar{P}_j < \epsilon$ with sum over $j \in C_t - C'_t$, we have

$$P_k^m = \sum_{j \in C_t} \bar{P}_j P_{jk}^m \leq p_t \sum_{j \in C'_t} P_{jk}^m / \tau_{jj} + \epsilon.$$

Upon replacing $\dfrac{1}{\tau_{jj}}$ by the limit of the mean in the asymptotic passage theorem with subscripts $h, j \in C'_t$, we obtain, by summing first over the j,

$$P_k^m \leqq p_t \lim_{n \to \infty} \frac{1}{n} \sum_{s=1}^{n} P_{hk}^{m+s} + \epsilon.$$

Hence

$$P_k^m \leqq \frac{p_t}{\tau_{kk}} + \epsilon,$$

so that, letting $\epsilon \to 0$, we have $P_k^m \leqq \dfrac{p_t}{\tau_{kk}}$.

If, for some k, the inequality is a strict one, then, since for null states $P_k^m = 0$, it follows, by summing over positive states k only, that

$$1 = \sum{}' P_k^m < \sum_t p_t \sum_{k \in Ct} \frac{1}{\tau_{kk}} = \sum_t p_t = 1.$$

Therefore, $P_k^m = \dfrac{p_t}{\tau_{kk}}$ for every m, and the "if" assertion is proved.

COMPLEMENTS AND DETAILS

I. Physical statistics. The problem is to determine the state of equilibrium of a physical system, of energy E, composed of a very large number N of "particles" of the same nature: electrons, protons, photons, mesons, neutrons, etc.

Hypotheses. There are g_1 microscopic states of energy e_1, g_2 of energy e_2, $\cdots$ and each particle is in one of these states. The macroscopic state, i.e., the state of the system, is specified by the number of particles at each energy level: ν_1 particles of energy e_1, ν_2 particles of energy e_2, $\cdots$. The set $\{\nu_1, \nu_2, \cdots\}$ is a set of random integers and the probability of a macroscopic state $\nu_1 = n_1$, $\nu_2 = n_2$, $\cdots$ is equal, up to a constant factor, to the number W of ways in which n_k particles can be distributed amongst g_k microscopic states of energy e_k, $k = 1$, 2, $\cdots$, provided

$$\sum_k n_k = N, \quad \sum_k n_k e_k = E.$$

The Maxwell-Boltzmann statistics (classical theory of gases) is that of distinguishable particles without exclusion, i.e., without any bound upon the possible number of particles in any of the microscopic states. The Bose-Einstein statistics (photons, mesons, deuterons, $\cdots$—particles with an integer "spin") is that of nondistinguishable particles without exclusion. The Fermi-Dirac statistics (electrons, protons, neutrons—particles with a semi-integer "spin") is that of nondistinguishable particles which obey the Pauli exclusion principle, that is, there cannot be more than one particle in any of the microscopic states.

Weights. Let w denote the weight of the macroscopic state $\{n_1, n_2, \cdots\}$ i.e., $W/N!$ in the distinguishable case and W in the nondistinguishable case. Prove that the combinatorial formulae give the following expressions for w, where it is assumed that $\sum_k n_k = N$, $\sum_k n_k e_k = E$ (in the case of photons N is not fixed and only the second condition remains):

	Distinguishable Particles	Nondistinguishable Particles
Without exclusion	$w = \prod g_i{}^{n_i}/\prod n_i!$ (Maxwell-Boltzmann)	$w = \prod \dfrac{(g_i + n_i - 1)!}{n_i!(g_i - 1)!}$ (Bose-Einstein)
With exclusion	$w = \prod \dfrac{g_i!}{n_i!(g_i - n_i)!}$ (corresponds to no physical reality)	$w = \prod \dfrac{g_i!}{n_i!(g_i - n_i)!}$ (Fermi-Dirac)

When $g_k \gg n_k$, then the expressions of the weights in B.-E. and F.-D. statistics are equivalent to w in M.-B. statistics. Assume distinguishability and let c be the "capacity" coefficient of the microscopic states, that is, if there are already n particles in the g_k states of energy $e_k(k = 1, 2, \cdots)$, the number of these g_k states which remains available for the $(n + 1)$th particle is $g_k - nc$—this is *Brillouin statistics*. The weights w of the macroscopic states, previously defined as $w = W/N!$, are given by

$$w = \prod_k \frac{1}{n_k!} g_k(g_k - c) \cdots [g_k - (n_k - 1)c]$$

and reduce to those of M.-B., B.-E., and F.-D. by giving to the parameter c the values 0, -1, $+1$ respectively.

Statistical equilibrium. For a very large N the equilibrium state of the macroscopic system is postulated to be the most probable one, that is, the one with the highest weight. Assume that Stirling's formula can be used for the factorials which figure in the table of weights above. Take the variation $\delta \log w$ which corresponds to the variation $\{\delta n_1, \delta n_2, \cdots\}$. Using the Lagrange multipliers method, the state which corresponds to the maximum of w is determined by solving the system (prove)

$$\delta \log w + \lambda \cdot \delta N + \mu \cdot \delta E = 0$$

$$\sum_k n_k = N, \quad \sum_k n_k e_k = E.$$

(In the case of photons take $\lambda = 0$ and suppress the second relation.) The equilibrium states for the various statistics are also obtained by replacing c by

0, -1, and 1 in the equilibrium state for Brillouin statistics, given by

$$n_k = g_k/(e^{\lambda + \mu e_k} + c)$$

where λ and μ are determined by the subsidiary conditions

$$N = \sum_k g_k/(e^{\lambda + \mu e_k} + c), \quad E = \sum_k g_k e_k/(e^{\lambda + \mu e_k} + c).$$

The Planck-Bose-Allard method. The macroscopic states can be described in a more precise manner. Instead of asking for the number n_k of particles in the states of energy e_k, we ask for the number g_{km} of states of energy e_k occupied by m particles. The particles are assumed to be nondistinguishable as required by modern physics. The combinatorial formulae give

$$w = \prod_k (g_k!/\prod_m g_{km}!) \text{ with } g_k = \sum_m g_{km}, \quad N = \sum_k \sum_m m g_{km}, \quad E = \sum_k \sum_m e_k m g_{km}.$$

To obtain the statistical equilibrium state use the procedure described above.

B.-E. statistics is obtained if no bounds are imposed upon the values of m. F.-D. statistics is obtained if m can take only the values 0 or 1; "intermediate" statistics is obtained if m can take only the values belonging to a fixed set of integers.

In the equilibrium state (with $c = -1$ or $+1$ when the statistics are B.-E.'s or F.-D.'s respectively), we have

$$g_{km} = g_k(1 + ca_k)^{-\frac{1}{c}} a_k{}^m, \text{ where } a_k = e^{-(\lambda + \mu e_k)}$$

and $g_k(u)$, determined by the usual subsidiary conditions, the generating function of the number of particles in a microscopic state of energy e_k, is

$$g_k(u) = (1 + ca_k)^{-g_k/c} \cdot (1 + ca_k u)^{g_k/c}.$$

II. The method of indicators.

1. Rule: In order to compute PB, $B = f(A_1, A_1{}^c, \cdots, A_m, A_m{}^c)$, take the following steps:

(a) Reduce the operations on events to complementations, intersections, and sums;

(b) Replace each event by its indicator, expand, and take the expectation. In this way find

$$P(\bigcup_{j=1}^m A_j) \text{ and } P(\bigcap_{j=1}^k A_j \bigcup_{j=k+1}^m A_j{}^c) \text{ in terms of } P(\bigcap_{j=1}^r A_j)\text{'s.}$$

Notations. Let $I_{A_j} = I_j$ and let $R = \sum_{j=1}^m I_j$ be the "repetition" of A_j's, that is, the number of events A_j which occur. Let $J_o = 1$, $J_r = \sum I_{j_1} \cdots I_{j_r}$, where the summation is over all combinations $1 \leqq j_1 < j_2 < \cdots < j_r \leqq m$. Let $I_{[r]}$ and $I_{(r)}$ be indicators of the events exactly r A's occur and at least r A's occur, respectively; set

$$S_r = EJ_r, \quad P_{[r]} = EI_{[r]}, \quad P_{(r)} = EI_{(r)}.$$

2. Prove

(a)
$$\sum_{r=0}^{m} u^r I_{[r]} = \sum_{s=0}^{m} (u-1)^s J_s = u^R,$$

and deduce

(b)
$$P_{[r]} = \sum_{k=r}^{m} (-1)^{k-r} C_k^r S_k,$$

(c)
$$S_k = \sum_{t=k}^{m} C_t^k P_{[t]} = \sum_{t=k}^{m} C_{t-1}^{k-1} P_{(t)},$$

(d)
$$R(R-1) \cdots (R-k+1) = k! J_k.$$

3. Let $k \leqq r \leqq m$. Using *2*(c) and the relations
$$I_{(m)} \leqq \cdots \leqq I_{(r)} \leqq I_{(r-1)} \leqq \cdots \leqq I_{(k)},$$

prove that
$$(S_k - C_{r-1}^k)/(C_m^k - C_{r-1}^k) \leqq P_{(r)} \leqq S_k/C_r^k.$$

Examine the special case $r = m$; the left-hand side becomes Gumbel's inequality; the right-hand side becomes Fréchet's inequality.

Let
$$J(k) = 1 - J_k/C_m^k, \quad \Delta f(k) = f(k+1) - f(k).$$

4. Prove

(a)
$$\Delta J(k) = \sum_{t=k}^{m-1} \frac{C_{m-k-1}^{t-k}}{C_m^t} I_{[t]},$$

(b)
$$I_{[r]} \leqq \frac{C_m^r}{C_{m-k-1}^{r-k}} \Delta J(k), \quad k \leqq r \leqq m-1,$$

(c)
$$\Delta J(k) \geqq 0;$$

deduce a scale of inequalities for the S_k's.

5.

(a)
$$-m\Delta\left(\frac{J(k)}{k}\right) = \sum_{t=k}^{m-1} \frac{C_{m-k-1}^{t-k}}{C_{m-1}^{t-1}} (1 - I_{(t)}),$$

(b)
$$1 - I_{(t)} \leqq \frac{C_{m-1}^{t-1}}{C_{m-k-1}^{r-k}} \left(-m\Delta \frac{J(k)}{k}\right),$$

(c)
$$-m\Delta \frac{J(k)}{k} \geqq 0;$$

deduce another scale of inequalities for S_k's.

6. The general symbolic method. The events $B_1, \cdots, B_m$ are called exchangeable if $P(B_{i_1} \cdots B_{i_r} B_{i_{r+1}}^c \cdots B^c{}_{i_{r+s}})$ depends only on the number r of events B_i and on the number s of events B_i^c.

Let
$$J_{r/s} = \sum I(A_{i_1}) \cdots I(A_{i_r}) I(A_{i_{r+1}}^c) \cdots I(A_{i_{r+s}}^c)$$
$$S_{r/s} = E(J_{r/s}) = \sum P(A_{i_1} \cdots A_{i_r} A_{i_{r+1}}^c \cdots A_{i_{r+s}}^c)$$
$$p_{r/s} = P(B_{i_1} \cdots B_{i_r} B_{i_{r+1}}^c \cdots B_{i_{r+s}}^c).$$

If we choose the B_i's such that

$$\sum_i P(A_{t_1} \cdots A_{t_r} A_{t_{r+1}}{}^c \cdots A_{t_{r+s}}{}^c) = \sum_i P(B_{t_1} \cdots B_{t_r} B_{t_{r+1}}{}^c \cdots B_{t_{r+s}}{}^c),$$

then

$$p_{r/s} = S_{r/s}/C_m^r C_{m-r}^s.$$

If we further introduce symbolic independent events having the same probability p, the complementary events having probability $q = 1 - p$, then, symbolically,

$$p_{r/s} = p^r q^s.$$

The symbolic method consists of the following steps:

(a) In any given identity (or identical inequality) for p, q ($0 \leqq p \leqq 1$ $q = 1 - p$) replace $p^r q^s$ by $p_{r/s}$.

(b) Replace $p_{r/s}$ by $S_{r/s}/C_m^r C_{m-r}^s$ and obtain an equality (or inequality resp.) for the $S_{r/s}$'s.

Examples:

(a) Starting from $p^r q^s = p^r (1 - p)^s$ obtain

$$\frac{S_{r/s}}{C_m^{r+s} C_{r+s}^r} = \sum_{i=0}^p (-1)^i \frac{C_s^i}{C_m^{r+i}} S_{r+i/0};$$

in the special case $r + s = m$, find

$$S_{r/m-r} = P_{[r]}.$$

(b) Starting from $p^r q^s = p^r q^s (p + q)^{m-r-s}$, obtain

$$S_{r/s} = \sum_{i=r}^{m-s} C_i^r C_{m-i}^s S_{i/m-i}.$$

In the special case $s = 0$, find

$$S_{r/0} = S_r = \cdots.$$

(c) Starting from $p^{r'} q^{s'} \geqq p^r q^s$, $r' \leqq r$, $s' \leqq s$, find

$$\frac{S_{r'/s'}}{C_m^{r'+s'} C_{r'+s'}^{r'}} \geqq \frac{S_{r/s}}{C_m^r C_{r+s}^r}, \quad r' \leqq r, \quad s' \leqq s,$$

and as a special case the scale of inequalities (4c).

(d) Starting from $1 \geqq \sum_{i=0}^r {}' C_i^n p^{r-i} q^i$ where $\sum'$ denotes a sum in which a certain number of terms is omitted, find

$$C_r^m \geqq \sum{}' S_{r-1/i}$$

and, taking only the terms $i = 0$ and $i = 1$, find the second scale of inequalities (5c).

7. *The classical problem of matching.* This problem (problème des rencontres) was studied first by Montmort (1708) and further treated by Lambert, Euler,

and others in different forms, all of which can be described by the following
setup: given m distinct numbers $X_1, X_2, \cdots, X_m$, choose at random a first X_{t_1},
then a second X_{t_2} from the remaining ones, etc. A match (coincidence, ren-
contre) is an event A_i which consists in choosing exactly X_i at the ith draw.

In the following, assume that each permutation $(X_{t_1} \cdots X_{t_m})$ has the same
probability of being chosen at random. Show that

(a)
$$P(A_{t_1}, \cdots, A_{t_r}) = \frac{(m-r)!}{m!} \quad \text{and} \quad S_r = \frac{1}{r!},$$

(b)
$$P_{[r]} = \frac{1}{r!} \sum_{s=0}^{m-r} \frac{(-1)^s}{s!},$$

(c) Find $\lim_{m \to \infty} P_{[r]}$; interpretation? Show that $P_{[m-1]} = 0$; interpretation?

(d) Show that $E(r) = \sum_{r=0}^{m} rP_{[r]} = S_1 = 1$ and $E[r - E(r)]^2 = 1$. (Use the
generating function $\sum_{r=0}^{m} u^r P_{[r]}$.)

III. Random walk. A particle starting at some point of an m-dimensional
space moves in such a way that its consecutive displacements can be represented by independent m-dimensional random vectors. Problems of the following type arise: find the probability that in time T or before time T the particle reaches a certain domain D, or that it reaches D without having reached
previously a domain D', or find the expected time for the particle to reach D,
etc. $\cdots$

We give a few examples which show the great variety of forms under which
this problem occurs, questions which can be asked, and methods of solution.
We restrict ourselves to the discontinuous case with every move taking one
unit of time.

1. Game of "heads or tails" and combinatorial method. To n tosses of a coin
with equal probabilities for heads and for tails we associate the score point whose
coordinates are respectively the number of heads and the number of tails which
occur. Thus, at every toss, the score point M moves by one unit either upwards
or to the right, and the game is represented by a two-dimensional one-sided
random walk on the lattice of points with integer coordinates.

The score points corresponding to the same number n of tosses lie on the line
$x + y = n$. The total number of paths between 0 and $M = (a, b)$ is $\dfrac{(a+b)!}{a!b!}$.

(a) If A and B ran for office, A got a votes and B got $b < a$ votes, find the
pr. P that in counting the votes A be always ahead of B.

(Equivalent to the pr. that the score point stays below the bisectrix until it
reaches the point $M = (a, b)$. Compute the pr. of the complementary event by
applying the symmetry principle of Désiré André as follows: the paths from 0
to M which intersect the bisectrix either go through $(1, 0)$ or through $(0, 1)$.
By reason of symmetry both classes contain the same number of paths. The
number of those which go from $(0, 1)$ to M is $(a + b - 1)!/a!(b - 1)!$, and
$$P = \frac{a - b}{a + b}.\Bigg)$$

(b) The probability that there be neither gain nor loss in exactly $2n$ tosses is
$$\frac{1\cdot 3\,\cdots\,(2n-3)}{2\cdot 4\cdot 6\,\cdots\,2n} \sim \frac{1}{2n\sqrt{\pi n}}.$$ (Start with the number of paths from 0 to $(n, n-1)$ which do not intersect the bisectrix.)

(c) The probability that the gambler who bets on heads and whose fortune is m times the stake loses his fortune in $m + 2n$ tosses is $m(m + n + 1)\,\cdots\,(m + 2n - 1)/2^{m+2n}n!$. (Reduce to (a) by taking for origin the point $(m + n, n)$.)

2. *Gambler's ruin.*

(a) *Method of difference equations.* Consider a one-dimensional random walk on the lattice $x = 0, \pm 1, \pm 2, \cdots$. At each step the particle at y has probability p_k to move from y to $y + k$, $k = 0, \pm 1, \pm 2, \cdots$. Let P_x be the probability of ruin, that is, starting at x with $0 < x < a$ to arrive at $y \leq 0$ before reaching $y \geq a$. Then $P_x = \sum_y P_y p_{x-y}$ with boundary conditions $P_y = 1$ if $y \leq 0$ and $P_y = 0$ if $y \geq a$.

The gambler has x dollars and wins or loses one dollar with respective probabilities p and $q = 1 - p$. Find the probability P_x of his ruin. Find the probability P_{xn} of his ruin at the nth game.

In the first case, $P_x = pP_{x+1} + qP_{x-1}$ with $P_o = 1$, $P_a = 0$. The solution is $P_x = \dfrac{(q/p)^a - (q/p)^x}{(q/p)^a - 1}$ for $p \neq q$ and $P_x = 1 - \dfrac{x}{a}$ for $p = q$.

In the second case $P_{x,n+1} = pP_{x+1,n} + qP_{x-1,n}$ with $P_{on} = P_{an} = 0$ and $P_{oo} = 1$, $P_{xo} = 0$. The solution is

$$P_{xn} = a^{-1}2^n p^{(n-x)/2}q^{(n+x)/2} \sum_{k=1}^{a-1} \cos^{n-1}\frac{\pi k}{a} \sin\frac{\pi k}{a} \sin\frac{\pi kx}{a}.$$

(b) *Method of matrices.* Same random walk but with $p_1 = p_{-1} = 1/2$. The particle starts from 0 and dies when it attains $a - 1 \leq 0$ or $b = a + c \geq -1$. Find the probability P_n that after n displacements the particle is still alive, as follows.

Set $g(k) = 1/2$ for $k = \pm 1$ and $g(k) = 0$ otherwise. Then $P_n = \sum g(k_1) \cdots g(k_n)$ where the sum is taken over all k's such that $a \leq \sum_{j=1}^{h} k_j \leq b$, $h = 1, 2, \cdots, n$. Set $d_j = k_1 + \cdots + k_j - a$. Then P_n is the sum of the elements of the $(1 - a)$-th column or row of the matrix A^n where

$$A = (g(j - h)) = \begin{bmatrix} 0 & \frac{1}{2} & 0 & 0 & \cdots \\ \frac{1}{2} & 0 & \frac{1}{2} & 0 & \cdots \\ 0 & \frac{1}{2} & 0 & \frac{1}{2} & \cdots \\ \cdot & \cdot & \cdot & \cdot & \cdot & \cdot \end{bmatrix}$$

The proper values λ_j of A are given by $\lambda_j = \cos\dfrac{\pi j}{c + 2}$, the proper values of A^n are λ_j^n, and

$$P_n = \frac{2}{c + 2} \sum_{j=1}^{c+1}{}' \cos^n \frac{\pi j}{c + 2} \sin\frac{\pi j(1 - a)}{c + 2} \cot\frac{\pi j}{c + 2},$$

where $\sum'$ denotes summation over the odd j's only.

IV. Geometric probabilities.

Elementary probabilities. Consider an n-dimensional space of points $U = (u_1, \cdots, u_n)$ and let G be a group of transformations of points into points. If there exists a differential element $d\mu = g(u_1, \cdots, u_n)du_1 \cdots du_n$ determined up to a constant factor by the property that its integral over domains is invariant with respect to the group G, $d\mu$ defines up to a constant factor an elementary probability. The constant factor is determined by fixing a domain D_0 within which all considered domains lie and by assigning to this domain the pr. one; that is, by setting $c \int_{D_0} d\mu = 1$. Then the points are said to be taken or thrown at random in D_0. To say that several points are taken or thrown at random means that the throws are stochastically independent; in other words, we make repeated trials.

Let M with or without affixes be points in an m-dimensional euclidean space and let $x_1, \cdots, x_m$ with same affixes, if any, be its cartesian coordinates with respect to a fixed orthogonal frame of reference. The group G which transforms points M into points M is the group of euclidean displacements (preserves euclidean lengths). This means that the probability is required to be independent of the choice of the frame of reference. Prove that $d\mu = c\, dx_1\, dx_2 \cdots dx_m$.

Let us now investigate straight lines in a euclidean plane determined by their equations $u_1 x_1 + u_2 x_2 = 1$ in rectangular coordinates, and let G_e be the group of euclidean displacements in the plane. Prove that $d\mu = c(u_1^2 + u_2^2)^{-3/2} du_1 du_2$ or, using the normal equations: $x_1 \cos \theta + x_2 \sin \theta - p = 0$, $d\mu = cdp\, d\theta$.

(The transformations of the group G_e are of the form $x'_1 = a_1 + x_1 \cos \alpha - x_2 \sin \alpha$, $x'_2 = a_2 + x_1 \sin \alpha + x_2 \cos \alpha$ and induce transformations of a group G on the plane (u_1, u_2) defined by

$$u_1 = (u'_1 \cos \alpha + u'_2 \sin \alpha)/(a_1 u'_1 + a_2 u'_2 + 1),$$

$$u_2 = (-u'_1 \sin \alpha + u'_2 \cos \alpha)/(a_1 u'_1 + a_2 u'_2 + 1).$$

The invariance condition yields

$$g(u'_1, u'_2) = g(u_1, u_2)\frac{D(u_1, u_2)}{D(u'_1, u'_2)} \quad \text{with} \quad \frac{D(u_1, u_2)}{D(u'_1, u'_2)} = \frac{(u_1^2 + u_2^2)^{3/2}}{(u'_1^2 + u'_2^2)^{3/2}}.$$

With the same group G_e there is no elementary probability for circles in the plane. But there is one for circles of fixed radius.)

Points on a line. The elementary probability for a point M on a segment $[0, l]$ is dx/l. Throw n points at random on the segment. The probability, say, that there be no thrown points on $[0, x]$ is $\left(1 - \dfrac{x}{l}\right)^n$. What is the expected distance of the nearest to 0 of the thrown points? What is the probability that k out of the n thrown points lie on a fixed subinterval of length a? Find what happens as $l \to \infty$ with $n/l \to \lambda > 0$. Denote then by $M_1, M_2, \cdots$ the points in the nondecreasing order of their distance to 0. What is the elementary probability for the length $M_j M_{j+1}$ to be between x and $x + dx$ and what is the expectation of this length?

Lines in a plane. The elementary probability of a straight line $x \cos \theta + y \sin \theta - p = 0$ thrown on a plane is $d\mu = cdp\, d\theta$. The integral $\int dp\, d\theta$ over a domain induced by a family of straight lines is said to be the measure of the family. The measure of the secants of a segment of length l is $2l$. (θ varies from $-\dfrac{\pi}{2}$ to $+\dfrac{\pi}{2}$ while p varies from 0 to $l \cos \theta$ for every fixed θ.) The measure of the secants of a polygonal line of length l is $2l$, provided every secant is counted as many times as there are points of intersections of the secant with the polygonal line; in particular, the measure of the secants of a closed convex polygon is its perimeter. The same is true for the secants of a curve formed by a finite number of analytic arcs. Prove it directly for the secants of a circle.

Let C and C_0 be two closed convex curves of respective lengths l and l_0 with C being interior to C_0. The probability that a secant of C_0 be secant of C is l/l_0.

Application to the needle problem. If C_0 is a circumference of radius $r/2$ and C is a segment of length l, then $p = 2l/\pi r$. Throw the figure formed by the circumference and the segment on a plane with parallel equidistant straight lines with common distance r. The probability that one of these lines intersects the segment is $2l/\pi r$. Prove it directly by throwing a needle of length l on this plane.

(The position of the needle AB is determined by the coordinates x, y of A and the angle α that AB makes with Ox, one of the equidistant lines. The elementary probability is $dx\, dy\, d\alpha$. It is not a restriction to assume θ between 0 and $\pi/2$, $x = 0$, and y between 0 and r. Then $p = \dfrac{2l}{\pi r} \displaystyle\int_0^{\pi/2} \sin \alpha\, d\alpha$.)

A differential method. Let D_0 be a domain of the plane on which are thrown at random n points. Intrinsic properties of the figure formed by the points are defined independently of D_0; for example, $M_1 M_2 < l$, triangle $M_1 M_2 M_3$ has acute angles, $\cdots$.

The probability of an intrinsic property is given by $P = a/s^n$ where s is the area of D_0 and a represents the measure of the set of favorable cases. Let D'_0 be a new domain containing D_0 and let $P + \Delta P = (a + \Delta a)/(s + \Delta s)^n$ be the new probability of the same property. If P_k is the probability of the property when $n - k$ points are in D_0 and k points are in $D'_0 - D_0$, then

$$a + \Delta a = a + a_1 + \cdots + a_n, \quad a_k = \frac{n!}{k!(n-k)!} P_k s^{n-k}(\Delta s)^k$$

and

$$(s + \Delta s)^n \Delta P = n(P_1 - P)s^{n-1}\, \Delta s + \cdots$$

$$+ \frac{n!}{k!(n-k)!} (P_k - P)s^{n-k}(\Delta s)^k + \cdots + (P_n - P)(\Delta s)^n.$$

Keeping infinitesimals of first order, we have

$$\delta P = n(P_1 - P) \frac{\delta s}{s}$$

where n is the number of points thrown at random on D_0, P is the probability of the property, P_1 is the probability of the same property when 1 point is thrown at random on an increment of D_0 of area δs, and $n - 1$ points are thrown at random on D_0. More generally,

$$\delta m = n(m_1 - m)\frac{\delta s}{s}$$

where m is the expectation of a function of the thrown points, and the other quantities are defined similarly to what precedes. The method and the formulae apply whatever be the number of dimensions of the space.

Application. Two points M_1 and M_2 are thrown at random on a segment of length l. The probability that $M_1 M_2 < x$ is $\frac{2x}{l} - \frac{x^2}{l^2}$. What happens when the segment is replaced by a circle of radius r? Find $\overline{EM_1 M_2^p}$ in both cases.

V. Bernoulli case and Weierstrass theorem. Consider the Bernoulli case (with $PA = x$ in lieu of p): $0 \leq x \leq 1$,

$$P(S_n = k) = p_{nk}(x) = \frac{n!}{k!(n-k)!}x^k(1-x)^{n-k}, k = 0, 1, \cdots, n.$$

(a)
$$\sum_{k=0}^{n} p_{nk}(x) = 1, \quad ES_n = \sum_{k=0}^{n} kp_{nk}(x) = 1,$$

$$\sigma^2 S_n = \sum_{k=0}^{n} (k - nx)^2 p_{nk}(x) = nx(1-x).$$

(b) Let f be a real or complex-valued continuous function on $[0, 1]$. It is bounded: $|f| \leq c < \infty$ and uniformly continuous: Given $\epsilon > 0$ there is a $\delta > 0$ such that $|x - x'| < \delta = |f(x) - f(x')| < \epsilon$. Form Bernstein polynomials

$$E(f(S_n/n)) = \sum_{k=0}^{n} f(k/n)p_{nk}(x),$$

that is,

$$P_n(x) = \sum_{k=0}^{n} f(k/n)\frac{n!}{k!(n-k)!} x^k(1-x)^{n-k}.$$

(c) *Weierstrass theorem* says that on $[0, 1]$ there are polynomials which converge uniformly to f.
Bernstein polynomials are such that

$$|E(f(x) - f(S_n/n))| = |f(x) - P_n(x)|$$

$$= |\sum_{k=0}^{n} (f(x) - f(k/n))p_{nk}(x)| \leq |\sum_{|k-nx| \leq n\delta}| + |\sum_{|k-nx| > n\delta}|$$

The first partial sum is bounded by $\epsilon \sum_{k=0}^{n} p_{nk}(x) = \epsilon$. The second partial sum is bounded by

$$2c \sum_{|k-nx|>n\delta} p_{nk}(x) \leqq \frac{2c}{n^2\delta^2} \sum_{k=0}^{n} (k - nx)^2 p_{nk}(x) = 2cx(1 - x)/n\delta^2 \leqq c/2n\delta^2;$$

note that the first inequality while algebraically immediate is due to Tchebichev's inequality:

$$E(|\ S_n/n - E(S_n/n)\ | > \delta) \leqq \sigma^2 S_n/n^2\delta^2.$$

Thus, for all $x \in [0, 1]$, as $n \to \infty$ then $\epsilon \to 0$,

$$|f(x) - P_n(x)| \leqq \epsilon + c/2n\delta^2 \to 0.$$

Leaving out all references to the Bernoulli case, the most elementary proof known of Weierstrass theorem obtains: It introduces explicit uniformly approximating polynomials and is primarily algebraic.

Part One

NOTIONS OF MEASURE THEORY

No rigorous presentation of probability theory is possible without using the notions of sets, measures, measurable functions, and integrals. Their first lineaments are already apparent in elementary probability theory. These notions are introduced and investigated systematically in this part.

The presentation is self-contained, and the material will suffice for later parts. It is organized—at the cost of a few repetitions—so as to make the unstarred portions independent of the starred ones and, at the same time, to make the sections on measurable functions, convergence, and integration independent of the remainder except for 1.1 to 1.5. This permits a reorganization of the course so as to proceed from the less abstract notions toward more abstract and more involved ones. The following order is possible: 1.1 to 1.5 with 5.1 to 7.2, then 3.1, 3.2 with 8.1, suffice for practically all of the unstarred portions of Parts II, III, then IV.

Chapter I

SETS, SPACES, AND MEASURES

§1. SETS, CLASSES, AND FUNCTIONS

1.1 Definitions and notations. A *set* is a collection of arbitrary elements. By an abuse of language, an *empty set* is a "set with no elements."

Unless otherwise stated, all sets will be sets of elements of a fixed non empty set Ω, to be called a *space*. Elements of Ω will be called *points* and denoted by ω, with or without affixes (such as subscripts, superscripts, primes, etc.). Capitals A, B, C, $\cdots$, with or without affixes, will denote sets of points, $\{\omega\}$ will denote a set consisting of the one point ω, and $\emptyset$ will denote the empty set, that is, the set "containing no points." If ω is a point of A, we write $\omega \in A$ and, if ω is not a point of A we write $\omega \notin A$.

A set of sets is called a *class* and classes will be denoted by $\mathcal{A}$, $\mathcal{B}$, $\mathcal{C}$, $\cdots$, with or without affixes. The class of all the sets in Ω is called the *space of sets* in Ω and will be denoted by $S(\Omega)$. Thus a class of sets in Ω is a set in $S(\Omega)$ and all set notions and operations apply to classes considered as sets in the corresponding space of sets.

A is said to be a *subset of B*, or *included in B*, or *contained in B*, if all points of A are points of B; we then write $A \subset B$ or, equivalently, $B \supset A$. In symbols, if $\omega \in A$ implies $\omega \in B$, then $A \subset B$, and conversely. Clearly, for every set A,

$$\emptyset \subset A \subset \Omega,$$

and *the relation of inclusion is reflexive and transitive:*

$$A \subset A; \quad A \subset B \quad \text{and} \quad B \subset C \quad \text{imply} \quad A \subset C.$$

A and B are said to be *equal* if $A \subset B$ and $B \subset A$; we then write $A = B$.

55

Clearly, *the relation of equality is reflexive, transitive, and symmetric:*

$$A = A; \quad A = B \quad \text{and} \quad B = C \quad \text{imply} \quad A = C;$$

$$A = B \quad \text{implies} \quad B = A.$$

1.2 Differences, unions, and intersections. The *difference* $A - B$ is the set of all points of A which do not belong to B; in symbols, if $\omega \in A$ and $\omega \notin B$, then $\omega \in A - B$, and conversely. The particular difference $\Omega - A$, that is, the set of all points which do not belong to A, is called the *complement of* A and is denoted by A^c.

The *intersection* $A \cap B$, or simply AB, is the set of all points common to A and B; in symbols, if $\omega \in A$ and $\omega \in B$, then $\omega \in AB$ and conversely. The *union* $A \cup B$ is the set of all points which belong to at least one of the sets A or B; in symbols, if $\omega \in A$ or $\omega \in B$, then $\omega \in A \cup B$ and conversely. If $AB = \emptyset$, then A and B are said to be *disjoint*, and their union is then denoted by $A + B$ and called a *sum*.

It follows from the definitions that *the operations of intersection and union are associative, commutative, and distributive:*

$$(A \cup B) \cup C = A \cup (B \cup C), \quad (AB)C = A(BC);$$

$$A \cup B = B \cup A, \quad AB = BA;$$

$$(A \cup B)C = AC \cup BC, \quad (A \cup B)(A \cup C) = A \cup BC.$$

Moreover, the operation of complementation has the following properties:

$$A \subset B \quad \text{implies} \quad A^c \supset B^c;$$

$$\Omega^c = \emptyset, \quad \emptyset^c = \Omega, \quad AA^c = \emptyset, \quad A + A^c = \Omega, \quad (A^c)^c = A;$$

$$A - B = AB^c, \quad (A \cup B)^c = A^c B^c, \quad (AB)^c = A^c \cup B^c.$$

The notions of intersection and union extend at once to arbitrary classes. Let T be a set, not necessarily in Ω, and to every $t \in T$ assign a set $A_t \subset \Omega$. The class $\{A_t, t \in T\}$ of all these sets, or simply $\{A_t\}$ if there is no confusion possible, is a class assigned to the *index set* T.

The *intersection*, or *infimum*, of all sets of $\{A_t\}$ is defined to be the set of all those points which belong to every A_t, and is denoted by $\bigcap_{t \in T} A_t$ or by $\inf_{t \in T} A_t$; we drop $t \in T$ if there is no confusion possible. In symbols, if $\omega \in A_t$ for every $t \in T$, then $\omega \in \bigcap A_t$ and conversely.

The *union*, or *supremum*, of all sets of the class $\{A_t\}$ is defined to be the set of all those points which belong to at least one A_t, and is denoted by $\bigcup_{t \in T} A_t$ or by $\sup_{t \in T} A_t$; we drop $t \in T$ if there is no confusion possible. In symbols, if $\omega \in A_t$ for at least one $t \in T$, then $\omega \in \bigcup A_t$ and conversely.

If all sets of $\{A_t\}$ are pairwise disjoint, $\{A_t\}$ is said to be a *disjoint class* and the union of its sets, denoted then by $\sum A_t$, is called a *sum*. Conversely, the term "sum" and the symbols $\sum$ and $+$ when used for sets of a class will imply that the class is disjoint.

If ω does not belong to at least one A_t, then it belongs to every A_t^c, and conversely; consequently (de Morgan rule),

$$(\bigcup A_t)^c = \bigcap A_t^c, \quad (\bigcap A_t)^c = \bigcup A_t^c.$$

When $\{A_t\}$ is empty, that is, T is empty, it is natural to make the convention that $\bigcup_{t \in \emptyset} A_t = \emptyset$. Then, in order to preserve the foregoing relations, we have to make the convention that $\bigcap_{t \in \emptyset} A_t = \Omega$. Thus, by convention,

$$\bigcup_{t \in \emptyset} A_t = \emptyset, \quad \bigcap_{t \in \emptyset} A_t = \Omega.$$

It is easily seen, collecting all the relations so far obtained, that the following *duality* rule holds:

Every valid relation between sets, obtained by taking complements, unions, and intersections, is transformed into a valid relation if, the symbols "$=$" and "^{c}" remaining unchanged, the symbols $\bigcap$, $\subset$, and $\emptyset$, are interchanged with the symbols $\bigcup$, $\supset$, and Ω, respectively.

Operations performed on elements of "countable" classes will play a prominent role later in connection with the notion of measure. A set, or a class, is said to be *finite*, or *denumerable*, according as its elements can be put in a one-to-one correspondence with the set $\{1, 2, \cdots, n\}$ of the first n positive integers, for some value of n, or with the set of all positive integers $\{1, 2, \cdots$ ad infinitum$\}$. It is said to be *countable* if it is either finite or denumerable. Similarly, operations performed on elements of finite, denumerable, or countable classes will be said to be *finite, denumerable, or countable operations*, respectively.

The following immediate transformation of countable unions into countable sums will prove useful in connection with the notion of measure:

$$\bigcup A_j = A_1 + A_1^c A_2 + A_1^c A_2^c A_3 + \cdots.$$

1.3 Sequences and limits. To every value of $n = 1, 2, \cdots$, assign a set A_n; these sets A_n, whether distinct or not, are distinguished by their indices. The ordered denumerable class A_1, A_2, $\cdots$, is called *sequence* A_n. The set of all those points which belong to almost all A_n (all but any finite number) is called the *inferior limit* of A_n, and is denoted by lim inf A_n. Clearly

$$\lim \inf A_n = \bigcup_{n=1}^{\infty} \bigcap_{k=n}^{\infty} A_k.$$

The set of all those points which belong to infinitely many A_n is called the *superior limit* of A_n and is denoted by lim sup A_n. Since every point which belongs to almost all $A_n{}^c$ belongs to a finite number of A_n only, and conversely, it follows, by duality, that

$$\lim \sup A_n = (\bigcup_{n=1}^{\infty} \bigcap_{k=n}^{\infty} A_k{}^c)^c = \bigcap_{n=1}^{\infty} \bigcup_{k=n}^{\infty} A_k.$$

Every point which belongs to almost all A_n belongs to infinitely many A_n, so that

$$\lim \inf A_n \subset \lim \sup A_n.$$

Thus, if the reverse inclusion is true, lim inf A_n and lim sup A_n are equal to the same set A. Then A is called the *limit* of A_n and is denoted by lim A_n; the sequence A_n is said to *converge* to A and we write $A_n \to A$. Clearly, limits (inferior or superior) of sequences of sets are formed by denumerable set operations.

Monotone sequences form a basic class of convergent sequences. A sequence A_n is said to be *monotone* if it is either *nondecreasing:* $A_1 \subset A_2 \subset \cdots$, and we then write $A_n \uparrow$; or if it is *nonincreasing:* $A_1 \supset A_2 \supset \cdots$, and we then write $A_n \downarrow$. From the expressions above of inferior and superior limits, it follows at once that

every monotone sequence is convergent, and lim $A_n = \bigcup A_n$ *or* $\bigcap A_n$ *according as* $A_n \uparrow$ *or* $A_n \downarrow$.

Moreover, if we consider this proposition as a definition of limits of monotone sequences then, since for an arbitrary sequence B_n,

$$\bigcap_{k=n}^{\infty} B_k = \inf_{k \geq n} B_k \uparrow \quad \text{and} \quad \bigcup_{k=n}^{\infty} B_k = \sup_{k \geq n} B_k \downarrow,$$

it follows that its inferior and superior limits can be defined by

$$\lim \inf B_n = \lim_{n} (\inf_{k \geq n} B_k) \quad \text{and} \quad \lim \sup B_n = \lim_{n} (\sup_{k \geq n} B_k).$$

1.4 Indicators of sets. Set operations can be replaced by equivalent but more familiar ones, in the following manner. To every set A assign a function I_A of ω, to be called the *indicator of A*, defined by

$$I_A(\omega) = 1 \quad \text{or} \quad 0 \quad \text{according as} \quad \omega \in A \quad \text{or} \quad \omega \notin A.$$

Conversely, every function of ω which can take only the values 0 and 1 is the indicator of the set for the points of which it takes the value 1. The one-to-one correspondences (denoted by $\Leftrightarrow$) and relations listed below are immediate.

$$I_A \leq I_B \Leftrightarrow A \subset B, \quad I_A = I_B \Leftrightarrow A = B, \quad I_{AB} = 0 \Leftrightarrow AB = \emptyset,$$

$$I_\emptyset = 0, \quad I_\Omega = 1, \quad I_A + I_{A^c} = 1,$$

$$I_{\inf A_t} = \inf I_{A_t}, \quad I_{\sup A_t} = \sup I_{A_t},$$

$$I_{\bigcap A_n} = \prod I_{A_n}, \quad I_{\sum A_n} = \sum I_{A_n},$$

$$I_{\bigcup A_n} = I_{A_1} + (1 - I_{A_1})I_{A_2} + (1 - I_{A_1})(1 - I_{A_2})I_{A_3} + \cdots$$

$$I_{\liminf A_n} = \liminf I_{A_n}, \quad I_{\limsup A_n} = \limsup I_{A_n}, \quad I_{\lim A_n} = \lim I_{A_n}.$$

1.5 Fields and σ-fields. Classes of sets in Ω are sets in the space $S(\Omega)$ of all sets in Ω and thus what precedes applies to classes. However, there is a notion specific to classes—that of closure under one or more set operations. A class $\mathcal{C}$ is said to be *closed* under a set operation if the sets obtained by performing this operation on sets of $\mathcal{C}$ are sets of $\mathcal{C}$. In particular, the class $S(\Omega)$ of all sets in Ω is closed under every set operation.

In connection with the notions of measurability and of measure, two species of classes play a prominent role—fields and σ-fields. A field is a (nonempty) class closed under all finite set operations; clearly, every field contains $\emptyset$ and Ω. A *σ-field* is a (nonempty) class closed under all countable set operations; clearly every σ-field is a field. We observe that, because of the duality rule, closure under complementations and finite (countable) intersections implies closure under finite (countable) unions. Also we can interchange in this property "intersections" and "unions."

Let $\mathcal{S}$-classes be species of classes closed under set operations $\mathcal{S}$; for example, the species of fields or the species of σ-fields. We observe that $S(\Omega)$ is an $\mathcal{S}$-class, whatever be the set operations $\mathcal{S}$.

a. *Arbitrary intersections of $\mathcal{S}$-classes are $\mathcal{S}$-classes. In particular, arbitrary intersections of fields or of σ-fields are fields or σ-fields, respectively.*

For the intersection of a collection of 𝖲-classes belongs to every one of these classes. Therefore, performing operations 𝖲 on sets of the intersection, we obtain sets belonging to every one of these classes, that is, to the intersection.

This property gives rise to the notion of a "minimal" 𝖲-class over a given class. An 𝖲-class $\mathcal{C}'$ containing $\mathcal{C}$ is a *minimal class over $\mathcal{C}$ or the 𝖲-class generated* by $\mathcal{C}$ if every 𝖲-class containing $\mathcal{C}$ contains $\mathcal{C}'$.

b. *There is one, and only one, minimal 𝖲-class over a class $\mathcal{C}$. In particular, there is one, and only one, minimal field and one, and only one, minimal σ-field over $\mathcal{C}$.*

For the intersection of all 𝖲-classes containing $\mathcal{C}$ contains $\mathcal{C}$ and is contained in every 𝖲-class containing $\mathcal{C}$.

A space Ω in which is selected a fixed σ-field $\mathcal{A}$ is called a *measurable space* $(\Omega, \mathcal{A})$. If there is no confusion possible, the sets of $\mathcal{A}$ are said to be *measurable*.

1.6 Monotone classes. We shall need the notion of monotone classes in connection with the problem of extending measures on a field to its minimal σ-field. A *monotone* class is a class closed under formation of limits of monotone sequences.

a. *A σ-field is a monotone field and conversely.*

The first assertion is obvious and the second follows from the fact that every countable intersection $\bigcap_n A_n$ and union $\bigcup_n A_n$ is a monotone limit of sequences $\bigcap_{k=1}^n A_k$ and $\bigcup_{k=1}^n A_k$ of finite intersections and unions. The property we shall require is as follows:

A. *The minimal monotone class $\mathfrak{M}$ and the minimal σ-field $\mathcal{A}$ over the same field $\mathcal{C}$ coincide.*

Proof. On account of **a** and minimality of $\mathfrak{M}$ and $\mathcal{A}$, it suffices to prove that $\mathfrak{M}$ is a field; for, a monotone field $\mathfrak{M}$ is a σ-field so that $\mathfrak{M} \supset \mathcal{A}$, and the σ-field $\mathcal{A}$ is monotone so that $\mathfrak{M} \subset \mathcal{A}$. Since $\mathfrak{M} \supset \mathcal{C} \ni \Omega$ and unions are reducible to intersections (by means of complementations), it suffices to prove that, if A and B belong to $\mathfrak{M}$, so do AB, A^cB, and AB^c.

For every fixed $A \in \mathfrak{M}$, let $\mathfrak{M}_A$ be the class of all $B \in \mathfrak{M}$ with the asserted property. Every $\mathfrak{M}_A$ is monotone for, if the sequence $B_n \in \mathfrak{M}_A$ is monotone, then $B = \lim B_n$ belongs to $\mathfrak{M}$ and so do the limits of monotone sequences

$$AB = \lim AB_n, \quad A^cB = \lim A^cB_n, \quad AB^c = \lim AB_n{}^c.$$

It follows that, for every $A \in \mathcal{C}$, the class $\mathfrak{M}_A$ coincides with $\mathfrak{M}$. For $\mathcal{C}$ being a field, every $B \in \mathcal{C}$ is $\in \mathfrak{M}_A$, so that $\mathcal{C} \subset \mathfrak{M}_A \subset \mathfrak{M}$ and, hence, $\mathfrak{M}$ being minimal over $\mathcal{C}$, $\mathfrak{M}_A = \mathfrak{M}$. In fact, $\mathfrak{M}_B = \mathfrak{M}$ for every $B \in \mathfrak{M}$. For, the conditions imposed upon pairs A, B being symmetric, $B \in \mathfrak{M}(= \mathfrak{M}_A$ for $A \in \mathcal{C})$ is equivalent to $A \in \mathfrak{M}_B$ for every $A \in \mathcal{C}$ so that $\mathcal{C} \subset \mathfrak{M}_B$ and hence as above, $\mathfrak{M}_B = \mathfrak{M}$. But this last property means that $\mathfrak{M}$ is a field, and the proof is complete.

*1.7 Product sets. We introduce now a different type of set operation and corresponding notions, for which we shall have need later. Let A_1 and A_2 be two arbitrary sets with elements ω_1 and ω_2, respectively. By the *product set $A_1 \times A_2$* we shall mean the set of all ordered pairs $\omega = (\omega_1, \omega_2)$ where $\omega_1 \in A_1$ and $\omega_2 \in A_2$. If A_1, B_1, $\cdots$ are sets in a space Ω_1 and A_2, B_2, $\cdots$ are sets in a space Ω_2, then $A_1 \times A_2$, $B_1 \times B_2$, $\cdots$ are sets in the *product space* $\Omega_1 \times \Omega_2$, called *intervals* or *rectangles* in $\Omega_1 \times \Omega_2$ and the properties below follow readily from the definition:

$$(A_1 \times A_2) \cap (B_1 \times B_2) = (A_1 \cap B_1) \times (A_2 \cap B_2)$$

$$(A_1 \times A_2) - (B_1 \times B_2) = (A_1 - B_1) \times (A_2 - B_2) + (A_1 - B_1)$$
$$\times (A_2 \cap B_2) + (A_1 \cap B_1) \times (A_2 - B_2)$$

In turn, it follows at once from these relations that

a. *If $\mathcal{C}_1$ and $\mathcal{C}_2$ are fields of sets in Ω_1 and Ω_2 respectively, then the class of all finite sums of intervals $A_1 \times A_2$, where $A_1 \in \mathcal{C}_1$ and $A_2 \in \mathcal{C}_2$, is a field of sets in $\Omega_1 \times \Omega_2$.*

This field will be called the *product field* of $\mathcal{C}_1$ and $\mathcal{C}_2$.

Yet, if $\mathcal{Q}_1$ and $\mathcal{Q}_2$ are σ-fields of sets in Ω_1 and Ω_2, respectively, then the product field of $\mathcal{Q}_1$ and $\mathcal{Q}_2$ is not necessarily a σ-field. The minimal σ-field over it will be called the *product σ-field* $\mathcal{Q}_1 \times \mathcal{Q}_2$. If $(\Omega_1, \mathcal{Q}_1)$ and $(\Omega_2, \mathcal{Q}_2)$ are measurable spaces, then their *product measurable space* is, by definition, $(\Omega_1 \times \Omega_2, \mathcal{Q}_1 \times \mathcal{Q}_2)$.

Let $\Omega = \Omega_1 \times \Omega_2$ and $\mathcal{Q} = \mathcal{Q}_1 \times \mathcal{Q}_2$. If $A \subset \Omega$ is measurable and $\omega_1 \in \Omega_1$ is a fixed point, then the set $A(\omega_1)$ of all points $\omega_2 \in \Omega_2$ such that $\omega = (\omega_1, \omega_2) \in A$ is called the *section of A at ω_1*; similarly for the section $A(\omega_2)$ at $\omega_2 \in \Omega_2$; by the definition, $A(\omega_1) \subset \Omega_2$ and $A(\omega_2) \subset \Omega_1$.

b. *Every section of a measurable set is measurable.*

For let $\mathcal{C}$ be the class of all measurable sets in Ω whose sections are measurable. It is easily seen that $\mathcal{C}$ is a σ-field. On the other hand, if $A = A_1 \times A_2$ is a measurable interval, that is, A_1 and A_2 are measurable, then every section of A is either empty or is A_1 or A_2, so that $A_1 \times A_2 \in \mathcal{C}$. Therefore, $\mathcal{C} = \mathcal{C}_1 \times \mathcal{C}_2$, being the minimal σ-field over all measurable intervals, is contained in $\mathcal{C}$, and the assertion is proved.

The foregoing definitions and properties extend at once to any finite number of sets and of measurable spaces. However, in the nonfinite case, some of these definitions have to be modified in order to preserve these properties.

Let $\{A_t, t \in T\}$ be an arbitrary collection of arbitrary sets A_t in arbitrary spaces Ω_t of points ω_t. The *product set* $A_T = \prod_{t \in T} A_t$ is the set of all the new elements $\omega_T = (\omega_t, t \in T)$ such that $\omega_t \in A_t$ for every $t \in T$. The product set A_T is in the *product space* $\Omega_T = \prod_{t \in T} \Omega_t$; we drop "$t \in T$" if there is no confusion possible. It follows from the foregoing definition that, for any set B, when the Ω_t are identical

$$\left(\bigcap A_t\right) \times B = \bigcap (A_t \times B), \quad \left(\bigcup A_t\right) \times B = \bigcup (A_t \times B).$$

Let $T_N = (t_1, \cdots, t_N)$ be a finite index subset and let A_{T_N} be a set in the product space Ω_{T_N}. The set $A_{T_N} \times \Omega_{T-T_N}$ is a *cylinder* in Ω_T with *base* A_{T_N}. If the base is a product set $\prod_{t \in T_N} A_t$, the cylinder becomes a *product cylinder* or an *interval* in Ω_T with *sides* $A_t, t \in T_N$. Let $\mathcal{C}_t$ be fields in Ω_t. It is easily seen that, as in the finite case,

A. *The class of all finite sums of all the intervals in Ω_T with sides $A_t \in \mathcal{C}_t$, is a field of sets in Ω_T.*

This field is the *product field* of the fields $\mathcal{C}_t$.

Let $(\Omega_t, \mathcal{C}_t)$ be measurable spaces. The minimal σ-field over the product field of the $\mathcal{C}_t$ is the *product σ-field* $\mathcal{C}_T = \prod \mathcal{C}_t$ *of measurable sets in Ω_T*, and the measurable space $(\Omega_T, \mathcal{C}_T)$ is the *product measurable space* $(\prod \Omega_t, \prod \mathcal{C}_t)$ of the measurable spaces $(\Omega_t, \mathcal{C}_t)$. It is easily seen, as in the finite case, that **b** remains valid:

B. *Sections at ω_{T_N} of measurable sets in Ω_T are measurable sets in Ω_{T-T_N}.*

*1.8 Functions and inverse functions. Perhaps the most important notion of mathematics is that of function (or transformation, or mapping, or correspondence). We have already encountered functions defined on an index set T whose "values" are sets in Ω. In general, a *function X on* a space Ω—the *domain* of X—to a space Ω'—the *range space of X*—is defined by assigning to every point $\omega \in \Omega$ a point $\omega' \in \Omega'$ called the *value of X at ω* and denoted by $X(\omega)$. Sets and classes of sets in Ω' will be denoted by A', B', $\cdots$, and $\mathcal{A}'$, $\mathcal{B}'$, $\cdots$, respectively. It will be assumed, once and for all, that functions are *single-valued*, that is, to every given $\omega \in \Omega$ corresponds one, and only one, value $X(\omega)$.

The set of values of X for all $\omega \in A$ is called the *image* $X(A)$ of A (by X) and the class of images $X(A)$ for all $A \in \mathcal{C}$ is called the *image* $X(\mathcal{C})$ of $\mathcal{C}$ (by X); in particular $X(\Omega)$ is the *range* (of all values) of X. Thus, a function X on Ω to Ω' determines a function on $S(\Omega)$ to $S(\Omega')$. While this new function is of no great interest, such is not the case for the inverse function that we shall introduce now.

By $[\omega; \cdots]$ where $\cdots$ stands for expressions and/or relations involving functions on Ω, we denote the set of points $\omega \in \Omega$ for which these expressions are defined and/or these relations are valid; if there is no confusion possible we drop "ω;". Thus, $[X = \omega']$, or *inverse image* of ω', is the set of all points ω for which $X(\omega) = \omega'$; $[X \in A']$, or *inverse image of A'*, is the set of all points ω for which $X(\omega) \in A'$; and $[A; X(A) \in \mathcal{C}']$, or *inverse image of $\mathcal{C}'$*, is the class of inverse images of all sets $A' \in \mathcal{C}'$. We observe that the inverse image of an ω' which does not belong to the range of X is the empty set $\emptyset$ in Ω.

The *inverse function* X^{-1} of X is defined by assigning to every A' its inverse image $[X \in A']$. In other words, X^{-1} is a function on $S(\Omega')$ to $S(\Omega)$ with values $X^{-1}(A') = [X \in A']$; if $A' = \{\omega'\}$, then we write $X^{-1}(\omega')$ for $X^{-1}(\{\omega'\}) = [X = \omega']$. Since X is single-valued, X^{-1} generates a partition of Ω into disjoint inverse images of points $\omega' \in \Omega'$. It follows readily that

$$X^{-1}(A' - B') = X^{-1}(A') - X^{-1}(B'),$$

$$X^{-1}(\bigcup A'_t) = \bigcup X^{-1}(A'_t), \quad X^{-1}(\bigcap A'_t) = \bigcap X^{-1}(A'_t), \cdots$$

Therefore,

A. BASIC PROPERTY OF INVERSE FUNCTIONS: *Inverse functions preserve all set and class inclusions and operations.*

It follows at once that

If $\mathcal{C}'$ is closed under a set operation so is $X^{-1}(\mathcal{C}')$. In particular, the inverse image of a σ-field is a σ-field, and the inverse image of the minimal σ-field over $\mathcal{C}'$ is the minimal σ-field over $X^{-1}(\mathcal{C}')$.

Moreover,

If $\mathcal{A}$ is a σ-field so is the class of all sets whose inverse images belong to $\mathcal{A}$.

The notion of function can be "iterated" as follows. Let X be a function on Ω to Ω' and let X' be a function on Ω' to Ω''. Then, the *function of function* $X'X$ defined by $(X'X)(\omega) = X'(X(\omega))$ is a function on Ω to Ω''. Clearly, its inverse function $(X'X)^{-1}$ is a function on $S(\Omega'')$ to $S(\Omega)$ such that, for every set $A'' \subset \Omega''$,

$$(X'X)^{-1}(A'') = X^{-1}(X'^{-1}(A''))$$

or, in a condensed form,

$$(X'X)^{-1} = X^{-1}X'^{-1}.$$

*1.9 Measurable spaces and functions. So far, we did not consider particular species of functions. There are two species which play a basic role in abstract analysis. We shall introduce them now. But first we examine, in more detail, the class of inverse images of points of the range space.

Let X be a function on Ω to Ω'. The partition of Ω formed by the inverse images $X^{-1}(\omega')$ of all points $\omega' \in \Omega'$ is said to be *induced* (or *determined*) *by* X and X is said to be *constant* $(=\omega')$ on $X^{-1}(\omega')$. Since the class of values $X^{-1}(A')$ of X^{-1} is the inverse image of the σ-field of all sets A' in Ω', it is a σ-field. If the partition induced by X is finite, or denumerable, or countable, then X is said to be *finitely*, or *denumerably*, or *countably valued*, respectively; in other words, X is, say, countably valued if the set of its values is countable. Setting $A_j = [X = \omega'_j]$, we can write every countably valued function X as a countable combination of indicators:

$$X = \sum_j \omega'_j I_{A_j}.$$

Conversely, we make the convention that every time such a "sum" is written, the sets A_j form a partition of the domain of the function X. If the ω'_j are distinct, then this partition is the one induced by the function represented by the "sum."

Now, let $\mathcal{Q}$ be a fixed σ-field in Ω. Ω, together with $\mathcal{Q}$, is called a *measurable space* $(\Omega, \mathcal{Q})$, and the sets of $\mathcal{Q}$ are then said to be *measurable* (although this terminology derives from the notion of measure, we emphasize that, nowadays, the notion of measurability is independent of that of measure). A countably valued function $X = \sum \omega'_j I_{A_j}$, where the sets A_j are measurable, is called a countably valued measurable function—for short, an *elementary function*; if X is finitely valued, then this elementary function is also called a *simple function*. Clearly

the sets of the σ-field induced by an elementary function are measurable.

We are now in a position to introduce the general notion of measurable functions. However, there are several ways for doing so, and the classes of measurable functions so defined are, in general, not the same.

One way of defining measurable functions is to extend a basic property of inverse functions of elementary functions, as follows: Let $(\Omega, \mathcal{Q})$ and $(\Omega', \mathcal{Q}')$ be two measurable spaces. The inverse images by elementary functions on Ω to Ω' of measurable sets are measurable. Extending this property, we say that a function X on Ω to Ω' is *measurable* if the inverse images by X of measurable sets ($\in \mathcal{Q}'$) are measurable ($\in \mathcal{Q}$). If, moreover, $(\Omega'', \mathcal{Q}'')$ is a measurable space and X' on Ω' to Ω'' is a measurable function, then $X'X$ is measurable, for

$$(X'X)^{-1}(\mathcal{Q}'') = (X^{-1}X'^{-1}(\mathcal{Q}'')) \subset X^{-1}(\mathcal{Q}') \subset \mathcal{Q}.$$

Thus, *with this definition, a measurable function of a measurable function is measurable.*

Another way of defining measurable functions is as follows: Let $(\Omega, \mathcal{Q})$ be a measurable space on which are defined simple (elementary) functions to a space Ω' (there are no measurable sets in Ω'). A notion of limit is introduced on Ω', and *measurable functions in the sense of this limit* are then defined to be limits of convergent sequences of simple (elementary) functions. This approach is particularly suited for the introduction of integrals of measurable functions. Later we shall see cases in which measurable sets and the notion of limit are selected in such a manner that the two definitions are equivalent.

*§ 2. TOPOLOGICAL SPACES

The selections of measurable sets and of concepts of limit in range-spaces are rooted in the properties of the euclidean line: real line $R = (-\infty, +\infty)$ with euclidean distance $|x - y|$ of points (numbers, reals) x, y. Species of spaces vary according to the preserved amount of these properties, an amount which increases as we pass from separated spaces to metric spaces, then to Banach spaces and to Hilbert spaces. We examine here the basic properties of these spaces and shall encounter them in various guises throughout the book. At the same time, the few notions of topology which follow are a recapitulation of the properties of the euclidean line and, more generally, of euclidean spaces. We urge the reader to keep this fact constantly in mind by illustrating the concepts and their relationships in terms of euclidean spaces; for this reason, we denote here the points by x, y, z, with or without affixes.

Points, sets, and classes will be those of the space $\mathfrak{X}$ under consideration, unless otherwise stated.

We use without comment the *axiom of choice*: given a nonempty class of nonempty sets, there exists a function which assigns to every set of the class a point belonging to this set; in other words, we can always "choose" a point from every one of the sets of the class.

2.1   Topologies and limits.   A class Θ is a *topology* or the class of *open sets* if it is closed under formation of arbitrary unions and finite intersections and contains $\emptyset$ and Ω (the last property follows from the closure property by the conventions relative to intersections and unions of sets of an empty class).   The dual class of complements of open sets is the class of *closed sets;* hence it is closed under formation of arbitrary intersections and finite unions and contains Ω and $\emptyset$.

A topological space $(\mathfrak{X}, \Theta)$ is a space $\mathfrak{X}$ in which is selected a topology Θ; from now on, all spaces under consideration will be topological and we shall frequently drop "Θ."   A *topological subspace* thereof (A, Θ_A) is a set A in which is selected its *induced topology* Θ_A which consists of all the intersections of open sets with A and is, clearly, a topology in A.   It is important to distinguish the properties of A considered as a set in $(\mathfrak{X}, \Theta)$ from those of A considered as a topological subspace of $(\mathfrak{X}, \Theta)$.

To every set A there are assigned an open set A^o and a closed set $\overline{A}$, as follows.   The *interior* A^o of A is the maximal open set contained in A, that is, the union of all open sets in A; in particular, if A is open, then $A^o = A$.   The *adherence* $\overline{A}$ of A is the minimal closed set containing A, that is, the intersection of all closed sets containing A; in particular, if A is closed, then $\overline{A} = A$.   The definitions of interiors and adherences of A and A^c are clearly dual, so that

$$(A^o)^c = (\overline{A^c}), \quad (A^c)^o = (\overline{A})^c.$$

In topological spaces relations between sets and points are described in terms of neighborhoods.   Every set containing a nonempty open set is a *neighborhood* of any point x of this open set; the symbol V_x will denote a neighborhood of x.   The points of the interior A^o of A are "interior" to A; in other words, x is *interior* to A if A is a V_x.   The points of the adherence $\overline{A}$ of A are adherent to A; in other words, x is *adherent* to A if no V_x is disjoint from A, that is, $x \notin (A^c)^o = (\overline{A})^c$.

Classical analysis is concerned primarily with continuous functions on euclidean lines to euclidean lines.   In general, a function X on a topological domain Ω to a topological range space $\mathfrak{X}$ is *continuous at*

$\omega \in \Omega$ if the inverse images of neighborhoods of $x = X(\omega)$ are neighborhoods of ω; X is *continuous* (on Ω) if it is continuous at every $\omega \in \Omega$. Since taking inverse images preserves all set operations, it follows readily that we can limit ourselves to open (closed) sets. Thus X is continuous if, and only if, the inverse images of open (closed) sets are open (closed) and, hence, a continuous function induces on its domain a topology contained in (no "finer" than) that of the domain. Therefore, *if in topological spaces the σ-fields of measurable sets are selected to be the minimal σ-fields over the topologies, then continuous functions are measurable.* The importance of the concept of continuity is emphasized by the fact that two spaces $\mathfrak{X}$ and $\mathfrak{X}'$ are considered to be "topologically equivalent" if, and only if, there exists a one-to-one correspondence X on $\mathfrak{X}$ to $\mathfrak{X}'$ such that X and X^{-1} are continuous.

The basic concept which distinguishes classical analysis from classical algebra and which gave rise to the various concepts examined in this section is that of limit of sequences of numbers. In a topological space it becomes: x is *limit of a sequence x_n* or the sequence x_n *converges to x* if, for every V_x, there exists an integer $n(V_x)$ such that $x_n \in V_x$ for all $n \geqq n(V_x)$. However, the need for a more general concept of limit is already apparent in the classical theory of integration where the partitions of the interval of integration form a "direction" and the Riemann sums form a "directed set" of numbers of which the Riemann integral, if it exists, is the "limit." It so happens that this type of limit is precisely the one required for general topological spaces, and we now define the foregoing terms; the role of sequences in some species of spaces (including the euclidean ones) will be better understood when considered within the general setup.

Let T be a set of points t, with or without indices. T is *partially ordered* if a partial ordering is defined on it. A *partial ordering* "$\prec$," to be read "precedes," is a binary relation which is transitive ($t \prec t'$ and $t' \prec t''$ imply $t \prec t''$), reflexive ($t \prec t$), and such that, if $t \prec t'$ and $t' \prec t$, then $t = t'$; upon writing $t' \succ t$ when $t \prec t'$, the relation "$\succ$," to be read "follows," is also a partial ordering. T is a *direction* if it is partially ordered and if every pair t, t' is followed by some t'' ($t \prec t''$, $t' \prec t''$). T is *linearly ordered*, and *a fortiori* is a direction, if every pair t, t' is ordered (either $t \prec t'$ or $t' \prec t$). For example, the sets in a space are partially ordered by the relation of inclusion and the neighborhoods of a point x form a direction (this is the root of the definition of limit as given below); the finite partitions of an interval of integration form a direction when ordered by the relation of refinement; integers and, in general, sets of numbers are linearly ordered by the relation "$\leq$," etc.

A function X on T to $\mathfrak{X}$ can be represented by the indexed set $\{x_t\}$ of its values which may or may not be distinct but which are always distinguished by their indices t. The indexed set $\{x_t\}$ is *directed* if T is a direction; sequences $\{x_n\}$ are special directed sets representing functions on the (linearly ordered) set of positive integers. We are now ready to define the general concept of limit.

The point x is the *limit* of a directed set $\{x_t\}$ and we write $x = \lim x_t$, or, equivalently, x_t *converges to x* and we write $x_t \rightarrow x$, if, for every V_x, there exists an index $t(V_x)$ such that $x_t \in V_x$ for all those indices which follow $t(V_x)$. However, the concept of limit is of use only if, when the limit exists, it is unique; this requirement leads to the introduction of "separated" or "Hausdorff" space as follows:

A. SEPARATION THEOREM. *The following three definitions are equivalent. A topological space is separated if*

(S_1) *every directed set has at most one limit,*
(S_2) *every pair of distinct points has disjoint neighborhoods,*
(S_3) *the intersection of all closed neighborhoods of a point reduces to this point.*

The term "separated" expresses property (S_2).

We observe that, according to (S_3), in a separated space every set reduced to a point is closed.

Proof. (S_1) *and* (S_2) *are equivalent.* Let $x \neq y$. If $x_t \rightarrow x$ and $x_t \rightarrow y$, then $x_t \in V_x \cap V_y$ for all those t which follow both $t(V_x)$ and $t(V_y)$; since T is a direction such t exist so that no pair V_x, V_y is disjoint.

Conversely, if no pair V_x, V_y is disjoint, then there exist points $z(V_x, V_y) \in V_x \cap V_y$ and, since these pairs form a direction when ordered by the relation $(V_x, V_y) \prec (V'_x, V'_y)$ if $V_x \supset V'_x$ and $V_y \supset V'_y$, these points form a directed set converging to both x and y.

(S_2) *and* (S_3) *are equivalent.* If for every $y \neq x$ there exists a V_x such that $y \notin \overline{V}_x$, then the intersection of all $\overline{V}_x$ reduces to x. Conversely, if the intersection of all $\overline{V}_x$ reduces to the set formed by x, then, for every $y \neq x$, there exists a V_x such that $y \notin \overline{V}_x$, and the open set $(\overline{V}_x)^c$ is a neighborhood of y disjoint from V_x. The proof is terminated.

From now on, all spaces will be separated spaces.

2.2 Limit points and compact spaces. Analysis of concepts or properties leads to the introduction of "weaker" ones. A property $\mathscr{P}$ is *weaker* than a property $\mathscr{P}'$ if $\mathscr{P}'$ implies $\mathscr{P}$; $\mathscr{P}$ is a necessary condition for $\mathscr{P}'$ and $\mathscr{P}'$ is a sufficient condition for $\mathscr{P}$.

Perhaps even more basic than the concept of limit is the weaker one of limit point. A point x is a *limit point* of the directed set $\{x_t\}$ if, for every pair t, V_x, there exists *some* $t' > t$ such that $x_{t'} \in V_x$. The definitions of limit and of limit point yield at once (i) and (ii) of the proposition below, and then (iii) follows.

a. *Let the sets A_t be formed by all those points $x_{t'}$ for which t' follows t:*
$A_t = \{x_{t'}, t' > t\}$.

(i) $x_t \to x$ *if, and only if, for every V_x there exists an $A_t \subset V_x$.*

(ii)   *x is a limit point of $\{x_t\}$ if, and only if, no pair A_t, V_x is disjoint.*

(iii) *the set of all limit points of $\{x_t\}$ coincides with the intersection of all $\overline{A}_t$, and if $x_t \to x$ then this set reduces to the single point x.*

The reason for the somewhat confusing terminology above is that every limit point of $\{x_t\}$ is *the* limit of some subset of $\{x_t\}$, in the following sense. A direction S of elements s, s', $\cdots$ is a *subdirection* of the direction T when there exists a function f on S to T with the property that, for every t, there is an s such that, if s' follows s, then $t' = f(s')$ follows t. The set $\{x_{f(s)}\}$ directed by the subdirection S of T is a *subdirected set.* Clearly, if $x_t \to x$, then every subdirected set $x_{f(s)} \to x$.

b. *A point x is a limit point of a directed set $\{x_t\}$ if, and only if, the set contains a subdirected set which converges to x.*

Proof. The "if" assertion follows at once from the definitions. As for the "only if" assertion, it suffices for every pair $s' = (t, V_x)$ to take $f(s') = t' > t$ such that $x_{t'} \in V_x$ and direct the pairs by $(t_1, V_x{}^1) > (t_2, V_x{}^2)$ when $t_1 > t_2$ and $V_x{}^1 \subset V_x{}^2$.

Compact spaces are separated spaces in which every directed set has at least one limit point; a set is *compact* if it is compact in its induced topology. Compactness plays a prominent role in analysis and it is important to have equivalent characterizations of compact spaces. We shall use repeatedly the following terminology: a subclass of open sets is an *open covering* of a set if every point of the set belongs to at least one of the sets of the subclass.

A. COMPACTNESS THEOREM. *The following three properties of separated spaces are equivalent:*

(C_1) BOLZANO-WEIERSTRASS PROPERTY: *every directed set has at least one limit point.*

(C_2) HEINE-BOREL PROPERTY: *every open covering of the space contains a finite covering of the space.*

(C_3) INTERSECTION PROPERTY: *every class of closed sets such that all its finite subclasses have nonempty intersections has itself a nonempty intersection.*

If *some* class has the property described in (C_3), we say that it has the *finite intersection property.*

Proof. The intersection property means by contradiction that every class of closed sets whose intersection is empty contains a finite subclass whose intersection is empty. Thus, it is the dual of the Heine-Borel property, and it suffices to show that it is equivalent to the Bolzano-Weierstrass one.

Let $\{x_t\}$ be a directed set and, for every $t_0 \in T$, consider the adherence of the set of all the x_t with t following t_0. Since T is a direction, these adherences form a class of closed sets with finite intersection property. Thus, if the intersection property is true, then there exists an x common to all these adherences and it follows that x is a limit point of $\{x_t\}$.

Conversely, consider a class of closed sets with the finite intersection property and adjoin all finite intersections to the class. The class so obtained is directed by inclusion so that, by selecting a point from every set of this class, we obtain a directed set. If the Bolzano-Weierstrass property is true, then this set has a limit point and this point belongs to every set of the class; hence the intersection of the class is not empty. This completes the proof.

COMPACTNESS PROPERTIES. 1° *In a compact space, a directed set $x_t \to x$ if, and only if, x is its unique limit point.*

Proof. We use **a** and its notations. The "only if" assertion holds by **a**(iii). As for the "if" assertion, if $x_t \nrightarrow x$ then, by **a**(i), there exists a V_x such that no A_t is disjoint from V_x^c; thus, for every t we can select a $t' > t$ such that $x_{t'} \in A_t \cap V_x^c$. Since the space is compact, the subdirected set $\{x_{t'}\}$, hence, by **b**, the directed set $\{x_t\}$, has a limit point $x' \in V_x^c$. Therefore, $x \neq x'$ and x cannot be the unique limit point of $\{x_t\}$.

$2°$ *Every compact set is closed, and in a compact space the converse is true.*

Proof. Let A be compact and let V_x, $V_y(x)$ be a disjoint pair of open neighborhoods of $x \in A$ and $y \in A^c$. By the Heine-Borel property, the open covering $\{V_x\}$ of A where x ranges over A contains a finite subcovering $\{V_{x_k}\}$, and the disjoint open sets $V = \bigcup_k V_{x_k}$, $V' = \bigcap_k V_y(x_k)$ are such that $A \subset V$ and $y \in V'$. Thus, the open neighborhood V' of y contains no points of A; hence $y \notin \overline{A}$. Since $y \in A^c$ is arbitrary, it follows that A^c and $\overline{A}$ are disjoint, and the first assertion is proved. The second assertion follows readily from the intersection property.

$3°$ *The intersection of a nonincreasing sequence of nonempty compact sets is not empty.*

Apply the intersection property.

$4°$ *The range of a continuous function on a compact domain is compact.*

Proof. Because of continuity of the function, the inverse image of every open covering of the range is an open covering of the compact domain; hence it contains a finite open subcovering which is the inverse image of a finite open subcovering of the range. Thus, the range has the Heine-Borel property, and the assertion is proved.

The euclidean line $R = (-\infty, +\infty)$ is not compact but, according to the Bolzano-Weierstrass or Heine-Borel theorems, every closed interval $[a, b]$ is compact. These theorems become valid for the whole line if it is "extended"—that is, if points $-\infty$ and $+\infty$ are added. Thus, the extended euclidean line $\overline{R} = [-\infty, +\infty]$ is compact. In fact, R is locally compact and every locally compact space can be compactified by adding one point only, as below.

A separated space is *locally compact* if every point has a compact neighborhood; it is easily shown that every neighborhood then contains a compact one. The one-point *compactification* of a separated space $(\mathfrak{X}, \mathcal{O})$ is as follows. Adjoin to the points of $\mathfrak{X}$ an arbitrary point $\infty \notin \mathfrak{X}$ and adjoin to the open sets all sets obtained by adjoining to the point ∞ those open sets whose complements are compact. Denote the topological space so obtained by $(\mathfrak{X}_\infty, \mathcal{O}_\infty)$.

5° *The one-point compactification of a locally compact but not compact space is a compact space, and the induced topology of the original space is its original topology.*

Proof. The last assertion follows at once from the definition of $\mathcal{O}_\infty$. As for the first assertion, observe that the new space is separated, since two distinct points belonging to the separated original space are separated and the point ∞ is separated from any $x \in \mathcal{X}$ by taking a compact and hence closed $V_x \subset \mathcal{X}$, so that $\infty \in V_x{}^c$. Also, the new space has the Heine-Borel property, since an open covering of it has a member $O + \{\infty\}$ with O^c compact and hence contains a finite subcovering of O^c which, together with $O + \{\infty\}$, is a finite subcovering of the new space.

2.3 Countability and metric spaces. The euclidean line possesses many countability properties, among them separability (the countable set of rationals is dense in it) and a countable base (the countable class of all intervals with rational extremities); this permits us to define limits in terms of sequences only. In general topological spaces, a set A is *dense in B* if $\overline{A} \supset B$; in other words, taking for simplicity $B = \mathcal{X}$, A is dense in $\mathcal{X}$ if no neighborhood is disjoint from A; and B is *separable* if there exists a countable set A dense in B. A *countable base at x* is a countable class $\{V_x(j)\}$ of neighborhoods of x such that every neighborhood of x contains a $V_x(j)$; and the space has a *countable base* $\{V(j)\}$ if, for every point x, a subclass of $V(j)$'s is a base at x.

a. *A space has a countable base only if it is separable and has a countable base at every point. Then every open covering of the space contains a countable covering of the space.*

Note that if a countable set $\{x_j\}$ is dense in a metric space, then at every x_j there is a countable base of spheres of rational radii, and the countable union of all these countable bases is a base for the space.

Proof. If the space has a countable base $\{V(j)\}$, then it has a countable base at every point. Moreover, if A is a set formed by selecting a point x_j from every $V(j)$, then, since any neighborhood of any point contains a $V(j)$, it contains the corresponding point x_j, so that no neighborhood is disjoint from A.

Finally, given an open covering of the space, every one of its sets contains a $V(j)$ so that, for every $V(j)$, we can select one set O_j of the covering containing it. The countable class $\{O_j\}$ is an open covering of the space, and the proof is terminated.

A basic type of space with a countable base at every point is that of metric spaces. In fact, topologies in euclidean spaces are determined by means of distances; this approach characterizes metric spaces. A *metric space* is a space with a *distance* (or *metric*) d on $\mathfrak{X} \times \mathfrak{X}$ to R such that, whatever be the points x, y, z, this function has

the triangle property: $d(x, y) + d(x, z) \geqq d(y, z)$,
the identification property: $d(x, y) = 0 \Leftrightarrow x = y$.

Upon replacing z by x and interchanging x and y, it follows that

$$d(x, y) = d(y, x), \quad d(x, y) \geqq 0.$$

It happens frequently, and we shall encounter repeatedly such cases, that, for some space, a function d with the two foregoing properties can be defined—except for the property $d(x, y) = 0 \Rightarrow x = y$. Then the usual procedure is to identify all points x, y such that $d(x, y) = 0$; the space is replaced by the space of "classes of equivalence" so obtained, and this new space is metrized by d.

The topology of a metric space $(\mathfrak{X}, d)$ is defined as follows: Let the *sphere* $V_x(r)$ with "center" x and "radius" $r(>0)$ be the set of all points y such that $d(x, y) < r$. A set A is *open* if, for every $x \in A$, there exists a sphere $V_x(r) \subset A$; it follows, by the triangle property, that every sphere is open. Clearly, the class of open sets so defined is a topology. Since, by the identification property, $d(x, y) > 0$ when $x \neq y$ and the spheres $V_x(r)$ and $V_y(s)$ are disjoint for $0 < r, s \leqq \frac{1}{2}d(x, y)$, it follows that with the *metric topology* so defined, the space is separated; we observe that $x_n \to x$ means that $d(x_n, x) \to 0$.

A basic property of the metric topology is that at every point x there is a countable base, say, the sequence of spheres $V_x\left(\dfrac{1}{n}\right), n = 1, 2, \cdots,$ and it is to be expected that properties of metric spaces can be characterized in countable terms. To begin with:

1. *Sequences can converge to at most one point.*
2. *A point* $x \in \overline{A}$ *if, and only if, A contains a sequence* $x_n \to x$, *so that a set is closed if, and only if, limits of all convergent sequences of its points belong to it.*
3. *Every closed (open) set is a countable intersection (union) of open (closed) sets.*
4. *A metric space has a countable base if, and only if, it is separable.*
5. *If X is a function on a metric domain (Ω, ρ) to a metric space $(\mathfrak{X}, d)$, then $X(\omega') \to X(\omega)$ as $\omega' \to \omega$ if, and only if, $X(\omega_n) \to X(\omega)$ whatever be the sequence $\omega_n \to \omega$.*

Proof. The first assertion follows from the separation theorem. The "if" part of the second assertion is immediate, and for the "only if" part it suffices to take $x_n \in A \cap V_x\left(\frac{1}{n}\right)$.

For the third assertion, form the open sets $O_n = \bigcup_{x \in A} V_x\left(\frac{1}{n}\right)$; those sets contain A, so that $A \subset \bigcap O_n$. On the other hand, for every $x \in \bigcap O_n$ there exist points $x_n \in O_n$ such that $x \in V_{x_n}\left(\frac{1}{n}\right)$, and hence $x_n \to x$; since A is closed, it follows by the second assertion that $x \in A$, and hence $A \supset \bigcap O_n$. Thus, closed $A = \bigcap O_n$ and the dual assertion for open sets follows by complementations.

The fourth assertion follows from **a**.

Finally, if $X(\omega') \to X(\omega)$ as $\omega' \to \omega$, then, clearly, $X(\omega_n) \to X(\omega)$ as $\omega_n \to \omega$. Since $X(\omega') \nrightarrow X(\omega)$ as $\omega' \to \omega$ implies that there exist points $\omega_n \in V_\omega\left(\frac{1}{n}\right)$ such that $X(\omega_n) \nrightarrow X(\omega)$, while $\omega_n \to \omega$, the last assertion follows.

Metric completeness and compactness. The basic criterion for convergence of numerical sequences is the (Cauchy) *mutual convergence criterion:* a sequence x_n is mutually convergent, that is, $d(x_m, x_n) \to 0$ as $m, n \to \infty$ if, and only if, the sequence x_n converges. In a metric space, if $x_n \to x$, then, by the triangle inequality, $d(x_m, x_n) \leqq d(x, x_m) + d(x, x_n) \to 0$ as $m, n \to \infty$, but the converse is not necessarily true (take the space of all rationals with euclidean distance); if it is true, that is, if $d(x_m, x_n) \to 0$ implies that $x_n \to$ some x, then the mutual convergence criterion is valid, and we say that the space is *complete*. Complete metric spaces have many important properties, which follow.

Call $\Delta(A) = \sup_{x,y \in A} d(x, y)$ the *diameter* of A; A is *bounded* if $\Delta(A)$ is finite.

A. Cantor's theorem. *In a complete metric space, every nonincreasing sequence of closed nonempty sets A_n such that the sequence of their diameters $\Delta(A_n)$ converges to 0 has a nonempty intersection consisting of one point only.*

Proof. Take $x_n \in A_n$ and $m \geqq n$. Since $d(x_m, x_n) \leqq \Delta(A_n) \to 0$, it follows that $x_n \to$ some x. Since $x_m \in A_m \subset A_n$ for all $m \geqq n$ and the set A_n is closed, x belongs to every A_n; hence $x \in \bigcap A_n$. If now $d(x, x') > 0$, then, from some k on, $d(x, x') > \Delta(A_k)$ so that $x' \notin A_k \supset \bigcap A_n$. The assertion is proved.

A set A is *nowhere dense* if the complement of $\overline{A}$ is dense in the space, or, equivalently, if $\overline{A}$ contains no spheres, that is, if the interior of $\overline{A}$ is empty. A set is of the *first category* if it is a countable union of nowhere dense sets, and it is of the *second category* if it is not of the first category.

B. BAIRE'S CATEGORY THEOREM. *Every complete metric space is of the second category.*

Proof. Let $A = \bigcup A_n$ where the A_n are nowhere dense sets. There exist a point $x_1 \notin \overline{A}_1$ and a positive $r_1 < 1$ such that the adherence of $V_{x_1}(r_1)$ is disjoint from A_1. Proceeding by recurrence, we form a decreasing sequence of spheres $V_{x_n}(r_n)$ such that $\overline{V_{x_n}(r_n)}$ is disjoint from A_n and $r_n < \dfrac{1}{n} \to 0$. Therefore, by Cantor's theorem, there exists a point $x \in \bigcap \overline{V_{x_n}(r_n)}$ and, because of the foregoing disjunction, $x \notin \bigcup A_n$. Thus $A \neq \mathfrak{X}$, and the theorem follows.

We investigate now compact metric spaces and require the two following propositions.

b. *If every mutually convergent sequence contains a convergent subsequence, then the space is complete.*

This follows from the fact that if a sequence x_n is mutually convergent and contains a convergent subsequence $x_{n'} \to x$, then, by the triangle inequality, $d(x_n, x) \leq d(x_{n'}, x_n) + d(x_{n'}, x) \to 0$ as $n, n' \to \infty$, so that $x_n \to x$.

A set is *totally bounded* if, for every $\epsilon > 0$, it can be covered by a finite number of spheres of radii $\leq \epsilon$. Clearly, a totally bounded set is bounded, and a subset of a totally bounded set is totally bounded.

c. *A metric space is totally bounded if, and only if, every sequence of points contains a mutually convergent subsequence. A totally bounded metric space has a countable base.*

Proof. Let the space be not totally bounded; there exists an $\epsilon > 0$ such that the space cannot be covered by finitely many spheres of radii $\leq \epsilon$. We can select by recurrence a sequence of points x_n whose mutual distances are $\geq \epsilon$; for, if there is only a finite number of points $x_1, \cdots, x_m$ with this property, then the spheres of radius ϵ centered at these points cover the space. Clearly, this sequence cannot contain a mutually convergent subsequence.

Conversely, let the space be totally bounded, so that every set is totally bounded. Then any sequence of points belonging to a set contains a subsequence contained in a sphere of radius $\leqq \epsilon$—member of a finite covering of the set by spheres of radii $\leqq \epsilon$. Thus, given a sequence $\{x_n\}$, setting $\epsilon = \frac{1}{2}, \frac{1}{3}, \cdots$, and proceeding by recurrence, we obtain subsequences such that each is contained in the preceding one and the kth one is formed by points $x_{1k}, x_{2k}, \cdots$ belonging to a sphere of radius $\leqq \frac{1}{k}$. The "diagonal" subsequence $\{x_{nn}\}$ is such that, from the kth term on, the mutual distances are $\leqq \frac{1}{k}$; hence this subsequence is mutually convergent.

The last assertion follows from the fact that given a totally bounded space, the class formed by all finite coverings by spheres of radii $\leqq \frac{1}{n}$, $n = 1, 2, \cdots$ is a countable base.

C. Metric compactness theorem. *The three following properties of a metric space are equivalent:*

(MC$_1$)   *every sequence of points contains a convergent subsequence;*
(MC$_2$)   *every open covering of the space contains a finite covering of the space (Heine-Borel property);*
(MC$_3$)   *the space is totally bounded and complete.*

Proof. It suffices to show that (MC$_2$) $\Rightarrow$ (MC$_1$) $\Rightarrow$ (MC$_3$) $\Rightarrow$ (MC$_2$).

(MC$_2$) $\Rightarrow$ (MC$_1$).   Apply the compactness theorem.

(MC$_1$) $\Rightarrow$ (MC$_3$).   Let every sequence of points contain a convergent (hence mutually convergent) subsequence. Then, by **b**, the space is complete and by **c**, it is also totally bounded.

(MC$_3$) $\Rightarrow$ (MC$_2$).   According to **a**, an open covering of a totally bounded space contains a countable covering $\{O_j\}$ of the space. If no finite union of the O_j covers the space, then, for every n, there exists a point $x_n \not\subset \bigcup_{j=1}^{n} O_j$, and, according to **c**, the sequence of these points contains a mutually convergent subsequence. Therefore, when the totally bounded space is also complete, this sequence has a limit point x which necessarily belongs to some set O_{j_o} of the open countable covering of the space. Since x is a limit point of the sequence $\{x_n\}$, there exists some $n > j_o$ such that $x_n \in O_{j_o} \subset \bigcup\limits_{j=1}^{n} O_j$, and we reach a contradiction. Thus, there exists a finite subcovering of the space.

COROLLARY 1. *A compact metric space is bounded and separable.*

COROLLARY 2. *A continuous function X on a compact metric space (Ω, ρ) to a metric space $(\mathfrak{X}, d)$ is uniformly continuous.*

By definition, X is *uniformly continuous* if for every $\epsilon > 0$ there exists a $\delta = \delta(\epsilon) > 0$, which depends only upon ϵ, such that $d(X(\omega), X(\omega')) < \epsilon$ for $\rho(\omega, \omega') < \delta$.

Proof. Let $\epsilon > 0$. Since X is continuous, for every $\omega \in \Omega$ there exists a δ_ω such that $d(X(\omega), X(\omega')) < \epsilon/2$ for $\rho(\omega, \omega') < 2\delta_\omega$. Since the domain is compact, it is covered by a finite number of spheres $V_{\omega_k}(\delta_{\omega_k})$, $k = 1, 2, \cdots, n$; let δ be the smallest of their radii. Any ω belongs to one of these spheres, say, $V_{\omega_k}(\delta_{\omega_k})$, and if $\rho(\omega, \omega') < \delta$, then $\rho(\omega_k, \omega') < 2\delta_{\omega_k}$. It follows, by the triangle inequality, that

$$d(X(\omega), X(\omega')) \leqq d(X(\omega_k), X(\omega)) + d(X(\omega_k), X(\omega')) < \frac{\epsilon}{2} + \frac{\epsilon}{2} = \epsilon$$

whenever $\rho(\omega, \omega') < \delta$, and the corollary is proved.

Let us indicate how a noncomplete metric space $(\mathfrak{X}, d)$ can be *completed*, that is, can be put in a one-to-one isometric correspondence with a set in a complete metric space—in fact, with a set dense in the latter space. The elementary computations will be left to the reader.

Consider all mutually convergent sequences $s = (x_1, x_2, \cdots)$, $s' = (x'_1, x'_2, \cdots)$, $\cdots$. The function ρ defined by $\rho(s, s') = \lim d(x_n, x'_n)$ exists and is finite and satisfies the triangular inequality. Let s, s' be *equivalent* if $\rho(s, s') = 0$; this notion is symmetric, transitive, and reflexive. It follows that the space (S, ρ) of all such equivalence classes is a metric space, and it is easily seen that it is complete. The one-to-one correspondence between $\mathfrak{X}$ and the set S' of classes of equivalence of all "constant sequences," defined by $x \leftrightarrow (x, x \cdots)$, preserves the distances. Moreover, S' is dense in S. Thus S may be considered as a "minimal completion" of $\mathfrak{X}$.

Distance of sets. In what follows the sets under consideration are nonempty subsets of a metric space $(\mathfrak{X}, d)$. The *distance of two sets* A and B is defined by

$$d(A, B) = \inf \{d(x, y) : x \in A, y \in B\}$$

and

$$d(x, B) = d(\{x\}, B) = \inf\{d(x, y): y \in B\}$$

is called the *distance of x to B*. Clearly there are sequences of points $x_n \in A$ and $y_n \in B$ such that $d(x_n, y_n) \to d(A, B)$ and in particular $d(x, y_n) \to d(x, B)$.

d. $d(x, A)$ *is uniformly continuous in x and, in fact,*

$$| d(x, A) - d(y, A) | < d(x, y).$$

For, upon taking infima in z in the triangle inequality $d(x, z) \leq d(x, y) + d(y, z)$, we obtain $d(x, A) \leq d(x, y) + d(y, A)$ and interchanging x and y the asserted inequality follows.

D. (i) $\bar{A} = \{x: d(x, A) = 0\}$.

(ii) *If disjoint sets A and B are closed then there are disjoint open sets $U \supset A$ and $V \supset B$ ($\mathfrak{X}$ is "normal") and there is a continuous function g with $0 \leq g \leq 1$, $g = 0$ on A, $g = 1$ on B ("Urysohn lemma").*

(iii) *If a compact A and a closed B are disjoint then $d(A, B) > 0$. If moreover B is also compact then $d(A, B) = d(x, y)$ for some $x \in A$ and $y \in B$.*

Proof. We use continuity in x of $d(x, A)$ without further comment. The set $A' = \{x: d(x, A) = 0\}$ contains A and is closed as inverse image of the closed singleton $\{0\}$ under a continuous mapping. Let a sequence of points x_n of A be such that $d(x, x_n) \to d(x, A)$. Then $d(x, x_n) \to 0$ for every $x \in A'$ so that $x \in \bar{A}$ hence A' is contained in A. Thus (i) is proved.

In (ii), the "normality" assertion follows by (i) and continuity in x of $d(x, A) - d(x, B)$ upon taking $U = \{x: d(x, A) - d(x, B) < 0\} \supset A$ and $V = \{x: d(x, A) - d(x, B) > 0\} \supset B$. "Urysohn lemma" obtains with $g(x) = \dfrac{d(x, A)}{d(x, A) + d(x, B)}$.

For (iii), let sequences of points x_n of A and y_n of B be such that $d(x_n, y_n) \to d(A, B)$. Since A is compact the sequence (x_n) contains a subsequence $x_{n'} \to x \in A$ hence $d(x, y_{n'}) \to d(A, B)$. If $d(A, B) = 0$ then $y_{n'} \to x$ so that, B being closed, $x \in B$ and A and B are not disjoint. Since they are disjoint, $d(A, B) > 0$. If, moreover, also B is compact then the sequence of points $y_{n'}$ of B contains a subsequence $y_{n''} \to y \in B$ hence $d(x, y) = d(A, B)$. The proof is terminated.

2.4 Linearity and normed spaces. Euclidean spaces are not only metric and complete but are also normed and linear as defined below. Unless specified, the "scalars" a, b, c, with or without subscripts, are *either* arbitrary real numbers *or* arbitrary complex numbers, and x, y, z, with or without subscripts, are arbitrary points in a space $\mathfrak{X}$.

A space $\mathfrak{X}$ is *linear* if a "linear operation" consisting of operations of "addition" and "multiplication by scalars" is defined on $\mathfrak{X}$ to $\mathfrak{X}$ with the properties:

(i) $$x + y = y + x, \quad x + (y + z) = (x + y) + z,$$
$$x + z = y + z \Rightarrow x = y;$$

(ii) $$1 \cdot x = x, \quad a(x + y) = ax + ay, \quad (a + b)x = ax + bx,$$
$$a(bx) = (ab)x.$$

By setting $-y = -1 \cdot y$, "subtraction" is defined by $x - y = x + (-y)$. Elementary computations show that (i) and (ii) imply uniqueness of the "zero point" or "null point" or "origin" θ, defined by $\theta = 0 \cdot x$, and with the property $x + \theta = x$. A set in a linear space generates a *linear subspace*—the *linear closure* of the set—by adding to its points x, y, $\cdots t$ all points of the form $ax + by + \cdots lt$.

A *metric linear space* is a linear space with a metric d which is invariant under translations and makes the linear operations continuous:

(iii) $$d(x, y) = d(x - y, \theta), \quad x_n \to \theta \Rightarrow ax_n \to \theta,$$
$$a_n \to 0 \Rightarrow a_n x \to \theta.$$

If

(iv) $$d(x, y) = d(x - y, \theta), \quad d(ax, \theta) = |a| d(x, \theta),$$

then (iii) holds, $d(x, \theta)$ is called norm of x and is denoted by $\| x \|$, and the metric linear space is then a "normed linear space."

Equivalently, a *normed linear space* is a linear space on which is defined a *norm* with values $\| x \| \geqq 0$ such that

(v) $$\| x + y \| \leqq \| x \| + \| y \|, \quad \| x \| = 0 \Leftrightarrow x = \theta,$$
$$\| ax \| = |a| \cdot \| x \|,$$

and the metric d is determined by the norm by setting

$$d(x, y) = \| x - y \|.$$

A *Banach space* is a normed linear space complete in the metric determined by the norm. For example, the space of all bounded continuous functions f on a topological space $\mathfrak{X}$ to the euclidean line is a Banach space with a norm defined by $\| f \| = \sup_x |f(x)|$. Real spaces with points $x = (x_1, \cdots, x_N)$ and norms $\| x \| = (|x_1|^r + \cdots + |x_N|^r)^{1/r}$, $r \geqq 1$, are Banach spaces, and we shall encounter similar but more general spaces L_r. If $r = 2$, then these (euclidean) spaces are Hilbert spaces.

A *Hilbert space* is a Banach space whose norm has the parallelogram property: $\| x + y \|^2 + \| x - y \|^2 = 2\| x \|^2 + 2\| y \|^2$; such a norm determines a scalar product. It is simpler to determine the Hilbert norm by means of a scalar product (corresponding to the scalar product defined by $(x, y) = \sum_{k=1}^{N} x_k y_k$ in a euclidean space R^N) as follows:

A *scalar product* is a function on the product of a linear space by itself to its space of scalars, with values (x, y) such that

(vi) $(ax + by, z) = a(x, z) + b(y, z), \quad (x, y) = \overline{(y, x)},$

$$x \neq \theta \implies (x, x) > 0.$$

Clearly (x, x) is real and nonnegative. The function with values $\| x \| = (x, x)^{\frac{1}{2}} \geq 0$ is the *Hilbert norm* determined by the scalar product. For, obviously, it has the two last properties (v) of a norm. And it also has the first property (v). This follows by using in the expansion of $(x + y, x + y)$ the Schwarz inequality

$$| (x, y) | \leq \| x \| \cdot \| y \| ;$$

when $(x, y) = 0$ this inequality is trivially true, and when $(x, y) \neq 0$ it is obtained by expanding $(x - ay, x - ay) \geq 0$ and setting $a = (x, x)/(y, x)$. Finally, the parallelogram property is immediate.

Linear functionals. The basic concept in the investigation of Banach spaces is the analogue of $f(x) = cx$—the simplest of nontrivial functions of classical analysis. A *functional f* on a normed linear space has for range space the space of the scalars (the scalars and the points below are arbitrary, unless specified). f is

linear if $f(ax + by) = af(x) + bf(y)$;
continuous if $f(x_n) \to f(x)$ as $x_n \to x$; if this property holds only for a particular x, then f is *continuous at* this x;
normed or *bounded* if $| f(x) | \leq c\| x \|$ where $c < \infty$ is independent of x; the *norm* of f is then the finite number $\|f\| = \sup_{x \neq \theta} \dfrac{| f(x) |}{\| x \|}$.

For example, a scalar product (x, y) is a linear continuous and normed functional in x for every fixed y. Clearly, if f is linear, then $f(\theta) = 0$, and a linear functional continuous at θ is continuous.

a. *Let f be a linear functional on a normed linear space. If f is normed, then it is continuous; and conversely.*

Proof. If f is normed, then it is continuous, since

$$\left|f(x_n) - f(x)\right| = \left|f(x_n - x)\right| \leq c\|x_n - x\| \to 0 \text{ as } \|x_n - x\| \to 0.$$

If f is not normed, then it is not continuous, since whatever be n there exists a point x_n such that $\left|f(x_n)\right| > n\|x_n\|$, and, setting $y_n = x_n/n\|x_n\|$, we have $\left|f(y_n)\right| > 1$ while $\|y_n\| = \dfrac{1}{n} \to 0$.

b. *The space of all normed linear functionals f on a normed linear space is a Banach space with norm $\|f\|$.*

Proof. Clearly the space is normed and linear and it remains to prove that it is complete.

Let $\|f_m - f_n\| \to 0$ as $m, n \to \infty$. For every $\epsilon > 0$ there exists an n_ϵ such that $\|f_m - f_n\| < \epsilon$ for $m, n \geq n_\epsilon$; hence $\left|f_m(x) - f_n(x)\right| < \epsilon\|x\|$ whatever be x. Since the space of scalars is complete, it follows that there exists a function f of x such that $f_n(x) \to f(x)$ and, clearly, f is linear and normed. By letting $m \to \infty$, we have, for $n \geq n_\epsilon$, $\left|f(x) - f_n(x)\right| \leq \epsilon\|x\|$ whatever be x, that is, $\|f_n - f\| \leq \epsilon$. Hence $f_n \to f$ and the proposition is proved.

What precedes applies word for word to more general functions (mappings, transformations) on a normed linear space to a normed linear space with the same scalars, and the foregoing proposition remains valid, provided the range space is complete; it suffices to replace every $\left|f(x)\right|$ by $\|f(x)\|$.

The Banach space of normed linear functionals on a Banach space is said to be its *adjoint;* a Hilbert space is adjoint to itself. However, *a priori*, the adjoint space may consist only of the trivial null functional f with $\|f\| = 0$. That it is not so will follow (see Corollary 1) from the basic Hahn-Banach

A. Extension theorem. *If f is a normed linear functional on a linear subspace A of a normed linear space, then f can be extended to a normed linear functional on the whole space without changing its norm.*

Proof. 1° We begin by showing that we can extend the domain of f point by point. Let $x_0 \notin A$ and let $\|f\| = 1$—this does not restrict the generality. First assume that the scalars, hence f, are real.

The linearity condition determines $f(x + ax_0)$, $x \in A$, by setting it equal to $f(x) + af(x_0)$, so that it suffices to show that there exists a number $f(x_0)$ such that $\left| f(x) + af(x_0) \right| \leq \left\| x + ax_0 \right\|$ for every $x \in A$ and every number a. Since A is a linear subspace, we can replace x by ax and, by letting x vary, the condition becomes

$$\sup_x \{-\left\| x + x_0 \right\| - f(x)\} \leq f(x_0) \leq \inf_x \{\left\| x + x_0 \right\| - f(x)\}.$$

Therefore, acceptable values of $f(x_0)$ exist if the above supremum is no greater than the above infimum, that is, if whatever be x', $x'' \in A$

$$-\left\| x' + x_0 \right\| - f(x') \leq \left\| x'' + x_0 \right\| - f(x'')$$

or

$$f(x'') - f(x') \leq \left\| x'' + x_0 \right\| + \left\| x' + x_0 \right\|.$$

Since by linearity of f and the triangle inequality

$$f(x'') - f(x') = f(x'' - x') \leq \left\| x'' - x' \right\| \leq \left\| x'' + x_0 \right\| + \left\| x' + x_0 \right\|,$$

acceptable values of $f(x_0)$ exist.

We can pass from real scalars to complex scalars, as follows: From $f(ix) = if(x)$ it follows that $f(x) = g(x) - ig(ix)$, $x \in A$, where $g = \Re f$ is a real-valued linear functional with $\left\| g \right\| \leq 1$; g extends first for all points $x + ax_0$ then for all points $(x + ax_0) + b \cdot ix_0 = x + (a + ib)x_0$, a, b real, and f extends by the foregoing relation. Now observe that f is linear on the so extended domain and that, for any given point x, upon setting $f(x) = re^{i\alpha}$, $r \geq 0$, α real, we obtain $\left| f(x) \right| = g(e^{-i\alpha}x) \leq \left\| x \right\|$.

2° We can extend the domain of f point by point. The family of all possible extensions of f to linear functionals without change of norm is partially ordered by inclusion of their domains. Any linearly ordered subfamily of extensions has a supremum in the family—the extension on the union of the domains. According to a consequence of the axiom of choice (Zorn's theorem), it follows that the whole family has a supremum which is a member of the family. It must have for domain the whole space, for otherwise, by 1°, it could be extended further. The theorem is proved.

COROLLARY 1. *Let x_0 be a nonzero point of a normed linear space, and let A be a closed linear subspace. There exist linear functionals f, f' on the space such that*

$$\left\| f \right\| = 1 \quad and \quad f(x_0) = \left\| x_0 \right\|,$$

$$f' = 0 \text{ on } A \quad and \quad f'(x_0) = d(x_0, A) = \inf_{x \in A} d(x_0, x).$$

Set $f(ax_0) = a\left\| x_0 \right\|$, $f'(ax_0 + x) = ad(x_0, A)$, $x \in A$, and extend.

COROLLARY 2. *A functional f on a set A in a normed linear space extends to a normed linear functional on the whole space with norm bounded by $c(<\infty)$ if, and only if,*

$$\left| \sum_k a_k f(x_k) \right| \leq c \left\| \sum_k a_k x_k \right\|$$

whatever be the finite number of arbitrary points $x_k \in A$ and of arbitrary scalars a_k.

Proof. The "only if" assertion is immediate. As for the "if" assertion, assume that the inequality is true, and observe that the linear closure of A consists of all points of the form $x = \sum_k a_k x_k$. Linearity of f on this closure implies that we must set $f(x) = \sum_k a_k f(x_k)$. Then, on the closure, $|f(x)| \leq c\|x\|$, and f is uniquely determined, since, for $x = \sum_k a_k x_k = \sum_{k'} a'_{k'} x'_{k'}$, we have

$$\left| \sum_k a_k f(x_k) - \sum_{k'} a'_{k'} f(x'_{k'}) \right| \leq c \left\| \sum_k a_k x_k - \sum_{k'} a'_{k'} x'_{k'} \right\| = 0.$$

The assertion follows by the extension theorem.

This corollary permits us to solve various moment problems as well as to find conditions for existence of solutions of systems of linear equations with an infinity of unknowns.

§ 3. ADDITIVE SET FUNCTIONS

3.1 Additivity and continuity. *A set function φ* is defined on a nonempty class $\mathcal{C}$ of sets in a space Ω by assigning to every set $A \in \mathcal{C}$ a single number $\varphi(A)$, finite or infinite, the *value of φ at A*. If all values of φ are finite, φ is said to be *finite*, and we write $|\varphi| < \infty$. If every set in $\mathcal{C}$ is a countable union of sets in $\mathcal{C}$ at which φ is finite, φ is said to be *σ-finite*. To avoid trivialities, we assume that every set function has at least one finite value. *Unless otherwise stated, φ denotes a set function and all sets considered are sets of the class on which this function is defined, so that the properties below are valid as long as φ is defined for the sets which appear there.*

φ is said to be *additive* if

$$\varphi(\sum A_j) = \sum \varphi(A_j)$$

either for every countable or only for every finite class of disjoint sets. In the first case φ is said to be *countably additive* or *σ-additive*, and in the second case φ is said to be *finitely additive*. In order that sums $\sum \varphi(A_j)$ be always meaningful we have to exclude the possibility of expressions of the form $+\infty - \infty$. In fact, if the sums always exist, φ is defined on a field, and $\varphi(A) = +\infty$ and $\varphi(B) = -\infty$, then $\varphi(\Omega) = \varphi(A) + \varphi(A^c) = +\infty$ and $\varphi(\Omega) = \varphi(B) + \varphi(B^c) = -\infty$, while the function φ is single-valued. Thus, by definition,

> *an additive set function has the additivity property above, and one of the values $+\infty$ or $-\infty$ is not allowed.*

To fix ideas we assume that the value $-\infty$ is excluded, unless otherwise stated.

A nonnegative additive set function is called a *content* or a *measure* according as it is finitely additive or σ-additive. Let φ be additive. If $A \supset B$, then, by additivity,

$$\varphi(A) = \varphi(B) + \varphi(A - B).$$

It follows, upon taking $A = B + \emptyset = B$ with $\varphi(B)$ finite, that $\varphi(\emptyset) = 0$.

A convergent series of terms, which are not necessarily of constant sign, may depend upon the order of the terms. This possibility is excluded in our case by

a. *If φ is σ-additive and $\left| \varphi(\sum A_n) \right| < \infty$, then the series $\sum \varphi(A_n)$ is absolutely convergent.*

Proof. Set $A_n{}^+ = A_n$ or $\emptyset$ according as $\varphi(A_n) \geq 0$ or $\varphi(A_n) < 0$, and set $A_n{}^- = A_n$ or $\emptyset$ according as $\varphi(A_n) \leq 0$ or $\varphi(A_n) > 0$. Then

$$\varphi(\sum A_n{}^+) = \sum \varphi(A_n{}^+), \quad \varphi(\sum A_n{}^-) = \sum \varphi(A_n{}^-),$$

and the terms of each series are of constant sign. Since the value $-\infty$ is excluded, the last series converges. Since the sum of both series converges, so does the first series. The assertion follows.

b. *If $\varphi(A)$ is finite and $A \supset B$, then $\varphi(B)$ is finite; in particular, if $\varphi(\Omega)$ is finite, then φ is finite. If $\varphi \geq 0$, then φ is nondecreasing: $\varphi(A) \geq \varphi(B)$ for $A \supset B$, and subadditive: $\varphi(\bigcup A_j) \leq \sum \varphi(A_j)$.*

Only the very last assertion needs verification and follows from

$$\varphi(\bigcup A_j) = \varphi(A_1 + A_1{}^c A_2 + A_1{}^c A_2{}^c A_3 + \cdots)$$
$$= \varphi(A_1) + \varphi(A_1{}^c A_2) + \varphi(A_1{}^c A_2{}^c A_3) + \cdots$$
$$\leq \varphi(A_1) + \varphi(A_2) + \varphi(A_3) + \cdots.$$

We intend to show that the difference between finite additivity and σ-additivity lies in continuity properties. φ is said to be *continuous from below* or *from above* according as

$$\varphi(\lim A_n) = \lim \varphi(A_n)$$

for every sequence $A_n \uparrow$, or for every sequence $A_n \downarrow$ such that $\varphi(A_n)$ is finite for some value n_0 of n (hence, by **b**, for all $n \geq n_0$). If φ is continuous from above and from below, it is said to be *continuous*. Continuity might hold at a fixed set A only, that is, for all monotone sequences which converge to A; continuity at $\emptyset$ reduces to continuity from above at $\emptyset$.

A. Continuity theorem for additive set functions. *A σ-additive set function is finitely additive and continuous. Conversely, if a set function is finitely additive and, either continuous from below, or finite and continuous at $\emptyset$, then the set function is σ-additive.*

Proof. Let φ be σ-additive and, *a fortiori*, additive. φ is continuous from below, for, if $A_n \uparrow$, then

$$\lim A_n = \bigcup A_n = A_1 + (A_2 - A_1) + (A_3 - A_2) + \cdots$$

so that

$$\varphi(\lim A_n) = \lim \{\varphi(A_1) + \varphi(A_2 - A_1) + \cdots + \varphi(A_n - A_{n-1})\}$$

$$= \lim \varphi(A_n).$$

φ is continuous from above, for, if $A_n \downarrow$ and $\varphi(A_{n_0})$ is finite, then $A_{n_0} - A_n \uparrow$ for $n \geq n_0$, the foregoing result for continuity from below applies and, hence,

$$\varphi(A_{n_0}) - \varphi(\lim A_n) = \varphi(\lim (A_{n_0} - A_n)) = \lim \varphi(A_{n_0} - A_n)$$

$$= \varphi(A_{n_0}) - \lim \varphi(A_n)$$

or

$$\varphi(\lim A_n) = \lim \varphi(A_n).$$

Conversely, let φ be finitely additive. If φ is continuous from below, then

$$\varphi(\sum A_n) = \varphi(\lim \sum_{k=1}^{n} A_k) = \lim \varphi(\sum_{k=1}^{n} A_k) = \lim \sum_{k=1}^{n} \varphi(A_k) = \sum \varphi(A_n),$$

so that φ is σ-additive. If φ is finite and continuous at $\emptyset$, then σ-additivity follows from

$$\varphi(\sum A_n) = \varphi(\sum_{k=1}^{n} A_k) + \varphi(\sum_{k=n+1}^{\infty} A_k) = \sum_{k=1}^{n} \varphi(A_k) + \varphi(\sum_{k=n+1}^{\infty} A_k)$$

and

$$\varphi(\sum_{k=n+1}^{\infty} A_k) \rightarrow \varphi(\emptyset) = 0.$$

The proof is complete.

The continuity properties of a σ-additive set function φ acquire their full significance when φ is defined on a σ-field. Then, not only is φ defined for all countable sums and monotone limits of sets of the σ-field but, moreover, φ attains its extrema at some sets of this σ-field. More precisely

    c. *If φ on a σ-field $\mathcal{C}$ is σ-additive, then there exist sets C and D of $\mathcal{C}$ such that $\varphi(C) = \sup \varphi$ and $\varphi(D) = \inf \varphi$.*

    Proof. We prove the existence of C; the proof of the existence of D is similar. If $\varphi(A) = +\infty$ for some $A \in \mathcal{C}$, then we can set $C = A$ and the theorem is trivially true. Thus, let $\varphi < \infty$, so that, since the value $-\infty$ is excluded, φ is finite.

    There exists a sequence $\{A_n\} \subset \mathcal{C}$ such that $\varphi(A_n) \rightarrow \sup \varphi$. Let $A = \bigcup A_n$ and, for every n, consider the partition of A into 2^n sets A_{nm} of the form $\bigcap_{k=1}^{n} A'_k$ where $A'_k = A_k$ or $A - A_k$; for $n < n'$, every A_{nm} is a finite sum of sets $A_{n'm'}$. Let B_n be the sum of all those A_{nm} for which φ is nonnegative; if there are none, set $B_n = 0$. Since, on the one hand, A_n is the sum of some of the A_{nm} and, on the other hand, for $n' > n$, every $A_{n'm'}$ is either in B_n or disjoint from B_n, we have

$$\varphi(A_n) \leqq \varphi(B_n) \leqq \varphi(B_n \cup B_{n+1} \cup \cdots \cup B_{n'}).$$

Letting $n' \rightarrow \infty$, it follows, by continuity from below, that

$$\varphi(A_n) \leqq \varphi(B_n) \leqq \varphi(\bigcup_{k=n}^{\infty} B_k).$$

Letting now $n \rightarrow \infty$ and setting $C = \lim \bigcup_{k=n}^{\infty} B_k$, it follows, by continuity from above (φ is finite), that $\sup \varphi \leqq \varphi(C)$. But $\varphi(C) \leqq \sup \varphi$ and, thus, $\varphi(C) = \sup \varphi$. The proof is complete.

    COROLLARY. *If φ on a σ-field $\mathcal{C}$ is σ-additive (and the value $-\infty$ is excluded), then φ is bounded below.*

3.2   Decomposition of additive set functions.   We shall find later that the "natural" domains of σ-additive set functions are σ-fields. We intend to show that on such domains σ-additive set functions coincide with *signed measures*, that is, differences of two measures of which one at least is finite. Clearly, a signed measure is σ-additive so that we need only to prove the converse.

Let φ be an additive function on a field $\mathfrak{C}$ and define φ^+ and φ^- on $\mathfrak{C}$ by

$$\varphi^+(A) = \sup_{B \subset A} \varphi(B), \quad \varphi^-(A) = -\inf_{B \subset A} \varphi(B), \quad A, B \in \mathfrak{C}.$$

The set functions φ^+, φ^- and $\bar{\varphi} = \varphi^+ + \varphi^-$ are called the *upper, lower,* and *total variation* of φ on $\mathfrak{C}$, respectively. Since $\varphi(\emptyset) = 0$, these variations are nonnegative.

A. JORDAN-HAHN DECOMPOSITION THEOREM.   *If φ on a σ-field $\mathfrak{A}$ is σ-additive, then there exists a set D such that, for every $A \in \mathfrak{A}$,*

$$-\varphi^-(A) = \varphi(AD), \quad \varphi^+(A) = \varphi(AD^c).$$

φ^+ and φ^- are measures and $\varphi = \varphi^+ - \varphi^-$ is a signed measure.

Proof.   According to 3.1c, there exists a set $D \in \mathfrak{A}$ such that $\varphi(D) = \inf \varphi$; since the value $-\infty$ is excluded, we have

$$-\infty < \varphi(D) = \inf \varphi \leqq 0.$$

For every set $A \in \mathfrak{A}$, $\varphi(AD) \leqq 0$ and $\varphi(AD^c) \geqq 0$, since $\varphi \geqq \varphi(D)$ while, if $\varphi(AD) > 0$, then

$$\varphi(D - AD) = \varphi(D) - \varphi(AD) < \varphi(D),$$

and if $\varphi(AD^c) < 0$, then

$$\varphi(D + AD^c) = \varphi(D) + \varphi(AD^c) < \varphi(D).$$

It follows that, for every $B \subset A$, $(A, B \in \mathfrak{A})$,

$$\varphi(B) \leqq \varphi(BD^c) \leqq \varphi(BD^c) + \varphi((A - B)D^c) = \varphi(AD^c),$$

and, hence, $\varphi^+(A) \leqq \varphi(AD^c)$. Since AD^c is one of the B's, the reverse inequality is also true. Therefore, for every $A \in \mathfrak{A}$, $\varphi^+(A) = \varphi(AD^c)$ and, similarly, $-\varphi^-(A) = \varphi(AD)$, so that

$$\varphi(A) = \varphi(AD^c) + \varphi(AD) = \varphi^+(A) - \varphi^-(A).$$

Moreover, φ^+ on $\mathfrak{A}$ is a measure since $\varphi^+ \geqq 0$ and

$$\varphi^+(\textstyle\sum A_j) = \varphi(\textstyle\sum A_j D^c) = \textstyle\sum \varphi(A_j D^c) = \textstyle\sum \varphi^+(A_j).$$

Similarly φ^- on α is a measure and, furthermore, it is bounded by $-\varphi(D)$ which is finite. Thus, $\varphi = \varphi^+ - \varphi^-$ is a signed measure, and the proof is complete.

JORDAN DECOMPOSITION. *If α is only a field but φ is also bounded, then it is still a signed measure.* Prove, proceeding directly from the definitions, showing first that $\varphi^\pm$ are bounded measures.

*§ 4. CONSTRUCTION OF MEASURES ON σ-FIELDS

4.1 Extension of measures. If two set functions φ on $\mathcal{C}$ and φ' on $\mathcal{C}'$ take the same values at sets of a common subclass $\mathcal{C}''$, we say that φ and φ' *agree* or *coincide* on $\mathcal{C}''$. If $\mathcal{C} \subset \mathcal{C}'$ and φ and φ' agree on $\mathcal{C}$, we say that φ is a *restriction* of φ' on $\mathcal{C}$, and φ' is an *extension* of φ on $\mathcal{C}'$. The general extension problem can be stated as follows: find extensions of φ which preserve some specified properties. If, given $\mathcal{C}' \supset \mathcal{C}$, there is one, and only one, such extension on $\mathcal{C}'$, we say that this extension is *determined*.

Here, we are concerned with the extension of measures to measures and shall denote extensions and restrictions of a measure μ by the same letter; as long as their domains are specified, there is no confusion possible. While any restriction of a measure is determined and is a measure, an extension of a measure to a measure on a given class may not exist, and if one exists it may not be unique. Our aim is to produce classes on which such extensions exist, and cases where they are determined. The results of the investigation are summarized by the Carathéodory

A. EXTENSION THEOREM. *A measure μ on a field $\mathcal{C}$ can be extended to a measure on the minimal σ-field over $\mathcal{C}$. If, moreover, μ is σ-finite, then the extension is determined and is σ-finite.*

We prove the extension theorem by means of an intermediate weaker extension which preserves a part only of the properties characterizing a measure. We shall need various notions that we collect here.

A set function μ^o on the class $S(\Omega)$ of all sets in the space Ω is called an *outer measure* if it is sub σ-additive, nondecreasing, and takes the value 0 at $\emptyset$:

$$\mu^o\left(\bigcup A_j\right) \leq \sum \mu^o(A_j) \text{ for every countable class } \{A_j\},$$

$$\mu^o(A) \leq \mu^o(B) \text{ for } A \subset B, \quad \mu^o(\emptyset) = 0.$$

A set A is called μ^o-*measurable* if, for every set $D \subset \Omega$,

$$\mu^o(D) \geq \mu^o(AD) + \mu^o(A^cD).$$

Since the relation is always true when $\mu^o(D) = \infty$, it suffices to consider sets D with $\mu^o(D) < \infty$. Since μ^o is sub σ-additive, the reverse inequality is always true and, hence, A is μ^o-measurable if, and only if,

$$\mu^o(D) = \mu^o(AD) + \mu^o(A^cD).$$

The class of all μ^o-measurable sets will be denoted by $\mathfrak{A}^o$ and, clearly, contains $\emptyset$ and Ω. The *outer extension* of a measure μ given on a field $\mathfrak{C}$ is defined for all sets $A \subset \Omega$ by

$$\mu^o(A) = \inf \sum \mu(A_j),$$

where the infimum is taken over all countable classes $\{A_j\} \subset \mathfrak{C}$ such that $A \subset \bigcup A_j$—*coverings* in $\mathfrak{C}$ of A, for short. Since $\Omega \in \mathfrak{C}$, there is at least one covering (consisting of Ω) in $\mathfrak{C}$ of every A so that the definition of an outer extension is justified. The use of the same symbol μ^o both for an outer measure and an outer extension is due to the property, to be proved first, that the outer extension of the measure μ on $\mathfrak{C}$ is an extension of μ to an outer measure. Next we shall prove that the restriction to $\mathfrak{A}^o$ of μ^o is a measure and that $\mathfrak{A}^o$ is a σ-field, and the extension theorem will follow.

a. *The outer extension μ^o of a measure μ on a field $\mathfrak{C}$ is an extension of μ to an outer measure.*

Proof. We prove first that μ^o is an extension of μ.

If $A \in \mathfrak{C}$, then $\mu^o(A) \leq \mu(A)$. On the other hand, since μ is a measure, $\mu(A) \leq \sum \mu(A_j)$ for every covering $\{A_j\}$ in $\mathfrak{C}$ of A, so that $\mu(A) \leq \mu^o(A)$ and, hence, $\mu^o(A) = \mu(A)$ for $A \in \mathfrak{C}$. It remains to prove that μ^o is an outer measure.

To begin with, $\mu^o(\emptyset) = 0$ since $\emptyset \in \mathfrak{C}$. Furthermore, $\mu^o(A) \leq \mu^o(B)$ for $A \subset B$, since every covering in $\mathfrak{C}$ of B is also a covering of A. Finally, we prove that μ^o is sub σ-additive.

Let $\epsilon > 0$ and let $\{A_j\}$ be an arbitrary countable class. For every A_j there is a covering $\{A_{jk}\}$ in $\mathfrak{C}$ such that

$$\sum_k \mu(A_{jk}) \leq \mu^o(A_j) + \frac{\epsilon}{2^j}.$$

Since $\bigcup_j A_j \subset \bigcup_{j,k} A_{jk}$, it follows that

$$\mu^o(\bigcup_j A_j) \leq \sum_{j,k} \mu(A_{jk}) \leq \sum_j \mu^o(A_j) + \epsilon,$$

and, $\epsilon > 0$ being arbitrarily close to zero, sub σ-additivity is proved.

b. *If μ^o is an outer measure, then the class $\mathcal{C}^o$ of μ^o-measurable sets is a σ-field and μ^o on $\mathcal{C}^o$ is a measure.*

Proof. We prove first that $\mathcal{C}^o$ is a field and μ^o on $\mathcal{C}^o$ is a content.

If $A \in \mathcal{C}^o$, then $A^c \in \mathcal{C}^o$, since the definition of μ^o-measurability is symmetric in A and A^c. If $A, B \in \mathcal{C}^o$, then $AB \in \mathcal{C}^o$, since

$$\mu^o(D) = \mu^o(AD) + \mu^o(A^cD)$$

$$= \mu^o(ABD) + \mu^o(AB^cD) + \mu^o(A^cBD) + \mu^o(A^cB^cD)$$

$$\geqq \mu^o(ABD) + \mu^o(AB^cD \cup A^cBD \cup A^cB^cD)$$

$$= \mu^o(ABD) + \mu^o(AB)^cD.$$

Thus $\mathcal{C}^o$ is closed under complementations and finite intersections and, hence, under finite unions, so that $\mathcal{C}^o$ is a field.

μ^o is finitely additive on $\mathcal{C}^o$ since, if $A, B \in \mathcal{C}^o$ and are disjoint,

$$\mu^o(A + B) = \mu^o((A + B)A) + \mu^o((A + B)A^c) = \mu^o(A) + \mu^o(B).$$

Since $\mu^o(A) \geqq \mu^o(\emptyset) = 0$, μ^o on $\mathcal{C}^o$ is a content.

To complete the proof, it suffices to show that, if the $A_n \in \mathcal{C}^o$ are disjoint, then $A = \sum A_n \in \mathcal{C}^o$ and $\mu^o(A) = \sum \mu^o(A_n)$.

Since $B_n = \sum_{k=1}^{n} A_k \in \mathcal{C}^o$, we have

$$\mu^o(D) = \mu^o(B_nD) + \mu^o(B_n{}^cD) \geq \sum_{k=1}^{n} \mu^o(A_kD) + \mu^o(A^cD)$$

and, letting $n \to \infty$,

$$\mu^o(D) \geqq \sum \mu^o(A_nD) + \mu^o(A^cD) \geq \mu^o(AD) + \mu^o(A^cD).$$

The inequality between the extreme sides shows that $A \in \mathcal{C}^o$. The first inequality with D replaced by A becomes

$$\mu^o(A) \geqq \sum \mu^o(A_n)$$

while the reverse inequality is always true.
Thus

$$\mu^o(A) = \sum \mu^o(A_n),$$

and the proof is complete.

REMARK. Most frequently, a measure μ is given on a class $\mathcal{D}$ whose closure under finite summations or under countable summations is a field $\mathcal{C}$. Then the requirement of σ-additivity determines the unique extension of μ on $\mathcal{C}$.

We are now in a position to prove the extension theorem.

1° For every $A \in \mathcal{C}$ and every D there is, for every $\epsilon > 0$, a covering $\{A_j\}$ in $\mathcal{C}$ of D such that

$$\mu^o(D) + \epsilon \geqq \sum \mu(A_j) = \sum \mu(AA_j) + \sum \mu(A^c A_j) \geqq \mu^o(AD) + \mu^o(A^c D).$$

Thus, $A \in \mathcal{C}^o$ and, hence, since the field $\mathcal{C}$ is contained in the σ-field $\mathcal{C}^o$, the minimal σ-field $\mathcal{C}$ over $\mathcal{C}$ is contained in $\mathcal{C}^o$. It follows, according to **a** and **b**, that the contraction on $\mathcal{C}$ of the measure μ^o on $\mathcal{C}^o$ is an extension of μ to a measure on $\mathcal{C}$. This proves the first part of the theorem.

2° Let μ on $\mathcal{C}$ be finite, let μ_1 and μ_2 be two extensions of μ to measures on $\mathcal{C}$, and let $\mathfrak{M} \subset \mathcal{C}$ be the class on which μ_1 and μ_2 agree. Since Ω belongs to $\mathcal{C}$, $\mu_1 (\Omega) = \mu_2(\Omega) = \mu(\Omega) < \infty$; hence μ_1 and μ_2 are finite. Since $\mathfrak{M}$ contains $\mathcal{C}$ and, for every monotone sequence $A_n \in \mathfrak{M}$,

$$\mu_1(\lim A_n) = \lim \mu_1(A_n) = \lim \mu_2(A_n) = \mu_2(\lim A_n),$$

$\mathfrak{M}$ is a monotone class. It follows, by 1.6**A**, that $\mathfrak{M}$ contains the minimal σ-field $\mathcal{C}$ over the field $\mathcal{C}$ and, therefore, μ_1 and μ_2 agree on $\mathcal{C}$.

Let now μ on $\mathcal{C}$ be σ-finite so that there is a countable class $\{A_j\} \subset \mathcal{C}$ with μA_j finite which covers Ω. Thus, the foregoing result applies to every subspace A_j, and the second part of the theorem follows.

Generalization. The extension theorem is valid for σ-finite *signed* measures $\varphi = \mu' - \mu''$. Extend μ' and μ'' and observe that 2° applies with φ instead of μ.

Completion. Given a measure μ on a σ-field $\mathcal{C}$, it is always possible to extend μ to a larger σ-field obtained as follows: For every $A \in \mathcal{C}$ and an arbitrary subset N of a *null* set of $\mathcal{C}$, that is, a set of measure zero, set $\mu(A \cup N) = \mu(A)$. Clearly, the class of all sets $A \cup N$ is a σ-field $\mathcal{C}_\mu \supset \mathcal{C}$ and μ on $\mathcal{C}_\mu$ is an extension of μ to a measure on $\mathcal{C}_\mu$. $\mathcal{C}_\mu$ is called the *completion* of $\mathcal{C}$ for μ and μ on $\mathcal{C}_\mu$ is called a *complete measure*. It is easily seen that $\mathcal{C}_\mu \subset \mathcal{C}^o$, so that the extension theorem provides us automatically with extensions to complete measures.

4.2 Product probabilities. A measure on a class containing the space is called a *normed measure* or a *probability* when its value for the whole space is one; we reserve the symbol P, with or without affixes, for such measures.

Let $(\Omega_t, \mathcal{C}_t, P_t)$, $t \in T$, be *probability spaces*, that is, triplets consisting of a space Ω_t of points ω_t, a σ-field $\mathcal{C}_t$ of measurable sets A_t (with or without superscripts) in Ω_t, and a probability P_t on $\mathcal{C}_t$. Let $\mathcal{C}_T$ be the class of all measurable cylinders of the form $\prod_{t \in T_N} A_t \times \prod_{t \in T - T_N} \Omega_t$ in the product measurable space $(\prod \Omega_t, \prod \mathcal{C}_t)$. The class $\mathcal{B}_T$ of all finite sums of these cylinders is a field, and the minimal σ-field α_T over $\mathfrak{B}_T$ is, by definition, the product σ-field $\prod \alpha_t$. The *product probability* $P_T = \prod P_t$ on the class $\mathfrak{C}_T$ is defined by assigning to every interval cylinder the product of the probabilities of its sides: in symbols,

$$P_T(\prod_{t \in T_N} A_t \times \prod_{t \in T - T_N} \Omega_t) = \prod_{t \in T_N} P_t A_t \cdot \prod_{t \in T - T_N} P_t \Omega_t = \prod_{t \in T_N} P_t A_t.$$

Clearly, $P_T \Omega_T = 1$ and P_T on $\mathfrak{C}_T$ is finitely additive and determines its extension to a finitely additive set function P_T on $\mathfrak{B}_T$. The defining term "product-probability" is justified by the following theorem (Andersen and Jessen).

A. PRODUCT PROBABILITY THEOREM. *The product probability P_T on $\mathfrak{B}_T$ is σ-additive and determines its extension to a probability P_T on the product σ-field α_T.*

Thus, the triplet $(\Omega_T, \alpha_T, P_T)$ is a probability space, to be called the *product probability space*.

Proof. 1° On account of the extension theorem, it suffices to prove that P_T on $\mathfrak{B}_T$ is σ-additive. Since it is obviously finitely additive on $\mathfrak{B}_T$, on account of the continuity theorem for additive set functions it suffices to prove that P_T on $\mathfrak{B}_T$ is continuous at $\emptyset$. *Ab contrario*, given $\epsilon > 0$ arbitrarily close to 0, it suffices to prove that, for every nonincreasing sequence of measurable cylinders $A^n \downarrow A$ with $P_T A^n > \epsilon$ for every n, the limit set A is not empty. Since every cylinder A^n depends only upon a finite subset of indices, the set of all indices involved in defining the sequence A^n is countable. By interchanging, if necessary, the indices, we can restrict ourselves to the product space $\Omega = \prod \Omega_n$ and sets $A^n = D^n \times \Omega'_n$ with $D^n \subset \Omega_1 \times \cdots \times \Omega_n$, $\Omega'_n = \Omega_{n+1} \times \Omega_{n+2} \times \cdots$.

If the set of all indices is finite, then there is an integer N such that, for every n, all the factors which follow the Nth one reduce to Ω_N, and the argument below applies with corresponding modifications.

2° Let $P'_1, P'_2, \cdots$ be the set functions defined on the fields $\mathfrak{B}_1'$, $\mathfrak{B}'_2, \cdots$ of all measurable cylinders in $\Omega'_1, \Omega'_2, \cdots$, as P_T is defined on $\mathfrak{B}_T$. Let $A^n(\omega_1), A^n(\omega_1, \omega_2), \cdots$ be the sections of A^n at $\omega_1 \in \Omega_1$, $(\omega_1, \omega_2) \in \Omega_1 \times \Omega_2$, etc. Clearly, $A^n(\omega_1) \in \mathfrak{B}'_1$. It is easily seen that, if B_1^n is the set of all ω_1 such that

$$P'_1 A^n(\omega_1) > \frac{\epsilon}{2},$$

then

$$P_1 B_1{}^n + \frac{\epsilon}{2}(1 - P_1 B_1{}^n) \geqq P_T A^n > \epsilon$$

and, hence,

$$P_1 B_1{}^n > \frac{\epsilon}{2}.$$

Since $A^n \downarrow$ implies that $B_1{}^n \downarrow$, it follows that, for $B_1 = \lim B_1{}^n$, $P_1 B_1$ $\geqq \frac{\epsilon}{2}$. Thus, B_1 is not empty; hence, there is a point $\bar\omega_1 \in \Omega_1$ common to all $B_1{}^n$ and, for every n, $P'_1(A^n(\bar\omega_1)) > \frac{\epsilon}{2}$. The same argument applied to $A^n(\bar\omega_1) \downarrow$ yields a point $\bar\omega_2 \in \Omega_2$ such that $P'_2(A^n(\bar\omega_1, \bar\omega_2)) > \frac{\epsilon}{2^2}$, and so on. Therefore, the point $\bar\omega = (\bar\omega_1, \bar\omega_2, \cdots)$ is common to all A^n, so that the limit set A is not empty, and the proof is complete.

We pass now to Borel spaces.

4.3 Consistent probabilities on Borel fields. We introduce the following terminology. The set $R = (-\infty, +\infty)$ of all finite numbers x is a *real line*, the minimal σ-field over the class of all intervals is the *Borel field* $\mathcal{B}$ in R, the elements of $\mathcal{B}$ are *Borel sets* in R, and the measurable space $(R, \mathcal{B})$ is a *Borel line*. Similarly, the product space $R_T = \prod R_t$, where every R_t is a real line with points x_t, is a *real space* with points $x_T = (x_t)$, the product σ-field $\mathcal{B}_T = \prod \mathcal{B}_t$, where every $\mathcal{B}_t$ is the Borel field in R_t, is the *Borel field* in R_T whose elements are *Borel sets* in R_T, and the measurable space $(R_T, \mathcal{B}_T)$ is a *Borel space*. If T is a finite set, we say that R_T is a finite product space. Cylinders with Borel bases are *Borel cylinders* and, clearly, the Borel field $\mathcal{B}_T$ is the minimal σ-field over the class of all Borel cylinders or, equivalently, over the class of all cylinders whose bases are product Borel sets.

Given a finite measure on $\mathcal{B}_T$ we can assume, by dividing it by its value for R_T, that it is a probability P_T. Let $T_N = \{t_1, \cdots t_N\}$ be a finite subset of indices and let $(R_{T_N}, \mathcal{B}_{T_N})$ be the corresponding Borel space. We define on $\mathcal{B}_{T_N}$ the *marginal probability* P_{T_N}, or *projection of P on* R_{T_N}, by assigning to every Borel set B_{T_N} in R_{T_N} the measure of the cylinder with basis B_{T_N}; in symbols

$$P_{T_N}(B_{T_N}) = P_T(B_{T_N} \times R'_{T_N}), \quad R'_{T_N} = \prod_{t \notin T_N} R_t.$$

Marginal probabilities are *consistent* in the following sense. If R' and R'' are two finite product subspaces of R_{T_N} with marginal measures P'

and P'', respectively, then the projections of P' and P'' on their common subspace, if any, coincide (with the projection of P_T on this subspace). We want to prove that the converse is true (Daniell, Kolmogorov).

A. Consistency theorem. *Consistent probabilities P_{T_N} on Borel fields of all finite product subspaces R_{T_N} of R_T determine a probability P_T on the Borel field in R_T such that every P_{T_N} is the projection of P_T on R_{T_N}.*

Proof. To every Borel cylinder with Borel base B_{T_N} in R_{T_N} we assign the probability value

$$P_T(B_{T_N} \times R'_{T_N}) = P_{T_N}(B_{T_N}).$$

It is easily seen that P_T on the class $\mathcal{C}_T$ of all Borel cylinders is finitely additive, and the theorem will follow from the extension theorem if we prove that P_T on $\mathcal{C}_T$ is continuous at $\emptyset$.

As in the proof of the product probability theorem, it suffices to prove that, given $\epsilon > 0$ arbitrarily close to zero, if a sequence $A_n \downarrow A$ of Borel cylinders with bases B_n formed by finite sums of intervals in $R_1 \times \cdots \times R_n$ is such that, for every n,

$$P_T(A_n) = P_{12\cdots n}(B_n) > \epsilon,$$

then A is not empty. To simplify the writing, set $P = P_T$ and $P_n = P_{12\cdots n}$. Since P_n is bounded and continuous from below, in every interval in $R_1 \times \cdots \times R_n$ we can find a bounded closed interval whose P_n-measure is as close as we wish to that of the original interval. Therefore, in every B_n, we can find a bounded closed Borel set B'_n—formed by a finite sum of bounded closed intervals—such that $P_n(B_n - B'_n) < \dfrac{\epsilon}{2^{n+1}}$ and, hence, if A'_n is the Borel cylinder with basis B'_n, then

$$P(A_n - A'_n) = P_n(B_n - B'_n) < \frac{\epsilon}{2^{n+1}}.$$

It follows, setting $C_n = A'_1 \cap \cdots \cap A'_n$, that $P(A_n - C_n) < \dfrac{\epsilon}{2}$ or, since $C_n \subset A'_n \subset A_n$,

$$P(C_n) > P(A_n) - \frac{\epsilon}{2} > \frac{\epsilon}{2}.$$

Thus every C_n is nonempty and we can select in it a point $x^{(n)} = (x_1^{(n)}, x_2^{(n)}, \cdots)$. It follows from $C_1 \supset C_2 \supset \cdots$ that for every $p = 0, 1, \cdots, x^{(n+p)} \in C_n \subset A'_n$ and hence $(x_1^{(n+p)}, \cdots, x_n^{(n+p)}) \in B'_n$.

Since every B'_n is bounded, we can select a subsequence n_{1k} of integers such that $x_1^{(n_{1k})} \to x_1$ as $k \to \infty$, then within it a subsequence n_{2k} such that $x_2^{(n_{2k})} \to x_2$, and so on. The diagonal subsequence of points $x^{(n_{kk})} = (x_1^{(n_{kk})}, x_2^{(n_{kk})}, \cdots)$ converges to the point $x = (x_1, x_2, \cdots)$ and $(x_1^{(n_{kk})}, \cdots, x_m^{(n_{kk})}) \to (x_1, \cdots, x_m) \in B'_m$ for every m. Therefore, $x \in A'_m \subset A_m$ whatever be m so that $x \in \bigcap_{m=1}^{\infty} A_m$. Thus this intersection is not empty, and the assertion is proved.

Extensions. The foregoing theorem can be extended, as follows: Let $\mathcal{C}_n$ be the σ-field of Borel cylinders with bases in $R_1 \times \cdots \times R_n$, and let $\mathcal{C}_\infty$ be the Borel field in $\prod R_n$.

1° *If uniformly bounded measures μ_n on $\mathcal{C}_n$ form a nondecreasing sequence, in the sense that $\mu_n A_n \leqq \mu_{n+1} A_n \leqq \cdots$ and hence $\mu_p A_n \uparrow \mu A_n$ as $p \to \infty$ whatever be n and $A_n \in \mathcal{C}_n$, then μ extends to a bounded measure on $\mathcal{C}_\infty$.*

The proof reduces to the previous one as follows. The set function μ so defined on the field $\bigcup \mathcal{C}_n$ of all Borel cylinders in $\prod R_n$ is, clearly, finitely additive and bounded. Therefore, it suffices to prove that on this field μ is continuous at $\emptyset$. Given $\epsilon > 0$ and $A_n \in \mathcal{C}_n$, we can find p sufficiently large so that $\mu_p A_n + \dfrac{\epsilon}{2^{n+2}} > \mu A_n$. Then we can select a Borel cylinder $A'_n \subset A_n$ whose basis is a closed and bounded Borel set in $R_1 \times \cdots \times R_n$ such that $\mu_p(A_n - A'_n) < \dfrac{\epsilon}{2^{n+2}}$. It follows that

$$\mu A'_n + \frac{\epsilon}{2^{n+1}} \geqq \mu_p A_n + \frac{\epsilon}{2^{n+1}} > \mu A_n$$

so that $\mu(A_n - A'_n) < \dfrac{\epsilon}{2^{n+1}}$. From here on, the end of the preceding proof applies word for word.

If φ_n on $\mathcal{C}_n$, $n = 1, 2, \cdots$, are such that $\varphi_n(A_n) = \varphi_{n+1}(A_n) = \cdots$, $A_n \in \mathcal{C}_n$, we say that the φ_n are *consistent*.

2° *If the uniformly bounded σ-additive set functions φ_n on $\mathcal{C}_n$ are consistent, hence $\varphi_p(A_n) \to \varphi(A_n)$ as $p \to \infty$ whatever be n and $A_n \in \mathcal{C}_n$, then φ extends to a σ-additive bounded set function on $\mathcal{C}_\infty$.*

The assertion follows from what precedes. For, clearly, the total variations $\bar{\varphi}_n$ on $\mathcal{C}_n$ form a nondecreasing bounded sequence on $\bigcup \mathcal{C}_n$, in the sense of 1°. Hence $\lim \bar{\varphi}_n$ is continuous at $\emptyset$ on $\bigcup \mathcal{Q}_n$ and, *a fortiori*, so is φ. Now use Jordan decomposition and generalization in **4.1.**

4.4 Lebesgue-Stieltjes measures and distribution functions. Complete measures on the Borel field in a real line $R = (-\infty, +\infty)$ did, and still do, play a prominent role. However, being set functions, they are not easy to handle with the tools of classical analysis, for methods of analysis were developed to deal primarily with finite point functions on R. It is, therefore, of the greatest methodological importance to establish a link between the modern notion of measure and the classical notions. This will be done by showing that there is a class of point functions on R which can be placed in a one-to-one correspondence with a very wide class of measures. In this manner, investigations of measures (and, thereafter, of integrals) will be reduced to investigations of the corresponding point functions and, thus, the familiar methods of analysis will apply. Whatever be these point functions they will be said *to represent* the corresponding measure.

Among possible representations of measures there are two which are fundamental: "distribution functions" which represent measures assigning finite values to finite intervals, to be called *Lebesgue-Stieltjes (L.S.) measures,* that we shall introduce now, and "characteristic functions" which represent the subclass of finite Lebesgue-Stieltjes measures required in connection with probability problems—that we shall introduce in Part II. Let $\mathfrak{B}$ be the Borel field in R and let μ be a Lebesgue-Stieltjes measure. The completion of $\mathfrak{B}$ for μ will be denoted by $\mathfrak{B}_\mu$, and called a *Lebesgue-Stieltjes field* in R, and its elements will be called *Lebesgue-Stieltjes sets* in R.

A function on R which is finite, nondecreasing, and continuous from the left is called a *distribution function* (d.f.). Two d.f.'s will be said to be *equivalent* if they differ by some fixed but arbitrary constant. This notion of equivalence has the usual properties of equivalence—it is reflexive, transitive, and symmetric. Thus, the class of all d.f.'s splits into equivalence classes. As the correspondence theorem below (Lebesgue, Radon) shows, the one-to-one correspondence between L.S.-measures and d.f.'s is not a correspondence between L.S.-measures and individual d.f.'s but a correspondence between L.S.-measures and classes of equivalent d.f.'s, each class to be represented by one of its elements, arbitrarily chosen.

Let F, with or without affixes, denote a d.f. and define its *increment function* by

$$F[a, b) = F(b) - F(a), \quad -\infty < a \leq b < +\infty.$$

Since two equivalent d.f.'s have the same increment function and conversely, it follows that every class of equivalent d.f.'s is characterized by its increment function. Moreover, the defining properties of d.f.'s are equivalent to the following:

$$\text{(i)} \quad 0 \leqq F[a, b) < \infty, \qquad \text{(ii)} \quad F[a, b) \to 0 \text{ as } a \uparrow b,$$

and

$$\text{(iii)} \quad \sum_{k=1}^{n} F[a_k, b_k) + \sum_{k=1}^{n-1} F[b_k, a_{k+1}) = F[a_1, b_n)$$

where $a < b, a_1 \leqq b_1 \leqq a_2 \leqq \cdots \leqq a_n \leqq b_n$ are arbitrary.

A. Correspondence theorem. *The relation*

$$\mu[a, b) = F[a, b), \quad -\infty < a \leqq b < +\infty$$

establishes a one-to-one correspondence between L.S.-measures μ and d.f.'s F defined up to an equivalence.

Proof. Let $\mathscr{B}_I$ be the class of all intervals $[a, b), -\infty < a \leqq b < +\infty$. $\mathscr{B}_I$ is closed under formation of finite intersections. The minimal field $\mathscr{B}_0$ over $\mathscr{B}_I$ is the class of all finite sums of elements of $\mathscr{B}_I$ and of intervals of the form $(-\infty, a), [b + \infty)$, and the minimal σ-field over $\mathscr{B}_0$ is the Borel field $\mathscr{B}$.

The proof of the correspondence theorem is summarized by the diagram below, where c represents an arbitrary constant:

$$F + c \text{ on } R \Leftrightarrow \mu \text{ on } \mathscr{B}_I \Leftrightarrow \mu \text{ on } \mathscr{B}_0 \Leftrightarrow \mu \text{ on } \mathscr{B} \Leftrightarrow \mu \text{ on } \mathscr{B}_\mu.$$

1° μ *on* $\mathscr{B}_\mu \Rightarrow F + c$ *on* R. For, μ on $\mathscr{B}_\mu$ determines its restriction to $\mathscr{B}_I$ and, from properties of L.S.-measures it follows that the relation

$$F[a, b) = \mu[a, b)$$

determines an increment function with properties (i), (ii), and (iii) given above.

2° μ *on* $\mathscr{B}_0 \Rightarrow \mu$ *on* $\mathscr{B}_\mu$. For, R being a denumerable sum of finite intervals, the measure μ on $\mathscr{B}_0$ is σ-finite and the extension theorem applies followed by completion.

3° μ *on* $\mathscr{B}_I \Rightarrow \mu$ *on* $\mathscr{B}_0$. It suffices to prove that if $A = \sum_k I_k \in \mathscr{B}_0, I_k \in \mathscr{B}_I$, then $\mu(A)$ is determined by the σ-additivity requirement $\mu(A) = \sum_k \mu(I_k)$, that is, if A can also be written as $\sum_j I'_j$, where I'_j

$\in \mathfrak{B}_I$, then $\sum_j \mu(I'_j) = \sum_k \mu(I_k)$. Since μ on $\mathfrak{B}_I$ is additive and

$$I'_j = AI'_j = \sum_k I_k I'_j, \quad I_k = AI_k = \sum_j I'_j I_k,$$

it follows that

$$\sum_j \mu(I'_j) = \sum_j \sum_k \mu(I_k I'_j) = \sum_k \sum_j \mu(I'_j I_k) = \sum_k^n \mu(I_k),$$

and the assertion is proved.

4° $F + c \Rightarrow \mu$ on $\mathfrak{B}_I$. We have to prove that the relation $\mu[a, b) = F[a, b)$ determines a measure μ on $\mathfrak{B}_I$, that is, if $I = \sum I_n$, where $I = [a, b)$ and $I_n = [a_n, b_n)$, then $\mu I = \sum \mu I_n$. By interchanging, if necessary, the subscripts, we can assume that, for every n,

$$a \leqq a_1 \leqq b_1 \leqq \cdots \leqq a_n \leqq b_n \leqq b.$$

It follows that

$$\sum_{k=1}^n \mu(I_k) = \sum_{k=1}^n F[a_k, b_k) \leqq \sum_{k=1}^n F[a_k, b_k) + \sum_{k=1}^{n-1} F[b_k, a_{k+1})$$

$$= F[a_1, b_n) \leqq F[a, b) = \mu(I),$$

and, letting $n \to \infty$, we get $\sum \mu(I_n) \leqq \mu(I)$.

It remains for us to prove the reverse inequality. We exclude the trivial case $a = b$, select $\epsilon > 0$ such that $\epsilon < b - a$ and set $I^\epsilon = [a, b - \epsilon]$. Because of the continuity from the left, for every n there is an $\epsilon_n > 0$ such that $F[a_n - \epsilon_n, a_n) < \dfrac{\epsilon}{2^n}$. If $I_n^\epsilon = (a_n - \epsilon_n, b_n)$, then, from $I^\epsilon \subset \bigcup_n I_n^\epsilon$ it follows, by the Heine-Borel lemma, that there is an n_0 finite such that $I^\epsilon \subset \bigcup_{k=1}^{n_0} I_k^\epsilon$. Let $k_1 \leqq n_0$ be such that $a \in I_{k_1}^\epsilon$ and, if $b_{k_1} < b$, then let $k_2 \leqq n_0$ be such that $b_{k_1} \in I_{k_2}$. Continue in this manner until some $b_{k_m} \geqq b - \epsilon$—the process necessarily stops for some $m \leqq n_0$. Omitting intervals that were not selected and, if necessary, changing the subscripts, it follows that $I^\epsilon \subset \bigcup_{k=1}^m I_k^\epsilon$ and

for

$$a_1 - \epsilon_1 < a < b_1, \quad a_{k+1} - \epsilon_{k+1} < b_k < b_{k+1}$$

$$k = 1, 2, \cdots m - 1, \quad a_m - \epsilon_m < b - \epsilon \leqq b_m.$$

Therefore,

$$F[a, b - \epsilon) \leqq F[a_1 - \epsilon_1, b_m) = F[a_1 - \epsilon_1, b_1) + \sum_{k=1}^{m-1} F[b_k, b_{k+1})$$

$$\leqq \sum_{k=1}^{m} F[a_k - \epsilon_k, b_k) \leqq \sum_{k=1}^{\infty} F[a_k, b_k) + \epsilon$$

and, letting $\epsilon \to 0$,

$$F[a, b) \leqq \sum F[a_n, b_n), \text{ that is, } \mu(I) \leqq \sum \mu(I_n),$$

which completes the proof of the final assertion and, hence, of the correspondence theorem.

Particular case. If F is defined, up to an additive constant, by $F(x) = x$, $x \in R$, then the corresponding measure of an interval is its "length." The extension of "length" to a measure μ on $\mathcal{B}$ and the completed measure μ on $\mathcal{B}_\mu$ are called *Lebesgue measure* on $\mathcal{B}$ or $\mathcal{B}_\mu$, respectively, and $\mathcal{B}_\mu$ will be called *Lebesgue field.* The Lebesgue measure is at the root of the general notion of measure.

REMARK. We can define a L.S.-measure on the Borel field $\bar{\mathcal{B}}$-minimal σ-field over the class of all intervals in $\bar{R} = [-\infty, +\infty]$ and, hence, on $\bar{\mathcal{B}}_\mu$, by adjoining to a L.S.-measure on $\mathcal{B}$, arbitrary measures for the sets $\{-\infty\}$ and $\{+\infty\}$.

Extension. The preceding definitions, proofs, and results, remain valid, word for word, if Borel lines are replaced by finite-dimensional Borel spaces $R^N = R_1 \times \cdots \times R_N$, provided the following interpretation of symbols is used: $a, b, x, \cdots$ are points in R^N, say, $a = (a_1, \cdots, a_N)$; $a < b(a \leqq b)$ means that $a_k < b_k(a_k \leqq b_k)$ for $k = 1, \cdots, N$. F on R^N is a function with values $F(a) = F(a_1, \cdots, a_N)$ and increments $F[a, b)$ are defined by

$$F[a, b) = \Delta_{b-a}F(a) = \Delta_{b_1 - a_1} \cdots \Delta_{b_N - a_N} F(a_1, a_2, \cdots a_N)$$

where, for every k, $\Delta_{b_k - a_k}$ denotes the difference operator of step $b_k - a_k$ acting on a_k. For instance, if $N = 2$,

$$\Delta_{b-a}F(a) = \Delta_{b_1 - a_1}\Delta_{b_2 - a_2}F(a_1, a_2) = \Delta_{b_1 - a_1}\{F(a_1, b_2) - F(a_1, a_2)\}$$

$$= F(b_1, b_2) - F(a_1, b_2) - F(b_1, a_2) + F(a_1, a_2)$$

and, in particular, if $F(a_1, a_2) = a_1 a_2$ is the area of the rectangle with sides 0 to a_1 and 0 to a_2, then $\Delta_{b-a}F(a) = (b_1 - a_1)(b_2 - a_2)$ is the area of the rectangle with sides a_1 to b_1 and a_2 to b_2.

The defining properties of a d.f. F on R^N become:

$$-\infty < F < +\infty, \quad F[a, b) = \Delta_{b-a}F(a) \geqq 0, \quad F[a, b) \to 0$$

as $a \uparrow b$, that is, $a_1 \uparrow b_1, \cdots, a_N \uparrow b_N$.

Product-d.f.'s and product-measures. A very important particular case is that of *product-d.f.'s*:

$$F(x_1, \cdots, x_N) = \prod_{k=1}^{N} F_k(x_k), \quad x_k \in R_k$$

where the F_k on R_k are d.f.'s. Then F on R^N is a d.f., for,

$$\Delta_{b-a}F(a) = \prod_{k=1}^{N} \Delta_{b_k-a_k}F_k(a_k) \geqq 0$$

and the other defining properties are clearly satisfied.

Every d.f. F_k determines a measure μ_k on the Borel field in R_k, by means of the relation $\mu_k[a_k, b_k) = F_k[a_k, b_k)$, and the measure μ on the product Borel field determined by means of the relation $\mu[a, b) = F[a, b)$, is clearly the product-measure $\prod_{k=1}^{N} \mu_k$.

Let now F_n be d.f.'s with $F_n(+\infty) - F_n(-\infty) = 1$, so that the measures μ_n are probabilities. Then, by the product-probability theorem or by the consistency theorem,

B. *A sequence F_n of d.f.'s corresponding to probabilities on R_n determines a product-probability on the Borel field in the product space $\prod R_n$.*

This result extends at once to any set $\{F_t, t \in T\}$, of such d.f.'s.

COMPLEMENTS AND DETAILS

In one guise or another, and especially when they are indefinite integrals, signed measures on a fixed σ-field are in constant use in measure theory and probability theory. Many of the properties established in this book are but properties of such set functions.

Notation. The measurable sets belong to a fixed σ-field on which the set functions and limits of their sequences are defined. Unless otherwise stated and with or without affixes, A, B, $\cdots$ denote sets, μ denotes a measure, φ denotes a signed measure.

1. If φ is σ-finite, then there are only countably many disjoint sets for which $\varphi \neq 0$ in every class.

2. For every A there exists a $B \subset A$ such that $\bar{\varphi}(A) \leqq 2|\varphi(B)|$.

3. If $\varphi_1 \leqq \varphi_2$, then $\varphi_1^+ \leqq \varphi_2^+$, $\varphi_1^- \geqq \varphi_2^-$. If $\varphi = \varphi_1 \pm \varphi_2$, then $\varphi^\pm \leqq \varphi_1^\pm + \varphi_2^\pm$.

4. Minimality of the Jordan-Hahn decomposition. If $\varphi = \mu^+ - \mu^-$, then $\varphi^\pm \leqq \mu^\pm$.

We say that A is a *φ-null set*, if $\varphi = 0$ on $\{AA', A' \in \mathfrak{A}\}$. We say that A and B are *φ-equivalent*, if they coincide up to a φ-null set. We say that a non-empty set is a *φ-atom*, if every measurable subset of A is φ-equivalent either to $\emptyset$ or to A.

5. The φ-null sets form a σ-ring; the φ-null sets of φ and of $\bar{\varphi}$ are the same. The φ-equivalence is an equivalence relation (reflexive, transitive, and symmetric), and $\mathfrak{A}$ splits into φ-equivalence classes.

6. Every φ-null set and every measurable set consisting of one point is a φ-atom; $\bar{\varphi}(A) = |\,\varphi(A)\,|$ for every φ-atom A. Atoms of φ and $\bar{\varphi}$ are the same; atoms of φ are atoms of φ^+ and φ^-, but the converse is not necessarily true.

If A is a φ-atom, then $\varphi = 0$ or $\varphi(A)$ on $A \cap \mathfrak{A}$; if φ is finite, then the converse is true. What if φ is σ-finite? What about $\varphi = \infty$ except for $\emptyset$?

7. If μ is finite, then $\Omega = \sum A_j + A$ where the A_j or A may be absent but, if present, then the A_j are μ-atoms of positive measure and, for every $B \subset A$ of positive measure, μ takes every value c between 0 and μB for measurable subsets of B. This decomposition of Ω is determined up to μ-null sets. Can μ be replaced by φ?

(There is only a countable number of μ-equivalence classes of such A_j's. Select representatives A_j of these classes and let $B \subset A = \Omega - \sum A_j$. Select inductively sets $C_n \in \mathfrak{C}_n$ such that $\mu C_n > \sup \mu C - \dfrac{1}{n}$ for all $C \in \mathfrak{C}_n$, where $\mathfrak{C}_n$ is the class of all $C \subset B - (C_1 \cup C_2 \cdots \cup C_{n-1})$ for which $\mu C \leq c - \mu(C_1 \cup C_2 \cup \cdots \cup C_{n-1})$. Then $\mu C = c$ for $C = \cup C_n$.)

8. If φ is finitely additive, μ is finite, and $\mu A_n \to 0$ implies $\varphi A_n \to 0$, then φ is σ-additive.

We say that φ is φ_0-*continuous* if $\varphi_0 A = 0$ implies $\varphi A = 0$.

9. If $\mu A_n \to 0$ implies $\varphi A_n \to 0(\bar{\varphi} A_n \to 0)$, then φ is μ-continuous. If φ is finite, then the converse is true.

(Assume the contrary of the converse; there exist $\epsilon > 0$ and A_n such that $\mu A_n < \dfrac{1}{2^n}$ and $\bar{\varphi} A_n \geq \epsilon$. Then $\mu B = 0$ and $\bar{\varphi} B \geq \epsilon$ for $B = \lim \sup A_n$.)

What if φ is σ-finite? What about $\mathfrak{A}$ consisting of all subsets of a denumerable space of points ω_n, and $\mu\{\omega_n\} = \dfrac{1}{2^n}$, $\varphi\{\omega_n\} = n$. What about μ replaced by φ_0?

10. If the μ_j are finite measures, then there exists a μ such that all the μ_j are μ-continuous. (Take $\mu = \sum \mu_j / 2^j \mu_j \Omega$.) What about μ_j's replaced by φ_j's?

Let $\mathfrak{B} \subset \mathfrak{A}$ be a σ-field such that the measurable subsets of elements of $\mathfrak{B}$ belong to $\mathfrak{B}$. Let $\mathfrak{B}(\varphi)$ be the class of sets such that their subsets which belong to $\mathfrak{B}$ are φ-null. Call the sets of $\mathfrak{B}$ "singular," and the sets of $\mathfrak{B}(\varphi)$ "regular." Call φ regular (singular) if every singular (regular) set is φ-null.

Let $\varphi_r = \varphi_r{}^+ - \varphi_r{}^-$, $\varphi_s = \varphi_s{}^+ - \varphi_s{}^-$, defined by

$$\varphi_r{}^\pm(A) = \sup \varphi^\pm(B) \quad \text{for all regular } B \subset A,$$

$$\varphi_s{}^\pm(A) = \sup \varphi^\pm(B) \quad \text{for all singular } B \subset A.$$

11. *Decomposition theorem.* φ_r is regular, φ_s is singular, and $\varphi = \varphi_r + \varphi_s$. If φ is finite, then the decomposition of φ into a regular and a singular part is unique. What if φ is σ-finite? What if $\mathfrak{A}$ consists of all subsets of a noncountable space, and $\varphi(A)$ equals the number of points of A? (Proceed as follows:

(i) $\mathfrak{B}(\varphi) = \mathfrak{B}(\bar{\varphi}) = \mathfrak{B}(\varphi^+) \cap \mathfrak{B}(\varphi^-)$ is a σ-field.

(ii) $\varphi_r(\varphi_s)$ is a regular (singular) signed measure.

(iii) Every A contains disjoint A_r regular and A_s singular such that $\varphi_r{}^{\pm}(A)$ = $\varphi^{\pm}(A_r)$, $\varphi_s{}^{\pm}(A) = \varphi^{\pm}(A_s)$.

(iv) If $A = A'_r + A'_s$ with A'_r regular and A'_s singular, then we can take $A_r = A'_r$ and $A_s = A'_s$.

(v) If φ is finite, every A can be so decomposed.)

12. We can take for singular sets:

(i) the μ-null sets—regular (singular) becomes μ-continuous (μ-discontinuous);

(ii) the countable measurable sets—regular (singular) becomes continuous (purely discontinuous);

(iii) the countable sums of atoms—regular (singular) becomes nonatomic (atomic).

In each case investigate the regular and singular parts.

13. Intermediate-value theorem (compare with continuous function on a connected set). If A is nonatomic and $A_n \uparrow A$ with φA_n finite, then φ takes every value between $-\varphi^- A$ and $+\varphi^+ A$ for measurable subsets in A. (See *7.*) What if $\mathfrak{A}$ consists of all sets in a noncountable space, $\varphi(A) = 0$ or ∞ according as A is countable or not?

In what follows, the φ_n are σ-additive but, unless otherwise stated, $\lim \varphi_n$ is not assumed to be σ-additive.

14. If $\varphi_n \to \varphi$ σ-additive, then $\varphi^{\pm} \leqq \lim \inf \varphi_n{}^{\pm}$. If, moreover, $\varphi_n \uparrow$ or $\varphi_n \downarrow$, then $\varphi^{\pm} = \lim \varphi_n{}^{\pm}$.

15. If $\varphi_n \uparrow (\downarrow)$ and $\varphi_1 > -\infty (< +\infty)$, then $\varphi_n \to \varphi$ σ-additive.

16. If $\varphi_n \to \varphi$ uniformly on $\mathfrak{A}$ and $\varphi > -\infty$ or $\varphi < +\infty$, then φ is σ-additive.

17. To a measure space $(\Omega, \mathfrak{A}, \mu)$ associate a complete metric space $(\mathfrak{X}, d)$ as follows: $\mathfrak{X}$ is the space of all sets A, B of finite measure, d is a metric defined by $d(A, B) = \mu(AB^c + A^c B)$. Prove that the metric space is complete.

(If A_n is a mutually convergent sequence in $\mathfrak{X}$, then the sequence I_{A_n} mutually converges in measure and hence converges in measure—see *6.3.*)

If ν on $\mathfrak{A}$ is a finite μ-continuous measure, then ν is defined and continuous on $(\mathfrak{X}, d)$.

We say that the φ_n are *uniformly μ-continuous* if $\mu A_m \to 0$ implies $\varphi_n A_m \to 0$ uniformly in n, as $m \to \infty$.

18. Let μ be σ-finite. If the finite φ_n are μ-continuous and $\lim \varphi_n$ exists and is finite, then the φ_n are uniformly μ-continuous and $\lim \varphi_n = \varphi$ is μ-continuous and σ-additive. (For every $\epsilon > 0$, set $A_k = \bigcap_{m=k}^{\infty} \bigcap_{n=k}^{\infty} \left[A \in \mathfrak{X}; \ |\varphi_m A - \varphi_n A| \right.$

$\left. \leqq \frac{\epsilon}{3} \right]$. By (*17*), every A_k is closed. By Baire's category theorem, there exists k_0, d_0 and $A_0 \in \mathfrak{X}$ such that $[A \in \mathfrak{X}; d(A, A_0) < d_0] \subset A_{k_0}$. Let $0 < \delta_0 < d_0$ such that $|\varphi_n A| < \epsilon$ whenever $\mu A < \delta_0$ and $n \leqq k_0$. If $\mu A < \delta_0$, then $d(A_0 - A, \ A_0) < d_0$, $d(A_0 \cup A, \ A_0) < d_0$, and $|\varphi_n A| \leqq |\varphi_{k_0} A| + |\varphi_n(A_0 \cup A) - \varphi_{k_0}(A_0 \cup A)| + |\varphi_n(A_0 - A) - \varphi_{k_0}(A_0 - A)|$.)

19. If finite $\varphi_n \to \varphi$ finite, then φ is σ-additive. (If $|\varphi_n| \leqq c_n$, set $\mu A = \sum \frac{1}{2^n c_n} |\varphi_n A|$ and apply *18.*)

Chapter II

MEASURABLE FUNCTIONS AND INTEGRATION

§ 5. MEASURABLE FUNCTIONS

5.1 Numbers. Spaces built with numbers are prototypes of all spaces, and functions whose values are numbers are prototypes of all functions.

By a *number x* we mean either a usual real number—*finite number*—or one of the symbols $+\infty$ and $-\infty$—*infinite numbers*. These symbols are defined by the following properties:

$$-\infty \leq x \leq +\infty,$$

$$\pm\infty = (\pm\infty) + x = x + (\pm\infty), \quad \frac{x}{\pm\infty} = 0 \quad \text{if} \quad -\infty < x < +\infty,$$

$$x(\pm\infty) = (\pm\infty)x = \begin{cases} \pm\infty & \text{if} \quad 0 < x \leq +\infty \\ 0 & \text{if} \quad x = 0 \\ \mp\infty & \text{if} \quad -\infty \leq x < 0. \end{cases}$$

The expression $+\infty - \infty$ is meaningless, so that, when speaking of a "sum" of two numbers, we assume that, if one of them is $\mp\infty$, the other one is not $\pm\infty$; then the sum exists.

The reason for the introduction of infinite numbers lies in the fact that, then, $\sup x_t$ and $\inf x_t = -\sup(-x_t)$, where t varies over an arbitrary set T, always exist (but may be infinite). Moreover, if inclusion, union, and intersection of numbers are defined by $x \leq y$, $\sup x_t$ and $\inf x_t$ respectively, then these operations have properties of the corresponding set operations; in particular, limits of monotone sequences of numbers always exist, but may be infinite.

103

If, as $n \to \infty$, the limit x of a sequence x_n of numbers exists, we write $x = \lim x_n$ or $x_n \to x$ and say that x_n *converges to x*; if x is infinite, say, $+\infty$, one also says that x_n *diverges to $+\infty$*. The Cauchy mutual convergence criterion is valid only for finite limits: x_n *converges to some finite x if, and only if, $x_m - x_n \to 0$ (as $m, n \to \infty$) or, equivalently, if $x_{n+v} - x_n \to 0$ uniformly in v*. On the other hand, the Bolzano-Weierstrass lemma remains valid without the usual restriction of boundedness: *every sequence of numbers is compact*, that is, contains a convergent subsequence, but if the sequence is not bounded then the limits may be infinite.

The set of all finite numbers is a *real line* $R = (-\infty, +\infty)$ and the set of all numbers is an *extended real line* $\overline{R} = [-\infty, +\infty]$. The basic class of sets in R is the class of *intervals*; there are four types of finite intervals of respective form:

$$[a, b): \text{set of all points } x \text{ such that } a \leq x < b;$$

$$(a, b]: \text{set of all points } x \text{ such that } a < x \leq b;$$

$$(a, b): \text{set of all points } x \text{ such that } a < x < b;$$

$$[a, b]: \text{set of all points } x \text{ such that } a \leq x \leq b.$$

The minimal σ-field over the class of all intervals in R is the *Borel field* in R and its elements are *Borel sets* in R. The Borel field in R coincides with the minimal σ-field over the subclass of all intervals of one of the foregoing four types, since countable operations performed upon elements of one of these subclasses yield any element of the other subclasses; for example, $(a, b) = \bigcup \left[a + \dfrac{1}{n}, b \right)$, $[a, b] = \bigcap \left[a, b + \dfrac{1}{n} \right)$, etc. Similarly, the Borel field in R is the minimal σ-field over the subclass of all infinite intervals of the form $(-\infty, x)$, $-\infty \leq x \leq +\infty$, since any finite interval $[a, b)$ is obtainable as a difference $\Delta_{b-a}(-\infty, a) = (-\infty, b) - (-\infty, a)$. The Borel field in $\overline{R}$ can be defined similarly by means of any of the foregoing types where $-\infty \leq a \leq b \leq +\infty$, or by means of the intervals $[-\infty, x)$, $-\infty \leq x \leq +\infty$; but, frequently, the most convenient way is to take the minimal σ-field over the class formed by the Borel field in R and the two sets $\{-\infty\}$, $\{+\infty\}$.

Extension. The preceding notions extend at once to finite-dimensional real spaces. The set of all ordered N-uples $x = (x_1, \cdots, x_N)$ of finite numbers is the *N-dimensional real space* R^N or, equivalently, the *product space* $\prod\limits_{v=1}^{N} R_v$ of N real lines $R_v = (-\infty < x_v < +\infty)$. If every R_v is replaced by $\overline{R}_\nu = [-\infty \leq x_\nu \leq +\infty]$, then we have the *extended N-dimensional real space* $\overline{R}^N$. If $a, b \in R^N$, then $a \leq b$ means that $a_\nu \leq b_\nu$ for $\nu = 1, 2, \cdots, N$, and, similarly, for $a < b$, $a = b$.

An interval, say $[a, b)$, will also be written more explicitly as $[a_1, a_2, \cdots, a_N; b_1, b_2, \cdots, b_N)$, and

$$[a, b) = \Delta_{b-a}(-\infty, a) = \Delta_{b_1-a_1}\Delta_{b_2-a_2}\cdots$$

$$\Delta_{b_N-a_N}(-\infty, -\infty, \cdots -\infty; a_1, a_2, \cdots a_N)$$

where $\Delta_{b_\nu-a_\nu}$ is the difference operator of step $b_\nu - a_\nu$ acting on a_ν. For example, if $N = 2$, then

$$[a_1, a_2; b_1, b_2) = \Delta_{b_1-a_1}\Delta_{b_2-a_2}(-\infty, -\infty; a_1, a_2)$$

$$= \Delta_{b_1-a_1}\{(-\infty, -\infty; a_1, b_2) - (-\infty, -\infty; a_1, a_2)\}$$

$$= (-\infty, -\infty; b_1, b_2) - (-\infty, -\infty; a_1, b_2) - (-\infty, -\infty;$$

$$b_1, a_2) + (-\infty, -\infty; a_1, a_2).$$

With this interpretation, the foregoing definitions of types of intervals and, thereafter, of Borel fields, remain the same.

5.2 Numerical functions. A *numerical function* X on a space Ω is a function on Ω to $\overline{R}$, defined by assigning to every point $\omega \in \Omega$ a single number $x = X(\omega)$, the *value* of X at ω. If infinite values are excluded, X is a *finite function* or, equivalently, a function on Ω to R. Ω is called the *domain* of X and $\overline{R}$ (or R) is called the *range space* of X. The functions $X^+ = XI_{[X \geq 0]}$ and $X^- = -XI_{[X < 0]}$ will be called the *positive part* and the *negative part* of X, respectively, and we have

$$X = X^+ - X^-, \quad |X| = X^+ + X^-.$$

Unless otherwise stated, all functions will be numerical functions and, in general, will be denoted by $X, Y, \cdots$, with or without affixes.

If definitions or relations between values of given functions hold for every ω belonging to a set $A \subset \Omega$, we say that these definitions or relations hold *on A* and drop "on A" if $A = \Omega$. For example,

$|X| < \infty$ means that X is finite;
$X \geq 0$ on A means that $X(\omega) \geq 0$ for every $\omega \in A$;
$X = \inf X_n$ means that $X(\omega) = \inf X_n(\omega)$ for every $\omega \in \Omega$;
$X_n \to X$ on A means that $X_n(\omega) \to X(\omega)$ for every $\omega \in A$, etc.

Conversely, the set of *all* $\omega \in \Omega$ on which definitions or relations hold is denoted by $[\omega; \cdots]$ or, if there is no confusion possible, by $[\cdots]$ where $\cdots$ stand for the definitions or relations. For example,

$[X]$ is the set on which X is defined;

$[X \geq Y]$ is the set of all $\omega \in \Omega$ for which $X(\omega) \geq Y(\omega)$;

$[X \in S]$, where $S \subset \bar{R}$, is the set of all $\omega \in \Omega$ for which the values $X(\omega)$ belong to the set S.

The set $[X = x]$ is called the inverse image of the set $\{x\}$ which consists of x only or, simply, of x. Since X is single-valued, the inverse images of distinct numbers x are disjoint, and the partition of Ω into inverse images of all $x \in \bar{R}$ is called the partition of the domain *induced* by X; we sometimes write $X = \sum_{x \in \bar{R}} x I_{[X = x]}$ where $I_{[X = x]}$ is the indicator of $[X = x]$. In particular, if X is *countably valued*, that is, takes only a countable number of values x_j, then, and only then,

$$X = \sum_j x_j I_{[X = x_j]}.$$

More generally, the set $[X \in S]$ is called *inverse image* of S and is also denoted by $X^{-1}(S)$. The symbol X^{-1}, which can be considered as representing a mapping of sets in $\bar{R}$ onto sets in Ω, is called the *inverse function* of X. Since inverse images of disjoint sets of $\bar{R}$ are disjoint, it follows easily that

X^{-1} *and set operations commute:*

$$X^{-1}(S - S') = X^{-1}(S) - X^{-1}(S'), \quad X^{-1}(\bigcup S_t) = \bigcup X^{-1}(S_t),$$
$$X^{-1}(\bigcap S_t) = \bigcap X^{-1}(S_t).$$

Similarly, $X^{-1}(\mathcal{C})$ or the *inverse image* of $\mathcal{C}$, where $\mathcal{C}$ is a class of sets in R, is the class of all inverse images of elements of $\mathcal{C}$. Since set operations commute with inverse functions, it follows that

a. *The inverse image of a σ-field is a σ-field, the inverse image of the minimal σ-field over a class is the minimal σ-field over the inverse image of the class, the class of all sets whose inverse images belong to a σ-field is a σ-field.*

The foregoing definitions and properties extend at once to functions $X = (X_1, \cdots, X_N)$ on Ω to an N-dimensional real space $\bar{R}^N$ (or R^N) or, equivalently, to N-uples of numerical functions $X_1, \cdots, X_N$. Classical analysis is concerned with functions from a real line to a real line or, more generally, from a finite-dimensional real space R^N to a finite-dimensional real space R^N. Still more generally, let X be a function on Ω to $\bar{R}^N$ and let g be a function on $\bar{R}^N$ to $\bar{R}^{N'}$. *The function of func-*

tion gX defined by $(gX)(\omega) = g(X(\omega))$ is a function on Ω to $\overline{R}^{N'}$. Clearly, its inverse function $(gX)^{-1}$ is a mapping of sets S' in $\overline{R}^{N'}$ onto sets in Ω such that

$$(gX)^{-1}(S') = X^{-1}(g^{-1}(S'))$$

or, in a condensed form,

$$(gX)^{-1} = X^{-1}g^{-1}.$$

5.3 Measurable functions. Classical analysis is concerned primarily with continuous functions on R to R' or, more generally, on R^N to $R^{N'}$. However, passages to the limit, which play such a basic role in analysis, do not, in general, preserve continuity (and also they cause the appearance of $\mp\infty$). The essential achievement of modern analysis, due to Borel, Baire, and Lebesgue, is the introduction of a wider class of functions which is closed under the "usual" operations of analysis: arithmetic operations and formation of infima, suprema, and limits of sequences. Those are the functions we intend to define now.

In the domain Ω of our functions we select a σ-field $\mathfrak{a}$ of sets, to be called $\mathfrak{a}$-*sets* or, if there is no confusion possible, *measurable sets;* the doublet $(\Omega, \mathfrak{a})$ is called a *measurable space.* In the range space $\overline{R}$ of our functions we select the σ-field $\overline{\mathfrak{B}}$ of Borel sets—the Borel field in $\overline{R}$; the doublet $(\overline{R}, \overline{\mathfrak{B}})$ is an (*extended*) *Borel line.* Thus, our functions are defined on a measurable space $(\Omega, \mathfrak{a})$ to the Borel line $(\overline{R}, \overline{\mathfrak{B}})$. More generally, if the range space is $\overline{R}^N$, then we select the Borel field $\overline{\mathfrak{B}}^N$, and the doublet $(\overline{R}^N, \overline{\mathfrak{B}}^N)$ is an *extended Borel space;* then the functions are defined on a measurable space $(\Omega, \mathfrak{a})$ to the Borel space $(\overline{R}^N, \overline{\mathfrak{B}}^N)$.

A countably valued function $X = \sum x_j I_{A_j}$ where the sets A_j are measurable is called an elementary measurable function or, simply, an *elementary function;* if the number of distinct values of X is finite, then X is also called a *simple function.*

(C) *Limits of convergent sequences of simple functions are called measurable functions.*

This is a constructive definition and, because of that, will play an essential role in the constructive definition of integrals. However, general properties of measurable functions are easier to discover and to prove when using the descriptive definition which follows.

(D) *Functions such that inverse images of all Borel sets are measurable sets are called measurable functions.*

Yet this definition is not the most economical one, since

(D') *In* (D), *it suffices to require measurability of inverse images of elements of any fixed class* $\mathcal{C}$ *such that the minimal σ-field over* $\mathcal{C}$ *is the Borel field.*

For example, we can take $\mathcal{C}$ to be the class of all intervals, or the class of all intervals $[-\infty, x]$, etc.

The proof is immediate. Since a mapping X^{-1} preserves all sets operations and the measurable sets form a σ-field, it follows that the class of all sets whose inverse images are measurable is a σ-field. Therefore, if, according to (D'), it contains $\mathcal{C}$, then it contains the minimal σ-field over $\mathcal{C}$ which, by assumption, is the Borel field.

Similarly, the constructive definition (C) is not the most economical one as we shall find in proving the basic theorem below.

A. MEASURABILITY THEOREM. *The constructive and descriptive definitions are equivalent, and the class of measurable functions is closed under the usual operations of analysis.*

Proof. 1° Let X_n be functions measurable (D), that is, measurable according to (D) or, equivalently, (D'). Then all sets

$$[\inf X_n < x] = \bigcup [X_n < x], \quad [-X_n < x] = [X_n > -x]$$

are measurable and, hence, the functions

$$\sup X_n = - \inf (-X_n), \quad \liminf X_n = \sup_n (\inf_{k \geq n} X_k),$$

$$\limsup X_n = - \liminf (-X_n)$$

are measurable (D). Thus, the class of functions measurable (D) is closed under formation of infima, suprema, and limits. But every simple function $X = \sum x_j I_{A_j}$ is measurable (D), since all sets $[X \leq x] = \sum_{x_j \leq x} A_j$ are measurable. Therefore, limits of convergent sequences of simple functions are measurable (D); in particular, functions measurable (C) are measurable (D).

2° Conversely, let X be measurable (D) so that the functions

$$X_n = -nI_{[X < -n]} + \sum_{-n2^n+1}^{n2^n} \frac{k-1}{2^n} I\left[\frac{k-1}{2^n} \leq X < \frac{k}{2^n}\right] + nI_{[X \geq n]},$$

$$n = 1, 2, \cdots$$

are simple. Since

$$\left| X_n(\omega) - X(\omega) \right| < \frac{1}{2^n} \quad \text{for} \quad |X(\omega)| < n$$

and

$$X_n(\omega) = \pm n \quad \text{for} \quad X(\omega) = \pm\infty,$$

it follows that $X_n \to X$ and this, together with what precedes, completes the proof of the equivalence of the two definitions of measurability.

We observe that if X is nonnegative, then the foregoing functions X_n become

$$X_n = \sum_{k=1}^{n2^n} \frac{k-1}{2^n} I\left[\tfrac{k-1}{2^n} \leq X < \tfrac{k}{2^n}\right] + nI_{[X \geq n]}$$

and we have $0 \leq X_n \uparrow X$. Also, if

$$X'_n = \sum_{k=-\infty}^{+\infty} \frac{k-1}{2^n} I\left[\tfrac{k-1}{2^n} \leq X < \tfrac{k}{2^n}\right] + (-\infty)I_{[X=-\infty]} + (+\infty)I_{[X=+\infty]},$$

then $\left| X'_n - X \right| < \dfrac{1}{2^n}$ on $[|X| < \infty]$ and $X'_n = X$ on $[|X| = \infty]$, so that $X'_n \to X$ uniformly.

3° It remains to prove closure under the arithmetic operations. Using definition (C) and the fact that arithmetic operations commute with passages to the limit by convergent sequences, it suffices to show that the class of simple functions is closed under the arithmetic operations. But much more is true, for if g on $\bar{R}^N$ is an arbitrary function and $X_k = \sum_j x_{kj} I_{A_{kj}}$, $k = 1, \cdots, N$, are simple (elementary) functions, then the function of functions

$$g(X_1, \cdots, X_N) = \sum g(x_{1j_1}, \cdots, x_{Nj_N}) I_{A_{1i_1}} \cdots I_{A_{Ni_N}}$$

is simple (elementary). This completes the proof.

According to this proof we have new *equivalent constructive definitions* of measurable functions that we state now.

(C′) *A nonnegative function is measurable if it is the limit of a nondecreasing sequence of nonnegative simple functions. A function X is measurable if its positive and negative parts X^+ and X^- are measurable.*

(C″) *A function is measurable if it is the limit of a uniformly convergent sequence of elementary functions. In particular, every bounded measurable function is limit of a uniformly convergent sequence of simple functions.*

Definition (C′) will play a central role in the theory of integration. Closure under the arithmetic operations is a very particular case of

a. *A Baire function of measurable functions is measurable.*

Proof. Let us recall a (constructive) definition of Baire functions (we consider only finite-dimensional Borel spaces). *Baire functions* are elements of the smallest class closed under passages to the limit containing all continuous functions. Therefore, since the class of measurable functions is closed under passages to the limit, it suffices to prove that

A continuous function of measurable functions is measurable.

Thus, let g on $\overline{R}^N$ be continuous; that is, for every point $(x_1, \cdots, x_N)$ $\in \overline{R}^N$,

$$g(x'_1, \cdots, x'_N) \rightarrow g(x_1, \cdots, x_N) \quad \text{as} \quad x'_1 \rightarrow x_1, \cdots, x'_N \rightarrow x_N.$$

Let X_k, $k = 1, 2, \cdots, N$, be measurable and let X_{nk} be sequences of simple functions such that $X_{nk} \rightarrow X_k$ for every k. We found (in 3°) that the functions $g(X_{n1}, \cdots, X_{nN})$(that we assumed tacitly to have meaning) are measurable and hence, by continuity and closure under passages to the limit, the function

$$g(X_1, \cdots, X_N) = \lim g(X_{n1}, \cdots, X_{nN})$$

is measurable. This completes the proof.

All the foregoing definitions and properties extend at once, and word for word, to functions on a measurable space to any finite-dimensional Borel space, provided we replace $\overline{R}$ by $\overline{R}^N$ and leave out the operations of multiplication and division that we do not define (at least here) for such functions. For example,

functions such that inverse images of Borel sets in their range space are measurable sets in their domain are called measurable functions.

This extension is useful but, in fact, brings nothing new, for

b. *A function $X = (X_1, \cdots, X_N)$ is measurable if, and only if, its components $X_1, \cdots, X_N$ are measurable.*

In other words such a function is merely an N-uple of numerical measurable functions.

Proof. If $X = (X_1, \cdots, X_N)$ is measurable, then, for every $k \leqq N$, the sets

$$[X_k \leqq x_k] = X_k^{-1}[-\infty, x_k]$$

$$= X^{-1}[-\infty, \cdots, -\infty; +\infty, \cdots, +\infty, x_k, +\infty, \cdots, +\infty]$$

are measurable, so that X_k is measurable.

Conversely, if all X_k are measurable, then the sets

$$[X \leqq x] = [X_1 \leqq x_1, \cdots, X_N \leqq x_N] = \bigcap_{k=1}^{N} [X_k \leqq x_k]$$

are measurable, so that $X = (X_1, \cdots, X_N)$ is measurable.

We give another (descriptive) definition of Baire functions. With this definition, it is customary to call these functions Borel functions. A measurable function on a finite-dimensional Borel space to a finite-dimensional Borel space is called a *Borel function*. In other words, g on $\overline{R}^N$ to $\overline{R}^{N'}$ is a Borel function if, and only if, the inverse images of Borel sets S' in $\overline{R}^{N'}$ are Borel sets S in $\overline{R}^N$. The proof of **a** in this more general case is then immediate and we have

a'. *A Borel function of a measurable function is measurable.*

For, if X is a measurable function (not necessarily numerical) and g is a Borel function on the range space of X, then, for every Borel set S' in the range space of g, the set $(gX)^{-1}(S') = X^{-1}(g^{-1}(S'))$ is measurable and, hence, gX is a measurable function.

§ 6.   MEASURE AND CONVERGENCES

6.1 Definitions and general properties. The notions of "measurable" sets and "measurable" functions are two out of a triplet of notions, due essentially to Lebesgue, the third being the notion of "measure" which gave its name to the two others, and which we shall introduce now.

A function φ on a σ-field $\mathcal{C}$ is said to be *σ-additive* if, for every countable disjoint class $\{A_j\} \subset \mathcal{C}$,

$$\varphi(\textstyle\sum A_j) = \sum \varphi(A_j).$$

To avoid trivialities, it is assumed that at least one value of φ, say, $\varphi(A_0)$, $A_0 \in \mathcal{C}$, is finite. Since

$$\varphi(A_0 + \emptyset) = \varphi(A_0) = \varphi(A_0) + \varphi(\emptyset),$$

this assumption is equivalent to $\varphi(\emptyset) = 0$. To avoid meaningless expressions of the form $+\infty - \infty$, it is assumed that at least one of the possible values $-\infty$ or $+\infty$ is excluded.

φ is said to be *finite* if its values are finite, and it is said to be *σ-finite* if the space in which $\mathcal{C}$ is defined can be partitioned into a countable number of sets in $\mathcal{C}$ for which the values of φ are finite.

A *measure* μ on a σ-field $\mathfrak{A}$ is a nonnegative and σ-additive function. In other words, μ is defined by the three following properties:

(i) $\mu(\sum A_j) = \sum \mu(A_j)$ for every countable disjoint class $\{A_j\} \subset \mathfrak{A}$;
(ii) $\mu(A) \geqq 0$ for every $A \in \mathfrak{A}$;
(iii) $\mu(\emptyset) = 0$.

The value $\mu(A)$ of μ at A is called the *measure of A* and, if there is no confusion possible, we drop the bracket following the symbol μ.

A *measure space* $(\Omega, \mathfrak{A}, \mu)$ is formed by the space Ω, the σ-field $\mathfrak{A}$ of measurable sets in this space, and the measure μ defined on this σ-field. Unless otherwise stated all sets under consideration will be measurable sets in our measure space. A set of measure 0 is said to be a *μ-null set* or, if there is no confusion possible, a *null set,* and definitions or relations valid outside a μ-null set are said to be valid *almost everywhere (a.e.).* The following properties of the measure μ are immediate:

a. *μ is nondecreasing, and μ is bounded if the space Ω is of finite measure.*

This follows from

$$\mu B = \mu A + \mu(B - A) \geqq \mu A \quad \text{for} \quad B \supset A.$$

b. *μ is sub σ-additive:* $\mu \bigcup A_j \leqq \sum \mu A_j.$

This follows from

$$\mu \bigcup A_j = \mu(A_1 + A_1{}^c A_2 + \cdots)$$
$$= \mu A_1 + \mu A_1{}^c A_2 + \cdots \leqq \mu A_1 + \mu A_2 + \cdots.$$

A. Sequences theorem. *If $A_n \uparrow A$, then $\mu A_n \uparrow \mu A$ and, in general,*

$$\liminf \mu A_n \geqq \mu(\liminf A_n).$$

If μ is finite, then, moreover,

$$A_n \downarrow A \text{ implies } \mu A_n \downarrow \mu A, \quad \limsup \mu A_n \leqq \mu(\limsup A_n),$$
$$A_n \rightarrow A \text{ implies } \mu A_n \rightarrow \mu A.$$

Proof. If $A_n \uparrow A$, then, by σ-additivity,

$$\mu A = \mu A_1 + \mu(A_2 - A_1) + \cdots$$
$$= \lim \{\mu A_1 + \mu(A_2 - A_1) + \cdots + \mu(A_n - A_{n-1})\}$$
$$= \lim \mu A_n.$$

If A_n is an arbitrary sequence, then, since $B_n = \bigcap_{k \geq n} A_k \uparrow \lim\inf A_n$ and $\mu A_n \geq \mu B_n$, it follows that

$$\lim\inf \mu A_n \geq \lim \mu B_n = \mu(\lim\inf A_n),$$

and the first assertion is proved.

Let now μ be finite and use the proved assertion. If $A_n \downarrow A$, then $A_1 - A_n \uparrow A_1 - A$ and, hence,

$$\mu A_1 - \mu A_n = \mu(A_1 - A_n) \uparrow \mu(A_1 - A) = \mu A_1 - \mu A,$$

so that $\mu A_n \downarrow \mu A$. If A_n is an arbitrary sequence, then $\mu\Omega - \lim\sup \mu A_n = \lim\inf \mu A_n^c \geq \mu(\lim\inf A_n^c) = \mu\Omega - \mu(\lim\sup A_n)$ and, hence, $\lim\sup \mu A_n \leq \mu(\lim\sup A_n)$. Finally, if $A_n \to A$, then the two inequalities proved above yield $\mu A_n \to \mu A$, and the proof is complete.

The introduction of measures yields new types of convergence founded upon the notion of measure and unknown in classical analysis. Before we introduce them, we recall the classical types of convergence; unless otherwise stated, we consider sequences X_n of measurable functions on a fixed measure space $(\Omega, \mathcal{C}, \mu)$ and limits taken as $n \to \infty$.

If X_n converges to X on A according to a definition "c" of convergence, we say that X_n converges "c" on A and write $X_n \xrightarrow{c} X$ on A. The Cauchy convergence criterion leads to the corresponding notion of mutual convergence: if $X_{n+\nu} - X_n$ converges "c" to 0 on A uniformly in ν (or $X_m - X_n$ converges "c" to 0 on A as $m, n \to \infty$), we say that X_n *mutually converges* "c" on A and write $X_{n+\nu} - X_n \xrightarrow{c} 0$ (or $X_m - X_n \xrightarrow{c} 0$). In defining mutual convergence, we naturally must assume that the differences exist, that is, meaningless expressions $+\infty - \infty$ do not occur. We drop "on A" if $A = \Omega$ and drop "c" if the convergence is ordinary pointwise convergence.

We recall that $X_n \to X$ on A means that, for every $\omega \in A$ and every $\epsilon > 0$, there is an integer $n_{\epsilon, \omega}$ such that, for $n \geq n_{\epsilon, \omega}$,

if $X(\omega)$ is finite, then $|X(\omega) - X_n(\omega)| < \epsilon$,

if $X(\omega) = -\infty$, then $X_n(\omega) < -\dfrac{1}{\epsilon}$,

if $X(\omega) = +\infty$, then $X_n(\omega) > +\dfrac{1}{\epsilon}\cdot$

If $n_{\epsilon,\omega} = n_\epsilon$ is independent of $\omega \in A$, then the convergence is *uniform* and, according to the preceding conventions, we write $X_n \overset{u}{\to} X$ on A. According to the closure property of measurable functions, if a sequence of measurable functions $X_n \to X$, then X is measurable. According to the Cauchy criterion, if X_n are finite, then

$X_n \to X$ finite if, and only if, $X_m - X_n \to 0$ or, equivalently $X_{n+\nu} - X_n \to 0$

$X_n \overset{u}{\to} X$ finite if, and only if, $X_m - X_n \overset{u}{\to} 0$ or, equivalently, $X_{n+\nu} - X_n \overset{u}{\to} 0$.

6.2   Convergence almost everywhere. A sequence X_n is said to *converge* a.e. to X, and we write $X_n \overset{a.e.}{\longrightarrow} X$, if $X_n \to X$ outside a null set; it *mutually converges* a.e., and we write $X_m - X_n \overset{a.e.}{\longrightarrow} 0$ or $X_{n+\nu} - X_n \overset{a.e.}{\longrightarrow} 0$, if it mutually converges outside a null set. It follows, by the Cauchy criterion and the fact that a countable union of null sets is a null set, that

a. *A sequence of a.e. finite functions converges a.e. to an a.e. finite function if, and only if, the sequence mutually converges a.e.*

Let $X_n \overset{a.e.}{\longrightarrow} X$. Since X_n are taken to be measurable, X is a.e. measurable, that is, X is the a.e. limit of a sequence of simple functions. Also, if X' is such that $X_n \overset{a.e.}{\longrightarrow} X'$, then $X = X'$ a.e., for X can differ from X' only on the null set on which X_n converges neither to X nor to X'. Thus, the limit of the sequence X_n is a.e. determined and a.e. measurable. Moreover, if every X_n is modified arbitrarily on a null set N_n, then the whole sequence is modified at most on the null set $\bigcup N_n$ and, therefore, the so modified sequence still converges a.e. to X.

These considerations lead to the introduction of the notion of "equivalent" functions: X and X' are *equivalent* if $X = X'$ a.e. Since the notion has the usual properties of an equivalence—it is reflexive, transitive, and symmetric—it follows that the class of all functions on our measure space splits into equivalence classes, and the discussion which precedes can be summarized as follows.

b. *Convergence a.e. is a type of convergence of equivalence classes to an equivalence class.*

In other words, as long as we are concerned with convergence a.e. of *sequences* of functions, these functions as well as the limit functions are to be considered as defined up to an equivalence. In particular, we can replace an a.e. finite and a.e. measurable function by a finite and measurable function, and conversely, without destroying convergence a.e.

Let us investigate in more detail the set on which a given sequence converges. To simplify, we restrict ourselves to the most important case of *finite* measurable functions, the study of the general case being similar. By definition of ordinary convergence, the set of convergence $[X_n \rightarrow X]$ of finite X_n to a finite measurable X is the set of all points $\omega \in \Omega$ at which, for every $\epsilon > 0$, $|X(\omega) - X_n(\omega)| < \epsilon$ for $n \geq n_{\epsilon,\omega}$ sufficiently large. Since, moreover, the requirement "for every $\epsilon > 0$" is equivalent to "for every term of a sequence $\epsilon_k \downarrow 0$ as $k \rightarrow \infty$," say, the sequence $\dfrac{1}{k}$, we have

$$[X_n \rightarrow X] = \bigcap_{\epsilon > 0} \bigcup_{n} \bigcap_{\nu} [|X_{n+\nu} - X| < \epsilon]$$

$$= \bigcap_{k} \bigcup_{n} \bigcap_{\nu} \left[|X_{n+\nu} - X| < \frac{1}{k}\right],$$

so that the set $[X_n \rightarrow X]$ is measurable. Similarly for the set of mutual convergence, since the set

$$[X_{n+\nu} - X_n \rightarrow 0] = \bigcap_{\epsilon > 0} \bigcup_{n} \bigcap_{\nu} [|X_{n+\nu} - X_n| < \epsilon]$$

$$= \bigcap_{k} \bigcup_{n} \bigcap_{\nu} \left[|X_{n+\nu} - X_n| < \frac{1}{k}\right]$$

is measurable. Thus

   c. *The sets of convergence (to a finite measurable function) and of mutual convergence of a sequence of finite measurable functions are measurable.*

In other words, to every sequence we can assign a "measure of convergence" and, the sets of divergence $[X_n \nrightarrow X]$ and $[X_{n+\nu} - X_n \nrightarrow 0]$ being complements of those of convergence and, hence, measurable, to every sequence we can assign a "measure of divergence." In particular, the definitions of a.e. convergence of a sequence X_n mean that

$$\mu[X_n \nrightarrow X] = 0 \quad \text{or} \quad \mu[X_{n+\nu} - X_n \nrightarrow 0] = 0.$$

Upon applying repeatedly the sequences theorem to the above-defined sets, we obtain the following

A. Convergence a.e. criterion. *Let* X, X_n *be finite measurable functions.*

$X_n \xrightarrow{\text{a.e.}} X$ *if, and only if, for every* $\epsilon > 0$,

$$\mu \bigcap_n \bigcup_\nu \left[|\, X_{n+\nu} - X\,| \geq \epsilon\right] = 0$$

and, if μ is finite, this criterion becomes

$$\mu \bigcup_\nu \left[|\, X_{n+\nu} - X\,| \geq \epsilon\right] \to 0.$$

$X_{n+\nu} - X_n \xrightarrow{\text{a.e.}} 0$ *if, and only if, for every* $\epsilon > 0$,

$$\mu \bigcap_n \bigcup_\nu \left[|\, X_{n+\nu} - X_n\,| \geq \epsilon\right] = 0$$

and, if μ is finite, this criterion becomes

$$\mu \bigcup_\nu \left[|\, X_{n+\nu} - X_n\,| \geq \epsilon\right] \to 0.$$

6.3   Convergence in measure.   A sequence X_n of finite measurable functions is said to *converge in measure* to a measurable function X and we write $X_n \xrightarrow{\mu} X$ if, for every $\epsilon > 0$,

$$\mu[|\, X_n - X\,| \geq \epsilon] \to 0.$$

The limit function X is then necessarily a.e. finite, since

$$\mu[|\, X\,| = \infty] = \mu[|\, X_n - X\,| = \infty] \leq \mu[|\, X_n - X\,| \geq \epsilon] \to 0.$$

Similarly, $X_{n+\nu} - X_n \xrightarrow{\mu} 0$ if, for every $\epsilon > 0$,

$$\mu[|\, X_{n+\nu} - X_n\,| \geq \epsilon] \to 0 \text{ (uniformly in } \nu\text{)}.$$

All considerations about equivalence classes in the case of convergence a.e. remain valid for convergence in measure. In particular, if $X_n \xrightarrow{\mu} X$ and $X_n \xrightarrow{\mu} X'$, then X and X' are equivalent, for

$$\mu[|\, X - X'\,| \geq \epsilon] \leq \mu\left[|\, X - X_n\,| \geq \frac{\epsilon}{2}\right] + \mu\left[|\, X_n - X'\,| \geq \frac{\epsilon}{2}\right] \to 0$$

and, hence,

$$\mu[X \neq X'] = \mu \bigcup_k \left[|\, X - X'\,| \geq \frac{1}{k}\right] = 0.$$

We compare now convergence in measure and convergence a.e.

A. COMPARISON OF CONVERGENCES THEOREM.   *Let X_n be a sequence of finite measurable functions.*

If X_n converges or mutually converges in measure, then there is a subsequence X_{n_k} which converges in measure and a.e. to the same limit function. If μ is finite, then convergence a.e. to an a.e. finite function implies convergence in measure to the same limit function.

Proof.   The second assertion is an immediate consequence of the a.e. convergence criterion, since μ finite and $X_n \xrightarrow{\text{a.e.}} X$ imply that, for every $\epsilon > 0$,

$$\mu[|X_{n+1} - X| \geqq \epsilon] \leqq \mu \bigcup_\nu [|X_{n+\nu} - X| \geqq \epsilon] \to 0.$$

As for the first assertion, let $X_{n+\nu} - X_n \xrightarrow{\mu} 0$. Then, for every integer k there is an integer $n(k)$ such that, for $n \geqq n(k)$ and all ν,

$$\mu\left[|X_{n+\nu} - X_n| \geqq \frac{1}{2^k}\right] < \frac{1}{2^k}.$$

Let $n_1 = n(1)$, $n_2 = \max(n_1 + 1, n(2))$, $n_3 = \max(n_2 + 1, n(3))$, etc., so that $n_1 < n_2 < n_3 < \cdots \to \infty$. Let $X'_k = X_{n_k}$ and

$$A_k = \left[|X'_{k+1} - X'_k| \geqq \frac{1}{2^k}\right], \quad B_n = \bigcup_{k \geqq n} A_k,$$

so that

$$\mu A_k < \frac{1}{2^k}, \quad \mu B_n \leqq \sum_{k \geqq n} \mu A_k < \frac{1}{2^{n-1}}.$$

Thus, for a given $\epsilon > 0$, n large enough so that $\frac{1}{2^{n-1}} < \epsilon$, and all ν, we have on $B_n{}^c$

$$|X'_{n+\nu} - X'_n| \leqq \sum_{k \geqq n} |X'_{k+1} - X'_k| < \frac{1}{2^{n-1}} < \epsilon.$$

Therefore,

$$\mu \bigcap_n \bigcup_\nu [|X'_{n+\nu} - X'_n| \geqq \epsilon] \leqq \mu \bigcup_\nu [|X'_{n+\nu} - X'_n| \geqq \epsilon]$$

$$\leqq \mu B_n < \frac{1}{2^{n-1}} \to 0$$

and, hence, by the convergence a.e. criterion, $X'_{n+\nu} - X'_n \xrightarrow{\text{a.e.}} 0$. Thus, by **6.2a**, there is a finite X' such that $X'_n \xrightarrow{\text{a.e.}} X'$. Since on

$B_n{}^c$ we have $| X'_{n+\nu} - X'_n | < \epsilon$ for all ν it follows, upon letting $\nu \to \infty$, that on $B_n{}^c$ we also have $| X' - X'_n | < \epsilon$ outside perhaps a null subset. Therefore, upon taking complements,

$$\mu[| X' - X'_n | \geq \epsilon] \leq \mu B_n < \frac{1}{2^{n-1}} \to 0,$$

so that $X'_n \xrightarrow{\mu} X'$. A similar argument shows that $X_n \xrightarrow{\mu} X$ implies $X'_n \xrightarrow{\text{a.e.}} X$. This completes the proof.

COROLLARY. *Convergence and mutual convergence in measure imply one another.*

Proof. If $X_n \xrightarrow{\mu} X$, then, for every $\epsilon > 0$ and all ν,

$$\mu[| X_{n+\nu} - X_n | \geq \epsilon] \leq \mu\left[| X_{n+\nu} - X | \geq \frac{\epsilon}{2}\right]$$

$$+ \mu\left[| X - X_n | \geq \frac{\epsilon}{2}\right] \to 0,$$

so that $X_{n+\nu} - X_n \xrightarrow{\mu} 0$. Conversely, if $X_{n+\nu} - X_n \xrightarrow{\mu} 0$, then, upon taking the subsequence X_{n_k} of the foregoing theorem, we obtain, for every $\epsilon > 0$, by letting $n_k, n \to \infty$,

$$\mu[| X - X_n | \geq \epsilon] \leq \mu\left[| X - X_{n_k} | \geq \frac{\epsilon}{2}\right] + \mu\left[| X_{n_k} - X_n | \geq \frac{\epsilon}{2}\right] \to 0,$$

so that $X_n \xrightarrow{\mu} X$, and the corollary is proved.

§ 7.   INTEGRATION

The concepts of σ-field, measure, and measurable function are born from the efforts, made in the nineteenth and the beginning of the twentieth centuries, to extend the concept of integration to wider and wider classes of functions. The decisive extension was accomplished by Lebesgue, after Borel opened the way. Lebesgue worked with the special "Lebesgue" measure. Radon applied the same approach working with Lebesgue-Stieltjes measures. Finally, Fréchet, still using Lebesgue's approach, got rid of the restrictions on the measure space on which the numerical functions to be integrated were defined.

Lebesgue had two equivalent definitions of the integral, a descriptive one and a constructive one. We shall use a constructive definition of the integral of which there are many variants, but the basic ideas are always the same and, in general, the integral is first defined for simple functions. Although infinite values are not excluded, nevertheless, the expression $+\infty -\infty$, being meaningless, must be avoided. Therefore, it behooves us to start with integrals of functions of constant sign, say, nonnegative ones. Furthermore, the central property of the integral, called "the monotone convergence theorem," says that for a nondecreasing sequence of nonnegative functions integration and passage to the limit can be interchanged. Therefore, we give here the approach aimed directly at this theorem, an approach which requires a minimum of notions and of effort. The reader will recognize in the central definition 2° below, a particular form of the monotone convergence theorem.

7.1 Integrals. We consider a fixed measure space $(\Omega, \mathcal{A}, \mu)$; A, B, $\cdots$, and X, Y, $\cdots$, with or without affixes, will denote measurable sets and (numerical) measurable functions, respectively.

DEFINITIONS 1° The *integral on Ω of a nonnegative simple function*

$X = \sum_{j=1}^{m} x_j I_{A_j}$ is defined by

$$\int_{\Omega} X \, d\mu = \sum_{j=1}^{m} x_j \mu A_j.$$

2° The *integral on Ω of a nonnegative measurable function X* is defined by

$$\int_{\Omega} X \, d\mu = \lim \int_{\Omega} X_n \, d\mu,$$

where X_n is a nondecreasing sequence of nonnegative simple functions which converges to X.

3° The *integral on Ω of a measurable function X* is defined by

$$\int_{\Omega} X \, d\mu = \int_{\Omega} X^+ \, d\mu - \int_{\Omega} X^- \, d\mu,$$

where $X^+ = XI_{[X \geq 0]}$ and $X^- = -XI_{[X < 0]}$ are the positive and negative parts of X respectively, provided the defining difference exists, that is, provided at least one of the terms of this difference is finite.

If $\int_{\Omega} X \, d\mu$ is finite, that is, if both of the terms of the difference are finite, X is said to be *integrable* on Ω.

Finally, if X is a.e. determined and measurable, that is, there exists a measurable function X' such that $X = X'$ outside a μ-null set, we set $\int X = \int X'$, provided the right-hand side exists.

Upon replacing, in the preceding definitions, Ω by a measurable set A (hence replacing, in 1°, every $A_j = \Omega A_j$ by AA_j), they become definitions of the *integral of X on A*, to be denoted by $\int_A X\, d\mu$. Since, for

$X = \sum_{j=1}^{m} x_j I_{A_j} \geqq 0$, we have

$$\int_\Omega X I_A\, d\mu = \sum_{j=1}^{m} x_j \mu A A_j \leqq \int_\Omega X\, d\mu,$$

it follows immediately that

if $\int_\Omega X\, d\mu$ exists so does $\int_A X\, d\mu$, and $\int_A X\, d\mu = \int_\Omega X I_A\, d\mu$.

To simplify the writing, we drop $d\mu$ and Ω in the foregoing symbols, unless confusion is possible; thus, the symbols $\int_\Omega X\, d\mu$ and $\int_A X\, d\mu$ will be replaced by $\int X$ and $\int_A X$, respectively.

Justification and additivity. We have to justify the three definitions 1°, 2°, 3°, that is, we have to show that the concepts as defined exist and are uniquely determined. In the course of the justification we shall have use for the elementary properties below; the first one is called the *additivity property* of the operation of integration.

A. ELEMENTARY PROPERTIES. *Let $\int X, \int Y, \int X + \int Y$ exist.*

I *Linearity:*

$$\int (X + Y) = \int X + \int Y, \quad \int_{A+B} X = \int_A X + \int_B X, \quad \int cX = c \int X.$$

II *Order-preservation:*

$$X \geqq 0 \Rightarrow \int X \geqq 0, \quad X \geqq Y \Rightarrow \int X \geqq \int Y,$$

$$X = Y \text{ a.e. } \Rightarrow \int X = \int Y.$$

III *Integrability:*

$$X \text{ integrable} \Leftrightarrow |X| \text{ integrable} \Rightarrow X \text{ a.e. finite;}$$

$$|X| \leq Y \text{ integrable} \Rightarrow X \text{ integrable;}$$

$$X \text{ and } Y \text{ integrable} \Rightarrow X + Y \text{ integrable.}$$

Assume that the additivity property is proved. Then the second of properties I follows by replacing in the first one X by XI_A and Y by XI_B. The third one follows directly by successive use of the definitions.

The successive use of the definitions also proves directly the first and third of properties II, and the second one follows by the additivity property upon setting $X = Y + Z$, where $Z \geq 0$.

Similarly for properties III, except for $|X|$ integrable $\Rightarrow X$ a.e. finite. But, if $\mu A > 0$ where $A = [|X| = \infty]$, then, on account of II,

$$\int |X| \geq \int |X| I_A \geq c\mu A \text{ whatever be } c > 0. \text{ It follows, by letting}$$

$c \to \infty$, that $\int |X| = \infty$, and the property is proved *ab contrario.* Thus

For each of the successive definitions, the elementary properties hold as soon as the additivity property is proved.

We use this fact repeatedly in proceeding to the successive justifications of the definitions and to the proof of the additivity property.

1° *Nonnegative simple functions.* Since $X = \sum_{j=1}^{m} x_j I_{A_j}$ is nonnegative, the defining sum in

$$\int X = \sum_{j=1}^{m} x_j \mu A_j \geq 0$$

exists; it may be infinite. Its value is independent of the way in which X is written. For, if X is written in some other form $\sum_{k=1}^{n} y_k I_{B_k}$, then $x_j = y_k$ if $A_j B_k \neq \emptyset$ and, from $\sum_{j=1}^{m} A_j = \sum_{k=1}^{n} B_k = \Omega$, it follows that

$$\sum_{j=1}^{m} x_j \mu A_j = \sum_{j,k} x_j \mu A_j B_k = \int \sum_{j,k} x_j I_{A_j B_k} = \int \sum_{j,k} y_k I_{A_j B_k}$$

$$= \sum_{j,k} y_k \mu A_j B_k = \sum_{k=1}^{n} y_k \mu B_k.$$

Thus, $\int X$ is unambiguously defined.

Let now $X = \sum_{j=1}^{m} x_j I_{A_j}$ and $Y = \sum_{k=1}^{n} y_k I_{B_k}$ be two nonnegative simple functions, so that $X + Y = \sum_{jk} (x_j + y_k) I_{A_j B_k}$. Proceeding as above, we have

$$\int (X + Y) = \sum_{jk} (x_j + y_k)\mu A_j B_k = \sum_{jk} x_j \mu A_j B_k + \sum_{jk} y_k \mu A_j B_k$$

$$= \sum_{j=1}^{m} x_j \mu A_j + \sum_{k=1}^{n} y_k \mu B_k = \int X + \int Y,$$

and the additivity property is proved.

2° *Nonnegative measurable functions.* In definition 2°, the sequence of simple functions $X_n \geq 0$ is nondecreasing, so that, by **AII** for simple functions, the sequence $\int X_n$ is nondecreasing and, hence, has a limit, finite or not. Moreover, for every nonnegative measurable function X there exists such a sequence $X_n \uparrow X$. Therefore, to justify the definition, it suffices to show that the defining limit is independent of the particular choice of the sequence X_n. In other words

a. *If two nondecreasing sequences X_n and Y_n of nonnegative simple functions have the same limit, then*

$$\lim \int X_n = \lim \int Y_n.$$

Proof. It suffices to prove that $0 \leq X_n \uparrow X$ and $\lim X_n \geq Y$, where Y is a nonnegative simple function, imply $\lim \int X_n \geq \int Y$. For, then, it follows from the assumptions that, for every integer p,

$$\lim \int X_n \overset{\cdot}{\geq} \int Y_p, \quad \lim \int Y_n \geq \int X_p,$$

and the asserted equality is obtained by letting $p \to \infty$.

First, we prove the asserted inequality under the supplementary restrictions

$$\mu\Omega < \infty, \quad m = \min Y > 0, \quad M = \max Y < \infty.$$

Let $\epsilon > 0$ be less than m. Since $\lim X_n \geq Y$, it follows that $A_n = [X_n > Y - \epsilon] \uparrow \Omega$. But, on account of the validity of **A** for simple functions and the finiteness of μ and Y, we have

$$\int X_n \geq \int X_n I_{A_n} \geq \int (Y - \epsilon) I_{A_n} = \int Y - \int Y I_{A_n^c} - \epsilon \mu A_n$$

$$\geq \int Y - M\mu A_n^c - \epsilon \mu A_n$$

and, hence, by letting $n \to \infty$ and then $\epsilon \to 0$, the asserted inequality follows. Now, we get rid of the supplementary restrictions.

If $\mu \Omega = \infty$, then

$$\int X_n \geq \int X_n I_{A_n} \geq \int (Y - \epsilon) I_{A_n} \geq (m - \epsilon)\mu A_n \to \infty,$$

and the asserted inequality is trivially true.

If $M = \infty$, then, the inequality being valid with X_n and $Y I_{[Y < \infty]} + c I_{[Y = +\infty]}$ where c is an arbitrary finite number, we have

$$\lim \int X_n \geq \int Y I_{[Y < \infty]} + c\mu[Y = +\infty]$$

and, letting $c \to \infty$, the right-hand side becomes $\int Y$.

Finally, if $m = 0$, then, since the functions X_n and Y are nonnegative and, by what precedes, the inequality is true for integrals on $[Y > 0]$, we have

$$\lim \int X_n \geq \lim \int_{[Y > 0]} X_n \geq \int_{[Y > 0]} Y = \int Y.$$

This completes the proof and the definition of the integral of a non-negative measurable function is justified.

Since the additivity property was proved for nonnegative simple functions X_n, Y_n, and $0 \leq X_n \uparrow X$, $0 \leq Y_n \uparrow Y$ imply $0 \leq X_n + Y_n \uparrow X + Y$, it follows, by letting $n \to \infty$ in

$$\int (X_n + Y_n) = \int X_n + \int Y_n,$$

that

$$\int (X + Y) = \int X + \int Y.$$

Thus, the additivity property remains valid for nonnegative measurable functions.

3° *Measurable functions.* The decomposition $X = X^+ - X^-$ of a measurable function into its positive and negative parts is unique, so that $\int X = \int X^+ - \int X^-$ is unambiguously defined, provided $\int X^+$ or $\int X^-$ is finite.

Finally, if X is determined and measurable outside a μ-null set N, then let X' be any measurable function such that $X = X'$ on N^c. The integral of X is defined by setting $\int X = \int X'$, provided $\int X'$ exists. By **AII** for nonnegative measurable functions, the integrals of such functions which coincide on N^c are equal. It follows, by definition 3°, that the same is true when the functions are not of constant sign. Therefore, $\int X$ is unambiguously defined.

It remains to prove the additivity property.

Since we assume that not only $\int X$ and $\int Y$ exist but also that $\int X + \int Y$ exists, that is, is not of the form $+\infty -\infty$, it follows that (excluding the trivial case of the three integrals infinite of the same sign) at least one of the functions, say Y, is integrable and, hence, by **AIII**, is a.e. finite. Therefore, $X + Y$ is a.e. determined, and we do not restrict the generality by taking determined X and Y, and changing Y to 0 on the μ-null event on which it is infinite and $X + Y$ may be not determined.

We decompose Ω into the six sets on each of which X, Y, and $X + Y$ are of constant sign ($\geqq 0$ or <0). Because of definition 3° and property **AI** for nonnegative functions, it suffices to prove the additivity property on each of these sets, say $A = [X \geqq 0, Y < 0, X + Y \geqq 0]$. But, on account of definition 3° and the additivity property for nonnegative functions $(X + Y)I_A$ and $-YI_A$, we have

$$\int_A X = \int_A (X + Y) + \int_A (-Y) = \int_A (X + Y) - \int_A Y$$

and, $\int_A Y$ being finite,

$$\int_A X + \int_A Y = \int_A (X + Y).$$

Similarly for the other sets, and the additivity property follows.

This completes the justification of the definitions and the proof of the elementary properties.

7.2 Convergence theorems. The central convergence property is as follows:

A. MONOTONE CONVERGENCE THEOREM. *If* $0 \leqq X_n \uparrow X$, *then* $\int X_n \uparrow \int X.$

Proof. Choose nonnegative simple functions $X_{km} \uparrow X_k$ as $m \rightarrow \infty$. The sequence $Y_n = \max_{k \leqq n} X_{kn}$ of nonnegative simple functions is non-decreasing, and

$$X_{kn} \leqq Y_n \leqq X_n, \quad \int X_{kn} \leqq \int Y_n \leqq \int X_n.$$

It follows, by letting $n \rightarrow \infty$, that

$$X_k \leqq \lim Y_n \leqq X, \quad \int X_k \leqq \int \lim Y_n \leqq \lim \int X_n$$

and, by letting $k \rightarrow \infty$, we obtain

$$X \leqq \lim Y_n \leqq X, \quad \lim \int X_n \leqq \int \lim Y_n \leqq \lim \int X_n.$$

Thus $\lim Y_n = X$ and $\int X = \lim \int X_n$. The assertion is proved.

COROLLARY 1. *The integral is σ-additive on the family of nonnegative measurable functions.*

This means that, if the X_n are nonnegative, then $\int \sum X_n = \sum \int X_n$. and follows by $0 \leqq \sum_{k=1}^{n} X_k \uparrow \sum X_n$.

COROLLARY 2. *If X is integrable, then* $\int_A |X| \rightarrow 0$ *as* $\mu A \rightarrow 0.$

For, if $X_n = X$ or n according as $|X| < n$ or $|X| \geqq n$, then $\int |X_n| \uparrow \int |X|$, so that, given $\epsilon > 0$, there exists an n_0 such that $\int |X| < \int |X_{n_0}| + \frac{\epsilon}{2}$. It follows that, for A with $\mu A < \epsilon/2n_0$,

$$\int_A |X| = \int_A |X_{n_0}| + \int_A (|X| - |X_{n_0}|) < \frac{\epsilon}{2} + \int |X| - \int |X_{n_0}| < \epsilon.$$

The monotone convergence theorem extends as follows:

B. Fatou-Lebesgue theorem. *Let Y and Z be integrable functions. If $Y \leq X_n$ or $X_n \leq Z$, then*

$$\int \lim \inf X_n \leq \lim \inf \int X_n, \quad resp. \lim \sup \int X_n \leq \int \lim \sup X_n.$$

If $Y \leq X_n \uparrow X$, or $Y \leq X_n \leq Z$ and $X_n \xrightarrow{a.e.} X$, then $\int X_n \rightarrow \int X$.

Proof. If the X_n are nonnegative, then

$$X_n \geq Y_n = \inf_{k \geq n} X_k \uparrow \lim \inf X_n,$$

so that, by the monotone convergence theorem,

$$\lim \inf \int X_n \geq \lim \int Y_n = \int \lim \inf X_n.$$

The asserted inequalities follow, by the additivity property, upon applying this result to the sequences $X_n - Y$ and $Z - X_n$ of nonnegative measurable functions, and the asserted equalities are immediate consequences.

Clearly, if the assumptions of this theorem hold only a.e., the conclusions continue to hold. In fact, the last assertion, frequently called the dominated convergence theorem, extends as follows:

C. Dominated convergence theorem. *If $|X_n| \leq Y$ a.e. with Y integrable and if $X_n \xrightarrow{a.e.} X$ or $X_n \xrightarrow{\mu} X$, then $\int X_n \rightarrow \int X$. In fact,*

$$\int_A X_n - \int_A X \rightarrow 0 \text{ uniformly in } A \text{ or, equivalently, } \int |X_n - X| \rightarrow 0.$$

Proof. Since

$$\left| \int_A (X_n - X) \right| \leq \int |X_n - X| = \int (X_n - X)^+ + \int (X_n - X)^-,$$

it follows that the last two assertions are equivalent and imply the first one. Thus, it suffices to prove that $\int |X_n - X| \rightarrow 0$. Set $Y_n = |X_n - X|$ and observe that $Y_n \leq 2Y$ a.e. and that the $\int Y_n$ remain the same when the Y_n are modified on null events. Therefore, it suffices to prove that, if $0 \leq Y_n \leq Z$ integrable and $Y_n \xrightarrow{\text{a.e.}} 0$ or $Y_n \xrightarrow{\mu} 0$, then $\int Y_n \to 0$.

The case $Y_n \xrightarrow{\text{a.e.}} 0$ follows from the last assertion in **B**. It implies the case $Y_n \xrightarrow{\mu} 0$, since, by selecting a subsequence $Y_{n'}$ ($\xrightarrow{\mu} 0$) such that $\int Y_{n'} \to \lim \sup \int Y_n$ and, within this subsequence, a sequence $Y_{n''} \xrightarrow{\text{a.e.}} 0$, it follows that $\int Y_{n''} \to 0$ and $\lim \sup \int Y_n = 0$. Hence $\int Y_n \to 0$, and the proof is complete.

Extension. In all the preceding convergence theorems the parameter $n \to \infty$ can be replaced by a parameter $t \to t_0$ along an arbitrary set $T \subset \bar{R}$ of values, the reason for this being that $a_t \to a$ as $t \to t_0$ along T is equivalent to $a_{t_n} \to a$ for every sequence t_n in T converging to t_0.

Applications I. We assume all functions X_t to be integrable.

The dominated convergence theorem yields at once

1° *If* $|X_t| \leq Y$ *integrable and* $X_t \to X_{t_0}$ *as* $t \to t_0 (t \in T)$, *then*

$$\int X_t \to \int X_{t_0}.$$

This proposition yields, by applying the definition of derivative,

2° *If, on* T, $\dfrac{dX_t}{dt}$ *exists at* t_0 *and* $\left| \dfrac{X_t - X_{t_0}}{t - t_0} \right| \leq Y$ *integrable, then*

$$\left(\frac{d}{dt} \int X_t \right)_{t_0} = \int \left(\frac{dX_t}{dt} \right)_{t_0}.$$

In turn, this proposition yields

3° *If, on a finite interval* $[a, b]$, $\dfrac{dX_t}{dt}$ *exists and* $\left| \dfrac{dX_t}{dt} \right| \leq Y$ *integrable, then, on* $[a, b]$,

$$\frac{d}{dt} \int X_t = \int \frac{dX_t}{dt}.$$

This follows from

$$X_t - X_{t'} = (t - t') \left(\frac{dX_t}{dt} \right)_{t''}$$

where t'' lies between t and t'. And in its turn, this proposition yields

4°   *If, on a finite interval* [a, b], X_t *is continuous and* $|X_t| \leq Y$ *integrable then,* for every $t \in [a, b]$,

$$\int_a^t \left(\int X_{t'} \right) dt' = \int \left(\int_a^t X_{t'}\, dt' \right).$$

Moreover, if the foregoing assumptions hold for every finite interval and $\int_{-\infty}^{+\infty} |X_t|\, dt \leq Z$ *integrable, then*

$$\int_{-\infty}^{+\infty} \left(\int X_t \right) dt = \int \left(\int_{-\infty}^{+\infty} X_t\, dt \right).$$

The integrals with respect to t are Riemann integrals.

The first assertion follows from the fact that the derivative of a Riemann integral $\int_a^t g(t)\, dt$ where g is continuous is $g(t)$ which is bounded on $[a, b]$, so that, upon applying 3° to the asserted equality, it follows that derivatives of both sides are equal and, since both sides vanish for $t = a$, the equality is proved. The second assertion follows by 1° from the first one, by letting $a \to -\infty$ and $t \to +\infty$.

II. *Integrals over the Borel line.* Let $\mathfrak{B}$ be the Borel field in $R = (-\infty, +\infty)$ and let μ be *a measure on* $\mathfrak{B}$ *which assigns finite values to finite intervals.* Let $\mathfrak{B}_\mu$ be the class of all sets which are unions of a Borel set and a subset of a μ-null Borel set. $\mathfrak{B}_\mu$ is closed under formation of complements and countable unions and, hence, is a σ-field. By assigning to every set of $\mathfrak{B}_\mu$ the measure of the Borel set from which it differs by a subset of a μ-null set, μ is extended to a σ-finite measure on $\mathfrak{B}_\mu$, that we continue to denote by μ. $\mathfrak{B}_\mu$ will be called a *Lebesgue-Stieltjes field* in R and μ on $\mathfrak{B}_\mu$ will be called a *Lebesgue-Stieltjes measure.* The relation

$$F(b) - F(a) = F[a, b) = \mu[a, b)$$

determines, up to an additive constant, a function F on R which is clearly finite, nondecreasing, and continuous from the left, called *a distribution function* corresponding to μ. (It was proved that, conversely, such a function *determines* a Lebesgue-Stieltjes measure μ.)

Let g be a $\mathfrak{B}_\mu$-measurable function. If g is integrable, the integral $\int g\, d\mu$ is called a *Lebesgue-Stieltjes integral.* If F is a distribution function corresponding to μ, this integral is also denoted by $\int g\, dF$, and the integral $\int_{[a, b)} g \, d\mu$ is also denoted by $\int_a^b g \, dF$. If $F(x) = x$, $x \in R$, the corresponding measure is called the *Lebesgue measure*; it assigns to every interval its "length" and, thus, is a direct extension of the notion of length. The corresponding σ-field, or *Lebesgue field*, is formed by *Lebesgue sets* and the corresponding integrals, say $\int g \, dx$, $\int_a^b g \, dx$, are called *Lebesgue integrals*. Lebesgue field, measure, and integral are prototypes of general σ-fields, measures, and integrals. One may say that the basic ideas and methods relative to measure spaces and integrals belong to Lebesgue.

Let g be continuous on [a, b]. The Lebesgue-Stieltjes integral $\int_a^b g \, dF$ becomes then a Riemann-Stieltjes integral and the Lebesgue integral $\int_a^b g \, dx$ becomes then a Riemann integral.

The proof is easy. We have to show that, g being continuous on $[a, b]$, $\int_a^b g \, dF$ is limit of Riemann-Stieltjes sums. This is possible because a continuous function on a closed interval is bounded and is the (uniform) limit of any sequence of step-functions

$$g_n = \sum g(x'_{nk}) I_{[x_{nk}, x_{n,k+1})}, \quad a = x_{n1} < \cdots < x_{n,k_n+1} = b,$$
$$x_{nk} \leqq x'_{nk} < x_{n,k+1},$$

such that $\max_{k \leqq k_n} (x_{n,k+1} - x_{nk}) \to 0$. Therefore, by the dominated convergence theorem or, more specifically, by the last assertion of the Fatou-Lebesgue theorem,

$$\int_a^b g \, d\mu = \lim \int_a^b g_n \, d\mu = \lim \sum_{k=1}^{k_n} g(x'_{nk}) \mu[x_{nk}, x_{n,k+1}),$$

that is,

$$\int_a^b g \, dF = \lim \sum_{k=1}^{k_n} g(x'_{nk}) F[x_{nk}, x_{n,k+1}),$$

where the right-hand side sums are precisely the usual Riemann-Stieltjes sums. Thus, in the case of g continuous on $[a, b]$, the integral $\int_a^b g \, dF$ can be defined directly in terms of F, or of measures assigned to intervals only.

However, when g is continuous on R, its Lebesgue-Stieltjes integral over R and its *improper* Riemann-Stieltjes integral do not necessarily coincide. In fact, the last integral is defined by

$$\int g \, dF = \lim_{\substack{a \to -\infty \\ b \to +\infty}} \int_a^b g \, dF,$$

provided the limit exists and is finite. It may happen that at the same time

$$\int |g| \, dF = \lim_{\substack{a \to -\infty \\ b \to +\infty}} \int_a^b |g| \, dF$$

is infinite so that $|g|$ not being Lebesgue-Stieltjes integrable, g is not Lebesgue-Stieltjes integrable. Such examples are familiar; one of the most classical ones is that of the improper Riemann-integral of $g(x) = \sin x/x$. However, if g is Lebesgue-Stieltjes integrable then, clearly, both integrals coincide. Thus, the class of continuous functions whose improper Riemann-Stieltjes integrals with respect to a distribution function F exist (and are finite) contains the class of continuous functions which are Lebesgue-Stieltjes integrable with respect to F.

§ 8.   INDEFINITE INTEGRALS; ITERATED INTEGRALS

8.1 Indefinite integrals and Lebesgue decomposition. We characterize now the indefinite integrals by using repeatedly the monotone convergence theorem. Let X be a measurable function whose integral exists—say, $\int X^-$ is finite. Then the indefinite integral φ on $\mathcal{C}$ defined by

$$\varphi(A) = \int_A X = \int X I_A$$

exists, for $\int X^- I_A$ is finite and $\int X^+ I_A$ exists. Since the integral of a function which vanishes a.e. is 0, *the indefinite integral is μ-continuous*, that is, vanishes for μ-null sets. Since for a countable measurable partition $\{A_j\}$, $X^{\pm} I_A = \sum X^{\pm} I_{AA_j}$, it follows that, by the monotone convergence theorem,

$$\int_{\sum A_j} X = \sum \int_{A_j} X,$$

and the *indefinite integral is σ-additive*.

If X is integrable, then it is a.e. finite, and the indefinite integral is finite. If X is not integrable but, still, *X is a.e. finite and μ is σ-finite, then the indefinite integral is σ-finite.* For, by decomposing Ω into sets A_n of finite measure, we have

$$\int X = \sum_{m=-\infty}^{+\infty} \sum_{n=1}^{\infty} \int_{A_n[m \leq X < m+1]} X$$

and every term of the double sum is finite.

The problem which arises is whether the foregoing properties characterize indefinite integrals and the answer lies in the celebrated Lebesgue (-Radon-Nikodym) decomposition theorem that we shall establish now. But first we introduce a notion in opposition to that of μ-continuity. A set function φ_s on $\mathcal{C}$ is said to be μ-*singular* if it vanishes outside a μ-null set; in symbols, there is a μ-null set N such that

$$\varphi_s(AN^c) = 0, \quad A \in \mathcal{C}.$$

A. Lebesgue decomposition theorem. *If, on $\mathcal{C}$, the measure μ and the σ-additive function φ are σ-finite, then there exists one, and only one, decomposition of φ into a μ-continuous and σ-additive set function φ_c and a μ-singular and σ-additive set function φ_s,*

$$\varphi = \varphi_c + \varphi_s,$$

and φ_c is the indefinite integral of a finite measurable function X determined up to a μ-equivalence.

φ_c and φ_s are called μ-continuous and μ-singular *parts* of φ, and X is called the *derivative $d\varphi/d\mu$ with respect* to μ; we emphasize that $d\varphi/d\mu$ is determined up to μ-equivalence.

Proof. 1° Since Ω is a countable sum of sets for which μ and φ are finite and since, by the Hahn decomposition theorem, φ is a difference of two measures, it suffices to prove the theorem for finite measures μ and φ. Furthermore, if there are two decompositions of φ into a μ-continuous and a μ-singular part:

$$\varphi = \varphi_c + \varphi_s = \varphi'_c + \varphi'_s,$$

then

$$\varphi_c - \varphi'_c = \varphi'_s - \varphi_s = 0,$$

for the μ-continuous function $\varphi_c - \varphi'_c$ vanishes for all μ-null sets while the μ-singular function $\varphi'_s - \varphi_s$ vanishes outside a μ-null set. Finally, an indefinite integral determines the integrand up to an equivalence:

if, for every $A \in \mathcal{Q}$,

$$\varphi_c(A) = \int_A X = \int_A X'$$

then $X = X'$ a.e.; for, if, say, $\mu A = \mu[X - X' > \epsilon] > 0$, then

$$\int_A (X - X') > 0.$$

Thus the uniqueness assertions hold if we prove the existence assertions under the assumption that μ and φ are finite measures.

2° Let Φ be the class of all nonnegative integrable functions X whose indefinite integrals are majorized by φ:

$$\int_A X \leq \varphi(A), \quad A \in \mathcal{Q}.$$

Φ is not empty, since $X = 0$ belongs to it; and there is a sequence $\{X_n\} \subset \Phi$ such that

$$\int X_n \rightarrow \sup_{X \in \Phi} \int X = \alpha \leq \varphi(\Omega) < \infty.$$

Let $X'_n = \sup_{k \leq n} X_k$, so that $0 \leq X'_n \uparrow X = \sup X_n$. Let

$$A_k = [X_k = X'_n], \quad A'_k = A_1{}^c A_2{}^c \cdots A_{k-1}{}^c A_k, \quad A'_1 = A_1,$$

so that

$$\sum_{k=1}^n A'_k = \bigcup_{k=1}^n A_k = \Omega$$

and, for every A,

$$\int_A X'_n = \sum_{k=1}^n \int_{AA'_k} X'_n = \sum_{k=1}^n \int_{AA'_k} X_k \leq \sum_{k=1}^n \varphi(AA'_k) = \varphi(A).$$

Upon letting $n \rightarrow \infty$ and applying the monotone convergence theorem, we get

$$\int_A X \leq \varphi(A), \quad \int X = \alpha.$$

Therefore X is a "maximal" element of Φ. This property will allow us to show that

$$\varphi_s = \varphi - \varphi_c \geq 0,$$

where φ_c is the indefinite integral of X, is μ-singular, and the proof will be complete.

3° Let $D_n + D_n{}^c$ be a Hahn decomposition for the finite and σ-additive set function $\varphi_n = \varphi_s - \dfrac{1}{n}\mu$, that is, $\varphi_n(AD_n) \leqq 0$ and $\varphi_n(AD_n{}^c) \geqq 0$ for every A. Let $D = \bigcap D_n$ (whence $D^c = \bigcup D_n{}^c$), so that, for every A and all n,

$$0 \leqq \varphi_s(AD) \leqq \frac{1}{n}\mu(AD).$$

Upon letting $n \to \infty$, it follows that $\varphi_s(AD) = 0$ and, hence, $\varphi_s(A) = \varphi_s(AD^c)$. Since

$$\varphi_c(A) = \varphi(A) - \varphi_s(AD^c) \leqq \varphi(A) - \varphi_s(AD_n{}^c),$$

it follows that

$$\int_A \left(X + \frac{1}{n}I_{D_n{}^c}\right) = \varphi_c(A) + \frac{1}{n}\mu(AD_n{}^c) \leqq \varphi(A) - \varphi_n(AD_n{}^c) \leqq \varphi(A),$$

so that $X + \dfrac{1}{n}I_{D_n{}^c} \in \Phi$. But this conclusion is contradicted by

$$\int \left(X + \frac{1}{n}I_{D_n{}^c}\right) = \alpha + \frac{1}{n}\mu D_n{}^c > \alpha$$

unless $\mu D_n{}^c = 0$. Therefore, all sets $D_n{}^c$ are μ-null sets and so is their countable union D^c. Since $\varphi_s(A) = \varphi_s(AD^c)$, it follows that φ_s is μ-singular, and the proof is complete.

In the particular case of a μ-continuous φ, the foregoing theorem reduces to

B. RADON-NIKODYM THEOREM. *If, on $\mathcal{Q}$, the measure μ and the σ-additive set function φ are σ-finite and φ is μ-continuous, then φ is the indefinite integral of a finite function determined up to an equivalence.*

We are now in a position to characterize indefinite integrals of finite functions on a σ-finite measure space.

C. *A set function φ on $\mathcal{Q}$ is the indefinite integral on a σ-finite measure space of a finite function X determined up to an equivalence, if, and only if, φ is σ-finite, σ-additive, and μ-continuous; and X is integrable if, and only if, this φ is finite.*

The "if" assertion is the Radon-Nikodym theorem and the "only if" assertion is contained in the discussion at the beginning of this subsection.

CorollaRY. *Let λ and μ be σ-finite measures on $\mathcal{Q}$. If μ is λ-continuous and X is a measurable function whose integral $\int X \, d\mu$ exists, then, for every $A \in \mathcal{Q}$,*

$$\int_A X \, d\mu = \int_A X \frac{d\mu}{d\lambda} \, d\lambda.$$

Proof. If $X = I_B$, $B \in \mathcal{Q}$, then the equality is valid, since

$$\int_A I_B \, d\mu = \mu AB = \int_{AB} \frac{d\mu}{d\lambda} \, d\lambda = \int_A I_B \frac{d\mu}{d\lambda} \, d\lambda.$$

It follows that the equality is valid for nonnegative simple functions and hence, by the monotone convergence theorem, for nonnegative measurable functions and, consequently, for measurable functions whose integral exists.

Extension. The indefinite integral of a measurable function X which is not necessarily finite is still σ-additive and μ-continuous, but it is not necessarily σ-finite. The question arises whether the Radon-Nikodym theorem can be extended to this case. The answer is in the affirmative.

D. *The Radon-Nikodym theorem remains valid if finiteness of X and σ-finiteness of φ are simultaneously suppressed therein.*

Proof. As usual, it suffices to consider a finite measure μ and a μ-continuous measure φ on $\mathcal{Q}$.

Let $\mathcal{B}$ be the class of all measurable sets such that φ on $\mathcal{B}$ is σ-finite, and let s be the supremum of μ on $\mathcal{B}$.

There exists a sequence $B_n \in \mathcal{B}$ such that $s = \lim \mu B_n$ and, hence, $B = \bigcup B_n \in \mathcal{B}$ with $\mu B = s$. If there exists a $C \in \{B^c A, \ A \in \mathcal{Q}\}$ such that $0 < \varphi(C) < \infty$, then $B + C \in \mathcal{B}$, $\mu C > 0$, and

$$s \geqq \mu(B + C) = \mu B + \mu C > s.$$

Therefore, while φ on $\{BA, \ A \in \mathcal{Q}\}$ is σ-finite, φ on $\{B^c A, \ A \in \mathcal{Q}\}$ can take values 0 and ∞ only.

Furthermore, whatever be $C \in \{B^c A, A \in \mathcal{Q}\}$, it is impossible to have $\mu C > 0$ and $\varphi(C) = 0$ since then $B + C \in \mathcal{B}$ and, as above, $s > s$. Since φ is μ-continuous, it is also impossible to have $\mu C = 0$ and $\varphi(C) > 0$. Thus, for every $C \in \{B^c A, \ A \in \mathcal{Q}\}$, either $\mu C > 0$ and $\varphi(C) = \infty \cdot \mu C = \infty$ or $\mu C = 0$ and $\varphi(C) = 0$. In other words, φ on $\{B^c A, A \in \mathcal{Q}\}$ is the indefinite integral of a function $X = \infty$ on B^c, determined up to an equivalence. On the other hand, by **B**, φ on $\{BA,\ A \in \mathfrak{A}\}$ is the indefinite integral of a function X on B, determined up to an equivalence. These values of X on B and on B^c determine it on Ω, up to an equivalence and, for every $A \in \mathfrak{A}$,

$$\int_A X = \int_{AB} X + \int_{AB^c} X = \varphi(AB) + \varphi(AB^c) = \varphi(A).$$

The extension follows.

8.2 Product measures and iterated integrals. Let $(\Omega_i,\ \mathfrak{A}_i,\ \mu_i)$, $i = 1, 2$, be two measure spaces. A space $(\Omega,\ \mathfrak{A},\ \mu)$ is their *product-measure space* if

$\Omega = \Omega_1 \times \Omega_2$ is the space of all points $\omega = (\omega_1, \omega_2)$, $\omega_i \in \Omega_i$;

$\mathfrak{A} = \mathfrak{A}_1 \times \mathfrak{A}_2$ is the minimal σ-field over the class of all measurable "rectangles" $A_1 \times A_2$, $A_i \in \mathfrak{A}_i$, where $A_1 \times A_2$ is the set of all points ω with $\omega_i \in A_i$;

$\mu = \mu_1 \times \mu_2$ is the "product-measure" on $\mathfrak{A}$, provided it exists, that is, is a measure on $\mathfrak{A}$ uniquely determined by the relations $\mu(A_1 \times A_2) = \mu_1 A_1 \times \mu_2 A_2$ for all measurable rectangles $A_1 \times A_2$.

We intend to find conditions under which the product-measure exists and conditions under which integrals with respect to this measure can be expressed in terms of integrals with respect to the factor measures μ_i. In what follows the subscripts 1 and 2 can be interchanged. We shall also frequently proceed to the usual abuse of notation which consists in the use of the same symbol for a function and for its values.

For every set $A \subset \Omega$, the *section* A_{ω_1} of A at ω_1 is the set of all points ω_2 such that $(\omega_1, \omega_2) \in A$. For every function X on Ω, the *section* X_{ω_1} of X at ω_1 is the function defined on Ω_2 by $X_{\omega_1}(\omega_2) = X(\omega_1, \omega_2)$.

a. *Every section of a measurable set or function is measurable.*

If $\mathcal{C}$ is the class of all the sets in Ω whose every section is measurable, then it is readily seen that $\mathcal{C}$ is a σ-field. But every section of a measurable rectangle $A_1 \times A_2$ is measurable, since it is either empty or is one of the sides. Therefore, $\mathcal{C} \supset \mathfrak{A}$ and the first assertion is proved. If X on Ω is measurable and $S \subset \overline{R}$ is an arbitrary Borel set, the second assertion follows by

$$X_{\omega_1}^{-1}(S) = [\omega_2;\ X_{\omega_1}(\omega_2) \in S] = [\omega_2;\ X(\omega_1, \omega_2) \in S]$$

$$= [\omega_2;\ (\omega_1, \omega_2) \in X^{-1}(S)] = (X^{-1}(S))_{\omega_1}.$$

A. Product-measure theorem. *If μ_1 on $\mathcal{Q}_1$ and μ_2 on $\mathcal{Q}_2$ are σ-finite, then, for every $A \in \mathcal{Q}_1 \times \mathcal{Q}_2$, the functions with values $\mu_1 A_{\omega_2}$ and $\mu_2 A_{\omega_1}$ are measurable, and the set function μ with values*

$$\mu A = \int (\mu_1 A_{\omega_2})\, d\mu_2 = \int (\mu_2 A_{\omega_1})\, d\mu_1,$$

is a σ-finite measure μ on $\mathcal{Q}_1 \times \mathcal{Q}_2$ uniquely determined by the relation

$$\mu(A_1 \times A_2) = \mu_1 A_1 \times \mu_2 A_2, \quad A_i \in \mathcal{Q}_i.$$

In other words, μ is the product-measure $\mu_1 \times \mu_2$.

Proof. The proof is based upon the fact that, by the monotone convergence theorem, the class $\mathfrak{M}$ of all those sets A for which the integrals are equal is closed under formation of countable sums.

Since the measures μ_1 and μ_2 are σ-finite, the product space is decomposable into a countable sum of rectangles with sides of finite measure. It follows that, without restricting the generality, we can suppose that these measures are finite. If $A = A_1 \times A_2$ is a measurable rectangle, then $\mu_1 A_{\omega_2} = \mu_1 A_1 \times I_{A_2}(\omega_2)$ and similarly by interchanging the subscripts 1 and 2. Thus, the functions with these values are measurable and both integrals reduce to $\mu_1 A_1 \times \mu_2 A_2$. The last asserted equality is proved and $\mathfrak{M}$ contains all measurable rectangles. It follows that $\mathfrak{M}$ contains the field of finite sums of these rectangles. But, $\mathfrak{M}$ is closed under nondecreasing passages to the limit, on account of the monotone convergence theorem, and, under nonincreasing ones, on account of the dominated convergence theorem and the finiteness of measures. Therefore, by 1.6, it contains the product σ-field $\mathcal{Q}_1 \times \mathcal{Q}_2$, and the equality of the integrals is proved. The finite set function μ on $\mathcal{Q}$ so defined is a measure, on account of the monotone convergence theorem, and it is uniquely determined by the stated relation, on account of the extension theorem. This terminates the proof.

Corollary. *$A \in \mathcal{Q}_1 \times \mathcal{Q}_2$ is a $(\mu_1 \times \mu_2)$-null set if, and only if, almost every section A_{ω_1} is a μ_2-null set.*

For the integral of a nonnegative function vanishes if, and only if, the integrand vanishes a.e.

We are now in a position to answer the second stated question. The result is due to Lebesgue and Fubini and is generally called the Fubini theorem.

B. ITERATED INTEGRALS THEOREM. *Let* $(\Omega_1, \mathfrak{A}_1, \mu_1)$ *and* $(\Omega_2, \mathfrak{A}_2, \mu_2)$ *be σ-finite measure spaces.*

If the $\mathfrak{A}_1 \times \mathfrak{A}_2$-measurable function X on $\Omega_1 \times \Omega_2$ is nonnegative or $\mu_1 \times \mu_2$-integrable, then

$$\int_{\Omega_1 \times \Omega_2} X d(\mu_1 \times \mu_2) = \int_{\Omega_1} d\mu_1 \int_{\Omega_2} X_{\omega_1} d\mu_2 = \int_{\Omega_2} d\mu_2 \int_{\Omega_1} X_{\omega_2} d\mu_1,$$

and in the integrability case almost every section of X is integrable.

The iterated integrals are to be read from right to left.

Proof. For $X = I_A$, the asserted equality reduces to that of the product-measure theorem. It follows that it holds for simple functions and hence holds for nonnegative measurable functions because of the monotone convergence theorem, since, if $0 \leqq X_n \uparrow X$, then $0 \leqq (X_n)_{\omega_i} \uparrow$ $(X)_{\omega_i}$. If $X \geqq 0$ is integrable, then the function $\int X_{\omega_1} d\mu_2$ of ω_1 is integrable and hence a.e. finite, so that the functions X_{ω_1} of ω_2 are almost all integrable. Therefore, if $X = X^+ - X^-$ is integrable, that is, X^+ and X^- are integrable, then $(X)_{\omega_i} = (X^+)_{\omega_i} - (X^-)_{\omega_i}$ are almost all integrable and a.e. finite. This terminates the proof.

Finite-dimensional case. What precedes extends in an obvious manner to the product of an arbitrary but finite number of measure spaces. The interesting case is the infinitely dimensional one, and we shall now investigate it from a somewhat more general point of view.

*8.3   Iterated integrals and infinite product spaces. In what follows we push the abuse of notation to its extreme.

We consider a sequence of measurable spaces $(\Omega_n, \mathfrak{A}_n)$ and denote by ω_n points of Ω_n and by A_n measurable sets in Ω_n (sets of $\mathfrak{A}_n$). The product measurable space $(\Omega_1 \times \cdots \times \Omega_n, \mathfrak{A}_1 \times \cdots \times \mathfrak{A}_n)$ is the space of points $(\omega_1, \cdots, \omega_n)$ together with the minimal σ-field over the *intervals* $A_1 \times \cdots \times A_n$. The product measurable space $(\prod \Omega_n, \prod \mathfrak{A}_n)$ is the space of points $(\omega_1, \omega_2, \cdots)$ and the minimal σ-field over all cylinders of the form $A_1 \times \cdots \times A_n \times \prod_{k=n+1}^{\infty} \Omega_k$ or, equivalently, over all cylinders of the form $C(B_n) = B_n \times \prod_{k=n+1}^{\infty} \Omega_k$ where the *base* B_n is a measurable set in $\Omega_1 \times \cdots \times \Omega_n$.

In the infinitely dimensional case, we must, for reasons of "consistency" (to be made clear later), limit ourselves to probabilities, that is, to measures which assign value one to the space, to be denoted by P, Q, $\cdots$, with or without affixes. Furthermore, in probability theory, the following more general concept plays a basic role (at least when "independence"—see Part III—is not assumed). Every function—to be denoted by $P_n(\omega_1, \cdots, \omega_{n-1}; A_n)$—which is a probability in A_n for every fixed point $(\omega_1, \cdots, \omega_{n-1})$ and a measurable function in this point for every fixed A_n will be called a *regular conditional probability*. For $n = 1$ it reduces to a probability P_1 on $\mathfrak{A}_1$ but for $n > 1$ it reduces to a probability on $\mathfrak{A}_n$ only when it is constant in $(\omega_1, \cdots, \omega_{n-1})$ for every fixed A_n, provided the ordered T has a first element. We observe that the functions $\mu_2 A_{\omega_1} = \mu_2(\omega_1; A_{\omega_1})$ are regular conditional probabilities when $\mu_2(\omega_1; \Omega_2) = 1$. On account of the monotone convergence theorem, iterated integrals of the form

$$Q_n B_n = \int P_1(d\omega_1) \int P_2(\omega_1; d\omega_2) \cdots$$
$$\int P_n(\omega_1, \cdots, \omega_{n-1}; d\omega_n) I_{B_n}(\omega_1, \cdots, \omega_n)$$

define probabilities Q_n on $\mathfrak{A}_1 \times \cdots \times \mathfrak{A}_n$. It follows by the same theorem that if a measurable function X on $\Omega_1 \times \cdots \times \Omega_n$ is nonnegative or Q_n-integrable, then

$$\int_{\Omega_1 \times \cdots \times \Omega_n} X \, dQ_n = \int P_1(d\omega_1) \int P_2(\omega_1; d\omega_2) \cdots$$
$$\int P_n(\omega_1, \cdots, \omega_{n-1}; d\omega_n) X(\omega_1, \cdots, \omega_n).$$

A. Iterated regular conditional probabilities theorem. *The iterated integrals*

$$QC(B_n) = \int P_1(d\omega_1) \int P_2(\omega_1; d\omega_2) \cdots$$
$$\int P_n(\omega_1, \cdots, \omega_{n-1}; d\omega_n) I_{B_n}(\omega_1, \cdots, \omega_n),$$

determine a probability Q on $\prod \mathfrak{A}_n$.
This extension of the product-probability theorem is due to Tulcea and, proceeding as therein (in 1°), permits one to determine Q on an arbitrary $\prod_{t \in T} \mathfrak{A}_t$ under obvious consistency conditions on the regular conditional pr.'s $P_{t_{n+1}}(\omega_{t_1}, \cdots, \omega_{t_n}; A_{t_{n+1}})$.

Proof. To begin with, the definition of Q_n on the class $\mathcal{C}$ of all cylinders of the form $C(B_n)$ is consistent. For, if $C(B_n) = C(B_m)$, $m < n$, then integrations with respect to the ω_k which do not belong to the product subspace where B_m lies yield factors one.

Since Q on $\mathcal{C}$ is finitely additive, the assertion will follow by the extension theorem if we prove that Q on $\mathcal{C}$ is continuous at $\emptyset$. We have to consider nonincreasing sequences of cylinders which converge to $\emptyset$.

Upon renumbering the indices, we can suppose that the sequences are of the form $C(B_n) \downarrow \emptyset$ with nonempty bases $B_n \in \mathcal{C}_1 \times \cdots \times \mathcal{C}_n$. We can write

(1)
$$QC(B_n) = \int P(d\omega_1)Q^{(1)}C(B_n)_{\omega_1}$$

where $(B_n)_{\omega_1}$ is the section of B_n at ω_1 and

$$Q^{(1)}C(B_n)_{\omega_1} = \int P_2(\omega_1; d\omega_2) \cdots \int P_n(\omega_1, \cdots, \omega_{n-1}; d\omega_n)I_{B_n}(\omega_1, \cdots, \omega_n).$$

In (1) the left-hand side is nonincreasing in n, and the integrand converges nonincreasingly to a certain limit $X_1(\omega_1) \geq 0$. By the dominated convergence theorem, the limit of the left-hand side is $\int P(d\omega_1)X_1(\omega_1)$.

Assume that this integral is positive. Then there exists a point $\bar{\omega}_1$ such that $X_1(\bar{\omega}_1) > 0$. It follows that we find ourselves in the same situation but with the sequence $Q^{(1)}C(B_n)_{\bar{\omega}_1}$ instead of $QC(B_n)$. Repeating the argument over and over again, we obtain a sequence $\bar{\omega} = (\bar{\omega}_1, \bar{\omega}_2, \cdots)$ such that $\bar{\omega}_n \in \Omega_n$ and $Q^{(n)}C(B_n)_{\bar{\omega}_1, \cdots, \bar{\omega}_n} \downarrow X_n(\bar{\omega}_n) > 0$. Therefore, every $C(B_n)$ contains at least one point of the form $(\bar{\omega}_1, \cdots, \bar{\omega}_n, \omega_{n+1}, \cdots)$. Since $C(B_n) = B_n \times \prod_{k=n+1}^{\infty} \Omega_k$, it contains the point $\bar{\omega}$ and, hence, $\bar{\omega} \in \bigcap C(B_n)$. Thus, when $QC(B_n) \not\to 0$ the intersection is not empty, and the theorem follows *ab contrario*.

Particular cases. 1° If $P_n(\omega_1, \cdots, \omega_{n-1}; A_n) = P_nA_n$ are constant for every fixed A_n, then we write $Q = \prod P_n$ and call it a *product-probability*. Then the theorem reduces to the *product-probability theorem* in the denumerable case (4.2A).

2° If the factor spaces are finite-dimensional Borel spaces, then, it follows from 27.2, Application 1, that the theorem yields the *consistency theorem*.

COMPLEMENTS AND DETAILS

Notation. Unless otherwise stated, the measure space $(\Omega, \mathcal{C}, \mu)$ is fixed, the (measurable) sets $A, B, \cdots$, with or without affixes, belong to $\mathcal{C}$, and the functions $X, Y, \cdots$, with or without affixes, are finite measurable functions.

1. The set C of convergence of a sequence X_n (to a finite or infinite limit function) is measurable.

$$(C = [\lim \inf X_n = \lim \sup X_n].)$$

2. If μ is finite, then given X, for every $\epsilon > 0$ there exists A such that $\mu A < \epsilon$ and X is bounded on A^c. If X is bounded, then there exists a sequence of simple functions which converges uniformly to X. Combine both propositions.

We say that a sequence X_n converges *almost uniformly* (a.u.) to X, and write $X_n \xrightarrow{\text{a.u.}} X$, if, for every $\epsilon > 0$, there exists a set A with $\mu A < \epsilon$ such that $X_n \xrightarrow{u} X$ on A^c.

3. If $X_n \xrightarrow{\text{a.u.}} X$, then $X_n \xrightarrow{\text{a.e.}} X$ and $X_n \xrightarrow{\mu} X$. (For the first assertion, form A_n where A_n is the A of the foregoing definition with $\epsilon = \dfrac{1}{n}$.)

4. If $X_n \xrightarrow{\mu} X$, then there exists a subsequence $X_{n'} \xrightarrow{\text{a.u.}} X$.

5. *Egoroff's theorem.* If μ is finite, then $X_n \xrightarrow{\text{a.e.}} X$. implies that $X_n \xrightarrow{\text{a.u.}} X$. Compare with 3. (Neglect the null set of divergence, and form $A = \bigcup\limits_{m=1}^{\infty} A_m$ with $A_m = \bigcup\limits_{k \geq n(m)} \left[\, | X_k - X | \geq \dfrac{1}{m} \,\right]$ and $n(m)$ such that $\mu A_m < \dfrac{\epsilon}{2^m}$.)

6. *Lusin's theorem.* If μ is σ-finite, then $X_n \xrightarrow{\text{a.e.}} X$ implies that $X_n \xrightarrow{u} X$ on every element A_j of some countable partition of $\Omega\text{-}N$ where N is some null set. (Neglect the null set of divergence, and start with μ finite. Use Egoroff's theorem to select inductively sets A_k such that $\mu \bigcap\limits_{k=1}^{n} A_k < \dfrac{1}{n}$ and $X_n \xrightarrow{u} X$ on $A_k{}^c$ for every k.)

7. If μ is finite, then $X_n \xrightarrow{\text{a.e.}} X$ implies existence of a set of positive measure on which the X_n are uniformly bounded. What if μ is σ-finite?

8. If μ is finite, then $X_{mn} \xrightarrow{\text{a.e.}} X_m$ as $n \to \infty$ and $X_m \xrightarrow{\text{a.e.}} X$ as $m \to \infty$ imply that there exists subsequences m_k, n_k such that $X_{m_k n_k} \xrightarrow{\text{a.e.}} X$ as $k \to \infty$. What if μ is σ-finite?

(Neglect the null sets of divergence. Select A_k and m_k such that $\mu A_k < \dfrac{1}{2^k}$ and $| X_{m_k} - X | < \dfrac{1}{2^k}$ on $A_k{}^c$. Select $B_k \subset A_k$ and n_k such that $\mu B_k < \dfrac{1}{2^k}$ and $| X_{m_k n_k} - X_{m_k} | < \dfrac{1}{2^k}$ on $A_k - B_k$.)

9. Let $X_n \xrightarrow{\mu} X$, $Y_n \xrightarrow{\mu} Y$. Do $aX_n + bY_n \xrightarrow{\mu} aX + bY$, $| X_n | \xrightarrow{\mu} | X |$, $X_n{}^2 \xrightarrow{\mu} X^2$, $X_n Y_n \xrightarrow{\mu} XY$? What about $1/X_n$? Let μ be finite and let g on R or on $R \times R$ be continuous. What about the sequences $g(X_n)$ and $g(X_n, Y_n)$?

10. Let the functions X_n, X on the measure space be complex-valued or vector-valued or, more generally, let them take their values in some fixed Banach space. Denote the norm of X by $| X |$, and denote $| X_n - X | \to 0$ by $X_n \to X$.

Transpose the constructive definitions of measurability and the definitions of various types of convergence. Investigate the validity of the transposed of the corresponding properties established in the text, as well as of those stated above.

11. *Examples and counterexamples of mutual implications of types of convergence.* Investigate convergences of the sequences defined below:

(i) The measure space is the Borel line with Lebesgue measure, $X_n = 1$ on $[n, n + 1]$ and $X_n = 0$ elsewhere.

(ii) The measure space is the Borel interval $(0, 1)$ with Lebesgue measure, $X_n = 1$ on $\left(0, \dfrac{1}{n}\right)$ and $X_n = 0$ elsewhere.

(iii) The measure space is the Borel interval $[0, 1]$ with Lebesgue measure, the sequence is $X_{11}, X_{21}, X_{22}, X_{31}, X_{32}, X_{33}, \cdots$ with $X_{nk} = 1$ on $\left[\dfrac{k-1}{n}, \dfrac{k}{n}\right]$ and $X_{nk} = 0$ elsewhere.

(iv) $\mathfrak{A}$ consists of all subsets of the set of positive integers, μA is the number of points of A, X_n is indicator of the set of the n first integers.

12. If X is integrable, then the set $[X \neq 0]$ is of σ-finite measure. What if $\int X$ exists? $\left(\mu[\,|\,X\,| \geq c] \leq \dfrac{1}{c} \int |\,X\,|. \right)$

13. Let $(T, \mathfrak{I}, \tau)$ be a measure space, to every point t of which is assigned a measure μ_t on $\mathfrak{A}$. Let the function on T defined by $\mu_t A$ for any fixed A be $\mathfrak{I}$-measurable.

The relation $\mu A = \displaystyle\int_T \mu_t A \, d\tau(t)$ defines a measure μ on $\mathfrak{A}$. If $\displaystyle\int_\Omega X(\omega) \, d\mu(\omega)$ exists, then the function defined on T by $U(t) = \displaystyle\int_\Omega X(\omega) \, d\mu_t(\omega)$ exists and is $\mathfrak{I}$-measurable, and $\displaystyle\int_\Omega X(\omega) \, d\mu(\omega) = \displaystyle\int_T U(t) \, d\tau(t)$.

14. Let φ be the indefinite integral of X. Express $\varphi^+, \varphi^-, \bar\varphi$ in terms of X.

15. If $\displaystyle\int_A X_n \to 0$ uniformly in n as $\mu A \to 0$ or as $A \downarrow \emptyset$, then the same is true of $\displaystyle\int_A |\,X_n\,|$; and conversely. Interpret in terms of signed measures.

$$\left(\int_A |\,X_n\,| = \int_{A[X_n \geq 0]} X_n - \int_{A[X_n < 0]} X_n. \right)$$

16. If finite $\displaystyle\int_A X_n \to \displaystyle\int_A X$ finite, uniformly in $A \,(\in \mathfrak{A})$, then $\displaystyle\int_\Omega |\,X_n - X\,| \to 0$; and conversely.

17. If $0 \leq X_n \xrightarrow{\mu} X$, then finite $\displaystyle\int_\Omega X_n \to \displaystyle\int_\Omega X$ finite implies that $\displaystyle\int_A X_n \to \displaystyle\int_A X$ uniformly in A (also if $\xrightarrow{\mu}$ is replaced by $\xrightarrow{\text{a.e.}}$)

$\left(0 \leq (X - X_n)^+ \leq X \text{ integrable, and } \int (X - X_n)^+ - \int (X - X_n) \to 0. \right)$

18. Rewrite in terms of integrals as many as possible of the complements and details of Chapter I.

19. If the X_n are integrable and $\lim \displaystyle\int_A X_n$ exists and is finite for every A, then the $\int |\,X_n\,|$ are uniformly bounded, $\displaystyle\int_A |\,X_n\,| \to 0$ uniformly in n as $\mu A \to 0$ and as $A \downarrow \emptyset$, and there exists an integrable X, determined up to an equivalence, such that $\displaystyle\int_A X_n \to \displaystyle\int_A X$ for every A. (Use *18.*)

20. If integrable $X_n \to X$ integrable, then existence and finiteness of $\lim \int_A X_n$ for every A are equivalent to the following properties:

(i)
$$\int_A X_n \to \int_A X \text{ uniformly in } A;$$

(ii)
$$\int_A X_n \to 0 \text{ uniformly in } n \text{ as } \mu A \to 0 \text{ and as } A \downarrow \emptyset.$$

If μ is finite, then "as $A \downarrow \emptyset$" can be suppressed. (Use the preceding propositions and the relations

$$\int_A |X_n| \le \int_A |X_n - X| + \int_A |X|,$$

$$\int_A |X_n - X| \le \epsilon + \int_{A[|X_n - X| \ge \epsilon]} (|X_n| + |X|).)$$

21. The differential formalism applies to Radon-Nikodym derivatives:

Let μ, ν be finite measures on $\mathcal{Q}$ and φ, φ' be σ-finite signed measures on $\mathcal{Q}$. Let φ be ν-continuous and ν, φ, φ' be μ-continuous. Then

$$\frac{d(\varphi + \varphi')}{d\mu} = \frac{d\varphi}{d\mu} + \frac{d\varphi'}{d\mu} \quad \mu\text{-a.e.}$$

$$\frac{d\varphi}{d\mu} = \frac{d\varphi}{d\nu} \frac{d\nu}{d\mu} \quad \mu\text{-a.e.}$$

(For the second assertion, it suffices to consider $\varphi \ge 0$, $X = \dfrac{d\varphi}{d\nu} \ge 0$ $Y = \dfrac{d\nu}{d\mu} \ge 0$. Take simple X_n with $0 \le X_n \uparrow X$ so that

$$\int_A X \, d\nu \leftarrow \int_A X_n \, d\nu = \int_A X_n Y \, d\mu \to \int_A XY \, d\mu.)$$

Let $\{\mu_t, t \in T\}$ and $\{\mu'_{t'}, t' \in T'\}$ be two families of measures on $\mathcal{Q}$; we drop $t \in T$ and $t' \in T'$ unless confusion is possible. We say that $\{\mu_t\}$ is $\{\mu'_{t'}\}$-continuous if every set null for all $\mu'_{t'}$ is null for all μ_t. If the converse is also true, we say that the two families are mutually continuous.

22. If $\{\mu_j\}$ is a countable family of finite measures, then there exists a finite measure μ such that $\{\mu_t\}$ and μ are mutually continuous. (Take $\mu = \sum \mu_j / 2^j \mu_j \Omega$.)

23. Let the μ_t and μ be finite measures. If $\{\mu_t\}$ is μ-continuous, then there exists a finite measure μ' such that $\{\mu_t\}$ and μ' are mutually continuous. (Select sets $A_t = \left[\dfrac{d\mu_t}{d\mu} > 0 \right]$. Denote by B, with or without affixes, sets such that, for some t, $B \subset A_t$ and $\mu_t B > 0$. Denote countable sums of sets B up to μ-null sets by C, with or without affixes. Every subset $C' \subset C$ with $\mu_t C' > 0$ is a set C; every countable union of sets C is a set C. Let $\mu C_n \to s$ where s is the supremum of values of μ over all the sets C. Then $s = \mu \bigcup C_n = \mu \bigcup B_m$ and to every m there corresponds a μ_t, say μ_m, such that $B_m \subset A_m$ and $\mu_m B_m > 0$. The families $\{\mu_t\}$ and $\{\mu_m\}$ are mutually continuous.)

24. Let $\bar{\mu}_n = \sum\limits_{k=1}^{n} \mu_k \rightarrow \bar{\mu}$ and $\bar{\nu}_n = \sum\limits_{k=1}^{n} \nu_k \rightarrow \bar{\nu}$, all the μ and ν with various affixes being finite measures on $\mathcal{C}$ and every $\bar{\nu}_n$ being $\bar{\mu}_n$-continuous.

(i) $\dfrac{d\mu_1}{d\bar{\mu}_n} \rightarrow \dfrac{d\mu_1}{d\bar{\mu}}$ μ-a.e.

(ii) if $\{\mu_n\}$ is ν-continuous, then $\dfrac{d\bar{\mu}_n}{d\nu} \rightarrow \dfrac{d\bar{\mu}}{d\nu}$ ν-a.e.

(iii) $\bar{\nu}$ is $\bar{\mu}$-continuous and $\dfrac{d\bar{\nu}_n}{d\bar{\mu}_n} \rightarrow \dfrac{d\bar{\nu}}{d\bar{\mu}}$ $\bar{\mu}$-a.e.

(For the last assertion, if $\bar{\mu}_n A_n = 0$ for all n, then $\bar{\mu}$ (lim sup A_n) $= 0$. It follows that it suffices to consider a particular choice of the $\dfrac{d\bar{\nu}_n}{d\bar{\mu}_n} = \sum\limits_{k=1}^{n} X_k / \sum\limits_{k=1}^{n} Y_k$ where $X_k = \dfrac{d\nu_k}{d\bar{\mu}}$, $Y_k = \dfrac{d\mu_k}{d\bar{\mu}}$. But $\sum X_n = \dfrac{d\bar{\nu}}{d\bar{\mu}}$ and $\sum Y_n = 1 \bar{\mu}$-a.e.)

The propositions which follow correspond to various definitions of the concept of integration. We shall assume that the measures and the functions are finite. Besides proving the statements, the reader should also examine removal of the restriction of finiteness as well as of other restrictions which may be introduced.

25. Set

$$\int X \, d\varphi = \int X \, d\varphi^+ - \int X \, d\varphi^-, \quad \int (X + iY) \, d\mu = \int X \, d\mu + i \int Y \, d\mu,$$

$$\int X d(\mu + i\nu) = \int X \, d\mu + i \int X \, d\nu$$

and investigate existence and properties of integrals so defined.

26. Descriptive approach. The Radon-Nikodym theorem characterizes an indefinite integral but not that of a given function. The following proposition answers this requirement.

φ on $\mathcal{C}$ is indefinite integral of X on Ω if, and only if, φ is σ-additive and, for every set $A = [a \leq X \leq b] B$, $B \in \mathcal{C}$,

$$a\mu A \leq \varphi(A) \leq b\mu A.$$

27. In the definition of the integral given in the text, start with (nonnegative) elementary functions instead of simple ones. The integral so defined coincides with the initial one.

28. Lebesgue's approach. The Cauchy-Riemann approach starts with arbitrary finite partitions of the interval of integration into intervals. The Lebesgue approach consists in partitioning the set of integration according to the function to be integrated so that the integral is tailored to order as opposed to the ready-to-wear Cauchy-Riemann one. Let $\mu < \infty$.

Set

$$\sum{}_n(X) = \sum\limits_{k=-\infty}^{+\infty} \frac{k-1}{2^n} \mu \left[\frac{k-1}{2^n} \leq X < \frac{k}{2^n} \right].$$

If X is bounded, these sums correspond to finite partitions and $\int X =$

$\lim \sum_n(X)$. If X is not bounded, set $X_{mn} = X$ if $-m \leq X \leq n$ and $X_{mn} = 0$ otherwise. If X is integrable, then $\int X_{mn} \to \int X$ as $m, n \to \infty$.

If X is not bounded, the series $\sum_n(X)$ correspond to countable partitions and $\int X = \lim \sum_n(X)$, in the sense that if X is integrable, then these series are absolutely convergent and the equality holds and, conversely, if one of these series is absolutely convergent, so are all of them and the equality holds.

(For the last assertion, it suffices to consider nonnegative elementary functions $X_n = \sum_1^\infty \frac{k-1}{2^n} I_{\left[\frac{k-1}{2^n} \leq x < \frac{k}{2^n}\right]}$. For the converse, use the relation $X \leq 2X_n + \mu\Omega$.)

29. *Darboux-Young approach.* Let X be measurable or not and set

$$\underline{\int} X = \sup \sum_{k=1}^n \inf_{\omega \in A_k} X(\omega)\mu A_k, \quad \overline{\int} X = \inf \sum_{k=1}^n \sup_{\omega \in A_k} X(\omega)\mu A_k$$

where the extrema of sums are taken over all finite measurable partitions $\sum_{k=1}^n A_k = \Omega$. If X is measurable and bounded, then

$$\underline{\int} X = \int X = \overline{\int} X.$$

If $\underline{\int} X$ and $\overline{\int} X$ exist and are equal, we say that $\int X$ exists and equals their common value.

We can also set

$$\underline{\int}' X = \sup \int Y, \quad \overline{\int}' X = \inf \int Z$$

where the extrema are taken over all integrable (and measurable) Y and Z such that $Y \leq X \leq Z$ and define $\int X$ as above. Compare the two definitions.

30. *Completion approach.* The Meray-Cantor method for completion of metric spaces adjoins to the given metric space elements which represent mutually convergent (in distance) sequences of its points. This method permits (Dunford) to define and study the integral of functions with values in an arbitrary Banach space (Bochner), as follows:

(i) Define the indefinite integral of a simple function as in the text. Since nonnegativity and infinite values may be meaningless, all simple functions under consideration are integrable.

(ii) Adjoin to the space of these integrable functions $X_m, X_n, \cdots$ all functions X such that $\int |X_m - X_n| \to 0$ and $X_n \to X$, by defining the indefinite integral of X as the limit of the indefinite integrals of the X_n. To justify this definition, prove for simple functions those elementary properties of integrals which continue to have content for an arbitrary Banach space: $\int |X_m - X_n| \to 0$ if, and only if, $X_n \xrightarrow{\mu} X$ where X is some measurable function, and $\int_A |X_n| \to 0$ uniformly in n as $\mu A \to 0$; $\int |X_m - X_n| \to 0$ implies that $\varphi_n \to \varphi$ where φ is σ-additive.

(iii) Extend the foregoing properties to all integrable functions and obtain the dominated convergence theorem.

31. Kolmogorov's approach. Let $\mathfrak{C}$ be a class closed under intersections. Let $\mathfrak{D}$, with or without affixes, be finite disjoint subclasses of $\mathfrak{C}$. Order them by the relation $\mathfrak{D}_1 \prec \mathfrak{D}_2$ if every set of $\mathfrak{D}_2$ is contained in some set of $\mathfrak{D}_1$. Fix $A \in \mathfrak{C}$ and consider all the $\mathfrak{D}$ which are partitions of A. They form a "direction" Δ in the sense that, if $\mathfrak{D}_1$ and $\mathfrak{D}_2$ are such partitions, then there exists such a partition which "follows" both, namely, $\mathfrak{D}_1 \cap \mathfrak{D}_2$.

Let φ on $\mathfrak{C}$ be a function, additive or not, single-valued or not. By definition,

$$\hat{\varphi}(A) = \int_A d\varphi = \lim \sum \varphi(A_j)$$

where the A_j are elements of partitions $\mathfrak{D}$ of A and the limit $\hat{\varphi}(A)$, if it exists, is "along the direction Δ," that is, to every $\epsilon > 0$ there corresponds a $\mathfrak{D}_\epsilon$ such that $|\hat{\varphi}(A) - \sum \varphi(A_j)| < \epsilon$ for all $\mathfrak{D} > \mathfrak{D}_\epsilon$ and all values of the $\varphi(A_j)$—if φ is multivalued. If $\hat{\varphi}(A)$ exists, it is unique. If $\hat{\varphi}$ on $\mathfrak{C}$ exists, then it is finitely additive.

Compare this integral to the Riemann-Stieltjes integral by selecting conveniently φ.

Compare $\int_{\alpha\beta} d\varphi$ with the length (if it exists) of the arc $\alpha\beta$ of a plane curve, by taking $\varphi(\alpha_{k-1}, \alpha_k) = \overline{\alpha_{k-1}\alpha_k}$, the length of the cord α_{k-1} to α_k, the $\alpha = \alpha_1$, $\cdots \alpha_{k-1}, \alpha_k, \cdots, \alpha_n = \beta$ being consecutive points on the arc $\alpha\beta$.

We say that φ and φ' on $\mathfrak{C}$ are "differentially equivalent" on A if, for every $\epsilon > 0$, there exists a partition $\mathfrak{D}_\epsilon$ of A such that $\sum |\varphi(A_j) - \varphi'(A_j)| < \epsilon$ for all $\mathfrak{D} > \mathfrak{D}_\epsilon$. If φ is finitely additive, then $\int_A d\varphi = \varphi(A)$. If not, then $\hat{\varphi}$ on $A \cap \mathfrak{C}$ (if it exists) is the unique additive function differentially equivalent on A to φ. Proceed as follows:

(i) φ and φ' are differentially equivalent on A if, and only if, $\hat{\varphi} = \hat{\varphi}'$.

(ii) φ and $\hat{\varphi}$ are differentially equivalent on A.

(iii) If finitely additive functions φ and φ' are differentially equivalent on A, then they coincide on A.

In all which precedes replace "finite" by "countable" and investigate the validity of the propositions so obtained. Compare the various definitions of the integral, by selecting conveniently φ.

Finally, take φ with values in a fixed but arbitrary Banach space, and go over what precedes.

32. A structure of the concept of integration. The concept of integration is constructed by means of the concepts of summations and of passage to the limit along a direction or, more generally, a cut-direction. A bipartition $\overline{\underline{\Delta}} = \underline{\Delta} + \overline{\Delta}$

of a set Δ with an order relation $\prec$ is a "cut-direction" if $\underline{\Delta}$ and $\overline{\Delta}$ are directions and every element of $\overline{\Delta}$ follows every element of $\underline{\Delta}$.

Let φ be a function, single-valued or not, on a direction Δ to a real line or a plane or, more generally, a Banach space. The element φ_Δ of the range space is "limit of φ along Δ" if, for every $\epsilon > 0$, there exists an $\alpha_\epsilon \in \Delta$ such that $|\varphi_\Delta - \varphi(\alpha)| < \epsilon$ for all $\alpha > \alpha_\epsilon$ and for all values of $\varphi(\alpha)$. If the direction Δ is replaced by a cut-direction $\underline{\Delta}$, then $\varphi_{\underline{\Delta}}$ is "limit of φ along $\underline{\Delta}$ if, for every $\epsilon > 0$ there exist $\underline{\alpha}_\epsilon \in \underline{\Delta}$ and $\overline{\alpha}_\epsilon \in \overline{\Delta}$ such that $|\varphi_{\underline{\Delta}} - \varphi(\alpha)| < \epsilon$ for all α such that $\underline{\alpha}_\epsilon \prec \alpha \prec \overline{\alpha}_\epsilon$ and for all values of $\varphi(\alpha)$. If φ_Δ or $\varphi_{\underline{\Delta}}$ exist, they are unique.

To every $\alpha \in \Delta$ assign some finite collection of points α_j of a Banach space, not necessarily distinct and not necessarily uniquely determined. Form $\varphi(\alpha) = \sum \varphi(\alpha_j)$. By definition, $\int_\Delta d\varphi$ is the limit, if it exists, of φ along Δ. If Δ is replaced by $\underline{\Delta}$, the definition continues to apply.

Investigate all definitions of the integral you know of from this structural point of view, that is, the selections of Δ or $\underline{\Delta}$, and of the functions φ.

33. Daniell approach. Let S be a family of bounded real-valued functions on Ω, closed under finite linear combinations and lattice operations $f \cup g = \max(f, g)$, $f \cap g = \min(f, g)$. Then $f \in L \Rightarrow |f| = f \cup 0 - f \cap 0 \in S$. Suppose that on S is defined an *integral* $\int$: a nonnegative linear functional continuous under monotone limits: $f \geq 0 \Rightarrow \int f \geq 0, \int (af + bg) = a\int f + b\int g, f_n \downarrow 0 \Rightarrow \int f_n \downarrow 0$.

a) Let U be the family of limits (not necessarily finite) of nondecreasing sequences in S. U contains S and is closed under addition, multiplication by nonnegative constants, and lattice operations. Extend the integral on U, setting $\int f = \lim \int f_n$ when $S \ni f_n \uparrow f$ (infinite values being permitted). The definition is justified, for if the nondecreasing sequences f_n and g_n in S are such that $\lim f_n \leq \lim g_n$, then $\lim \int f_n \leq \lim \int g_n$.

If $U \ni f_n \uparrow f$ then $f \in U$ and $\int f_n \uparrow \int f$.

b) Let $-U$ be the family of functions f such that $-f \in U$, and set $\int f = -\int(-f)$.

If $g \in -U, h \in U$ and $g \leq h$, then $h - g \in U$ and $\int h - \int g = \int (h - g) \geq 0$.

By definition, f is *integrable* if, for every $\epsilon > 0$, there exist $g_\epsilon \in -U$ and $h_\epsilon \in U$ such that $g_\epsilon \leq f \leq h_\epsilon, \int g_\epsilon$ and $\int h_\epsilon$ are finite, and $\int h_\epsilon - \int g_\epsilon < \epsilon$. Then $\inf_\epsilon \int h_\epsilon = \sup_\epsilon \int g_\epsilon$ and $\int f$ is defined to be this common value.

Let L be the family of integrable functions. L and the integral on L have all the properties of S and of the integral on S.

If $L \ni f_n \uparrow f$ and $\lim \int f_n < \infty$, then $f \in L$ and $\int f_n \uparrow \int f$.

Let $\mathcal{F}$ be the smallest monotone family over S (closed under monotone passages to the limit by sequences). $\mathcal{F}$ is closed under algebraic and lattice operations.

Let $L_1 = L \cap \mathfrak{F}$. $f \in L_1$ if and only if $f \in \mathfrak{F}$ and there exists $g \in L_1$ such that $|f| \leq g$.

e) Let $\mathfrak{F}^+$ be the smallest monotone family over S^+ (consisting of all nonnegative functions of S). Set $\int f = \infty$ if $f \in \mathfrak{F}^+$ is not integrable. By definition, for

$f \in \mathfrak{F}, \int f = \int f^+ - \int f^-$ exists if f^+ or f^- is integrable.

If $\int f$ and $\int g$ exist and they are not infinite with opposite sign, then $\int (f + g)$ exists and equals $\int f + \int g$.

If $\int f_n$ exist, $\int f_1 > -\infty$, and $f_n \uparrow f$, then $\int f$ exists and $\int f_n \uparrow \int f$.

f) If $I_A \in \mathfrak{F}$, then, by definition, the *measure* of A is $\mu A = \int I_A$.

If I_A, $I_B \in \mathfrak{F}$, then $I_{A \cup B}$, $I_{A \cap B}$, $I_{A-B} \in \mathfrak{F}$ and if the $I_{A_n} \in \mathfrak{F}$, then $I_{\Sigma I A_n} \in \mathfrak{F}$ and $\mu \sum A_n = \sum \mu A_n$.

g) Suppose that $f \in S \Rightarrow f \cap 1 \in S$. Then $f \in \mathfrak{F} \Rightarrow f \cap 1 \in \mathfrak{F}$ and if $a > 0$, then $I_{[f>a]} \in \mathfrak{F}$.

If $f \geq 0$, $I_{[f>a]} \in \mathfrak{F}$ for every $a > 0$, then $f \in \mathfrak{F}$.

h) Suppose that $1 \in S$. Then $f \in \mathfrak{F}^+ \Rightarrow \int f = \int f d\mu$ where the right side is taken in the customary sense. What if $f \in \mathfrak{F}$?

i) The family S is a real linear normed space with the uniform norm $\|f\| = \sup f$. Every bounded linear functional $\varphi(f)$ on this space is difference of two bounded nonnegative linear functionals $\varphi(f) = \varphi^+(f) - \varphi^-(f)$: Take $\varphi^+(f) = \sup \{\varphi(f'), 0 \leq f' \leq f\}$ on S^+, then extend to S by linearity.

34. Riesz representation. Let $\mathfrak{X}$ be a locally compact space with points x, compacts K, and the σ-field S of topological Borel sets S, with or without subscripts. Let C be the space of bounded continuous functions g, with or without affixes, with the uniform norm $\|g\| = \sup g$. $C_0 \subset C$ consists of those g which vanish or infinity: Given $\epsilon > 0$ there exists a K_ϵ such that $|g| < \epsilon$ on $K_\epsilon{}^c$. $C_{00} \subset C$ consists of those g which vanish off compacts and $C_K \subset C_{00}$ of those g which vanish off K. If $\mathfrak{X}$ is compact, then $C_{\mathfrak{X}} = C_{00} = C_0 = C$.

a) *Dini.* If $g_n \in C_{00}$ and $g_n \downarrow 0$, then $g_n \downarrow 0$ uniformly, that is, $\|g\| \downarrow 0$.

b) Nonnegative linear functionals $\mu(g)$ on C_{00} are bounded on every C_K and are integrals on C_{00}: Bounded, since there exists $g_0 \in C_{00}{}^+$ with $g_0 \geq 1$ on C_K, hence $g \in C_K$ implies $|g| \leq g_0 \|g\|$ and $|\mu(g)| \leq \mu(g_0) \|g\|$. Integrals, since $g_1 \in C_K$ and $g_n \downarrow 0$ imply $g_n \in C_K$, $\|g\| \downarrow 0$, hence $|\mu(g_n)| \leq \mu(g_0) \|g_n\| \downarrow 0$.

c) There is a one-to-one correspondence between nonnegative linear functionals $\mu(g)$ on C_{00} and measures $\mu(S)$ bounded on compacts, given by $\mu(g) = \int \mu(dx)g(x)$: By b) and 33, $\mu(g)$ determines the measure $\mu(S)$.

d) There is a one-to-one correspondence between bounded linear functionals $\varphi(g)$ on C_{00} and bounded signed measures $\varphi(S)$ on S given by $\varphi(g) = \int \varphi(dx)g(x)$ with $\|g\| = \text{Var } \varphi$: Apply c) and 35i).

e) There is a one-to-one correspondence between bounded linear functionals on C_0 and bounded signed measures on S. Compactify and apply d.

Part Two

GENERAL CONCEPTS AND TOOLS OF PROBABILITY THEORY

Probability concepts can be defined in terms of measure-theoretic concepts. Since probability is a normed measure and random variables are finite measurable functions, the properties of sequences of random variables are more precise than those of measurable functions on a general measure space. Since in probability theory probability spaces are but frames of reference for families of random variables, probability properties are to be expressed in terms of the laws of the families only. These laws are expressed in terms of distributions which are set functions on the Borel fields in the range spaces. The distributions are expressed in terms of distribution functions which are point functions on the range spaces. In turn, to distribution functions correspond their Fourier-Stieltjes transforms (called characteristic functions) which are easier to deal with.

The following Parts utilize the tools so developed to investigate probability problems. These problems are centered about the concepts of independence and of conditioning introduced in Parts III and IV, respectively. The corresponding sections 15 and 24 may be read immediately after section 9.

149

Chapter III

PROBABILITY CONCEPTS

§ 9. PROBABILITY SPACES AND RANDOM VARIABLES

9.1 Probability terminology. Probability theory has its own terminology, born from and directly related and adapted to its intuitive background; for the concepts and problems of probability theory are born from and evolve with the analysis of random phenomena. As a branch of mathematics, however, probability theory partakes of and contributes to the whole domain of mathematics and, at present, its general set-up is expressible in terms of measure spaces and measurable functions. We give below a first table of correspondences between the probability and measure theoretic terms. Within parentheses appear the abbreviations to be used throughout this book.

probability space (pr. space)	normed measure space
elementary event	point belonging to the space
event	measurable set
sure event	whole space
impossible event	empty set
probability (pr.)	normed measure
almost sure, almost surely (a.s.)	almost everywhere
random variable (r.v.)	finite numerical measurable function
expectation E	integral $\int$

We shall use the pr. theory terms or the measure theory terms according to our convenience. We summarize below in pr. terms the properties which are specializations of those established in Part I.

I. A *pr. space* $(\Omega, \mathcal{a}, P)$ consists of the *sure event* Ω, the (nonempty) σ-field $\mathcal{a}$ of *events* and the *pr.* P on $\mathcal{a}$. Unless otherwise stated, the pr. space $(\Omega, \mathcal{a}, P)$ is fixed and $A, B, \cdots$, with or without affixes, represent events. If so required, the pr. space can always be *completed*, so that every subset of a null event becomes an event—necessarily null.

1° $\mathcal{a}$ *is a σ-field: for all A's, A^c, $\bigcup_{j=1}^{\infty} A_j$, $\bigcap_{j=1}^{\infty} A_j$ are events.*

It follows that, for every sequence A_n, $\lim\inf A_n$, $\lim\sup A_n$, and $\lim A_n$ (if it exists) are events.

2° P *is defined on $\mathcal{a}$ and, for all A's,*

$$PA \geqq 0, \quad P(\textstyle\sum A_j) = \sum PA_j, \quad P\Omega = 1.$$

It follows that

$$P\emptyset = 0, \quad PA \leqq PB \quad \text{when} \quad A \subset B, \quad P(\textstyle\bigcup A_j) \leqq \sum PA_j,$$

$$P(\lim\inf A_n) \leqq \lim\inf PA_n \leqq \lim\sup PA_n \leqq P(\lim\sup A_n),$$

and, if $\lim A_n$ exists, then $P(\lim A_n) = \lim PA_n$.

II. A r.v. X is a function on Ω to $R = (-\infty, +\infty)$ such that the inverse images under X of all Borel sets in R are events; it suffices to require the same of all intervals, or of all intervals $[a, b)$, or of all intervals $(-\infty, b)$, etc.

An *elementary r.v.* is a function on Ω to R of the form $X = \sum x_j I_{A_j}$ where x_j's are finite numbers, A_j's are disjoint events, and $\sum A_j = \Omega$; if there is only a finite number of distinct x_j's, then X is a *simple r.v.*

1° *Every r.v. is the finite limit of a sequence of simple r.v.'s and the finite uniform limit of a sequence of elementary r.v.'s; and conversely.*

Every nonnegative r.v. is the finite limit of a nondecreasing sequence of nonnegative simple r.v.'s; and conversely.

2° *The class of all r.v.'s is closed under the usual operations of analysis, provided these operations yield finite functions.*

3° *Every finite Borel function of a finite number of r.v.'s is a r.v.*

A *random function* is a family of r.v.'s; if the family is finite, it is a *random vector*, and, if the family is denumerable, it is a *random sequence*, that is, a sequence of r.v.'s.

III. Unless otherwise stated, $X, Y, \cdots$, with or without affixes, will represent r.v.'s and, as usual, limits will be taken for $n \to \infty$.

X_n *converges in pr.* to X, and we write $X_n \xrightarrow{P} X$, if, for every $\epsilon > 0$,

$$P[|X_n - X| \geq \epsilon] \to 0.$$

X_n *converges a.s.* to X, and we write $X_n \xrightarrow{\text{a.s.}} X$, if $X_n \to X$, except perhaps on a null event (event of pr. 0) or, *equivalently*, if for every $\epsilon > 0$,

$$P \bigcup_{k \geq n} [|X_k - X| \geq \epsilon] \to 0.$$

Mutual convergence in pr. $(X_n - X_m \xrightarrow{P} 0)$ and a.s. $(X_n - X_m \xrightarrow{\text{a.s.}} 0)$ are defined by replacing above $X_n - X$ by $X_n - X_m$ and $X_k - X$ by $X_k - X_l$ with $k, l \geq n$, and taking limits as $m, n \to \infty$.

1° $X_n \xrightarrow{P} X$ *if, and only if,* $X_n - X_m \xrightarrow{P} 0$. $X_n \xrightarrow{\text{a.s.}} X$ *if, and only if,* $X_n - X_m \xrightarrow{\text{a.s.}} 0$.

2° *If* $X_n \xrightarrow{\text{a.s.}} X$ *then* $X_n \xrightarrow{P} X$. *If* $X_n \xrightarrow{P} X$, *then there is a subsequence* $X_{n_k} \xrightarrow{\text{a.s.}} X$ *as* $k \to \infty$, *with*

$$\sum_{k=1}^{\infty} P\left[|X_{n_k} - X| \geq \frac{1}{2^k}\right] < \infty.$$

The terms "integral" and "expectation" and the notations $\int$ and E will be considered as equivalent. In the case of r.v.'s, we have

IV. The *expectation of a simple r.v.* $X = \sum_{k=1}^{n} x_k I_{A_k}$ is defined by

$$EX = \sum_{k=1}^{n} x_k P A_k.$$

The *expectation of a nonnegative r.v.* $X \geq 0$ is the limit of expectations of nonnegative simple r.v.'s X_n which converge nondecreasingly to X:

$$EX = \lim EX_n, \quad 0 \leq X_n \uparrow X.$$

The *expectation of a r.v.* $X = X^+ - X^-$ is given by

$$EX = EX^+ - EX^-,$$

provided the right-hand side is not of the form $+\infty - \infty$, and if EX exists and is finite, X is *integrable*.

1° X is integrable if, and only if, $|X|$ is integrable.

If X_1 and X_2 are integrable and a_1 and a_2 are finite numbers, then $a_1X_1 + a_2X_2$ is integrable and $E(a_1X_1 + a_2X_2) = a_1EX_1 + a_2EX_2$; if, moreover, $X_1 \leqq X_2$, then $EX_1 \leqq EX_2$.

If $|X_1| \leqq X_2$ and X_2 is integrable, then X_1 is integrable; in particular, every bounded r.v. is integrable, and if X degenerates at a $(X = a$ a.s.), then $EX = a$.

The indefinite expectation φ_X of a r.v. X whose expectation exists is defined on the σ-field $\mathfrak{A}$ of events A by $\varphi_X(A) = EXI_A$.

2° φ_X on $\mathfrak{A}$ is σ-finite, σ-additive, and P-continuous; if X is integrable, then φ_X is bounded by $E|X|$, and $\varphi_X(A) \to 0$ as $PA \to 0$.

3° MONOTONE CONVERGENCE THEOREM. If $0 \leqq X_n \uparrow X$ finite or not, then $EX_n \uparrow EX$; if EX is finite, then the measurable function X is a.s. a r.v.

DOMINATED CONVERGENCE THEOREM. If $X_n \xrightarrow{\text{P}} X$ and $|X_n| \leqq Y$ integrable, then X is integrable, and $EX_n \to EX$.

FATOU-LEBESGUE THEOREM. If Y and Z are integrable r.v.'s and $Y \leqq X_n$ or $X_n \leqq Z$, then

$$E(\liminf X_n) \leqq \liminf EX_n \quad \text{or} \quad \limsup EX_n \leqq E(\limsup X_n).$$

If, moreover, $\liminf EX_n$ or $\limsup EX_n$ is finite, then, respectively, $\liminf X_n$ or $\limsup X_n$ is a.s. a r.v.

EQUIVALENCE. Two functions on Ω are equivalent if they agree outside a null event. Convergences in pr. and a.s., integrals and integrability are, in fact, defined for equivalence classes and not for individual functions. Therefore, as long as we are concerned with a sequence of r.v.'s we can consider every r.v. of the sequence as defined up to an equivalence. In particular, we can then extend the notion of a r.v. as follows: a r.v. is an a.s. defined, a.s. finite and a.s. measurable function.

Let us observe, once and for all, that when the measurable functions under consideration are by definition $\mathfrak{B}$-measurable where $\mathfrak{B}$ is a sub σ-field of events, then almost sure relations are $P_{\mathfrak{B}}$-equivalences, that is, valid up to null $\mathfrak{B}$-measurable sets.

THE COMPLEX-VALUED CASE. A complex r.v. X is of the form $X = X' + iX''$ where X' and X'' are "ordinary" or "real-valued" r.v.'s as defined at the beginning of this section and where $i^2 = -1$; X takes its values in the complex plane of points $x' + ix''$, that is, in the plane $R \times R$, and its expectation is the point $EX = EX' + iEX''$. In other words, a complex r.v. X is a representation of the random vector $\{X', X''\}$. Similarly, a complex Borel function $g = g' + ig''$ is a representation of the Borel vector $\{g', g''\}$. The definitions and properties given below of random vectors, random sequences and, in general, random functions extend at once to the complex case where the components instead of being ordinary r.v.'s are complex-valued r.v.'s or, equivalently, two-dimensional random vectors. The relation $|EX| \leqq E|X|$ is still true; it suffices to use polar coordinates, setting $X = \rho e^{i\alpha}$, $EX = re^{it}$, and observe that

$$r = e^{-it}E\rho e^{i\alpha} = E\rho \cos(\alpha - t) \leqq E\rho$$

*9.2   Random vectors, sequences, and functions.   A *random vector* $X = (X_1, \cdots, X_n)$ is a finite family of r.v.'s called *components* of the random vector. Every component X_k induces a sub σ-field $\mathcal{B}(X_k)$ of events—inverse image of the Borel field in the range-space R_k of X_k. The random vector has for range space the n-dimensional real space $R^n = \prod\limits_{k=1}^{n} R_k$ with points $x = (x_1, \cdots, x_n)$ and it induces a σ-field $\mathcal{B}(X)$ $= \mathcal{B}(X_1, X_2, \cdots, X_n)$—inverse image of the Borel field in R^n. The inverse images of intervals $(-\infty, x) \subset R^n$ are events

$$[X < x] = [X_1 < x_1, \cdots, X_n < x_n] = \bigcap_{k=1}^{n} [X_k < x_k]$$

and, hence, are intersections of events belonging to the $\mathcal{B}(X_k)$. Since the Borel field $\mathcal{B}^n$ in R^n is the minimal σ-field over the class of these intervals, the σ-field $\mathcal{B}(X)$ is the minimal σ-field over these intersections or, equivalently, over the union of the $\mathcal{B}(X_k)$—a *compound or union σ-field* $\mathcal{B}(X_1, \cdots X_n)$ with *component σ-fields* $\mathcal{B}(X_k)$. Thus, the elements of $\mathcal{B}(X)$ are events and the random vector X can be defined as a measurable function on the pr. space to the n-dimensional Borel space $(R^n, \mathcal{B}^n)$. We define EX to be $(EX_1, EX_2, \cdots, EX_n)$—a point in the space R^n.

A *random sequence* $X = (X_1, X_2, \cdots)$ is a sequence of r.v.'s called its *components*; it takes its values in the space $R^\infty = \prod\limits_{n=1}^{\infty} R_n$ of points $x = (x_1, x_2, \cdots)$, that is, the space of numerical sequences. To every point x with an arbitrary but *finite number of finite coordinates* $x_{k_1}, \cdots x_{k_n}$ there corresponds the interval $(-\infty, x)$ of all points y such that $y_{k_1} < x_{k_1}, \cdots y_{k_n} < x_{k_n}$, and the minimal σ-field over the class of these intervals is the Borel field $\mathcal{B}^\infty$ in R^∞. Exactly as for random vectors, it follows that the inverse image under X of $\mathfrak{B}^{\infty}$ is the minimal σ-field over the class of all finite intersections of events $A_n \in \mathfrak{B}(X_n)$—the *compound* or *union σ-field* $\mathfrak{B}(X)$ with *component σ-fields* $\mathfrak{B}(X_n)$—then we write $\mathfrak{B}(X) = \mathfrak{B}(X_1, X_2, \cdots)$ and the random sequence can be defined as a measurable function on the pr. space to the *Borel space* $(R^{\infty}, \mathfrak{B}^{\infty})$. Similarly, the definition of the expectation of the random sequence is $EX = \{EX_1, EX_2, \cdots\}$—when $EX_1, EX_2, \cdots$ exist.

A random function $X_T = (X_t, t \in T)$ is a family of r.v.'s X_t where t varies over an arbitrary but fixed index set T. Exactly as above, the range space of X_T is the *real space* $R_T = \prod_{t \in T} R_t$ of points $x_T = (x_t, t \in T)$—the space of numerical functions; *intervals* $(-\infty, x_T)$ are defined for points x_T with an arbitrary but finite number of finite coordinates to be sets of all points $y_T < x_T$, that is, $y_t < x_t$, $t \in T$; the *Borel field* $\mathfrak{B}_T$ is the minimal σ-field over the class of these intervals. The random function X_T induces the *compound* or *union σ-field* $\mathfrak{B}(X_T)$ with *component σ-fields* $\mathfrak{B}(X_t)$—the minimal σ-field over the class of all finite intersections of events $A_t \in \mathfrak{B}(X_t)$ as t varies on T or, equivalently, the inverse image under X_T of the Borel field $\mathfrak{B}_T$; and the random function X_T can be defined as a measurable function on the pr. space to the *Borel space* $(R_T, \mathfrak{B}_T)$. By definition, $EX_T = \{EX_t, t \in T\}$ is a numerical function—when the EX_t exist.

A *Borel function* $g_{T'}$ is a function on a Borel space $(R_T, \mathfrak{B}_T)$ to a Borel space $(R_{T'}, \mathfrak{B}_{T'})$ such that the inverse image under $g_{T'}$ of the Borel field in the range space is contained in the Borel field $\mathfrak{B}_T$ in the domain R_T. Therefore, if X_T is a random function to R_T, then the function of function $g_{T'}(X_T)$ on the pr. space to the Borel space $(R_{T'}, \mathfrak{B}_{T'})$ induces a sub σ-field of events—inverse image under X_T of the inverse image under $g_{T'}$ of the Borel field $\mathfrak{B}_{T'}$. Thus, $\mathfrak{B}(g_{T'}(X_T)) \subset \mathfrak{B}(X_T)$; in other words, $g_{T'}(X_T)$ is $\mathfrak{B}(X_T)$-measurable and, hence, is a random function. We state this conclusion as a theorem.

A. Borel functions theorem. *A Borel function of a random function is a random function which induces a sub σ-field of events contained in the one induced by the original random function.*

Loosely speaking, a Borel function of a random function induces a "coarser" sub σ-field of events and has "fewer" values.

9.3 Moments, inequalities, and convergences. Expectations of powers of r.v.'s are called *moments* and play an essential role in the investigations of pr. theory. They appear in the simple but powerful Markov inequality and in the definition of the very useful notion of convergence "in the rth mean," that we shall introduce in this subsection. They appear in the expansions of "characteristic functions" that we shall examine in the next chapter. They play a basic role in the study of sums of "independent" r.v.'s to which the next part is devoted. Furthermore, the powerful "truncation" method—to be used extensively in the following parts—expands tremendously the domain of applicability of the methods of investigation based upon the use of moments.

EX^k ($k = 1, 2, \cdots$) and $E|X|^r$ ($r > 0$) are called, respectively, the kth *moment* and the rth *absolute moment* of the r.v. X. We may also consider 0th moments but, for all r.v.'s, the 0th moments are 1, and we shall limit ourselves to kth moments where k is a positive integer, and to rth absolute moments where r is a positive number, unless otherwise stated.

We establish now a few simple properties of moments. While a kth moment may not exist, absolute moments always exist but may be infinite. Since integrability is equivalent to absolute integrability, if the kth absolute moment of X is finite, then its kth moment exists and is finite; and conversely. More generally, since $|X|^{r'} \leq 1 + |X|^r$ for $0 < r' < r$, we have

a. *If* $E|X|^r < \infty$, *then* $E|X|^{r'}$ *is finite for* $r' \leq r$ *and* EX^k *exists and is finite for* $k \leq r$.

In other words, finiteness of a moment of X implies existence and finiteness of all moments of X of lower order.

Upon applying the elementary inequality

$$|a + b|^r \leq c_r|a|^r + c_r|b|^r, \quad r > 0,$$

where $c_r = 1$ or 2^{r-1} according as $r \leq 1$ or $r \geq 1$, replacing a by X, b by Y and, taking expectations of both sides, we obtain the

c_r-INEQUALITY. $E|X + Y|^r \leq c_r E|X|^r + c_r E|Y|^r$, *where* $c_r = 1$ *or* 2^{r-1} *according as* $r \leq 1$ *or* $r \geq 1$.

This inequality shows that if the rth absolute moments of X and Y exist and are finite, so is the rth absolute moment of $X + Y$.

Similarly, excluding the trivial case of vanishing $E|X|^r$ or $E|Y|^s$ (in which case the Hölder inequality below is trivially true), and replacing a by $X/E^{\frac{1}{r}}|X|^r$, b by $Y/E^{\frac{1}{s}}|Y|^s$ in the elementary inequality

$$|ab| \leq \frac{|a|^r}{r} + \frac{|b|^s}{s}, \quad r > 1, \quad \frac{1}{r} + \frac{1}{s} = 1,$$

we obtain the

HÖLDER INEQUALITY. $E|XY| \leq E^{\frac{1}{r}}|X|^r \cdot E^{\frac{1}{s}}|Y|^s$, *where* $r > 1$ *and*
$\dfrac{1}{r} + \dfrac{1}{s} = 1$.

From this inequality follows the

MINKOWSKI INEQUALITY. *If* $r \geq 1$, *then*

$$E^{\frac{1}{r}}|X + X'|^r \leq E^{\frac{1}{r}}|X|^r + E^{\frac{1}{r}}|X'|^r.$$

In fact, upon excluding the trivial case $r = 1$, and applying the Hölder inequality with $Y = |X + X'|^{r-1}$ to the right-hand side terms in the obvious inequality

$$E|X + X'|^r \leq E(|X| \cdot |X + X'|^{r-1}) + E(|X'| \cdot |X + X'|^{r-1}),$$

we find

$$E|X + X'|^r \leq (E^{\frac{1}{r}}|X|^r + E^{\frac{1}{r}}|X'|^r)E^{\frac{1}{s}}|X + X'|^{(r-1)s},$$

where $\dfrac{1}{r} + \dfrac{1}{s} = 1$. Upon excluding the trivial case of vanishing $E|X + X'|^r$, noticing that $(r - 1)s = r$, and dividing both sides by $E^{\frac{1}{s}}|X + X'|^r$, the asserted inequality follows.

Hölder's inequality with $r = s = 2$, is called the

SCHWARZ INEQUALITY: $E^2|XY| \leq E|X|^2 \cdot E|Y|^2$.

Replacing X by $|X|^{\frac{r-r'}{2}}$ and Y by $|X|^{\frac{r+r'}{2}}$, with $r' \leq r$, and, taking logarithms of both sides, we obtain the inequality

$$\log E|X|^r \leq \tfrac{1}{2} \log E|X|^{r-r'} + \tfrac{1}{2} \log E|X|^{r+r'}$$

b. $\log E|X|^r$ *is a convex function of* r.

Hölder's inequality with X, Y, r, s replaced respectively by $|X|^p$, 1^p, p/r, q/r $\left(\text{hence } \dfrac{1}{r} = \dfrac{1}{p} + \dfrac{1}{q}\right)$ becomes $E^{1/r}|X|^r \leq E^{1/p}|X|^p$ for $r < p$.

Hence,

c. $E^{1/r}|X|^r$ *is nondecreasing in* r.

In fact, $E^{1/r}|X|^r \uparrow E^{1/p}|X|^p$ as $r \uparrow p$. For, if $E|X|^p < \infty$ then $|X|^r \leq \max(1, |X|^p)$ and the dominated convergence theorem applies. If $E|X|^p = \infty$ apply what precedes to $Y_n = |X|I_{[|X|<n]}$ then let $n \uparrow \infty$.

We introduce now convergence in the rth mean. Let X_n and X be r.v.'s with finite rth absolute moments, so that, by the c_r-inequality, the same is true of $X_n - X$. We say that the sequence X_n converges to X *in the* rth *mean*, and write $X_n \overset{r}{\to} X$, if $E| X_n - X |^r \to 0$.

Let $X_n \overset{r}{\to} X$. If $r \le 1$ then it follows, by the c_r-inequality, that

$$| E| X_n |^r - E| X |^r | \le E| X_n - X |^r \to 0,$$

and, if $r > 1$, then it follows, by the Minkowski inequality, that

$$| E^{\frac{1}{r}}| X_n |^r - E^{\frac{1}{r}}| X |^r | \le E^{\frac{1}{r}}| X_n - X |^r \to 0.$$

This proves that

d. *If* $X_n \overset{r}{\to} X$, *then* $E| X_n |^r \to E| X |^r$.

We conclude this subsection with a simple but basic inequality and a few of its applications.

A. BASIC INEQUALITY. *Let X be an arbitrary r.v. and let g on R be a nonnegative Borel function.*

If g is even and is nondecreasing on $[0, +\infty)$ then, for every $a \ge 0$

$$\frac{Eg(X) - g(a)}{\text{a.s. sup } g(X)} \le P[| X | \ge a] \le \frac{Eg(X)}{g(a)}.$$

If g is nondecreasing on R, then the middle term is replaced by $P [X \ge a]$, where a is an arbitrary number.

The proof is immediate. Since g is a Borel function on R, it follows that $g(X)$ is a measurable function on Ω and, since g is nonnegative on R, its integral exists. If g is even and is nondecreasing on $[0, +\infty)$, then, setting $A = [| X | \ge a]$, from the obvious relations

$$Eg(X) = \int_A g(X) + \int_{A^c} g(X)$$

and

$$g(a)PA \le \int_A g(X) \le \text{a.s. sup } g(X) \cdot PA, \quad 0 \le \int_{A^c} g(X) \le g(a),$$

it follows that

$$g(a)PA \le Eg(X) \le \text{a.s. sup } g(X) \cdot PA + g(a).$$

This proves the first assertion and the second is similarly proved.

Applications. (1) Upon taking $g(x) = e^{rx}(r > 0)$, we obtain

$$\frac{Ee^{rX} - e^{ra}}{\text{a.s. sup } e^{rX}} \leq P[X \geq a] \leq e^{-ra}Ee^{rX}$$

(2) Upon taking $g(x) = |x|^r(r > 0)$ we obtain

$$\frac{E|X|^r - a^r}{\text{a.s. sup } |X|^r} \leq P[|X| \geq a] \leq \frac{E|X|^r}{a^r};$$

the right-hand side inequality is called the *Markov inequality*, and for $r = 2$ it reduces to the celebrated *Tchebichev inequality*.

Upon applying Markov's inequality with X replaced by $X_n - X$, it follows that

If $X_n \xrightarrow{r} X$, then $X_n \xrightarrow{P} X$, and if the X_n are a.s. uniformly bounded, then, conversely, $X_n \xrightarrow{P} X$ implies that $X_n \xrightarrow{r} X$.

(3) Upon taking $g(x) = \dfrac{|x|^r}{1 + |x|^r}$ $(r > 0)$, we obtain

$$E\frac{|X|^r}{1 + |X|^r} - \frac{a^r}{1 + a^r} \leq P[|X| \geq a] \leq \frac{1 + a^r}{a^r} E\frac{|X|^r}{1 + |X|^r};$$

replacing X by $X_n - X$ and by $X_m - X_n$, it follows that, as $m, n \to \infty$,

$$X_n \xrightarrow{P} X \quad \textit{if, and only if,} \quad E\frac{|X_n - X|^r}{1 + |X_n - X|^r} \to 0;$$

$$X_m - X_n \xrightarrow{P} 0 \quad \textit{if, and only if,} \quad E\frac{|X_m - X_n|^r}{1 + |X_m - X_n|^r} \to 0.$$

REMARK. Observe that the function defined by $d(X, Y) = E\dfrac{|X - Y|}{1 + |X - Y|}$ has the triangular and identification properties of a distance, except that $d(X, Y) = 0$ implies only that $X = Y$ a.s. It follows from the foregoing proposition that

*The space of the equivalence classes of the r.v.'s defined in a **pr.** space is a complete metric space with distance d defined by*

$$d(X, Y) = E\frac{|X - Y|}{1 + |X - Y|},$$

*and convergence in distance is equivalent to convergence in **pr.***

*CONVEX FUNCTIONS. The relations between moments established at the beginning of this subsection are essentially convexity properties. Let us recall a few classical properties of convex functions.

Let g be a (numerical) Borel function defined on a finite or an infinite *open* interval $I \subset R$. g is said to be *convex* if, for every pair of points x, x' of I,

$$g\left(\frac{x + x'}{2}\right) \leq \frac{1}{2} g(x) + \frac{1}{2} g(x');$$

if g is twice differentiable on I, then the convexity property is equivalent to $g'' \geq 0$ on I. The same definition applies to g on an N-dimensional interval I^N and is equivalent to the convexity of the function $g(x + ux')$ of the numerical argument u for all values of u for which $x + ux' \in I^N$, so that it suffices to consider convex functions on $I \subset R$. A convex function on I is either continuous on I or is not a Borel function. Thus, from now on, a convex function will be assumed to be continuous on its domain. In that case, g is convex on I if, and only if, to every $x_0 \in I$ there corresponds a number $\lambda(x_0)$ such that, for all $x \in I$,

$$\lambda(x_0)(x - x_0) \leq g(x) - g(x_0).$$

Let X be a r.v. whose values lie a.s. in I and whose expectation EX exists and is finite. Replacing x_0 by EX and x by X, and taking the expectation of both sides of the foregoing inequality, it follows that

e. *If g is convex and EX is finite, then*

$$g(EX) \leq Eg(X).$$

If g is strictly monotone, then this relation can be written

$$EX \leq g^{-1}(Eg(X)).$$

For example, for $r \geq 1$, $g(x) = x^r (x \in (0, +\infty))$ being convex, we have

$$E|X| \leq E^{1/r}|X|^r.$$

More generally, let G_1 and G_2 be two continuous and strictly increasing functions such that $g = G_2 G_1^{-1}$ is convex; we say then that G_2 *is convex in* G_1. Since $Y = G_1(X)$ implies that $X = G_1^{-1}(Y)$, it follows by **e**, upon assuming that EX and EY are finite, that

$$G_2 G_1^{-1}(EY) \leq E G_2 G_1^{-1}(Y)$$

and, hence,

e'. *If G_2 is convex in G_1, then*

$$G_1^{-1}(EG_1(X)) \leq G_2^{-1}(EG_2(X)).$$

For example, since on $(0, +\infty)$, x^{r_2} is convex in x^{r_1} for $r_2 \geq r_1$, that is, the function x^{r_2/r_1} is convex, we have

$$E^{\frac{1}{r_1}}|X|^{r_1} \leq E^{\frac{1}{r_2}}|X|^{r_2} \quad \text{for} \quad r_2 \geq r_1.$$

*9.4 Spaces L_r. The r.v.'s whose rth absolute moments are finite are said to form the *space L_r* over the pr. space $(\Omega, \mathcal{A}, P)$; in symbols, $X \in L_r$ if $E|X|^r < \infty$; we drop r if $r = 1$. We shall find later that the space L_2 is a very important tool in the investigation of pr. problems, especially those relative to sums of "independent" r.v.'s. It will be convenient to introduce two boundary cases. The first is the trivial space L_0 of all r.v.'s X since $E|X|^0 = 1$ is finite. The second is the space L_∞ of all a.s. bounded r.v.'s. Since $\lim_{r \to \infty} E|X|^r < \infty$ if, and only if, $|X| \leq 1$ a.s., it seems that only the subspace $L'_\infty \subset L_\infty$ of r.v.'s a.s. bounded by 1 ought to be introduced. However, for $r \to \infty$ it is $\lim E^{\frac{1}{r}}|X|^r$ which counts, and this limit is finite if, and only if, X is a.s. bounded. In fact, let s be the a.s. supremum of $|X|$, defined by $P[|X| > s] = 0$ and $P[|X| \geq c] > 0$ for every $c < s$; we have $s \leq \infty$. The foregoing assertion is implied by

a. $E^{\frac{1}{\infty}}|X|^\infty \equiv \lim_{r \to \infty} E^{\frac{1}{r}}|X|^r = \text{a.s. sup}|X| \equiv s.$

For

$$s \geq E^{\frac{1}{r}}|X|^r \geq E^{\frac{1}{r}}(|X|^r I_{[|X| \geq c]}) \geq c P^{\frac{1}{r}}[|X| \geq c] \to s$$

as $r \to \infty$, then $c \uparrow s$.

The foregoing definitions permit us to state 9.3a as follows:

b. $L_0 \supset L_r \supset L_s \supset L_\infty \supset L'_\infty, 0 \leq r \leq s \leq \infty.$

Let us observe that the space of all simple r.v.'s is a subspace of L_∞ and, hence, of all the spaces L_r.

Since, by the c_r- and Minkowski inequalities and by **a**,

$$E|X + Y|^r \leq E|X|^r + E|Y|^r, \quad 0 < r < 1,$$

$$E^{\frac{1}{r}}|X + Y|^r \leq E^{\frac{1}{r}}|X|^r + E^{\frac{1}{r}}|Y|^r, \quad 1 \leq r \leq \infty,$$

and $E|X - Y|^r = 0$ if, and only if, X and Y are equivalent, we have, according to the definitions relative to metric and normed spaces, the following theorem.

A. *The spaces L_r are linear metric spaces with metric defined by*

$$d(X, Y) = E|X - Y|^r \quad for \quad 0 < r < 1$$

and norm

$$\|X\| = E^{\frac{1}{r}}|X|^r \quad for \quad 1 \le r \le \infty,$$

provided equivalent r.v.'s are identified.

The problem arises whether the spaces L_r are complete and what are the convergence theorems in these spaces. Unless otherwise stated, *from now on $0 < r < \infty$* (the reader is invited to examine in each case the boundary spaces L_0 and L_∞).

First we observe that on account of **A** and 9.3d we have

c. *Convergence in distance $d(X_n, X) \to 0$ in L_r is equivalent to convergence in the rth mean $X_n \overset{r}{\to} X$ and implies convergence of distances $d(X_n, X_0) \to d(X, X_0)$ to any fixed $X_0 \in L_r$.*

Also, if $X_n \in L_r$, then, for a r.v. X, $E|X_n - X|^r$, which always exists, can converge to 0 only if, from some value of n on, $E|X_n - X|^r$ is finite and, hence, only if $X \subset L_r$, so that

d. *If X_n is a sequence in L_r and $E|X_n - X|^r \to 0$, then $X \in L_r$.*

We are now in a position to prove the

B. L_r-COMPLETENESS THEOREM. *Let the $X_n \in L_r$. Then $X_n \overset{r}{\to}$ some X if, and only if, $X_m - X_n \overset{r}{\to} 0$, as $m, n \to \infty$.*

Proof. If $X_n \overset{r}{\to} X$, then $X_m - X_n \overset{r}{\to} 0$, since, by the c_r-inequality,

$$E|X_m - X_n|^r \le c_r E|X_m - X|^r + c_r E|X - X_n|^r \to 0.$$

Conversely, if $X_m - X_n \overset{r}{\to} 0$, then, by the Markov inequality, for every $\epsilon > 0$,

$$P[|X_m - X_n| \ge \epsilon] \le \frac{1}{\epsilon^r} E|X_m - X_n|^r \to 0 \quad as \quad m, n \to \infty,$$

so that $X_m - X_n \overset{P}{\to} 0$. Therefore, there is a subsequence $X_{n'} \overset{a.s.}{\longrightarrow}$ some X as $n' \to \infty$ and, for every fixed m, $X_m - X_{n'} \overset{a.s.}{\longrightarrow} X_m - X$ as $n' \to \infty$. Since $E|X_m - X_{n'}|^r \to 0$ as $m, n' \to \infty$, it follows, by the Fatou-Lebesgue theorem and the hypothesis, that

$$E|X_m - X|^r \le \liminf_{n'} E|X_m - X_{n'}|^r \to 0 \quad as \quad m \to \infty.$$

Thus, $X_n \overset{r}{\to} X$, and the proof is complete.

If a r.v. X is integrable, then the (indefinite) integral of X is P-absolutely continuous: $\int_A |X| \to 0$ as $PA \to 0$. Let $B = [|X| \geq a]$. Since $PB \to 0$ as $a \to \infty$, it follows that $\int_B |X| \to 0$ as $a \to \infty$. Conversely, this implies that

$$\int_A |X| = \int_{AB} |X| + \int_{AB^c} |X| \leq \int_B |X| + aPA \to 0$$

as $PA \to 0$ then $a \to \infty$, and thus implies that X is integrable, since, given $\epsilon > 0$,

$$\int |X| \leq \int_B |X| + a < \epsilon + a_\epsilon < \infty$$

for $a = a_\epsilon$ sufficiently large.

The integrals of r.v.'s X_n are *uniformly P-absolutely continuous* or simply *uniformly continuous* if $\int_A |X_n| \to 0$ uniformly in n as $PA \to 0$; in other words, for every $\epsilon > 0$ there exists a δ_ϵ independent of n such that $\int_A |X_n| < \epsilon$ for any set A with $PA < \delta_\epsilon$. Let $B_n = [|X_n| \geq a]$. The r.v.'s $|X_n|$ are *uniformly integrable*, if $\int_{B_n} |X_n| \to 0$ uniformly in n, as $a \to \infty$. Observe that if the $\int |X_n|$ are uniformly bounded, say, by $c(< \infty)$, then, by Markov's inequality, $PB_n \leq c/a \to 0$ as $a \to \infty$. Upon replacing X by X_n and B by B_n in the foregoing discussion, it follows that

e. *The r.v.'s X_n are uniformly integrable if, and only if, their integrals are uniformly bounded and uniformly continuous.*

Let $X_n \xrightarrow{r} X$ hence $X_n I_A \xrightarrow{r} X I_A$. It follows, by 9.3d and the above lemma (take $A = \Omega$, and take A such that $PA \to 0$)

f. *If $X_n \xrightarrow{r} X$, then the $|X_n|^r$ are uniformly integrable.*

For use on the forthcoming theorem, note that (Young)
The Fatou-Lebesgue theorem and the dominated convergence theorem remain valid if therein Y and Z are replaced by U_n and V_n with $U_n \xrightarrow{a.c.} U$, $V_n \xrightarrow{a.e.} V$ and $\int U_n \to \int U$ finite, $\int V_n \to \int V$ finite.

For then, the argument pp. 125–6 remains valid. Furthermore, by selecting $\{n''\}$ p. 126 so that also $U_{n''} \xrightarrow{\text{a.e.}} U$, we have

If $|X_n| \leqq U_n$ with $U_n \xrightarrow{\mu} U$ and $\int U_n \to \int U$ finite, then $X_n \xrightarrow{\mu} X$

implies that $\int X_n \to \int X$ in fact $\int |X_n - X| \to 0$.

C. L_r-CONVERGENCE THEOREM. *Let the $X_n \in L_r$. Then*
(i) $X_n \xrightarrow{r} X$ *if and only if* (ii) $X_n \xrightarrow{P} X$
and one of the following conditions holds:

(iii) $\int |X_n|^r \to \int |X|^r < \infty$; (iv) *the $|X_n|^r$ are uniformly integrable;*
(v) *the $|X_n|^r$, or* (vi) *the $|X_n - X|^r$, have uniformly continuous integrals.*

Proof. Let $\epsilon > 0$ be arbitrary, set $A_n = [|X_n - X| \geqq \epsilon]$, $A_{mn} = [|X_m - X_n| \geqq \epsilon]$, and let $m, n \to \infty$. We use the c_r-inequality without further comment. Note that (iv) implies $X_n \in L_r$.

Condition (i) implies (ii) by Markov inequality $(PA_n \leqq E|X_n - X|^r/\epsilon^r \to 0)$ and implies (iii) by 9.3d. Conversely, (ii) and (iii) imply (i), since then $|X_n - X|^r \leqq c_r |X_n|^r + c_r |X|^r = U_n$ with $U_n \xrightarrow{P} 2c_r |X|^r$ and $\int U_n \to 2c_r \int |X|^r < \infty$.

As for the remaining assertions, (i) implies (iv) by **f**, and (iv) implies (v) by **e** applied to the $|X_n|^r$ in lieu of the X_n. Also, clearly (i) implies (vi), and (vi) implies (v), since it implies integrability of $|X_n - X|^r$ hence of $|X|^r$ (because $X_n \in L_r$) so that $\int_A |X_n| \leqq c_r \int_A |X_n - X|^r + c_r \int_A |X|^r < \epsilon$ for PA sufficiently small.

Thus, to complete the proof, it suffices to show that (ii) and (v) imply (i). Since convergence in pr. (in the rth mean) is equivalent to mutual convergence in pr. (in the rth mean) and $X_n \xrightarrow{P} X$, $X_n \xrightarrow{r} Y$ imply that $Y = X$ a.s., we can replace (i) and (ii) by (i') $E|X_m - X_n|^r \to 0$ and (ii') $PA_{mn} \to 0$. The assertion follows since, upon integrating $|X_m - X_n|^r$ on A_{mn} and on $A_{mn}{}^c$, (i') and (v) imply that as $m, n \to 0$ then $\epsilon \to 0$, $E|X_m - X_n|^r \leqq c^r \int_{A_{mn}} |X_m|^r + c_r \int_{A_{mn}} |X_n|^r + \epsilon^r \to 0$.

Corollary 1. $X_n \xrightarrow{r} X$ implies $X_n \xrightarrow{r'} X$ for $r' < r$.

Set $A_n = [|X_n - X| \geq 1]$ and observe that

$$\int_A |X_n - X|^{r'} = \int_{AA_n} |X_n - X|^{r'}$$

$$+ \int_{AA_n^c} |X_n - X|^{r'} \leq \int_A |X_n - X|^r + PA.$$

Corollary 2. If $\sup E|X_n|^r = c < \infty$, then $X_n \xrightarrow{P} X$ implies $X_n \xrightarrow{r'} X$ for $r' < r$.

Let $A_n = [|X_n| \geq a]$ and observe that

$$\int_A |X_n|^{r'} = \int_{AA_n} |X_n|^{r'} + \int_{AA_n^c} |X_n|^{r'} \leq ca^{r'-r} + a^r PA < \epsilon$$

by taking a sufficiently large to have $ca^{r'-r} < \dfrac{\epsilon}{2}$ and, then, PA sufficiently

small to have $a^r PA < \dfrac{\epsilon}{2}$.

Corollary 3. If $|X_n| \leq Y \in L_r$ for large n, then $X_n \xrightarrow{P} X$ implies $X_n \xrightarrow{r} X \in L_r$.

Observe that for large n, $\displaystyle\int_A |X_n|^r \leq \int_A Y^r$.

We proved in 9.3 a particular case of this corollary, with $Y = c < \infty$.

We summarize below the relations between various types of convergence:

$$X_n \xrightarrow{a.s.} X \Rightarrow X_n \xrightarrow{P} X \Rightarrow X_{n_k} \xrightarrow{a.s.} X \text{ with } \sum_k P\left[|X_{n_k} - X| \geq \frac{1}{2^k}\right] < \infty$$

$$\Uparrow$$

$$X_n \xrightarrow{r} X \Rightarrow X_n \xrightarrow{r'} X, \quad r' < r.$$

The operation of integration on the complete normed linear space L_r with $r \geq 1$ can be characterized as a functional of the integrand, as follows:

D. Integral representation theorem. *Let* $\dfrac{1}{r} + \dfrac{1}{s} = 1$ *with* $1 \leq r < \infty$.

A functional f on L_r is linear and continuous if, and only if, there exists a r.v. $Y \in L_s$ such that $f(X) = EXY$ for every $X \in L_r$; then f determines Y up to an equivalence and $\|f\| = E^{\frac{1}{s}}|Y|^s$.

Proof. Since $\dfrac{1}{r} + \dfrac{1}{s} = 1$ and $1 \leq r < \infty$, it follows that $1 < s \leq \infty$, and we apply repeatedly Hölder's inequality $E|XY| \leq \|X\|_r \|Y\|_s$, where $\|X\|_r = E^{\frac{1}{r}}|X|^r$ and $\|Y\|_s = E^{\frac{1}{s}}|Y|^s$ with $\|Y\|_\infty = \lim\limits_{s \to \infty} E^{\frac{1}{s}}|Y|^s = \text{a.s.} \sup |Y|$.

If $\|X\|_r \|Y\|_s$ is finite, then $f(X) = EXY$ exists, is finite, and defines a normed functional f on L_r with $\|f\| \leq \|Y\|_s$. Since EXY is linear in $X \in L_r$, so is $f(X)$. Being normed and linear, f is continuous.

Conversely, let a functional f on L_r be continuous and linear; linearity implies additivity and additivity implies $f(\theta) = 0$, where θ is the zero-point of L_r, that is, the class of r.v.'s degenerate at 0. Therefore, the set function φ on $\mathcal{C}$ defined by $\varphi(A) = f(I_A)$ is continuous and additive, hence σ-additive, and vanishes for null events, hence is P-continuous. Thus, the Radon-Nikodym theorem applies and φ on $\mathcal{C}$ determines up to an equivalence a r.v. Y such that

$$f(I_A) = \varphi(A) = EI_AY.$$

Since $f(X)$ and EXY are both linear in X, it follows that $f(X) = EXY$ for all simple finite $X(\in L_r)$. If $Y \in L_s$ and $L_r \ni X_n \xrightarrow{r} X$ hence $X_nY \xrightarrow{1} XY$, then, by continuity of f and of E on L_r, this equality extends to all $X \in L_r$. Since f has finite norm $\|f\| \leq \|Y\|_s$, to complete the proof it suffices to show that the reverse inequality $\|f\| \geq \|Y\|_s$ is true.

Let $r > 1$. If the X_n are simple finite and $0 \leq X_n \uparrow |Y|$, then

$$E|X_n|^s \leq E(X_n{}^{s-1} \operatorname{sign} Y)Y \leq \|f\| E^{\frac{1}{r}}|X_n|^{(s-1)r}$$

yields

$$\|Y\|_s \leftarrow \|X_n\|_s \leq \|f\|.$$

Let $r = 1$. If there exists an $\epsilon > 0$ such that $\|Y\|_\infty \geq \|f\| + 2\epsilon$ and we set $A = [|Y| \geq \|f\| + \epsilon]$, then $PA > 0$ while

$$(\|f\| + \epsilon)PA \leq E|I_AY| = E(I_A \operatorname{sign} Y)Y \leq \|f\|PA,$$

and we reach a contradiction. This completes the proof.

REMARK. The definitions and results of this subsection extend at once to complex-valued r.v.'s.

§ 10.   Probability Distributions

10.1   Distributions and distribution functions. Let X be a r.v. on our pr. space $(\Omega, \, \mathcal{C}, \, P)$. The nonnegative set function P_X defined on the Borel field $\mathcal{B}$ in R by

$$P_X S = P[X \in S], \quad S \in \mathcal{B}$$

is called the *pr. distribution* or, simply, *distribution* of X. Since X is finite, the inverse image under X of R is Ω and, since the inverse image of a sum of Borel sets is the sum of their inverse images, we have

$$P_X R = 1, \quad P_X(\textstyle\sum S_j) = \sum P_X S_j, \quad S_j \in \mathcal{B}.$$

Therefore, P_X on $\mathcal{B}$ is a probability. Thus, the r.v. X induces on its range space a new pr. space $(R, \mathcal{B}, P_X)$, to be called a pr. space *induced* by X on its range space or the *sample pr. space* of X. Moreover,

a. *The distribution P_X of X determines the distributions of all r.v.'s $g(X)$ where g is a finite Borel function on R; and $Eg(X) = \int_R g \, dP_X$ in the sense that, if either side of this expression exists, so does the other, and then they are equal.*

Proof. Every finite Borel function $g(X)$ of a r.v. X is a r.v. and, by definition,

$$[g(X) \in S] = [X \in g^{-1}(S)]$$

where S and $g^{-1}(S)$ are Borel sets. Therefore

$$P_{g(X)}(S) = P_X g^{-1}(S), \quad S \in \mathcal{B},$$

and the first assertion is proved.

The second assertion will follow if we prove it for nonnegative functions g. Because of the monotone convergence theorem, it suffices to prove it for nonnegative simple functions g and, because of the additivity property of integrals, it suffices to prove the assertion for indicators. Thus, let $g = I_S$, so that $g(X) = I_{[X \in S]}$. But, then, the left-hand side of the asserted equality becomes

$$\int_\Omega I_{[X \in S]} dP = P[X \in S],$$

while the right-hand side becomes $\displaystyle\int_R I_S \, dP_X = P_X S$. Therefore, by

definition of P_X, the asserted equality holds, and the proof is complete.

Distributions are *set* functions and are not easy to handle by means of classical analysis developed primarily to deal with point functions. Thus, in order to be able to use analytical methods and tools, it is of the greatest importance to find, and learn to use, *point* functions which "represent" distributions, that is, which are in a one-to-one correspondence with distributions. Such functions are obtained by the correspondence theorem according to which, to the finite measure P_X corresponds one, and only one, interval function defined by

$$F_X[a, b) = P_X[a, b) = P[a \leqq X < b), \quad [a, b) \subset R.$$

In turn, to this interval function corresponds one, and only one, class of point functions on R defined up to an additive constant, by

$$F_X(b) - F_X(a) = F_X[a, b), \quad a < b \in R.$$

Recalling that P_X is the distribution of a r.v. X, we select among all those functions the function F_X defined on R by

$$F_X(x) = P_X(-\infty, x) = P[X < x], \quad x \in R,$$

and call it the *distribution function (d.f.) of* X. Then, according to the usual notational convention, the equality in **a** can be written $Eg(X) = \int_R g \, dF_X$ and, if g is integrable and continuous on R, then the right-hand side L.-S.-integral becomes an improper R.-S.-integral.

b. *The d.f.* F_X *of a r.v.* X *is nondecreasing and continuous from the left on* R, *with* $F_X(-\infty) = 0$ *and* $F_X(+\infty) = 1$. *Conversely, every function* F *with the foregoing properties is the d.f. of a r.v. on some pr. space.*

Proof. The first assertion follows from the fact that $P[X < x]$ does not decrease as x increases, approaches $P[X < x']$ as $x \uparrow x'$, and approaches $P[X = -\infty] = 0$ or $P[X < +\infty] = 1$ according as $x \rightarrow -\infty$ or $x \rightarrow +\infty$. The converse follows by taking, say, for pr. space $(R, \mathfrak{B}, P)$ where P is the pr. determined, according to the correspondence theorem, by F. Then F is the d.f. of the r.v. X defined on this pr. space by $X(x) = x, x \in R$.

REMARK. *There are pr. spaces on which there can be defined r.v.'s for every function* F *with the stated properties.*

For example, take for the space Ω the interval $(0, 1)$, for the σ-field of events the σ-field of all Borel sets in this interval, and for pr. the Lebesgue measure on this σ-field. Then any function F with the stated properties is the d.f. of an inverse function X of F.

The weakest type of convergence of sequences of r.v.'s considered so far is convergence in pr. In turn, it implies a type of convergence of d.f.'s, as follows:

c. *If $X_n \xrightarrow{P} X$, then $F_{X_n} \to F_X$ on the continuity set $C(F_X)$ of F_X.*

Proof. Since

$$[X < x'] = [X_n < x, X < x'] + [X_n \geq x, X < x']$$

$$\subset [X_n < x] + [X_n \geq x, X < x'],$$

we have

$$P[X < x'] \leq F_{X_n}(x) + P[X_n \geq x, X < x'].$$

If $X_n - X \xrightarrow{P} 0$, then, for $x' < x$,

$$P[X_n \geq x, X < x'] \leq P[|X_n - X| \geq x - x'] \to 0$$

and, hence,

$$F_X(x') \leq \liminf F_{X_n}(x), \quad x' < x.$$

Similarly, interchanging X and X_n, x and x', we obtain

$$\limsup F_{X_n}(x) \leq F_X(x''), \quad x < x''.$$

Therefore, for $x' < x < x''$,

$$F_X(x') \leq \liminf F_{X_n}(x) \leq \limsup F_{X_n}(x) \leq F_X(x'')$$

and, if $x \in C(F_X)$, it follows, letting $x' \uparrow x$ and $x'' \downarrow x$, that

$$F_X(x) = \lim F_{X_n}(x).$$

The same argument with X'_n in lieu of X and $x', x'' \in C(F_X)$ yields

d. *If $X_n - X'_n \xrightarrow{P} 0$ and $F_{X'_n} \to F_X$ on $C(F_X)$, then $F_{X_n} \to F_X$ on $C(F_X)$.*

Particular case. There is an important case in which convergence in pr. and convergence of d.f.'s are equivalent:

$$X_n \xrightarrow{P} c \text{ if, and only if, } F_{X_n} \to 0 \text{ or } 1 \text{ according as } x < c \text{ or } x > c.$$

Follows by **c** and **d**.

FIRST EXTENSION. Let $X = (X_1, \cdots, X_N)$ be a random vector or, equivalently, a finite class of r.v.'s $X_1, \cdots, X_N$. The *distribution* of X

is defined on the Borel field $\mathscr{B}^N$ in the N-dimensional space $R^N = \prod_{k=1}^{N} R_k$, by

$$P_X(S) = P[X \in S], \quad S \in \mathscr{B}^N.$$

As for a r.v., P_X is a pr. and the *induced* pr. space is $(R^N, \mathscr{B}^N, P_X)$. Proposition **a**, with its proof, continues to be valid: the first part holds for every finite Borel function g on R^N to some $R^{N'}$ and the second part holds for every component of g.

The *distribution function* (d.f.) F_X on R^N of X is still defined by

$$F_X(x) = P_X(-\infty, x) = P[X < x], \quad x \in R^N,$$

or, more explicitly, by

$$F_{X_1,\cdots,X_N}(x_1, \cdots, x_N) = P[X_1 < x_1, \cdots, X_N < x_N].$$

P_X determines the increment function of F_X and, conversely, by

$$P_X[a, b) = F_X[a, b) = \Delta_{b-a}F_X(a), \quad a < b \in R^N$$

or, more explicitly, by

$$P[a_1 \leqq X_1 < b_1, \cdots, a_N \leqq X_N < b_N]$$
$$= \Delta_{b_1 - a_1} \cdots \Delta_{b_N - a_N} F_{X_1,\cdots,X_N}(a_1, \cdots, a_N),$$

where $\Delta_{b_k - a_k}$, $k = 1, \cdots, N$, is the difference operator of step $b_k - a_k$ operating on a_k.

Proposition **b** and its proof, as well as the remark, remain valid, provided F_X "nondecreasing" means that $\Delta_h F_X \geqq 0$ for $h > 0$, that is, $h_1 > 0, \cdots, h_N > 0$, and $x \rightarrow -\infty$ or $x \rightarrow +\infty$ means that one at least of the $x_k \rightarrow -\infty$ or that all the $x_k \rightarrow +\infty$, respectively.

Proposition **c** and its proof remain valid, provided $X_n \overset{P}{\rightarrow} X$ means that every one of the components $X_{nk} \overset{P}{\rightarrow} X_k$, $k = 1, \cdots, N$.

*Let $X = \{X_t, t \in T\}$ be an arbitrary random function or, equivalently, an arbitrary class of r.v.'s X_t, $t \in T$. Then X *induces* the pr. space $(R^T, \mathscr{B}^T, P_X)$—its *sample pr. space*—where $R^T = \prod_{t \in T} R_t$ is the range space of X, $\mathscr{B}^T$ is the Borel field in R^T, and P_X is the *distribution* of X defined by

$$P_X(S) = P[X \in S], \quad S \in \mathscr{B}^T.$$

According to the consistency theorem, P_X determines the consistent family of the distributions $P_{X_{t_1},\cdots,X_{t_N}}$ of all finite subfamilies $(X_{t_1}, \cdots, X_{t_N})$ of the family X and, conversely, a consistent family of distributions on Borel fields of all finite subspaces $R_{t_1,\ldots,t_N}$ of R^T determines a distribution on $\mathfrak{B}^T$. Similarly, the d.f. F_X on R^T is defined by the consistent family of the d.f.'s $F_{X_{t_1},\ldots,X_{t_N}}$ of all finite subfamilies of the family X and, conversely, a consistent family of d.f.'s on all finite subspaces of R^T defines a d.f. on R^T.

REMARK. So far, the numerical functions under consideration were r.v.'s, that is, finite (or a.s. finite) measurable functions. However, the preceding definitions remain valid for nonfinite measurable functions, provided the range-spaces are extended, that is, R, R_k, $R_t = (-\infty, +\infty)$ are replaced by $\bar{R}$, $\bar{R}_k$, $\bar{R}_t = [-\infty, +\infty]$. Thus, say, R^N is replaced by $\bar{R}^N = \prod_{k=1}^{N} \bar{R}_k$ and, at the same time, $\mathfrak{B}^N$ is replaced by $\bar{\mathfrak{B}}^N$—the Borel field in $\bar{R}^N$, and P_X on $\mathfrak{B}^N$ is replaced by P_X on $\bar{\mathfrak{B}}^N$.

To fix the ideas, let X be a numerical measurable function, not necessarily finite. Since $\bar{\mathfrak{B}}$ is determined by $\mathfrak{B}$ and the sets $\{-\infty\}$ and $\{+\infty\}$, P_X on $\bar{\mathfrak{B}}$ is determined by P_X on $\mathfrak{B}$ and the values

$$P_X(-\infty) = P[X = -\infty], \quad P_X(+\infty) = P[X = +\infty].$$

In fact, P_X on $\bar{\mathfrak{B}}$ is determined by the d.f. F_X of X, defined by

$$F_X(x) = P[X < x] = P_X[-\infty, x), \quad x \in R,$$

since

$$F_X(-\infty) = \lim_{x \to -\infty} F_X(x) = P[X = -\infty] \geqq 0$$

and

$$F_X(+\infty) = \lim_{x \to +\infty} F_X(x) = P[X < +\infty] = 1 - P[X = +\infty] \leqq 1.$$

10.2   The essential feature of pr. theory. We are now in a position to describe the essential feature of pr. theory as distinct from measure theory.

While pr. concepts are born from experience and, in their rough form, are perhaps older than the measure-theoretic ones, yet their rigorous formulation was given in this chapter in terms of and by specializing the measure-theoretic concepts. Thus, it looks as if, nowadays, pr. theory were a part of measure theory or, conversely, as if measure theory were a generalized and rigorous pr. theory. Therefore, it is important to point out the basic distinction between these two interlocking branches of mathematics. The fact is that the distinction does not lie in the greater or lesser generality of the concepts, but in the properties investigated in these branches of mathematics.

Let us start with an analogy. Geometry, say, euclidean plane geometry, appears to be a part of algebra and analysis, since we can consider a point in a plane as an ordered pair (x, y) of reals or as a complex number, a straight line as a linear equation in x and y, etc. Yet, geometry remains a science *per se*, not because it has its own terminology or is older than algebra and analysis, but because geometry studies those properties of sets of points that remain invariant under all the transformations which, say, preserve the distances; for example, euclidean displacements in the case of the euclidean geometry. And geometric terminology developed, frequently unconsciously, for this specific purpose is, on the whole, well adapted to the geometrical intuition, problems, and methods.

Now, measure theory investigates families of functions on a measure space to other spaces, distinct or not from the first. On the other hand, pr. theory has developed and continues to develop the intuition, problems, and methods of its own in exploring those properties of families of functions which remain invariant under all the transformations which preserve their joint distributions—the reason being that the primary datum in random phenomena is not the pr. space but the joint distributions of the families of r.v.'s which describe the characteristics of the phenomena. Since the measurable characteristics are finite, pr. theory limited itself to r.v.'s (which, by definition, are finite). This explains the historical reason for the restrictions imposed on the measure-theoretic setup of pr. theory. However, today pr. theory is sufficiently mature mathematically to show signs of getting rid of those restrictions, by considering more general families of functions on measure spaces (normed or not) to more and more abstract spaces. We can summarize the essential feature of pr. theory as follows:

A PROPERTY IS PR.-THEORETICAL IF, AND ONLY IF, IT IS DESCRIBABLE IN TERMS OF A DISTRIBUTION.

In other words,

A property of a family of functions on a measure space is pr.-theoretical if, and only if, the property remains the same when the family is replaced by any other family with the same distribution.

In particular, since in the numerical case a distribution is represented by the corresponding d.f.'s, we can say that

—*the pr.-theoretic properties of a r.v. X are those which can be expressed in terms of its d.f. F_X,*

—the pr.-theoretic properties of a finite family $(X_1, X_2, \cdots, X_N)$ of r.v.'s are those which can be expressed in terms of the joint d.f. $F_{X_1, X_2, \cdots, X_N}$,

—the pr.-theoretic properties of any family $(X_t, t \in T)$ of r.v.'s are those which can be expressed in terms of the joint d.f.'s of its finite subfamilies.

More generally, consider a function X on a pr. space $(\Omega, \mathcal{A}, P)$ to some abstract space Ω'. The class of all sets in Ω' whose inverse images under X are events is a σ-field $\mathcal{A}'$ in Ω'; assign to $A' \in \mathcal{A}'$ the number $P'A' = P(X^{-1}A')$. This defines the induced pr. space $(\Omega', \mathcal{A}', P')$. *The pr.-theoretic properties of X are those which can be expressed in terms of P' on $\mathcal{A}'$.* If we limit ourselves to these properties only, we can speak of a "*stochastic variable*" X described by a "*pr. law*" represented by P'. Those are the mathematical beings we are concerned with, and the function X, the measure P' (or the d.f.'s in the preceding cases) are only various ways of talking about those beings in various languages. It is important to realize fully that measurements of a stochastic variable are relative to the induced pr. space; the original pr. space is but a mathematical fiction. Yet it is basic, for it permits the use of a "common frame of reference" for the families of stochastic variables we investigate—the families of sub σ-fields of events they induce on the original pr. space. However, precisely because of the existence of a common frame of reference in the present setup, modern physics forces us to introduce a different setup that we shall see in the next volume.

COMPLEMENTS AND DETAILS

Notation. Unless otherwise stated, the pr. space $(\Omega, \mathcal{A}, P)$ is fixed, the spaces L_r, $L_s(r, s > 0)$ are defined over the pr. space, and, with or without affixes, A, B, $\cdots$ denote events, while X, Y, $\cdots$ denote r.v.'s.

1. Rewrite in pr. terms as many as possible of the complements and details of Part I.

2. The convex function $\log E|X|^r$ of r is linear if, and only if, X is a degenerate r.v.

3. Liapounov's inequality. Let $\mu_r = E|X|^r$. If $r \geqq s \geqq t \geqq 0$, then $\mu_r^{s-t} \mu_s^{t-r} \mu_t^{r-s} \geqq 1$. When does this inequality become an equality? Prove Hölder's inequality by means of properties of convex functions. When does this inequality become an equality?

4. Investigate the possible behaviors of $E^{\frac{1}{r}}|X|^r$ as r varies from $-\infty$ to 0.

5. Apply Markov's inequality to $X - \dfrac{a+b}{2}$ to obtain a bound for $P[a \leqq X \leqq b]$. Also use the method of proof of the basic inequalities to obtain various bounds for this pr.

6. If g_0 on $[0, +\infty)$ is a nonnegative Borel function such that $g_0(x) \geq g_0(\epsilon)$ for $x \geq \epsilon$, then $P[|X| \geq \epsilon] \leq Eg_0(|X|)/g_0(\epsilon)$. Construct a function g on $[0, +\infty)$ with $g(0) = 0$, $g(\epsilon) = g_0(\epsilon)$, which is nondecreasing, continuous where g_0 is continuous, and such that $Eg(|X|) \leq Eg_0(|X|)$. Then the above bound is at least as sharp with g instead of g_0.

(Form $g_i(x) = \inf g(x')$ for $x' \geq x$ and $g(x) = \min (g_i(x), \frac{x}{\epsilon} g_0(x))$.)

7. Let g with $g(0) = 0$ be a continuous and nondecreasing function on $[0, +\infty)$. If there exists an $h = h(Eg(|X|), \epsilon)$ such that $P[|X| \geq \epsilon] \leq h \leq Eg(|X|)/g(\epsilon)$ for all r.v.'s X, then $h = Eg(|X|)/g(\epsilon)$ for those $\epsilon > 0$ for which the bound is of interest, that is, for which $Eg(|X|) < g(\epsilon)$. Loosely speaking, the bound $Eg(|X|)/g(\epsilon)$ is the sharpest of all bounds which depend upon $Eg(|X|)$ and ϵ.
(Take $|X| = \epsilon$ or 0 with pr. p and $q = 1 - p$ $(pq \neq 0)$, respectively.)

8. For $\epsilon > 0$ sufficiently small, the bound $E|X|^r/\epsilon^r$ is at least as sharp as the bound $E|X|^s/\epsilon^s$ with $s > r$.

9. Let
$d_0(X, Y) = \inf \{P[|X - Y| \geq \epsilon] + \epsilon\}$ for all $\epsilon > 0$;
$d_1(X, Y) = \inf \epsilon$ such that $P[|X - Y| \geq \epsilon] < \epsilon$;
$d_2(X, Y) = Eg(|X - Y|), g$ on $[0, +\infty)$ is bounded continuous and increasing with $g(0) = 0$ and $g(x + x') \leq g(x) + g(x')$; for instance, take $g(x) = \dfrac{cx}{1 + cx}$ with $c > 0$, $g(x) = 1 - e^{-x}$, or $g(x) = \tanh x$.

Each of the three functions d_0, d_1, d_2 is a metric on the space of all r.v.'s, provided equivalent r.v.'s are identified. Convergence in pr. is equivalent to convergence in any of the corresponding metric spaces.

10. (a) $\sum |X_n| < \infty$ a.s. if, and only if, the sequence of d.f.'s of consecutive sums converges to the d.f. of a r.v.

(b) If $E \sum |X_n|^r < \infty$, then $\sum |X_n|^r < \infty$ a.s.

(c) Let $s = 1$ or $\dfrac{1}{r}$ according as $r < 1$ or $r \geq 1$. If $\sum E^s|X_n|^r < \infty$, then $\sum |X_n| < \infty$ a.s.

11. $X_n \xrightarrow{P} X$ if, and only if, given $\epsilon > 0$ and $\delta > 0$, there exists $n(\epsilon, \delta)$ such that $P[|X_n - X| \geq \epsilon] < \delta$ for $n \geq n(\epsilon, \delta)$.

(a) $X_n \xrightarrow{a.s.} X$ if, and only if, given $\epsilon > 0$ and $\delta > 0$, there exists $n(\epsilon, \delta)$ such that $P[|X_n - X| \geq \epsilon$ for some $n \geq n(\epsilon, \delta)] < \delta$.

(b) $X_n \xrightarrow{u} X$ except on a null event if, and only if, given $\epsilon > 0$ there exists $n(\epsilon)$ such that $P[|X_n - X| \geq \epsilon] = 0$ for $n \geq n(\epsilon)$ or, equivalently, $P[|X_n - X| \geq \epsilon$ for some $n \geq n(\epsilon)] - 0$.

12. $P[X_n \nrightarrow X] = \lim_{\epsilon \to 0} \lim_{n \to \infty} P \bigcup_{k \geq n} [|X_k - X| \geq \epsilon]$.

(a) If $\sum P[|X_n - X| \geq \epsilon] < \infty$ for every $\epsilon > 0$, then $X_n \xrightarrow{a.s.} X$.

(b) If $\sum E|X_n - X|^r < \infty$ for some $r > 0$, then $X_n \xrightarrow{a.s.} X$.

13. $X_n \xrightarrow{a.s.} X$ if, and only if, there exists a sequence $\epsilon_n \to 0$ such that $P \bigcup_{k \geq n} [|X_k - X| \geq \epsilon_k] \to 0$. (For the "only if" assertion select $n_m \uparrow \infty$ by
$P \bigcup_{k \geq n_m} \left[|X_k - X| \geq \dfrac{1}{m}\right] < \dfrac{1}{2^m}$ and take $\epsilon_n = \dfrac{1}{m}$ for $n_m \leq n < n_{m+1}$.) Let D be the set where the sequence X_n does not converge to a finite function.

$$PD = \lim_{\epsilon \to 0} \lim_{m \to \infty} \lim_{n \to \infty} P \bigcup_{k=m}^{n} [|\, X_k - X\,| \geqq \epsilon]$$

$$PD = \lim_{\epsilon \to 0} \lim'_{m \to \infty} \lim_{n \to \infty} P \bigcup_{k=m}^{n} [|\, X_k - X_m\,| \geqq \epsilon]$$

where lim' denotes lim inf or lim sup indifferently. Can lim' be replaced by lim?

14. (a) If $\sum P[X_{n+1} - X_n \,|\, \geqq \epsilon_n] < \infty$ and $\sum \epsilon_n < \infty$, then the sequence X_n converges a.s. to a r.v.

(b) If $\sum \sup_{p} P[|\, X_{n+p} - X_n\,| \geqq \epsilon] < \infty$ for every $\epsilon > 0$, or

$$\sup_{p} P[|\, X_{n+p} - X_n\,| \geqq \epsilon] \to 0 \quad \text{and} \quad \sum \liminf_{p} P[|\, X_{n+p} - X_n\,| \geqq \epsilon] < \infty$$

for every $\epsilon > 0$, then the sequence X_n converges a.s. to a r.v. (In the last two cases, $X_n \xrightarrow{P}$ some r.v. X and $P[|\, X_n - X\,| \geqq 2\epsilon]$ is bounded by the corresponding term of each of the two series.)

15. Take $X_n = n^c$ or 0 with pr. $\frac{1}{n}$ and $1 - \frac{1}{n}$, respectively, and investigate convergences of the sequences X_n and $E|\, X_n\,|^r$ according to the choice of c and of r.

16. If $F_{X_n} \to F_X$ on $C(F_X)$ and $Y_n \xrightarrow{P} c$, then $F_{X_n+Y_n} \to F_{X+c}$ on $C(F_{X+c})$ (*Slutsky*).
What about $X_n Y_n$, X_n/Y_n and in general $g(X_n, Y_n)$ where g is continuous? (Use 10.1d.)

17. Take $X_{2n-1} = \frac{1}{n}$, $X_{2n} = -\frac{1}{n}$ and investigate the sequences X_n and F_{X_n}.

Take $X_n = 0$ or 1, each with pr. $\frac{1}{2}$, and $X = 1$ or 0, each with pr. $\frac{1}{2}$. Then $|\, X_n - X\,| = 1$ but $F_{X_n} = F$. To what converse is it a counterexample?

18. If the sequence X_n converges a.s. to a nonfinite function, what can be said about the sequence F_{X_n}?

19. Let $\{F_n\}$ be a denumerable family of d.f.'s with $F_n(-\infty) = 0$ and $F_n(+\infty) = 1$. The family of all functions $F_{n_1 \cdots n_m} = F_{n_1} \times \cdots \times F_{n_m}$ is a consistent family of d.f.'s. Construct as many pr. spaces as you can, on which are defined r.v.'s X_n such that $F_{X_{n_1}, \ldots, X_{n_m}} = F_{n_1 \cdots n_m}$ for all finite index sets.

Extend what precedes to a family $\{F_t\}$ where t ranges over an arbitrarily given set T.

20. There is no universal pr. space for all possible r.v.'s on all possible pr. spaces.

21. Extend as much as possible of this chapter and of the foregoing complements and details to complex-valued r.v.'s and to complex vectors, by suitably interpreting the symbols used.

Chapter IV

DISTRIBUTION FUNCTIONS AND
CHARACTERISTIC FUNCTIONS

§ 11. DISTRIBUTION FUNCTIONS

11.1 Decomposition. In pr. theory, a *distribution function* (*d.f.*), to be denoted by F, with or without affixes, is a nondecreasing function, continuous from the left and bounded by 0 and 1 on R. This definition entails at once that the quantities,

$$F(-\infty) = \lim_{x \to -\infty} F(x) = \inf F, \quad F(+\infty) = \lim_{x \to +\infty} F(x) = \sup F,$$

$$F(x) = F(x - 0) = \lim_{x_n \uparrow x} F(x_n) = \sup_{x' < x} F(x'),$$

$$F(x + 0) = \lim_{x_n \downarrow x} F(x_n) = \inf_{x' > x} F(x'),$$

exist and are bounded by 0 and 1, and x is a continuity or a discontinuity point of F according as $F(x + 0) - F(x - 0) = 0$ or > 0. As we have seen, a d.f. is always the d.f. of a measurable function on a pr. space, and if $F(-\infty) = 0$, $F(+\infty) = 1$, then it is the d.f. of a r.v.

The requirement of continuity from the left is of no importance, since every nondecreasing function F_1 on R bounded by 0 and 1 determines a d.f. F by setting $F(x) = F_1(x)$ or $F(x) = F_1(x - 0)$ according as x is a continuity or a discontinuity point of F_1. In fact, even less is necessary to determine a d.f.

Let D denote a set dense in R (for example, the set of all rationals) and let F_D denote a nondecreasing function on D bounded by 0 and 1. We can assume, without loss of generality, that it is continuous from the left on D. Since, for every $x \in R$, there exists a sequence $\{x_n\} \subset D$

such that $x_n \uparrow x$, $x_n < x$, it follows easily that, according to the definition of d.f.'s,

a. *The function F defined on R by*

$$F(x) = \lim_{x_n \uparrow x} F_D(x_n), \quad x_n \in D, \quad x_n < x$$

is a d.f.

It follows that, if two d.f.'s coincide on a set dense in R, they coincide everywhere. Furthermore, monotoneity of d.f.'s leads to the

A. Decomposition theorem. *Every d.f. F has a countable set of discontinuity points and determines two d.f.'s F_c and F_d such that F_c is continuous, F_d is a step-function, and $F = F_c + F_d$.*

Proof. If F has at least n discontinuity points x_k

$$a \leqq x_1 < x_2, \cdots, < x_n < b$$

in a finite interval $[a, b)$, then, from

$$F(a) \leqq F(x_1) < F(x_1 + 0) \leqq \cdots \leqq F(x_n) < F(x_n + 0) \leqq F(b),$$

it follows, setting $p(x_k) = F(x_k + 0) - F(x_k)$, that

$$\sum_{k=1}^{n} p(x_k) = \sum_{k=1}^{n} \{F(x_k + 0) - F(x_k)\} \leqq F(b) - F(a).$$

Therefore, the number of discontinuity points x in $[a, b)$ with jumps $p(x) > \epsilon > 0$ is bounded by $\dfrac{1}{\epsilon} \{F(b) - F(a)\}$. Thus, for every integer m, the number of discontinuity points with jumps greater than $\dfrac{1}{m}$ is finite and, hence, there is no more than a countable set of discontinuity points in every finite interval $[a, b)$. Since R is a denumerable sum of such intervals, the same is true of the set of all discontinuity points, and the first assertion is proved. Furthermore, denoting the discontinuity set by $\{x_n\}$, we have, for every interval $[a, b)$, finite or not,

$$\sum_{a \leqq x_n < b} p(x_n) \leqq F(b) - F(a).$$

Upon defining F_d by

$$F_d(x) = \sum_{x_n < x} p(x_n), \quad x \in R,$$

and setting $F_c = F - F_d$, it follows at once that F_d and F_c are d.f.'s.

But, for $x < x'$,

$$F_c(x') - F_c(x) = F(x') - F(x) - \sum_{x \leq x_n < x'} p(x_n)$$

$$= F(x') - F(x + 0) - \sum_{x < x_n < x'} p(x_n),$$

so that, letting $x' \downarrow x$, we obtain

$$F_c(x + 0) - F_c(x) = 0;$$

thus F_c is also continuous from the right and hence continuous.

Finally, if there are two such decompositions of F,

$$F = F_c + F_d = F'_c + F'_d,$$

then $F_c - F'_c = F'_d - F_d$, and both sides must vanish since the left-hand side is continuous while the right-hand side is discontinuous, except when it vanishes identically. This completes the proof.

REMARK. Since the discontinuity set of a d.f. is countable, its continuity set is *always* dense in R. However, the discontinuity set *can* also be dense in R. For example, let $\{r_n\}$ be the set of all rationals in R (it is dense in R); if $p(r_n) = \dfrac{6}{\pi^2} \cdot \dfrac{1}{n^2}$, then the function F defined by

$$F(x) = \sum_{r_n < x} p(r_n), \quad x \in R,$$

is a d.f. and, in fact, is the d.f. of a r.v., since $F(-\infty) = 0$ and $F(+\infty) = \dfrac{6}{\pi^2} \sum_1^\infty \dfrac{1}{n^2} = 1.$

FURTHER DECOMPOSITION. F_c determines, by $\mu_c(-\infty, x) = F_c(x) - F_c(-\infty)$, a finite measure μ_c on the Borel field $\mathfrak{B}$ in R. Upon applying to μ_c the Lebesgue decomposition theorem with respect to the Lebesgue measure on $\mathfrak{B}$ we obtain

$$\mu_c = \mu_{ac} + \mu_s, \quad \mu_{ac}(S) = \int_S g(x)\, dx, \quad S \subset \mathfrak{B},$$

where $g \geq 0$ is a Borel function and $\mu_s = 0$ on the complement of some Lebesgue-null set N_s. It follows that there are d.f.'s F_{ac} and F_s which correspond to the measures μ_{ac} and μ_s, respectively, such that

$$F_c = F_{ac} + F_s, \quad F_{ac}(x) = \int_{-\infty}^x g(x)\, dx, \quad g \geq 0,$$

and F_s is a continuous d.f. whose points of increase all lie in N_s. Thus

A′. *Every d.f. F determines three d.f.'s of which F is the sum:*
—*the step part F_d which is a step function,*
—*the absolutely continuous part F_{ac} such that*

$$F_{ac}(x) = \int_{-\infty}^{x} g(x)\,dx, \quad g \geqq 0, \quad x \in R,$$

—*the singular part F_s which is a continuous function with points of increase all belonging to a Lebesgue-null set.*

11.2   Convergence of d.f.'s.   As 10.1c and 11.1a suggest, convergence of d.f.'s to a d.f. F ought to be defined without taking into account what happens on the discontinuity set of F.

We say that a sequence F_n of d.f.'s converges *weakly* to a d.f. F and write $F_n \overset{w}{\to} F$, if $F_n \to F$ on the continuity set $C(F)$ of F. This definition is justified—that is, the weak limit, if it exists, is unique, since $F_n \overset{w}{\to} F$ and $F_n \overset{w}{\to} F'$ imply $F = F'$ on the set $C(F) \cap C(F')$ and, on the remaining set, which, by 11.1A, is countable, $F = F'$ by continuity from the left.

We say that a sequence F_n of d.f.'s converges *completely* and write $F_n \overset{c}{\to} F$, if $F_n \overset{w}{\to} F$ and $F_n(\mp\infty) \to F(\mp\infty)$. Weak convergence does not imply complete convergence. For example, given a d.f. F_0 with at least one point of increase so that $F_0(-\infty) \neq F_0(+\infty)$, let $F_n(x) = F_0(x + n)$. Then $F_n \to F_0(+\infty)$ and the weak convergence holds but not the complete convergence. However, in the case of weak convergence we have

a. Let $F_n \overset{w}{\to} F$.   *Then*

$$\limsup F_n(-\infty) \leqq F(-\infty) \leqq F(+\infty) \leqq \liminf F_n(+\infty),$$

$$\operatorname{Var} F \leqq \liminf \operatorname{Var} F_n$$

and $F_n \overset{c}{\to} F$ if, and only if, $\operatorname{Var} F_n \to \operatorname{Var} F$ or $\operatorname{Var} F_n - F_n[-a, +a] \to 0$ uniformly in n as $a \to \infty$.

For, from
$$F_n(-\infty) \leqq F_n(x) \leqq F_n(+\infty),$$
it follows that, for $x \in C(F)$,

$$\limsup F_n(-\infty) \leqq F(x) \leqq \liminf F_n(+\infty)$$

and, letting $x \to \mp\infty$ along $C(F)$, the first inequalities are proved.

Thus

$$\text{Var } F = F(+\infty) - F(-\infty) \leqq \lim \inf (F_n(+\infty) - F_n(-\infty))$$

$$= \lim \inf \text{Var } F_n,$$

and the second assertion follows from the same inequalities.

We still have to find a way to recognize whether a given sequence F_n of d.f.'s converges, weakly or completely.

b. *A sequence F_n of d.f.'s converges weakly if, and only if, it converges on a set D dense in R.*

Proof. The "only if" assertion follows from the fact that the continuity set of a d.f. is dense in R. As for the "if" assertion, let $F_D = \lim F_n$ on D. The relation of 11.1a determines a d.f. F on R. Since, for $x' < x < x''$,

$$F_n(x') \leqq F_n(x) \leqq F_n(x''),$$

it follows that, for $x', x'' \in D$

$$F_D(x') \leq \lim \inf F_n(x) \leq \lim \sup F_n(x) \leq F_D(x'').$$

Taking $x \in C(F)$ and letting $x' \uparrow x$ and $x'' \downarrow x$ along D, we obtain

$$F(x) = \lim F_n(x), \quad x \in C(F),$$

and the "if" assertion is proved.

We are now in a position to prove the basic Helly

A. WEAK COMPACTNESS THEOREM. *Every sequence of d.f.'s is weakly compact.*

We recall that (at least here) a set is *compact* in the sense of a type of convergence if every infinite sequence in the set contains a subsequence which converges in the same sense.

Proof. It suffices to show that, if F_n is a sequence of d.f.'s, then there is a subsequence which converges weakly. According to **b**, it suffices to prove that there is a subsequence which converges on a set D dense in R.

Let $D = \{x_n\}$ be an arbitrary countable set dense in R, say, the set of all rationals. All terms of the numerical sequence $F_n(x_1)$ lie between 0 and 1 and, therefore, by the Bolzano-Weierstrass compactness lemma, this sequence contains a convergent subsequence $F_{n1}(x_1)$. Similarly, the numerical sequence $F_{n1}(x_2)$ contains a convergent subsequence $F_{n2}(x_2)$ and the sequence $F_{n2}(x_1)$ converges, and so on. It follows that the "diagonal" sequence F_{nn} of d.f.'s, contained in all the subsequences $\{F_{n1}\}$, $\{F_{n2}\}$, $\cdots$, converges on D, and the proof is complete.

B. Complete compactness criterion. *A sequence F_n of d.f.'s is completely compact if, and only if, it is equicontinuous at infinity:* $\mathrm{Var}\ F_n - F_n[-a, +a) \to 0$ *uniformly in n as $a \to +\infty$.*

Proof. The "if" assertion is immediate. As for the "only if" assertion, if the F_n are not equicontinuous at infinity, then, by **a** and **A**, there exists a subsequence $F_{n'}$ which converges weakly but not completely. Note that our "complete" convergence is frequently called "weak" and our "weak" is sometimes replaced by "vague."

11.3   Convergence of sequences of integrals.   Let g denote a function continuous on R and let F, with or without affixes, denote a d.f. We intend to investigate conditions under which weak or complete convergence of a sequence F_n implies convergence of the corresponding sequence of integrals $\int g\ dF_n$, when these integrals exist. Let us observe that these integrals do not change if arbitrary constants are added to the d.f.'s. The investigation is centered upon the basic

a. Helly-Bray lemma. *If $F_n \overset{w}{\to} F$ up to additive constants, then, for every pair $a < b$ such that $F_n(a) \to F(a)$ and $F_n(b) \to F(b)$,*

$$\int_a^b g\ dF_n \to \int_a^b g\ dF.$$

Proof. Setting $g_m = \sum_{k=1}^{k_m} g(x_{mk}) I_{[x_{mk},\ x_{m,k+1})}$, where

$$a = x_{m1} < x_{m2} < \cdots < x_{m,k_m+1} = b$$

and $\Delta_m = \sup_k (x_{m,k+1} - x_{mk}) \to 0$ as $m \to \infty$, we have, according to the definition of R.-S. integrals,

$$\int_a^b g_m\ dF_n \to \int_a^b g\ dF_n, \quad \int_a^b g_m\ dF \to \int_a^b g\ dF, \quad m \to \infty.$$

Upon selecting all subdivision points x_{mk} to be continuity points of F, it follows from $F_n \overset{w}{\to} F$ that, for every m and every k, as $n \to \infty$

$$F_n[x_{mk}, x_{m,k+1}) \to F[x_{mk}, x_{m,k+1}),$$

and, hence,

$$\int_a^b g_m \, dF_n = \sum_{k=1}^{k_m} g(x_{mk}) F_n[x_{mk}, x_{m,k+1}) \to \sum_{k=1}^{k_m} g(x_{mk}) F[x_{mk}, x_{m,k+1})$$

$$= \int_a^b g_m \, dF.$$

Since

$$\int_a^b g \, dF_n - \int_a^b g \, dF$$

$$= \int_a^b (g - g_m) \, dF_n + \int_a^b g_m \, dF_n - \int_a^b g_m \, dF + \int_a^b (g_m - g) \, dF$$

and the first and last integrals on the right-hand side are bounded by $\sup\limits_{a \leq x \leq b} |g(x) - g_m(x)| \to 0$ as $m \to \infty$, the assertion follows by letting $n \to \infty$ and then $m \to \infty$.

The extensions of this lemma will be based upon the obvious inequality

$$(\text{I}) \quad \left| \int g \, dF_n - \int g \, dF \right| \leq \left| \int g \, dF_n - \int_a^b g \, dF_n \right|$$

$$+ \left| \int_a^b g \, dF - \int_a^b g \, dF_n \right| + \left| \int_a^b g \, dF - \int g \, dF \right|$$

with a and b continuity points of F, provided the integrals exist and are finite.

A. EXTENDED HELLY-BRAY LEMMA. *If* $g(\mp\infty) = 0$, *then* $F_n \xrightarrow{w} F$ *up to additive constants, implies* $\int g \, dF_n \to \int g \, dF$.

Proof. Since g is continuous and its limits as $x \to \mp\infty$ exist and are finite, g is bounded on R and the integrals $\int g \, dF_n$ and $\int g \, dF$ exist and are finite. Letting $n \to \infty$ and then $a \to -\infty$, $b \to +\infty$, it follows that, out of the three right-hand side terms in (I), the second converges to 0 by the Helly-Bray lemma, whereas the first and the third ones are bounded by $\sup\limits_{x \notin (a,b)} |g(x)| \to 0$. The assertion is proved.

B. HELLY-BRAY THEOREM. *If g is bounded on R, then $F_n \overset{c}{\to} F$ up to additive constants implies $\int g \, dF_n \to \int g \, dF$.*

Proof. Since $|g| \leq c < \infty$, the integrals exist and are finite. Letting $n \to \infty$ and then $a \to -\infty$, $b \to +\infty$, it follows that, out of the three terms on the right-hand side of (I), the second converges to 0 by the Helly-Bray lemma, whereas the first and the third ones are bounded, respectively, by

$$c\{\operatorname{Var} F_n - F_n[a, b)\} \to 0 \quad \text{and} \quad c\{\operatorname{Var} F - F[a, b)\} \to 0;$$

and the assertion follows.

REMARK. All the results of these subsections extend, without further ado, to d.f.'s F on R^N and continuous functions g on R^N, with the usual conventions for the symbols used above.

*11.4 Further extension and convergence of moments. Let g on R be continuous and F on R, with or without affixes, be a d.f. The integrals we are interested in, are finite Lebesgue-Stieltjes integrals of the form $\int g \, dF$, that is, such that $\int |g| \, dF < \infty$; they are, therefore, absolutely convergent improper Riemann-Stieltjes integrals.

We say that $|g|$ is *uniformly integrable in F_n* if, as $a \to -\infty$, $b \to +\infty$, $\int_a^b |g| \, dF_n \to \int |g| \, dF_n < \infty$ uniformly in n; in other words, given $\epsilon > 0$,

$$\int |g| \, dF_n - \int_a^b |g| \, dF_n < \epsilon$$

for $a \leq a_\epsilon$ and $b \geq b_\epsilon$ independent of n. Since $\int_a^b |g| \, dF_n$ does not decrease as $a \downarrow -\infty$ and/or $b \uparrow +\infty$, it suffices to require the foregoing conditions for some set of values of $|a|$ and b going to infinity; for example, that $\int_{|x| \geq c_m} |g| \, dF_n \to 0$ uniformly in n as $c_m \to \infty$ with $m \to \infty$.

We consider now properties of the foregoing integrals which follow from the weak convergence of d.f.'s F_n; they contain the extensions of the Helly-Bray lemma of the preceding subsection (we leave the verification to the reader).

A. Convergence theorem. *If $F_n \xrightarrow{w} F$ up to additive constants, then*

(i) $$\lim \inf \int |g| \, dF_n \geqq \int |g| \, dF$$

(ii) $\quad |g|$ *is uniformly integrable in $F_n \Rightarrow \int g \, dF_n \to \int g \, dF$*

(iii) $\int |g| \, dF_n \to \int |g| \, dF < \infty \Leftrightarrow |g|$ *is uniformly integrable in F_n.*

Proof. Let $\pm c$ be continuity points of F, and use repeatedly the Helly-Bray lemma.

(i) follows, by letting $n \to \infty$ and then $c \to +\infty$, from

$$\int |g| \, dF_n \geqq \int_{-c}^{+c} |g| \, dF_n \to \int_{-c}^{+c} |g| \, dF \to \int |g| \, dF.$$

(ii) is proved as follows:

Given $\epsilon > 0$, let $\int_{|x| \geqq c} |g| \, dF_n < \epsilon$ for $c \geqq c_\epsilon$ whatever be n. By the Helly-Bray lemma, if $c' > c$ and $\pm c'$ (like $\pm c$) are continuity points of F, then $\int_{c \leqq |x| < c'} |g| \, dF < \epsilon$ and, letting $c' \to \infty$, we have $\int_{|x| \geqq c} |g| \, dF < \epsilon$ and hence $\int |g| \, dF < \infty$. Furthermore, by taking $c \geqq c_\epsilon$ and letting $n \to \infty$ and then $\epsilon \to 0$,

$$\left| \int g \, dF_n - \int g \, dF \right| \leqq \int_{|x| \geqq c} |g| \, dF_n + \left| \int_{-c}^{+c} g \, dF_n \right.$$

$$\left. - \int_{-c}^{+c} g \, dF \right| + \int_{|x| \geqq c} |g| \, dF \to 0.$$

(iii) $\Rightarrow$ follows from

$$\int_{|x| \geqq c} |g| \, dF_n \leqq \left| \int |g| \, dF_n - \int |g| \, dF \right|$$

$$+ \int_{|x| \geqq c} |g| \, dF + \left| \int_{-c}^{+c} |g| \, dF - \int_{-c}^{+c} |g| \, dF_n \right|$$

by taking $c = c_0$ such that the second right-hand side term is less than $\epsilon/3$, then $n \geqq n_0$ such that the first and the third right-hand side terms are less than $\epsilon/3$, and finally $c_\epsilon = \max(c_0, c_1, \cdots, c_{n_0-1})$ where c_k $(k = 1, \cdots n_0 - 1)$ are such that $\int_{|x| \geq c_k} |g|\, dF_k < \epsilon$; thus,

$$\int_{|x| \geq c} |g|\, dF_n < \epsilon$$

for $c \geq c_\epsilon$ whatever be n.

(iii) $\Leftarrow$ follows by (ii) where g is replaced by $|g|$.

This proves the last assertion and terminates the proof.

Application. Let

$$m^{(k)} = \int x^k\, dF(x), \quad k = 0, 1, 2, \cdots, \quad \mu^{(r)} = \int |x|^r\, dF(x), \quad r \geq 0$$

define, respectively, the kth *moment* (if it exists) and the rth *absolute moment* of the d.f. F or, equivalently, of the finite part of a measurable function X with d.f. F; if X is a r.v., then this definition coincides with that given in 9.3. If F possesses subscripts, we affix the same subscripts to its moments.

B. Moment convergence theorem. *If, for a given $r_0 > 0$, $|x|^{r_0}$ is uniformly integrable in F_n, then the sequence F_n is completely compact and, for every subsequence $F_{n'} \xrightarrow{c} F$ and all $k, r \leq r_0$,*

$$m_{n'}{}^{(k)} \longrightarrow m^{(k)} \quad \text{finite}, \quad \mu_{n'}{}^{(r)} \longrightarrow \mu^{(r)} \quad \text{finite}.$$

Proof. According to the weak compactness theorem, there is a subsequence $F_{n'}$ and a d.f. F such that $F_{n'} \xrightarrow{w} F$. On the other hand, the uniformity condition for $|x|^{r_0}$ implies that, for every $r \leq r_0$,

$$\int_{|x| \geq c} |x|^r\, dF_{n'}(x) \leq c^{r-r_0} \int_{|x| \geq c} |x|^{r_0}\, dF_{n'}(x) \to 0 \quad \text{as} \quad c \to +\infty$$

uniformly in n', so that the uniformity condition holds for $|x|^r$. Therefore, the preceding convergence theorem applies to every sequence $m_{n'}{}^{(k)}$ and $\mu_{n'}{}^{(r)}$ with $k, r \leq r_0$. In particular, taking $r = 0$, we obtain $\operatorname{Var} F_{n'} \to \operatorname{Var} F$, so that $F_{n'} \xrightarrow{c} F$. The theorem is proved.

Corollary. *If the sequence $\mu_n{}^{(r_0+\delta)}$ is bounded for some $\delta > 0$, then the conclusion of the foregoing theorem holds.*

For $\mu_n{}^{(r_0+\delta)} \leq a < \infty$ implies that, as $c \to +\infty$,

$$\int_{|x| \geq c} |x|^{r_0}\, dF_n(x) \leq c^{-\delta} \int_{|x| \geq c} |x|^{r_0+\delta}\, dF_n(x) \leq c^{-\delta} a \to 0,$$

so that the uniformity condition holds for $|x|^{r_0}$.

This corollary yields at once the following solution of the celebrated "moment convergence problem" (Fréchet and Shohat).

C. *If, for $k \geqq k_0$ arbitrary but fixed, the sequences $m_n^{(k)} \to m^{(k)}$ finite, then these sequences converge for every value of k, and their limits $m^{(k)}$ are finite and are the moments of a d.f. F such that there exists a subsequence $F_{n'} \xrightarrow{c} F$.*

If, moreover, these limits determine F up to an additive constant, then $F_n \xrightarrow{c} F$ up to an additive constant.

It suffices to apply the foregoing corollary and to observe that, if the $m^{(k)}$ determine F up to an additive constant, then all completely convergent subsequences $F_{n'}$ have the same limit d.f. F up to additive constants.

*11.5. Discussion. A d.f. F determined up to additive constants corresponds biunivoquely to an interval function F determined by $F[a, b) = F(b) - F(a)$ which in turn corresponds biunivoquely to a measure F on the Borel field in R (4.4a)—a *subprobability* (*subpr.*) since $F(R) \leqq 1$.

Weak convergence of d.f.'s F_n to F—all determined up to additive constants, is equivalent to convergence of interval functions defined by $F_n[a, b) \to F[a, b)$ for every F-continuity interval $[a, b)$, that is with $F\{a\} + F\{b\} = 0$, and we can still write $F_n \xrightarrow{w} F$. The above appearance of subpr.'s permits to extend propositions in 11.3 and 11.4 to noncontinuous functions g. Since these propositions derive from Helly–Bray lemma 11.3a, it will suffice to generalize it and the others will follow as before. Denote by D_g the set of discontinuities of a function g on R to R; it is a Borel set (see §12). If $F(D_g) = 0$ we say that g is F—*a.e. continuous*.

a. Generalized Helly–Bray lemma. *If $F_n \xrightarrow{w} F$ then $\int_a^b g \, dF_n \to \int_a^b g \, dF_n$ for every F-continuity interval $[a, b)$ and every F-a.e. continuous function g bounded on every bounded interval.*

Proof. The method of proof of the Helly–Bray lemma in 11.3 applies but for one necessary change due to the fact that our integrals are now Lebesgue-Stieltjes ones so that instead of Riemann sums we use Darboux sums: Instead of g_m we need $\underline{g}_m$ and $\bar{g}_m$ defined by

$$\underline{g}_m = \sum_{k=1}^{k_m} \underline{g}_{mk} I_{mk}, \quad \bar{g}_m = \sum_{k=1}^{k_m} \bar{g}_{mk} I_{mk},$$

where I_{mk} are indicators of F-continuity intervals $J_{mk} = [x_{mk}, x_{m,k+1})$ of length $|J_{mk}|$ with $\sum_{k=1}^{k_m} J_{mk} = [a, b)$, $\sup_k |J_{mk}| \to 0$ as $m \to \infty$, and where

$$\underline{g}_{mk} = \inf\{g(x): x \in J_{mk}\}, \quad \bar{g}_{mk} = \sup\{g(x): x \in J_{mk}\}.$$

Since as $n \to \infty$, by hypothesis, $F_n(J_{mk}) \to F(J_{mk})$ so that

$$\int_a^b \underline{g}_m \, dF_n \to \int_a^b \underline{g}_m \, dF, \quad \int_a^b \bar{g}_m \, dF_n \to \int_a^b \bar{g}_m \, dF,$$

while $F(D_g) = 0$ implies that F-a.e., as $m \to \infty$,

$$\underline{g}_m \uparrow g \downarrow \bar{g}_m;$$

letting $n \to \infty$ then $m \to \infty$ in

$$\int_a^b \underline{g}_m \, dF_n \leqq \int_a^b g \, dF_n \leqq \int_a^b \bar{g}_m \, dF_n,$$

it follows that

$$\int_a^b g \, dF_n \to \int_a^b g \, dF.$$

The lemma is proved.

So far we considered only numerical functions g. But all propositions in 11.3 and 11.4 as well as the one above remain valid for complex valued $g = \Re g + i \Im g$ by, say, $\int g \, dF = \int (\Re g) \, dF + i \int (\Im g) \, dF$. In fact, then, the inverses of the Helly–Bray lemma and of the Helly–Bray theorem are valid because of the weak and complete convergence criteria in 13.2. We shall leave these immediate extensions to the reader.

Several questions arise at once: Since Borel fields are generated by the class of open (of closed) sets, are subpr.'s determined by their values on such a class? Is weak convergence determined by the behaviour of subpr.'s on open (on closed) sets? Since weak and complete convergence are determined by convergence of integrals of some families of functions are there other such families?

It will be convenient to discuss these questions for subpr.'s on Borel fields of metric spaces. First, because this generality is needed for "functional limit theorems" (see Chapter XII) and second, because the proofs are not more involved than for the real line. However, this generality creates two difficulties: First, we do not have intervals in general metric spaces hence no interval functions and are reduced to work directly with subpr.'s. Second, nontrivial continuous functions g vanishing at infinity (that is, such that given $\epsilon > 0$ there is á compact K with $|g| < \epsilon$ on K^c) may not exist.   In fact, on the separable Banach space $C[a, b]$ of continuous functions on $[a, b]$ to R with the supremum norm, the only continuous function vanishing at infinity is the zero function.   Or this space is central to Ch. XII.   Thus the extended Helly–Bray lemma is useless.   However, the Helly–Bray theorem, with integrals of bounded continuous functions, with respect to subpr.'s μ_n, μ, remains meaningful.   But, in the case of the real line, it corresponds to complete convergence $\mu_n \overset{c}{\to} \mu$ or, equivalently, weak convergence of pr.'s $\mu_n/\mu_n(R)$ to a pr. $\mu/\mu(R)$ (excluding the trivial case of $\mu(R) = 0$).   Thus, in the general case we are led to consider only weak convergence of pr.'s to a pr. and the corresponding "relative compactness": As is easily seen, 11.2b implies that a sequence of pr.'s F_n on R contains a subsequence which converges weakly to a pr. if and only if for every $\epsilon > 0$ there is a compact K_ϵ in R with $F_n(K_\epsilon^c) < \epsilon$ for all n.   Is there a similar criterion for metric spaces?   Answers to the foregoing questions are to be found in the next section.

*§ 12.   CONVERGENCE OF PROBABILITIES ON METRIC SPACES

Throughout this section and unless otherwise stated, with or without affixes

1. $\mathfrak{X}$ is a space with metric d and Borel field $\mathcal{S}$ generated by the class of its open (of its closed) sets, U, C, K are its open, closed, compact sets, respectively, and $\partial A = \bar{A} - A^\circ$ is the boundary of a set A in $\mathfrak{X}$.   Properties of metric spaces in 5.3 are to be used without further comment.

2. P is a pr. on $\mathcal{S}$ and A in $\mathfrak{X}$ is a P-continuity set when $P(\partial A) = 0$, g, h are Borel functions on the Borel space $(\mathfrak{X}, \mathcal{S})$ to the Borel line or Borel space $(\mathfrak{X}', \mathcal{S}')$, respectively. D_g is the discontinuity set of g and g is P-a.e. continuous when $P(D_g) = 0$; similarly for h.   If $g = I_A$ then clearly $D_g = \partial A$.   Note that for any function h on $(\mathfrak{X}, d)$ to $(\mathfrak{X}', d')$, D_h is a Borel set, since $D_h = \underset{r}{\cup} \underset{s}{\cap} D_{rs}$ where r and s vary over the rationals and D_{rs} are the open sets

$$D_{rs} = \{x: d(x, y) < s, \quad d(x, z) < s, \quad d'(h(y), h(z)) \geq r\}.$$

For later use, we observe that except for a change of notation the same proof as for 10.1a yields

CHANGE OF VARIABLE FORMULA. *Let P on $\mathcal{S}$ be a pr. Let h be a Borel function on $\mathcal{X}$ to $\mathcal{X}'$ and g be a Borel function on $\mathcal{X}'$ to R. The distribution Ph^{-1} of h defined by $Ph^{-1}(A') = P(h^{-1}(A'))$, $A' \in \mathcal{S}'$-Borel field in $\mathcal{X}'$ determines the distribution of random variables $g(h)$, and*

$$\int g(h) \, dP = \int_{x'} g \, d(Ph^{-1}),$$

in the sense that if either integral exists so does the other one and then both are equal.

The main concepts and results of this section originated with Alexandrov and their final form is primarily due to Prohorov.

*12.1 Convergence. The basic theorem below is essentially due to Alexandrov. Any of its six equivalent properties defines *weak convergence on $\mathcal{S}$ of pr.'s P_n to a pr. P*, and we write $P_n \overset{w}{\to} P$. The usual definition is

(ii): $\int g \, dP_n \to \int g \, dP$ for all bounded continuous functions g. Since $1 = P_n(\mathcal{X}) \to P(\mathcal{X}) = 1$, *this "weak" convergence is in fact complete convergence.*

A.   CONVERGENCE CRITERIA. *Let P_n, P be pr.'s on the Borel field $\mathcal{S}$ of a metric space $(\mathcal{X}, d)$. Let g be functions on $\mathcal{X}$ to R and the integrals be over $\mathcal{X}$.*

The following six properties are equivalent and define $P_n \overset{w}{\to} P$:
I:

$$\int g \, dP_n \to \int g \, dP$$

  (i) *for all bounded P-a.e. continuous g*
  (ii) *for all bounded continuous g*
  (iii) *for all bounded uniformly continuous g*
II:
  (iv) limsup $P_n C \leqq PC$ *for all closed sets C*
  (v) liminf $P_n U \geqq PU$ *for all open sets U*
  (vi) $P_n A \to PA$ *for all P-continuity sets A*

Proof. Clearly (i) $\Rightarrow$ (ii) $\Rightarrow$ (iii).

(iii) $\Rightarrow$ (iv): The function g_m defined by $g_m(x) = e^{-md(x,C)}$ is bounded by 1 and uniformly continuous with $I_C \leqq g_m \downarrow I_C$ as $m \to \infty$. Thus $P_n C \leqq \int g_m \, dP_n$ and, by Fatou–Lebesgue theorem, as $n \to \infty$ then $m \to \infty$,

$$\text{limsup } P_n C \leqq \int g_m dP \to PC.$$

(iv) $\Rightarrow$ (v): The two properties are dual: each implies the other one by complementation.

(v) $\Rightarrow$ (vi): Since (iv) and (v) are equivalent, by using both and the fact that $A° \subset A \subset \bar{A}$, we obtain
$$PA° \leqq \liminf P_nA° \leqq \liminf P_nA \leqq \limsup P_nA \leqq \limsup P_n\bar{A} \leqq P\bar{A}.$$
Since $P(\bar{A} - A°) = P(\partial A) = 0$ by hypothesis in (vi), we have $PA° = P\bar{A}$ so that in the above inequalities the extreme terms hence all the terms coincide and $PA = \lim PA_n$.

(vi) $\Rightarrow$ (i): The method of proof of Helly–Bray lemma in 11.5 still applies but with another necessary change due to the fact that it is the range space, and not the domain, of g which is R. The sets $g^{-1}(c) = \{x: g(x) = c\}$ are disjoint for distinct $c \in R$. Since $P(\mathfrak{X})$ is finite, it follows that $P(g^{-1}(c)) > 0$ only for a countable set of values of c. Since g is bounded there is a bounded interval $[a, b)$ with $g(\mathfrak{X}) \subset [a, b)$. We can take $a = x_{m1} < \ldots < x_{m,k_m+1} = b \notin D$ with no $x_{mk} \in D$ for $k \leqq k_m$, $m = 1, 2, \ldots$, and $\max_k(x_{m,k+1} - x_{mk}) \to 0$ as $m \to \infty$.

Let I_{mk} be indicators of the $J_{mk} = g^{-1}[x_{mk}, x_{m\ k+1})$, omit the empty J_{mk}, set $g_{mk} = \inf\{g(x): x \in J_{mk}\}$, $\bar{g}_{mk} = \sup\{g(x): x \in J_{mk}\}$, and
$$\underline{g}_m = \sum_k \underline{g}_{mk} I_{mk}, \quad \bar{g}_m = \sum_k \bar{g}_{mk} I_{mk}.$$
Since, by (vi), $P_n(J_{mk}) \to P(J_{mk})$, it follows that, as $n \to \infty$,
$$\int \underline{g}_m\, dP \leftarrow \int \underline{g}_m\, dP_n \leqq \int \bar{g}_m\, dP_n \to \int \bar{g}_m\, dP,$$
while $P(D_g) = 0$, by hypothesis in (i), implies that, as $m \to \infty$, P-a.e.
$$\underline{g}_m \uparrow g \downarrow \bar{g}_m.$$
Therefore, letting $n \to \infty$ then $m \to \infty$ in
$$\int \underline{g}_m\, dP_n \leqq \int g\, dP_n \leqq \int \bar{g}_m\, dP_n,$$
we obtain (i):
$$\int g\, dP_n \to \int g\, dP.$$
The proof is terminated.

COROLLARY 1. *If* $P_n \overset{w}{\to} P$ *then* $P_nh^{-1} \overset{w}{\to} Ph^{-1}$ *for every P-a.e. continuous h on* $\mathfrak{X}$ *to* $\mathfrak{X}'$, *equivalently* $\int g(h)\, dP_n \to \int g\, d(P_nh^{-1})$ *for all bounded continuous g on* $\mathfrak{X}'$ *to R.*

For, $PH_h = 0$ and $\overline{(h^{-1}C)} \subset (h^{-1}C) \cup D_h$ for every closed C, imply $P\overline{(h^{-1}C)} = P(h^{-1}C)$ hence

limsup $P_n(h^{-1}C) \leq$ limsup $P_n\overline{(h^{-1}C)} \leq P\overline{(h^{-1}C)} = P(h^{-1}C)$ and, by A(iv), $P_n h^{-1} \xrightarrow{w} Ph^{-1}$. The equivalence assertion results at once from the change of variable formula by A(ii).

COROLLARY 2.   *If $P_n \to P$ on $\mathcal{C} \subset \mathcal{S}$ where $\mathcal{C}$ is closed under finite intersections and each open set is a countable union of members of $\mathcal{C}$, then $P_n \xrightarrow{w} P$.*

Proof.   Let $U = \bigcup_{k=1}^{\infty} A_k$, $A_k \in \mathcal{C}$.   By hypothesis,

$$P_n(A_1 \cup A_2)$$
$$= P_n(A_1) + P_n(A_2) - P_n(A_1 A_2) \to P(A_1) + P(A_2) - P(A_1 A_2)$$
$$= P(A_1 \cup A_2)$$

and, by induction, for every integer m,

$$P_n(A_1 \cup \cdots \cup A_m) \to P(A_1 \cup \cdots \cup A_m).$$

Since $U_m = \bigcup_{k=1}^{m} A_k \uparrow U$ as $m \to \infty$, there is an $m = m_\epsilon$ such that $PU - \epsilon \leq PU_m$.   Therefore,

$$PU - \epsilon \leq PU_m = \lim_n P_n U_m \leq \liminf P_n U$$

and, letting $\epsilon \downarrow 0$,

$$\liminf P_n U \geq PU$$

so that $A(v)$ holds, and $P_n \xrightarrow{w} P$.

COROLLARY 3.   *Let $\mathfrak{X}$ be separable and let $P_n \to P$ on $\mathcal{C} \subset \mathcal{S}$.   Then $P_n \xrightarrow{w} P$ if*

(i) $\mathcal{C}$ is closed under finite intersections and, given $\epsilon > 0$ and open U, for every $x \in U$ there is an $A \in \mathcal{C}$ with $x \in A^\circ \subset A \subset U$.
or

(ii) $\mathcal{C}$ consists of those finite intersections of open spheres which are P-continuity sets.

Proof.   (i): Since $\mathfrak{X}$ is separable, given open U, there is a sequence (A_n) in $\mathcal{C}$ with $U = \bigcup_n A_n^\circ$ and $A_n \subset U$ so that $U = \bigcup A_n$ andCorollary 2 applies.

Note that the second condition on $\mathcal{C}$ in (i) is implied by: for every x and every $r > 0$ there is an $A \in \mathcal{C}$ with

$$x \in A^\circ \subset A \subset S_x \ (r)\text{-open } r\text{- sphere about } x.$$

(ii): Since $\partial(AB) \subset \partial A \cup \partial B$ while $\partial S_x \ (r) \subset \{x : d(x, y) = r\}$ has P-measure 0 except for countably many values of r, (i) applies, and the proof is terminated.

*12.2   Regularity and tightness. Since the Borel field of the metric space $\mathcal{X}$ is generated by the class of open (of closed) sets, it is to be expected that a *pr.* P on $\mathcal{S}$ would be determined by its restriction to such a class.

a. REGULARITY LEMMA. *Every pr.* P *on* $\mathcal{S}$ *is regular: given* $A \in \mathcal{S}$ *and* $\epsilon > 0$, *there are open* U_ϵ *and closed* C_ϵ *such that*

$$C_\epsilon \subset A \subset U_\epsilon \text{ and } P(U_\epsilon - C_\epsilon) < \epsilon,$$

equivalently,

$$PA = \sup_{C \subset A} PC = \inf_{U \supset A} PU.$$

Proof. The equivalence assertion is immediate. To prove the ϵ-assertion, let $\mathcal{C} \subset \mathcal{S}$ be the subclass of those Borel sets for which the assertion holds.

$\mathcal{C}$ contains the class of closed sets C since open $U_r = \{x : d(x, C) < r\} \downarrow$ C as $r \downarrow 0$. It is clearly closed under complementations. Also it is closed under countable unions: Given $A_n \in \mathcal{C}$ and $\epsilon > 0$, there are $C_n \subset A \subset U_n$ with $P(U_n - C_n) < \epsilon/2^{n+1}$; take $U_\epsilon = \bigcup U_n$ and $C_\epsilon = \bigcup_{n \leq m} C_n$ with m such that $P(\bigcup C_n - C_\epsilon) < \epsilon/2$, so that $C_\epsilon \subset A \subset U_\epsilon$ and $P(U_\epsilon - C_\epsilon) < \epsilon$. Thus $\mathcal{C} \subset \mathcal{S}$ is a σ-field containing the class of closed sets hence $\mathcal{C} = \mathcal{S}$.

COROLLARY. *The set* $\left\{ \int g \ dP : g \text{ bounded uniformly continuous} \right\}$ *determines* P.

For, the functions g_m defined by $g_m(x) = e^{-md(x, C)}$ are bounded and uniformly continuous with $g_m = 1$ on C and $g_m \downarrow 0$ on C^c as $m \to \infty$, so that $\int g_m \ dP \to PC$.

The concept of "tightness" below was named by Le Cam in a memoir which followed within a year that of Prohorov and extended the whole theory to much more general topological spaces than the metric ones.

A family $\mathcal{P}$ of pr.'s on $\mathcal{S}$ is said to be *tight* if for every $\epsilon > 0$ there is a compact K_ϵ such that $PK_\epsilon{}^c < \epsilon$ for all $P \in \mathcal{P}$. We say that $\mathcal{P}$ *lives* on a Borel set $\mathfrak{X}_0$ if $P(\mathfrak{X}_0) = 1$ for all $P \in \mathcal{P}$, equivalently, if $PA = PA\mathfrak{X}_0$ for every $A \in \mathcal{S}$ and for all $P \in \mathcal{P}$. If $\mathcal{P} = \{P\}$ is a singleton we replace above the family $\mathcal{P}$ by the *pr. P*. Given a Borel set $\mathfrak{X}_0$, the σ-field $\mathcal{S}_0 = \{A : A \subset \mathfrak{X}_0, A \in \mathcal{S}\}$ is the Borel field of the metric space $\mathfrak{X}_0$ with its relative topology. Thus the above definitions apply to families of *pr.'s* on $\mathcal{S}_0 \subset \mathcal{S}$.

b. TIGHTNESS LEMMA. (i) *If a pr. P on $\mathcal{S}$ is tight then it lives on a σ-compact $\mathfrak{X}_0$ and $PA = \sup\limits_{K \subset A} PK$ for every $A \in \mathcal{S}$.*

(ii) *Conversely, if P on $\mathcal{S}$ lives on a σ-compact $\mathfrak{X}_0$ or if $PA = \sup\limits_{K \subset A} PK$ for every $A \in \mathcal{S}$ then P is tight.*

(iii) *Every pr. P on $\mathcal{S}$ is tight when $\mathfrak{X}$ is separable and complete.*

Proof. 1°. If P is tight then for every n there is a compact K_n with $PK_n{}^c < 1/n$, so that $P(\bigcap K_n{}^c) = 0$ and P lives on the σ-compact $\mathfrak{X}_0 = \bigcup K_n$. Note that $\mathfrak{X}_0$ is separable since compacts in metric spaces are separable.

By **a**, P is regular so that, given $A \in \mathcal{S}$ and $\epsilon > 0$, there is a closed $C \subset A$ with $P(A - C) < \epsilon/2$. But for n sufficiently large, $PK_n{}^c < \epsilon/2$ and $K_\epsilon = CK_n$ is compact with $K_\epsilon \subset C \subset A$. Since

$$P(A - K_\epsilon) \leqq P(A - C) + P(C - K_\epsilon) < \epsilon/2 + P(\mathfrak{X} - K_\epsilon) < \epsilon$$

and $\epsilon > 0$ is arbitrarily small, it follows that $PA = \sup\limits_{K \subset A} PK$, and (i) is proved.

Conversely, if P lives on $\mathfrak{X}_0 = \bigcup K_n$, that is, $P(\bigcup K_n) = 1$ then, given $\epsilon > 0$, there is an m such that $PK_\epsilon{}^c < \epsilon$ for compact $K_\epsilon = \bigcup\limits_{n \leqq m} K_n$, and P is tight. This proves the first assertion in (ii) and the second is immediate.

2°. When $\mathfrak{X}$ is separable then, for every n, open $1/n$-spheres U_{n1}, $U_{n2}, \cdots$ cover $\mathfrak{X}$. Therefore, given a pr. P on $\mathcal{S}$ and $\epsilon > 0$, for k_n sufficiently large $PU_n{}^c < \epsilon/2^{n+1}$ with $U_n = \bigcup\limits_{k \leqq k_n} U_{nk}$. When moreover $\mathfrak{X}$ is complete then the closure K_ϵ of the totally bounded set $\bigcap U_n$ is compact. Since

$$PK_\epsilon{}^c \leqq P(\bigcup U_n{}^c) < \sum \epsilon/2^{n+1} = \epsilon$$

P is tight and (iii) is proved.

A. TIGHTNESS THEOREM. *Let the family $\mathcal{P}$ of pr.'s on $\mathcal{S}$ be tight. Then (i) $\mathcal{P}$ lives on a σ-compact set $\mathfrak{X}_0$ and $PA = \sup\limits_{K \subset A} PK$ for every $A \in \mathcal{S}$.*

(ii) *The family* $\mathcal{P}h^{-1} = \{Ph^{-1}: P \in \mathcal{P}\}$ *is tight for every continuous function h on* $\mathcal{X}$ *to a metric space* $\mathcal{X}'$.

Proof. The proof of (i) is exactly the same as that of **b**(i); it suffices to observe that the compacts K_n therein are the same for all $P \in \mathcal{P}$. Note that, in general, **b**(ii) does not hold for families $\mathcal{P}$.

For (ii), given $\epsilon > 0$ there is a compact K_ϵ with $PK_\epsilon^c < \epsilon$ for all $P \in \mathcal{P}$. Since h on $\mathcal{X}$ to $\mathcal{X}'$ is continuous, $K_\epsilon' = h(K_\epsilon)$ is compact in $\mathcal{X}'$ and $K_\epsilon \subset h^{-1}(K_\epsilon')$ implies that for all $P \in \mathcal{P}$

$$Ph^{-1}(K_\epsilon')^c = P(h^{-1}(K_\epsilon')^c) = P(h^{-1}K_\epsilon')^c \leq PK_\epsilon^c < \epsilon,$$

and $\mathcal{P}h^{-1}$ is tight. The proof is terminated.

Let $\mathcal{S}_0$ be the σ-field of Borel sets on a Borel set $\mathcal{X}_0 \subset \mathcal{X}$. Given a family $\mathcal{P}^0$ of pr.'s on $\mathcal{S}_0$, its *extension* to $\mathcal{S}$ is defined by

$$\mathcal{P} = \{P: PA = P^0(A\mathcal{X}_0), P^0 \in \mathcal{P}^0, A \in \mathcal{S}\};$$

note that $P\mathcal{X}_0 = 1$ for all $P \in \mathcal{P}$,

COROLLARY. (i) *If* $\mathcal{P}^0$ *on* $\mathcal{S}_0$ *is tight so is its extension* $\mathcal{P}$ *to* $\mathcal{S}$.

(ii) *If* $P_n^0 \overset{w}{\to} P^0$ *on* $\mathcal{S}_0$ *then their extensions* $P_n \overset{w}{\to} P$ *on* $\mathcal{S}$.

For, upon taking h to be the (continuous) identity mapping of $\mathcal{X}_0$ into $\mathcal{X}$, **A**(ii) yields (i) and 12.1**A** Corollary 1 yields (ii).

*12.3   Tightness and relative compactness. We say that a family $\mathcal{P}$ of pr.'s on $\mathcal{S}$ is *relatively compact* if every sequence of members of $\mathcal{P}$ contains a subsequence which converges weakly to a pr. on $\mathcal{S}$. Thus "relative compactness" is, in fact, relative sequential complete compactness. Prohorov theorem below is the second basic theorem of this section.

A. RELATIVE COMPACTNESS CRITERION. *Let* $\mathcal{X}$ *be a separable complete metric space. Then a family* $\mathcal{P}$ *of pr.'s on its Borel field* $\mathcal{S}$ *is relatively compact if and only if* $\mathcal{P}$ *is tight. In fact, the "if" part holds for general metric spaces* $\mathcal{X}$.

Proof. 1°. Let $\mathcal{P}$ be relatively compact. Since $\mathcal{X}$ is separable for every $r > 0$, there are open r-spheres $U_1, U_2, \cdots$ which cover $\mathcal{X}$ so that $V_n = U_1 \cdots U_n \uparrow \mathcal{X}$. Given $\epsilon > 0$, there is an n such that $PV_n^c < \epsilon$ for all $P \in \mathcal{P}$: Otherwise, for every n there is some $P_n \in \mathcal{P}$ with $P_nV_n \leq 1 - \epsilon$ and, by relative compactness, the sequence (P_n) contains a subsequence $P_{n'} \overset{w}{\to}$ some pr. P on $\mathcal{S}$; thus, by 12.1 **A**(v), for every n

$$PV_n \leqq \liminf_{n'} P_{n'} V_n \leqq \liminf_{n'} P_{n'} V_{n'} \leqq 1 - \epsilon,$$

while $PV_n \uparrow 1$—contradiction.

2°. For the "if" part we follow Billingsley who bypasses Prohorov's use of integral representation of linear functionals by hewing closely to Halmos' generation of Borel measures from "content" to "inner content" to "outer extension." The difference is that "content" is defined by Halmos on the class of all compacts while here the corresponding set function has the same properties but only on a subclass of compacts.

For the time being, assume that $\mathfrak{X}$ is separable so that it has a countable base of open spheres $U_1, U_2, \cdots$; include $\mathfrak{X}$ in this base.

Let $\mathcal{P}$ be tight so that for every n there is a compact $K(n)$ with $P(K(n))^c < 1/n$ for all $P \in \mathcal{P}$. Let $\mathcal{K}$ consist of all finite unions of sets of the form $\bar{U}_m K(n)$. Thus the class $\mathcal{K}$ is countable, closed under finite unions, and its members—to be denoted by K with or without affixes, are compact.

Given a sequence (P_n) of members of $\mathcal{P}$, Cantor's diagonal procedure yields a subsequence $P_{n'} \to$ some λ on $\mathcal{K}$. We have to prove that $P_{n'} \overset{w}{\to}$ some pr. P on $\mathcal{S}$. Let

$$\lambda_0 U = \sup_{K \subset U} \lambda K, \quad \lambda^0 A = \inf_{U \supset A} \lambda_0 U,$$

so that λ is defined on $\mathcal{K}$, λ_0 on the class $\mathcal{U}$ of open sets U, and λ^0 on the class of all subsets. We shall show that the restriction of λ^0 to $\mathcal{S}$ is precisely the pr. P.

Clearly, λ on $\mathcal{K}$ is nondecreasing, additive, and subadditive: $K_1 \subset K_2 \Rightarrow \lambda K_1 \leqq \lambda K_2, \lambda(K_1 + K_2) = \lambda K_1 + \lambda K_2, \lambda(K_1 \cup K_2) \leqq \lambda K_1 + \lambda K_2$, λ_0 and λ^0 are nondecreasing, and $\lambda^0 = \lambda_0$ on $\mathcal{U}$. We shall use these properties without further comment.

3°. λ_0 on $\mathcal{U}$ *is σ-subadditive*:
Let $K \subset U_1 \cup U_2$ and set

$$C_1 = \{x \in K : d(x, U_1^c) \geqq d(x, U_2^c)\},$$

$$C_2 = \{x \in K : d(x, U_2^c) \geqq d(x, U_1^c)\}.$$

These closed sets, being contained in compact K, are compact and so are $C_1 U_1^c$ and $C_2 U_2^c$. If $x \in C_1 U_1^c \neq \emptyset$ belongs to U_2, then $d(x, U_1^c) = 0 < d(x, U_2^c)$ hence $x \notin C_1$—contradiction. Thus $C_1 \subset U_1$ and, by definition of $\mathcal{K}$, $C_1 \subset K_1 \subset U_1$ for some K_1; similarly $C_2 \subset K_2 \subset U_2$.

Therefore, upon taking the supremum in K in

$$\lambda K \leqq \lambda(K_1 \cup K_2) \leqq \lambda K_1 + \lambda K_2 \leqq \lambda_0 U_1 + \lambda_0 U_2,$$

we obtain $\lambda_0 \, (U_1 \cup U_2) \leqq \lambda_0 U_1 + \lambda_0 U_2$, so that λ_0 on $\mathfrak{U}$ is subadditive and, by induction, is finitely subadditive. Now, if $K \subset \bigcup U_n$ then, by compactness, $K \subset V_m = \bigcup_{n \leqq m} U_n$ for some m. Therefore, upon taking the supremum in K in

$$\lambda K \leqq \lambda_0 V_m \leqq \sum_{m \leqq n} \lambda_0 U_n \leqq \sum_{n} \lambda_0 U_n,$$

we obtain $\lambda_0(\bigcup_n U_n) \leqq \sum_n \lambda_0 U_n$ so that λ_0 on $\mathfrak{U}$ is σ-subadditive.

For closed C and open U, $\lambda_0 U \geqq \lambda_0 UC + \lambda_0 UC^c$: Given $\epsilon > 0$, there is a $K_1 \subset UC^c$ with $\lambda K_1 > \lambda_0 UC^c - \epsilon/2$, and then there is a $K_2 \subset UK_1^c$ with $\lambda K_2 > \lambda UK_1^c - \epsilon/2$. Since K_1 and K_2 are disjoint and contained in U,

$$\lambda_0 U \geqq \lambda(K_1 + K_2) = \lambda K_1 + \lambda K_2 > \lambda_0(UC^c)$$
$$+ \lambda_0(UK_1^c) - \epsilon \geqq \lambda_0(UC^c) + \lambda_0(UC) - \epsilon$$

hence, letting $\epsilon \to 0$, the assertion is proved.

4°. *λ_0 is an outer measure and Borel sets are λ_0-measurable*:
Given $\epsilon > 0$ and $A_n \subset \mathfrak{X}$ there are $U_n \supset A_n$ with $\lambda_0 U_n < \lambda^0 A_n + \epsilon/2^{n+1}$. Since λ_0 is σ-subadditive,

$$\lambda^0(\bigcup_n A_n) \leqq \lambda_0 \, (\bigcup_n U_n) \leqq \sum_n \lambda_0 U_n < \sum_n \lambda^0 A_n + \epsilon$$

so that, letting $\epsilon \to 0$, λ^0 is σ-subadditive. Since λ^0 is also nondecreasing, λ^0 is an outer measure. Furthermore, for closed C and open $U \supset A$, upon taking the infimum in U in

$$\lambda_0 U \geqq \lambda_0 UC + \lambda_0 UC^c \geqq \lambda^0(AC) + \lambda^0(AC^c),$$

we obtain

$$\lambda^0 A \geqq \lambda^0(AC) + \lambda^0(AC^c),$$

so that closed sets are λ^0-measurable. Therefore, the Borel field $\mathcal{S}$ (that the class of closed sets generates) is contained in the σ-field of λ^0-measurable sets.

5°. Let P be the restriction of λ^0 to $\mathcal{S}$, so that P on $\mathcal{S}$ is a measure; in fact, P is a pr. since

$$1 \geqq P\mathfrak{X} = \lambda_0 \mathfrak{X} = \sup_n \lambda(K(n)) \geqq \sup_n \left(1 - \frac{1}{n}\right) = 1.$$

Since for all open U

$$PU = \lambda_0 U = \sup_{K \subset U} \lambda K,$$

upon taking the supremum in $K \subset U$ in

$$\lambda K = \lim_{n'} P_{n'} K \leqq \liminf_{n'} P_{n'} U,$$

we obtain

$$PU \leqq \liminf P_{n'} U.$$

Thus, by 12.1 **A**(v), $P_{n'} \overset{w}{\to} P$ and the "if" part is proved but under the restriction of separability of $\mathfrak{X}$.

Now, let $\mathfrak{X}$ be a general metric space. By 12.2 **A**, $\mathcal{P}$ on $\mathcal{S}$, being tight, lives on a σ-compact $\mathfrak{X}_0$—a separable metric space in its relative topology. Thus what precedes applies to the restriction $\mathcal{P}^0$ of $\mathcal{P}$ to the Borel field $\mathcal{S}_0$ of $\mathfrak{X}_0$. But, by 12.2A Corollary (ii), $P_{n'}{}^0 \overset{w}{\to} P^0$ on $\mathcal{S}_0$ implies $P_{n'} \overset{w}{\to} P$ on $\mathcal{S}$. The proof is terminated.

CoROLLARY. *Let $\mathfrak{X}$ be separable and complete. Then $\mathcal{P}$ on $\mathcal{S}$ is relatively compact if and only if, for every $\epsilon > 0$ and $r > 0$, there is a finite union V_n of r-open spheres with $PV_n{}^c < \epsilon$.*

§ 13. characteristic functions and distribution functions

Pr. properties are properties describable in terms of distributions— and those are set functions. The introduction of d.f.'s makes it possible to describe pr. properties in terms of point functions, easier to handle with the tools of classical analysis. Yet, to a distribution corresponds not a single d.f. F but the family of all functions $F + c$ where c is an arbitrary constant. The selection of one of them is somewhat arbitrary, and we have constantly to bear this fact in mind. The introduction of characteristic functions (ch.f.) assigned to the family $F + c$ by the relation

$$f(u) = \int e^{iux} \, dF(x), \quad u \in R$$

obviates this difficulty and, moreover, is of the greatest practical importance for the following reasons.

1° To the family $F + c$ corresponds a unique ch.f., and conversely. Therefore, there is a one-to-one correspondence between distributions and ch.f.'s.

2° The methods and results of classical analysis are particularly well suited to the handling of ch.f.'s. In fact, ch.f.'s are continuous and uniformly bounded (by 1) functions. Moreover, to complete and weak convergence of d.f.'s (defined up to additive constants) correspond, respectively, ordinary convergence of ch.f.'s and ordinary convergence of their indefinite integrals.

3° The oldest and, until recent years, almost the only general problem of pr. theory is the "Central Limit Problem," concerned with the asymptotic behavior of d.f.'s of sequences of sums of independent r.v.'s. Much of Part III will be devoted to this problem. The d.f.'s of such sums are obtained by "composition" of the d.f 's of their summands, and this "composition" involves repeated integrations and results in unwieldly expressions, whereas the ch.f.'s of these sums are simply the products of the ch.f.'s of the summands. The Central Limit Problem was satisfactorily solved in the 15 years (1925–1940) which followed the establishment by P. Lévy of the properties of ch.f.'s.

13.1 Uniqueness. *The characteristic function (ch.f.) f of a d.f. F is* defined on R by

$$f(u) = \int e^{iux} \, dF(x) = \int \cos ux \, dF(x) + i \int \sin ux \, dF(x), \quad u \in R.$$

Since, for every $u \in R$, the function of x with values e^{iux} is continuous and bounded by 1, f exists and is continuous and bounded by 1 on R. Moreover, to all functions $F + c$, where c is an arbitrary constant, corresponds the same function f. The converse (and, thus, the one-to-one correspondence between distributions and ch.f.'s) follows from the formula below.

A. INVERSION FORMULA.

$$F[a, b] = \lim_{U \to \infty} \frac{1}{2\pi} \int_{-U}^{+U} \frac{e^{-iua} - e^{-iub}}{iu} f(u) \, du,$$

provided $a < b$ are continuity points of F.

The inversion formula holds for all $a < b \in R$, provided F is normalized.

We say that F is *normalized* if the values of F at its discontinuity points x are taken to be $\dfrac{F(x - 0) + F(x + 0)}{2}$. Normalization destroys the continuity from the left of F at its discontinuity points. However, according to 11.1, the normalized d.f. determines the original one, so that nothing is lost by normalization.

We observe that, in the integral which figures on the right-hand side of the inversion formula, the integrand is defined at $u = 0$ by continuity, so that it is continuous on R; also it is bounded on R by its value $(b - a)$ $f(0)$ at $u = 0$. Thus, for every finite U, this integral is an ordinary Riemann integral and, in proving the inversion formula, we shall find that the limit of this integral, as $U \to \infty$, exists.

Proof. The proof uses repeatedly the dominated convergence theorem applied to an interchange of integrations and is based on the classical Dirichlet formula

$$\frac{1}{\pi} \int_a^b \frac{\sin v}{v} \, dv \to 1 \quad \text{as} \quad a \to -\infty, \quad b \to +\infty,$$

so that the left-hand side is bounded uniformly in a and b. Let

$$I_U = \frac{1}{2\pi} \int_{-U}^{+U} \frac{e^{-iua} - e^{-iub}}{iu} f(u) \, du, \quad a < b \in R,$$

and replace $f(u)$ by its defining integral $\int e^{iux} \, dF(x)$. We can interchange the integrations, so that, by elementary computations,

$$I_U = \int J_U(x) \, dF(x),$$

where

$$J_U(x) = \frac{1}{\pi} \int_{U(x-b)}^{U(x-a)} \frac{\sin v}{v} \, dv.$$

Since J_U is bounded uniformly in U, integration and passage to the limit as $U \to \infty$ can be interchanged in

$$\lim_{U \to \infty} I_U = \lim_{U \to \infty} \int J_U(x) \, dF(x).$$

Therefore

$$\lim_{U \to \infty} I_U = \int J(x) \, dF(x)$$

where

$$J(x) = \lim_{U \to \infty} J_U(x) = \begin{cases} 1 & \text{for} \quad a < x < b \\ \frac{1}{2} & \text{for} \quad x = a, \quad x = b \\ 0 & \text{for} \quad x < a, \quad x > b, \end{cases}$$

and, hence,

$$\lim_{U \to \infty} I_U = \tfrac{1}{2}\{F(a + 0) - F(a - 0)\} + \{F(b - 0) - F(a + 0)\}$$
$$+ \tfrac{1}{2}\{F(b + 0) - F(b - 0)\}$$
$$= \frac{F(b - 0) + F(b + 0)}{2} - \frac{F(a - 0) + F(a + 0)}{2}.$$

Thus, if F is normalized or if $a < b \in C(F)$, then

$$\lim_{U \to \infty} I_U = F[a, b),$$

and the inversion formula is proved.

REMARK.   If an improper Riemann integral

$$\int_{-\infty}^{+\infty} g \, dx = \lim_{\substack{a \to -\infty \\ b \to +\infty}} \int_a^b g \, dx$$

exists and is finite, then

$$\lim_{U \to \infty} \int_{-U}^{+U} g \, dx = \int_{-\infty}^{+\infty} g \, dx.$$

However, the left-hand side limit may exist and be finite (as in the inversion formula), whereas the right-hand side improper integral does not exist. Yet the inversion formula can be written in terms of an improper Riemann integral as follows:

$$F[a, b) = \frac{1}{\pi} \int_0^\infty \frac{\mathscr{I}\{(e^{-iua} - e^{-iub})f(u)\}}{u} \, du$$

where $\mathscr{I}$ stands for "imaginary part of," so that

$$\mathscr{I}\{(e^{-iua} - e^{-iub})f(u)\} =$$

$$(\cos ua - \cos ub)\mathscr{I}f(u) - (\sin ua - \sin ub)\mathscr{R}f(u).$$

It suffices to write $\displaystyle\int_{-U}^{+U} = \int_{-U}^0 + \int_0^U$, change u into $-u$ in the first

right-hand side integral, and take into account the fact that then the integrand changes into its complex-conjugate.

COROLLARY.   *F is differentiable at a and its derivative $F'(a)$ at a is given by*

(1) $$F'(a) = \lim_{h \to 0} \lim_{U \to \infty} \frac{1}{2\pi} \int_{-U}^{+U} \frac{1 - e^{-iuh}}{iuh} e^{-iua} f(u) \, du$$

if, and only if, the right-hand side exists.

In particular, if f is absolutely integrable on R, then F' exists and is bounded and continuous on R and, for every $x \in R$,

(2) $$F'(x) = \frac{1}{2\pi} \int_{-\infty}^{+\infty} e^{-iux} f(u) \, du.$$

Proof. The first assertion follows directly from the inversion formula by the definition of the derivative. The second assertion follows from the first and from the assumption that $\int |f|\, du < \infty$ since, the integrand in (1) being bounded by $|f|$, we have, in (1),

$$\lim_{U \to \infty} \int_{-U}^{+U} = \int_{-\infty}^{+\infty} \quad \text{and} \quad \lim_{h \to 0} \int_{-\infty}^{+\infty} = \int_{-\infty}^{+\infty} \lim_{h \to 0}.$$

REMARK. Thus, if the ch.f.'s f_n of d.f.'s F_n are uniformly Lebesgue-integrable on R, and if $f_n \to f$ ch.f. of F, then f is Lebesgue-integrable on R, and $F'_n \to F'$.

B. *For every $x \in R$,*

$$F(x + 0) - F(x - 0) = \lim_{U \to \infty} \frac{1}{2U} \int_{-U}^{+U} e^{-iux} f(u)\, du.$$

For we can interchange below the integrations and the passage to the limit, so that

$$\lim_{U \to \infty} \frac{1}{2U} \int_{-U}^{+U} e^{-iux} f(u)\, du = \lim_{U \to \infty} \frac{1}{2U} \int_{-U}^{+U} du \left\{ \int e^{iu(y-x)}\, dF(v) \right\}$$

$$= \lim_{U \to \infty} \int \frac{\sin U(y - x)}{U(y - x)}\, dF(y)$$

$$= F(x + 0) - F(x - 0).$$

13.2   Convergences. Since there is a one-to-one correspondence between d.f.'s defined up to additive constants and ch.f.'s, it has to be expected that a one-to-one correspondence also exists between the weak and complete convergence, up to additive constants, of sequences of d.f.'s and certain types of convergence—to be found—of ch.f.'s. For this purpose we introduce the *integral ch.f. $\hat{f}$* of F defined on R by

$$\hat{f}(u) = \int_0^u f(v)\, dv = \int \frac{e^{iux} - 1}{ix}\, dF(x).$$

The last integral is obtained upon replacing $f(v)$ by its defining integral and noting that the interchange of integrations is permissible. Since there is a one-to-one correspondence between $\hat{f}$ and its continuous derivative f, it follows, by 12.1, that there is a one-to-one correspondence between $\hat{f}$ and F defined up to an additive constant.

We are now in a position to show that the weak and the complete convergence up to additive constants of sequences of d.f.'s correspond to the ordinary convergence of the corresponding sequences of integral ch.f.'s and of ch.f.'s, respectively. Unless otherwise stated, a d.f., its ch.f., and its integral ch.f. will be denoted by $F, f, \hat{f}$ respectively, with the same affixes if any.

A. Weak convergence criterion. *If $F_n \overset{w}{\to} F$ up to additive constants, then $\hat{f}_n \to \hat{f}$. Conversely, if $\hat{f}_n$ converges to some function $\hat{g}$, then there exists a d.f. F with $F_n \overset{w}{\to} F$ up to additive constants and $\hat{f} = \hat{g}$.*

Proof. Since $\dfrac{e^{iux} - 1}{ix} \to 0$ as $x \to \mp\infty$, the first assertion follows at once, by the extended Helly-Bray lemma, from the definition of the integral ch.f.'s.

Conversely, let $\hat{f}_n \to \hat{g}$. According to the weak compactness theorem, there is a d.f. F and a subsequence $F_{n'} \overset{w}{\to} F$ as $n' \to \infty$. Therefore, by the extended Helly-Bray lemma, for every $u \in R$,

$$\hat{g}(u) = \lim_{n'} \hat{f}_{n'}(u) = \lim_{n'} \int \frac{e^{iux} - 1}{ix} \, dF_{n'}(x) = \int \frac{e^{iux} - 1}{ix} \, dF(x) = \hat{f}(u).$$

Since $\hat{f}$ determines F up to an additive constant, it follows that weakly convergent subsequences of the sequence F_n have the same limit F up to additive constants, with $\hat{f} = \hat{g}$. This proves the second assertion.

Corollary 1. *Every sequence $\hat{f}_n$ of integral ch.f.'s is compact in the sense of ordinary convergence on R.*

For, in view of the above criterion, this statement is equivalent to the weak compactness theorem for d.f.'s.

Corollary 2. *If $f_n \to g$ a.e., then $F_n \overset{w}{\to} F$ up to additive constants, with $f = g$ a.e.*

Here "a.e." is taken with respect to the Lebesgue measure on R.

Proof. Since $f_n \to g$ a.e. and the f_n are continuous and uniformly bounded by 1, it follows that g is measurable and bounded a.e. so that, by the dominated convergence theorem, $\hat{f}_n \to \hat{g}$ where $\hat{g}$ is defined on R by the Lebesgue integral

$$\hat{g}(u) = \int_0^u g(v) \, dv, \quad u \in R.$$

Therefore, by the foregoing criterion, $F_n \xrightarrow{w} F$ up to additive constants, and $\hat{f} = \hat{g}$. Since the derivative of $\hat{f}$ is f, whereas that of the indefinite Lebesgue integral $\hat{g}$ exists and equals g a.e., it follows that $f = g$ a.e.

B. COMPLETE CONVERGENCE CRITERION. *If $F_n \xrightarrow{c} F$ up to additive constants, then $f_n \to f$. Conversely, if $f_n \to g$ continuous at $u = 0$, then $F_n \xrightarrow{c} F$ up to additive constants, and $f = g$.*

When the F_n and f_n are d.f.'s and ch.f.'s of r.v.'s, the converse becomes the celebrated P. Lévy's *continuity theorem* for ch.f.'s.

Proof. Let $F_n \xrightarrow{c} F$ up to additive constants. Then, by the Helly-Bray theorem, for every $u \in R$,

$$f_n(u) = \int e^{iux} \, dF_n(x) \to \int e^{iux} \, dF(x) = f(u).$$

Conversely, let $f_n \to g$ continuous at $u = 0$. Then, for every $u \in R$,

$$\hat{f}_n(u) = \int_0^u f_n(v) \, dv \to \int_0^u g(v) \, dv = \hat{g}(u),$$

and, hence, by the weak convergence criterion, for some d.f. F with ch.f. f, $F_n \xrightarrow{w} F$ up to additive constants, and $\hat{f} = \hat{g}$. Therefore,

$$\frac{1}{u} \int_0^u f(v) \, dv = \frac{1}{u} \int_0^u g(v) \, dv$$

and, letting $u \to 0$, we obtain $f(0) = g(0)$ on account of continuity of f and of g at the origin. Thus,

$$\text{Var } F_n = f_n(0) \to g(0) = f(0) = \text{Var } F,$$

and the proof is completed by taking into account the direct assertion.

C. UNIFORM CONVERGENCE THEOREM. *If a sequence f_n of ch.f.'s converges to a ch.f. f, then the convergence is uniform on every finite interval $[-U, +U]$.*

Proof. On account of **B**, $F_n \xrightarrow{c} F$ up to additive constants. Let $\epsilon > 0$ and $U > 0$ be arbitrarily fixed. We have

$$|f_n(u) - f(u)| \leq \left| \int_a^b e^{iux} \, dF_n(x) - \int_a^b e^{iux} \, dF(x) \right|$$

$$+ \text{Var } F_n - F_n[a, b) + \text{Var } F - F[a, b)$$

where we take a, b to be continuity points of F. Let $|a|, b$ and then n be so large that Var $F - F[a, b) | < \dfrac{\epsilon}{6}$,

$$\text{Var } F_n - F_n[a, b) < \text{Var } F - F[a, b) + \frac{\epsilon}{6} < \frac{\epsilon}{3}.$$

It suffices to show that, for n sufficiently large and all $u \in [-U, +U]$,

$$\Delta_n = \Big| \int_a^b e^{iux} \, dF_n(x) - \int_a^b e^{iux} \, dF(x) \Big| < \frac{\epsilon}{2}.$$

Let

$$a = x_1 < x_2, \cdots < x_{N+1} = b$$

where the subdivision points are continuity points of F and $\alpha = \max_{k \leq N} (x_{k+1} - x_k) < \epsilon/8U$. Since, by the mean value theorem,

$$| e^{iux} - e^{iux'} | \leq | x - x' | U \quad \text{for} \quad | u | \leq U,$$

it follows that, upon replacing x by x_k in every interval $[x_k, x_{k+1})$, Δ_n is modified by at most

$$\alpha U \int_a^b dF_n(x) + \alpha U \int_a^b dF(x) \leq 2\alpha U < \frac{\epsilon}{4}.$$

Thus, it remains to show that, for n sufficiently large,

$$\Big| \sum_{k=1}^{N} e^{iux_k} \{ F_n[x_k, x_{k+1}) - F[x_k, x_{k+1}) \} \Big|$$

$$\leq \sum_{k=1}^{N} | F_n[x_k, x_{k+1}) - F[x_k, x_{k+1}) | < \frac{\epsilon}{4}.$$

Since $F_n[x_k, x_{k+1}) \to F[x_k, x_{k+1})$ for every $k \leq N$, the last assertion follows and the proof is complete.

REMARK. In fact, we proved, with a supplementary detail, the first assertion of the complete convergence criterion without using the Helly-Bray theorem.

COROLLARY 1. *If $f_n \to f$ and $u_n \to u$ finite, then $f_n(u_n) \to f(u)$.*

This follows, by **C** and continuity of f, from

$$| f_n(u_n) - f(u) | \leq | f_n(u_n) - f(u_n) | + | f(u_n) - f(u) |.$$

COROLLARY 2. *A set $\{F_t\}$ of d.f.'s is completely compact (up to additive constants) if, and only if, the corresponding set $\{f_t\}$ of ch.f.'s is equicontinuous at $u = 0$.*

Proof. By 13.4B equicontinuity of $\{f_t\}$ at $u = 0$ is equivalent to equicontinuity on R.

On the other hand, Ascoli's theorem and its converse say that a set of continuous functions is compact in the sense of uniform convergence on a finite closed interval if, and only if, it is uniformly bounded and equicontinuous on this interval. Since the f_t are uniformly bounded, the assertion follows by **B** and **C**.

REMARK. If the d.f.'s F_n, F of r.v.'s, are differentiable and $F_n' \to F'$ on R, then $f_n \to f$ uniformly on R. It suffices to use 17 in Complements and Details of Ch. II.

13.3 Composition of d.f.'s and multiplication of ch.f.'s. A function F on $R = (-\infty, +\infty)$ is said to be *composed* of d.f.'s F_1 and F_2, and written $F_1 * F_2$, if

$$F(x) = \int F_1(x - y)\, dF_2(y), \quad x \in R$$

where we assume, for simplicity, that $F_1(-\infty) = F_2(-\infty) = 0$; otherwise, to avoid trivial complications, we would have to replace F_1 by $F_1 - F_1(-\infty)$.

Since, for every fixed y, $F_1(x - y)$ are values of a d.f., nondecreasing, continuous from the left and bounded by $F_1(-\infty) = 0$ and $F_1(+\infty) \leqq 1$, it follows, upon applying the dominated convergence theorem, that F has the same properties and that $\operatorname{Var} F = \operatorname{Var} F_1 \cdot \operatorname{Var} F_2$.

A. COMPOSITION THEOREM. *If $F = F_1 * F_2$, then $f = f_1 f_2$, and conversely.*

Proof. Let $F = F_1 * F_2$ and let $a = x_{n1} < \cdots < x_{n,k_n+1} = b$ with $\sup_k (x_{n,k+1} - x_{nk}) \to 0$ as $n \to \infty$. Since, for every $u \in R$,

$$\int_a^b e^{iux}\, dF(x) = \lim \sum_k e^{iux_{nk}} F[x_{nk}, x_{n,k+1})$$

$$= \lim \int \sum_k e^{iu(x_{nk}-y)} F_1[x_{nk} - y, x_{n,k+1} - y) e^{iuy}\, dF_2(y),$$

it follows that

$$\int_a^b e^{iux} \, dF(x) = \int \left\{ \int_{a-y}^{b-y} e^{iux} \, dF_1(x) \right\} e^{iuy} \, dF_2(y)$$

and, letting $a \to -\infty$ and $b \to +\infty$,

$$\int e^{iux} \, dF(x) = \int e^{iux} \, dF_1(x) \int e^{iuy} \, dF_2(y),$$

so that $f = f_1 f_2$ and the first assertion is proved.

Conversely, according to the first assertion, $f_1 f_2$ is the ch.f. of $F_1 * F_2$ and, hence, on account of the one-to-one correspondence between f and $F + c$, $F = F_1 * F_2$ up to an additive constant. The converse is proved.

COROLLARY 1.   *A product of ch.f.'s is a ch.f. and, in particular, if f is a ch.f. so is $|f|^2$.*

For $f = f_1 f_2$ is the ch.f. of the d.f. $F = F_1 * F_2$, and the particular case follows from the fact that, if f is a ch.f., so is its complex-conjugate $\bar{f}$ which corresponds to the d.f. $F(+\infty) - F(-x + 0)$.

COROLLARY 2.   *Composition of d.f.'s is commutative and associative.*

For the corresponding multiplication of ch.f.'s has these properties.

13.4   Elementary properties of ch.f.'s and first applications. In the sequel, the elementary properties we establish now will play an important ancillary role, and the first applications will be used, improved, and generalized.

We denote by F and f, with same subscripts if any, corresponding d.f.'s and ch.f.'s; in general, the corresponding d.f.'s F are defined up to additive constants, but if f is ch.f. of a r.v., then, as usual, we take $F(-\infty) = 0$, $F(+\infty) = 1$. We say that a r.v. X is *symmetric* if X and $-X$ have the same d.f., that is, for every $x \in R$, $P[X < x] = P[X > -x]$.

A. GENERAL PROPERTIES.   *Every ch.f. f is uniformly continuous and*

$$|f| \le f(0) = \text{Var } F \le 1, \quad f(-u) = \bar{f}(u).$$

If f is the ch.f. of a r.v. X, then the function with values $e^{iua}f(bu)$ is the ch.f. of the r.v. $a + bX$. In particular, $\bar{f}$ is the ch.f. of $-X$ and f is real if, and only if, X is symmetric.

Elementary inequality:$f(0) - Rf(2u) \le 4(f(0) - Rf(u))$.

Proof. The first assertion follows from $f(u) = \int e^{iux} \, dF(x)$. The second assertion follows from $E e^{iu(a+bX)} = e^{iua} E e^{ibuX}$. Finally, if X is symmetric, then $f(u) = Ee^{iuX} = Ee^{-iuX} = f(-u) = \overline{f}(u)$ so that f is real; conversely, if f is real, then changing the signs of a and b in the inversion formula is equivalent to taking the complex-conjugate of the integrand and changing its sign, so that $F[a, b] = F[-b, -a]$ and, hence, by letting $a \to -\infty$ and $b \uparrow x$, we have $P[X < x] = F(x) = 1 - F(-x + 0) = P[X > -x]$.

The elementary inequality obtains upon integrating $1 - \cos 2ux \le 4(1 - \cos ux)$ with respect to F.

B. INCREMENTS INEQUALITY: *for any* $u, h \in R$

$$|f(u) - f(u + h)|^2 \le 2f(0)\{f(0) - \Re f(h)\}.$$

INTEGRAL INEQUALITY: *for* $u > 0$ *there exist functions* $0 < m(u) < M(u) < \infty$ *such that*

$$m(u) \int_0^u \{f(0) - \Re f(v)\} \, dv \le \int \frac{x^2}{1 + x^2} \, dF(x)$$

$$\le M(u) \int_0^u \{f(0) - \Re f(v)\} \, dv;$$

if $f(0) = 1$, *then, for* u *sufficiently close to* 0,

$$\int \frac{x^2}{1 + x^2} \, dF(x) \le -M(u) \int_0^u (\log \Re f(v)) \, dv.$$

Proof. The increments inequality follows, by Schwarz's inequality, from

$$|f(u) - f(u + h)|^2 = \left| \int e^{iux}(1 - e^{ihx}) \, dF(x) \right|^2$$

$$\le \int dF(x) \int |1 - e^{ihx}|^2 \, dF(x)$$

$$= 2f(0) \int (1 - \cos hx) \, dF(x)$$

$$= 2f(0)\{f(0) - \Re f(h)\}.$$

The integral inequality follows, by the elementary inequality with $u \ne 0$

$$0 < M^{-1}(u) \le |u| \left(1 - \frac{\sin ux}{ux}\right) \frac{1 + x^2}{x^2} \le m^{-1}(u) < \infty, \quad x \in R,$$

from

$$\int_0^u dv \int (1 - \cos vx) \, dF(x) = u \int \left(1 - \frac{\sin ux}{ux}\right) \frac{1 + x^2}{x^2} \cdot \frac{x^2}{1 + x^2} \, dF(x).$$

The case $f(0) = 1$ follows then from the elementary inequality $1 - a \leq - \log a$ for $a \geq 0$.

The integral inequality permits us in turn to find bounds for $\int_{|x|<c} x^2 \, dF(x)$ and $\int_{|x|\geq c} dF(x)$, $(c > 0)$, by

(I) $\quad \dfrac{1}{1 + c^2} \displaystyle\int_{|x|<c} x^2 \, dF(x) + \dfrac{c^2}{1 + c^2} \int_{|x|\geq c} dF(x)$

$$\leq \int \frac{x^2}{1 + x^2} \, dF(x)$$

$$\leq \int_{|x|<c} x^2 \, dF(x) + \int_{|x|\geq c} dF(x).$$

However, it is sometimes more convenient to use the direct

B′. Truncation inequality: *for $u > 0$:*

$$\int_{|x|<1/u} x^2 \, dF(x) \leq \frac{3}{u^2} \{f(0) - \Re f(u)\},$$

$$\int_{|x|\geq 1/u} dF(x) \leq \frac{7}{u} \int_0^u \{f(0) - \Re f(v)\} \, dv.$$

If $f(0) = 1$ and u is sufficiently close to 0, then we can replace $1 - \Re f$ in the foregoing by $- \log \Re f$.

These inequalities follow, respectively, from

$$\int (1 - \cos ux) \, dF(x) \geq \int_{|x|<1/u} \frac{u^2 x^2}{2} \left(1 - \frac{u^2 x^2}{12}\right) dF(x)$$

$$\geq \frac{11 u^2}{24} \int_{|x|<1/u} x^2 \, dF(x)$$

and from

$$\frac{1}{u} \int_0^u dv \int (1 - \cos vx) \, dF(x) = \int \left(1 - \frac{\sin ux}{ux}\right) dF(x)$$

$$\geq (1 - \sin 1) \int_{|x|\geq 1/u} dF(x).$$

The case $f(0) = 1$ follows as in **B**.

Applications. 1° *If $f_n \to g$ continuous at $u = 0$, then g is continuous on R.*

This follows from the fact that the increments inequality with f_n becomes, as $n \to \infty$, the same inequality with g.

2° *If the sequence f_n is equicontinuous at $u = 0$, then it is equicontinuous at every $u \in R$.*

For, then, as $h \to 0$,

$$\left| f_n(u) - f_n(u + h) \right|^2 \leqq 2\{f_n(0) - \Re f_n(h)\} \to 0$$

uniformly in n.

3° *If $f_n \to 1$ on $(-U, +U)$, then $f_n \to 1$ on R.*

This follows by induction as $f_n(2u) \to 1$ for $|u| < U$ follows from

$$\left| f_n(u) - f_n(2u) \right|^2 \leqq 2\{f_n(0) - \Re f_n(u)\} \to 0 \quad \text{for} \quad |u| < U.$$

If we take into account the fact that the set of all differences of numbers belonging to a set of positive Lebesgue measure contains a nondegenerate interval $(-U, +U)$, this proposition can be improved as follows:

If $f_n \to 1$ on a set A of positive Lebesgue measure, then $f_n \to 1$ on R.

For, we can assume that the set A is symmetric with respect to the origin and contains it, since, for $u \in A$,

$$f_n(-u) = \bar{f}_n(u) \to 1, \quad 1 \geqq f_n(0) \geqq |f_n(u)| \to 1,$$

and, then, $f_n(u - u') \to 1$ for $u, u' \in A$ on account of

$$\left| f_n(u) - f_n(u - u') \right|^2 \leqq 2\{f_n(0) - \Re f_n(-u')\} \to 0.$$

4° We shall now prove an elegant proposition (slightly completed) due to Kawata and Ugakawa. We use repeatedly Corollary 2 of the weak convergence criterion which says that, if a sequence of ch.f.'s $g_n \to g$ a.e., then the corresponding sequence of d.f.'s $G_n \overset{w}{\to} G$ up to additive constants and the ch.f. of G coincides a.e. with g.

Let $g_n = \prod\limits_{k=1}^{n} f_k \to g$ a.e.   Either $g = 0$ a.e., and then $G_n \overset{w}{\to} 0$ up to additive constants.   Or $g \neq 0$ on a set A of positive Lebesgue measure, and then $G_n \overset{c}{\to} G$ up to additive constants.

Proof. In both cases $G_n \overset{w}{\to} G$ up to additive constants. The first case follows from the recalled proposition. In the second case, we have to prove that $\mathrm{Var}\, G_n \to \mathrm{Var}\, G$. Since $\mathrm{Var}\, F^s = \mathrm{Var}\, (F * F) = (\mathrm{Var} F)^2$ and $f^s = |f|^2$, it suffices to consider real-valued nonnegative ch.f.'s. But then $\lim\limits_{m \to \infty} \prod\limits_{n+1}^{m} f_k$ exists on R and coincides a.e. with a ch.f., while, for m, n sufficiently large, $g_m g_n \neq 0$ a.e. on A, and, as $m \to \infty$ and then $n \to \infty$,

$$\prod_{k=n+1}^{m} f_k = g_m/g_n \to g/g_n = \prod_{k=n+1}^{\infty} f_k \to 1 \quad \text{a.e. on } A.$$

It follows, by 3°, that $\prod\limits_{k=n+1}^{\infty} f_k \to 1$ a.e. on R. Therefore, if H_n is the d.f. whose ch.f. coincides a.e. with $\prod\limits_{k=n+1}^{\infty} f_k$, then $\mathrm{Var}\, H_n \to 1$. But, by 11.2a and the composition theorem 13.3A,

$$\liminf \mathrm{Var}\, G_n \geq \mathrm{Var}\, G = \mathrm{Var}\, G_n \cdot \mathrm{Var}\, H_n.$$

It follows, by letting $n \to \infty$, that $\mathrm{Var}\, G_n \to \mathrm{Var}\, G$. The proof is completed.

5° Let F_{nk} be d.f.'s of r.v.'s, $k = 1, \cdots, k_n \to \infty$, $\gamma_n = \sum\limits_{k} (1 - f_{nk})$.

Set $\Psi_n(x) = \sum\limits_{k} \int_{-\infty}^{x} \dfrac{y^2}{1 + y^2} \, dF_{nk}(y)$ and $\alpha(c)_{-} = \sup\limits_{n} \sum\limits_{k} \int_{|x| \geq c} dF_{nk}(x)$,

$\beta(c) = \sup\limits_{n} \sum\limits_{k} \int_{|x| < c} x^2 \, dF_{nk}(x)$, $c > 0$ finite.

If $f_n = \prod\limits_{k} f_{nk}$ with f_{nk} real-valued, then the following properties are equivalent:

(C_1) *the sequence F_n is completely compact.*
(C_2) *the sequence γ_n is equicontinuous at $u = 0$.*
(C_3) *$\alpha(c) \to 0$ as $c \to \infty$ and $\alpha(c) + \beta(c) < \infty$ for every (some) c.*
(C_4) *the sequence Ψ_n is bounded and completely compact.*

Proof. $(C_1) \Leftrightarrow (C_2)$ by 13.2 C Cor. 2 and the inequality $1 - \sum a_k \leq \Pi(1 - a_k) \leq \exp\{-\sum a_k\}, 0 \leq a_k \leq 1$. $(C_2) \Rightarrow (C_3)$ by B' and $(C_3) \Rightarrow (C_2)$ by $\gamma_n(u) \leq 2\alpha(c) + \beta(c)u^2/2$. Finally, $(C_3) \Leftrightarrow (C_4)$ and "some c" $\Leftrightarrow$ "every c" by (I), $\alpha(c)c^2/(1 + c^2) \leq \int_{|x| \geq c} d\Psi_n(x) \leq \alpha(c)$ and 11.2B.

Let

$$m^{(k)} = \int x^k \, dF(x), \quad \mu^{(r)} = \int |x|^r \, dF(x), \quad k = 0, 1, 2, \cdots, \quad r \geq 0,$$

be, respectively, the kth moments and the rth absolute moments of F. Let $f^{(k)}$ be the kth derivative of $f(f^{(0)} = f)$ and, as usual, let θ, θ' be quantities with modulus bounded by 1.

C. DIFFERENTIABILITY PROPERTIES. *If $f^{(2n)}(0)$ exists and is finite, then $\mu^{(r)} < \infty$ for $r \leq 2n$.*

If $\mu^{(n+\delta)} < \infty$ for a $\delta \geq 0$, then for every $k \leq n$

$$f^{(k)}(u) = i^k \int e^{iux} x^k \, dF(x), \quad u \in R,$$

and $f^{(k)}$ is continuous and bounded by $\mu^{(k)}$; moreover

$$f(u) = \sum_{k=0}^{n-1} m^{(k)} \frac{(iu)^k}{k!} + \rho_n(u), \quad u \in R$$

where

$$\rho_n(u) = u^n \int_0^1 \frac{(1-t)^{n-1}}{(n-1)!} f^{(n)}(tu) \, dt = m^{(n)} \frac{(iu)^n}{n!} + o(u^n) = \theta \mu^{(n)} \frac{|u|^n}{n!},$$

and if $0 < \delta \leq 1$, then

$$\rho_n(u) = m^{(n)} \frac{(iu)^n}{n!} + 2^{1-\delta} \theta' \mu^{(n+\delta)} \frac{|u|^{n+\delta}}{(1+\delta)(2+\delta) \cdots (n+\delta)}.$$

Proof. To begin with, we observe that, since $|x|^{r'} \leq 1 + |x|^r$ for $r' < r$, finiteness of $\mu^{(r)}$ implies that of $\mu^{(r')}$.

The first assertion follows from the existence and finiteness of the $2n$th symmetric derivative by using the Fatou-Lebesgue theorem in

$$|f^{(2n)}(0)| = \lim_{h \to 0} \int \left(\frac{\sin hx}{hx} \right)^{2n} x^{2n} \, dF(x) \geq \int x^{2n} \, dF(x).$$

The second assertion follows from the fact that, by differentiating $\int e^{iux} \, dF(x)$ k times under the integral sign, the integral so obtained is absolutely convergent and, hence, this differentiation and the integration can be interchanged.

The limited expansions follow by integrating the limited expansions of e^{iux} with corresponding forms of its remainder term. The last and less usual corresponding form of its remainder is obtained upon observing that $\left| e^{ia} - 1 \right| \leq 2 \left| a/2 \right|^\delta$ (since, for $0 < \delta \leq 1$, if $\left| a/2 \right| < 1$, then $\left| e^{ia} - 1 \right| \leq \left| a \right| \leq 2 \left| a/2 \right|^\delta$ and, if $\left| a/2 \right| \geq 1$, then $\left| e^{ia} - 1 \right| \leq 2 \leq 2 \left| a/2 \right|^\delta$), and using successive integrations by parts in

$$\left| \int_0^1 \frac{(1-t)^{n-1}}{(n-1)!} (e^{itux} - 1) \, dt \right| \leq 2^{1-\delta} \left| ux \right|^\delta \int_0^1 \frac{(1-t)^{n-1}}{(n-1)!} t^\delta \, dt$$

$$= \frac{2^{1-\delta} \left| ux \right|^\delta}{(1+\delta)(2+\delta) \cdots (n+\delta)}.$$

COROLLARY. *If all moments of F exist and are finite, then* $f^{(k)}(0) = i^k m^{(k)}$ *for every k, and*

$$f(u) = \sum_{n=0}^\infty m^{(n)} \frac{(iu)^n}{n!}$$

in the interval of convergence of the series.

Applications. We consider d.f.'s F and ch.f.'s f of r.v.'s X, with the same subscripts if any. If $m^{(1)} = EX = 0$, we write σ^2 instead of $m^{(2)} = EX^2$.

1° NORMAL DISTRIBUTION. A "reduced normal" d.f. is defined by $F'(x) = e^{-x^2/2}/\sqrt{2\pi}$. It is the d.f. of a r.v., since $m^{(0)} = 1$ by

$$\left(\frac{1}{\sqrt{2\pi}} \int e^{-x^2/2} \, dx \right) \left(\frac{1}{\sqrt{2\pi}} \int e^{-y^2/2} \, dy \right) = \frac{1}{2\pi} \iint e^{-(x^2+y^2)/2} \, dx \, dy$$

$$= \frac{1}{2\pi} \int_0^{2\pi} d\theta \int_0^\infty e^{-\rho^2/2} \rho \, d\rho = 1$$

Since $F'(-x) = F'(x)$, it follows at once that the odd moments vanish, while, by integration by parts, we obtain

$$m^{(2n)} = (2n - 1)m^{(2n-2)} = \cdots = (2n)!/2^n n!.$$

Therefore, by the foregoing corollary, the "reduced normal" ch.f. is

$$f(u) = \sum_{n=0}^\infty \frac{(-u^2/2)^n}{n!} = e^{-u^2/2}, \quad u \in R.$$

2° BOUNDED LIAPOUNOV THEOREM. *Let* $\left| X_n \right| \leq c < \infty$ *and* $EX_n = 0$.

If $s_n^2 = \sum_{k=1}^n \sigma_k^2 \to \infty$, *then* $\prod_{k=1}^n f_k(u/s_n) \to e^{-u^2/2}$ *for every* $u \in R$.

Since $E|X_n|^3 \leq cEX_n^2$ and $\sigma_n^2 = EX_n^2 \leq c^2$, it follows, upon fixing u arbitrarily, that

$$f_k\left(\frac{u}{s_n}\right) = 1 - \frac{\sigma_k^2}{2s_n^2}u^2 + \theta_{nk}\frac{c\sigma_k^2}{6s_n^3}|u|^3 \to 1$$

uniformly in $k \leq n$. Therefore, for n sufficiently large,

$$\sum_{k=1}^{n}\log f_k\left(\frac{u}{s_n}\right) = -\frac{u^2}{2}(1 + o(1)) + \theta_n\frac{c|u|^3}{6s_n}(1 + o(1)) \to -\frac{u^2}{2},$$

and the assertion is proved.

§ 14.   PROBABILITY LAWS AND TYPES OF LAWS

14.1   Laws and types; the degenerate type.   Since there is a one-to-one correspondence between distributions, d.f.'s defined up to an additive constant, and ch.f.'s, they are different but equivalent "representations" of the same mathematical concept which we shall call *pr. law* or, simply, *law*.   Moreover, to a given distribution on the Borel field $\mathcal{B}$ we can always make correspond the finite part of a measurable function X on some pr. space $(\Omega, \mathcal{A}, P)$, and the restriction of P to $X^{-1}(\mathcal{B})$ with values $P[X \in S]$, $S \in \mathcal{B}$, is still another representation of the law defined by the given distribution; there are many such measurable functions and many such spaces.   Nevertheless, the various representations of a given law have their own intuitive value.   Thus, for every law we have a multiplicity of representations and we shall use them according to convenience.

A law will be denoted by the symbol $\mathcal{L}$, with the same affixes if any as the d.f. or the ch.f. which represents this law, and the terminology and notations for operations on laws will be those introduced for d.f.'s; in particular, if $F_n \xrightarrow{w} F$ we write $\mathcal{L}_n \xrightarrow{w} \mathcal{L}$, and if $F_n \xrightarrow{c} F$ we write $\mathcal{L}_n \xrightarrow{c} \mathcal{L}$.   The case of laws of r.v.'s (with d.f.'s of variation 1) is by far the most important.   The law of a r.v. X will be denoted by $\mathcal{L}(X)$, and if a sequence $\mathcal{L}(X_n)$ of laws of r.v.'s converges completely—necessarily to the law $\mathcal{L}(X)$ of a r.v. X—we shall drop "complete" and write $\mathcal{L}(X_n) \to \mathcal{L}(X)$. *From now on a law will be law of a r.v., unless otherwise stated.*

The origin and the scale of values of measured quantities, say a r.v. X, are more or less arbitrarily chosen.   By modifying them we modify linearly the results of measurements, that is, we replace X by $a + bX$

where a and $b > 0$ are finite numbers. If, moreover, the orientation of values can be modified, then the only restriction on the finite numbers a and b is that $b \neq 0$. This leads us to assign to a law $\mathcal{L}(X)$ the family $\mathfrak{I}(X) = \{\mathcal{L}(a + bX)\}$ of all laws obtainable by changes of origin, scale, and orientation, to be called a *type* of laws. If b is restricted either to positive or to negative values, the corresponding families of laws will be called *positive*, resp. *negative* types of laws.

Letting $b \to 0$ we encounter a boundary case—the simplest and at the same time the everywhere pervading *degenerate type* $\{\mathcal{L}(a)\}$ of laws of r.v.'s which degenerate at some arbitrary but finite value a, that is, such that $X = a$ a.s. The corresponding family of "degenerate" d.f.'s is that of d.f.'s with one, and only one, point of increase $a \in R$ with $F(a + 0) - F(a - 0) = 1$. The corresponding family of "degenerate" ch.f.'s is that of all ch.f.'s of the form $f(u) = e^{iua}$, $u \in R$, so that their moduli reduce to 1. The converse is also true and, more precisely,

a. *A ch.f. is degenerate if, and only if, its modulus equals* 1 *for two values* $h \neq 0$ *and* $\alpha h \neq 0$ *of the argument whose ratio* α *is irrational. In particular, a ch.f. f is degenerate if* $|f(u)| = 1$ *in a nondegenerate interval.*

Proof. Since $|f(h)| = 1$, there is a finite number a such that $f(h) = e^{iha}$ and, hence,

$$e^{-iha}f(h) = \int e^{ih(x-a)} \, dF(x) = 1.$$

Thus

$$\int [1 - \cos h(x - a)] \, dF(x) = 0$$

and, since the integrand is nonnegative, it follows that, for points $\bar{x}$ of increase of F, $\cos h(\bar{x} - a) = 1$ so that $\bar{x}' - \bar{x}''$ is a multiple of $\dfrac{2\pi}{h}$ when the points of increase $\bar{x}'$, $\bar{x}''$ are distinct. Replacing h by αh, we find that $\bar{x}' - \bar{x}''$ is also a multiple of $\dfrac{2\pi}{\alpha h}$, which is impossible when α is irrational unless there is only one point of increase. The particular case follows.

REMARK. The foregoing argument proves that, if $|f(h)| = 1$ for an $h \neq 0$, then $f(u) = \sum\limits_{k=0}^{\infty} p_k e^{iux_k}$, $u \in R$, where $p_k \geqq 0$, $\sum\limits_{k=0}^{\infty} p_k = 1$ and $x_k = a + k \cdot \dfrac{2\pi}{h}$; the converse is immediate.

14.2   Convergence of types. If $\mathcal{L}(X_n) \to \mathcal{L}(X)$, then, for every a, $b \neq 0$, $\mathcal{L}(a + bX_n) \to \mathcal{L}(a + bX)$, since $f_n \to f$ implies that $e^{iua}f_n(bu) \to e^{iua}f(bu)$, $u \in R$. Thus, we may say that convergence of sequences of laws to a law is, in fact, convergence of sequences of types to a type. It may even happen that, given a sequence $\mathcal{L}(X_n)$ convergent or not, we can proceed to changes of origin and of scale varying with n and giving rise to a convergent sequence $\mathcal{L}(a_n + b_n X_n)$. In the particular case of consecutive sums X_n of "independent" r.v.'s, a special form of the problem of finding the sequences of laws which converge for given changes of origin and of scale is the oldest and, until recently, was the only limit problem of pr. theory; we shall investigate it in Part III. Meanwhile there is an immediate question to answer: given a sequence $\mathcal{L}(X_n)$ of laws, do all the limit laws of convergent sequences of the form $\mathcal{L}(a_n + b_n X_n)$ belong to a same type? The answer, due to Khintchine for positive types, is as follows:

A. CONVERGENCE OF TYPES THEOREM. *If $\mathcal{L}(X_n) \to \mathcal{L}(X)$ nondegenerate and $\mathcal{L}(a_n + b_n X_n) \to \mathcal{L}(X')$ nondegenerate, then the laws $\mathcal{L}(X)$ and $\mathcal{L}(X')$ belong to the same type. More precisely, $\mathcal{L}(X') = \mathcal{L}(a + bX)$ with $|b_n| \to |b|$, and if $b_n > 0$ then $b_n \to b$, $a_n \to a$.*

However, for every finite a and for every sequence $\mathcal{L}(X_n)$ of laws, there exist numbers a_n and $b_n \neq 0$ such that $\mathcal{L}(a_n + b_n X_n) \to \mathcal{L}(a)$.

In other words, given a sequence of laws, the changes of origin, scale, and orientation can yield in the limit no more than one nondegenerate type and can always yield in the limit the degenerate type. This shows once more that the degenerate type is to be considered as the "degenerate part" of every type.

Proof. The second assertion is immediate. For, by taking the numbers c_n sufficiently large so as to have $P[|X_n| \geq c_n] < \dfrac{1}{n} \to 0$, we obtain

$$P\left[\frac{|X_n|}{nc_n} \geq \frac{1}{n}\right] < \frac{1}{n} \to 0$$

and, it follows at once, that $\mathcal{L}\left(\dfrac{X_n}{nc_n}\right) \to \mathcal{L}(0)$, so that $\mathcal{L}\left(a + \dfrac{X_n}{nc_n}\right) \to \mathcal{L}(a)$.

The first assertion means that $f_n \to f$ nondegenerate and $e^{iua_n}f_n(b_n u) \to f'(u)$ nondegenerate, $u \in R$, imply existence of two finite numbers a

and $b \neq 0$ such that $f'(u) = e^{iua}f(bu)$, $u \in R$. We can always select from the sequence b_n a convergent subsequence $b_{n'}$, but its limit b may be 0 or $\pm\infty$. If $b = 0$, then, since the convergence of ch.f.'s to a ch.f. is uniform in every finite interval, we have, for every fixed $u \in R$,

$$|f'(u)| = \lim_{n'} |f_{n'}(b_{n'}u)| = |f(0)| = 1,$$

so that, by 14.1a, f' is degenerate and this contradicts the assumption. Similarly, if $b_{n'} \to \pm\infty$, then, replacing u by $\dfrac{u}{b_{n'}}$, it follows that

$$|f(u)| = \lim_{n'} \left| f'_{n'}\left(\frac{u}{b_{n'}}\right) \right| = |f'(0)| = 1,$$

so that f is degenerate and this contradicts the assumption. Thus $b_{n'} \to b$ finite and different from 0. On the other hand, for all u sufficiently close to 0, the continuous functions $f(bu)$ and $f'(u)$ (with values 1 for $u = 0$) differ from 0; and we have, for n' sufficiently large,

$$e^{iua_{n'}} = \frac{e^{iua_{n'}}f_{n'}(b_{n'}u)}{f_{n'}(b_{n'}u)} \to \frac{f'(u)}{f(bu)} \neq 0, \quad n' \to \infty,$$

so that $\lim e^{iua_{n'}}$ exists and is finite for $|u| \leq$ some $u_0 > 0$. But then $\limsup |a_{n'}| < \infty$. Therefore, for any convergent subsequences of (a_n), $a'_n \to a'$ and $a''_n \to a''$, we have $e^{iu(a'_n - a''_n)} \to e^{iu(a' - a'')} = 1$ for $|u| \leq u_0$. It follows, by 13.4 Application 3°, that $a' - a'' = 0$ hence $a_{n'} \to$ some $a \in R$ and $f'(u) = e^{iua}f(bu)$, $u \in R$.

Clearly, it remains only to prove that $|b_n| \to |b|$. Let $b_{n'} \to b$ and $b_{n''} \to b'$ hence $a_{n'} \to a$ and $a_{n''} \to a'$; it suffices to prove that if, for every u, $e^{iua}f(bu) = e^{iua'}f(b'u)$, then $|b| = |b'|$. Upon replacing $b'u$ by u and $\dfrac{b}{b'}$ by c, it suffices to prove that, if $|c| \leq 1$ and, for every u, $|f(u)|^2 = |f(cu)|^2$, then $|c| = 1$. But $|c| < 1$ entails, upon replacing repeatedly u by cu,

$$|f(u)|^2 = |f(cu)|^2 = \cdots = \lim |f(c^n u)|^2 = 1.$$

Thus, the nondegeneracy assumption excludes the possibility $|c| < 1$, so that $|c| = 1$ and the proof is complete.

REMARK. It is immediately seen that if we limit ourselves to, say, positive types only, then, under the foregoing assumptions, $a_n \to a$ and $b_n \to b$. We leave to the reader to find conditions under which this property remains valid for types.

COROLLARY. *If, for every u,*

$$e^{iua_n}f_n(b_nu) \to f(u) \quad and \quad e^{iua'_n}f_n(b'_nu) \to f(u)$$

where f is a nondegenerate ch.f. and $b_nb'_n > 0$ *for every n, then*

$$\frac{a_n - a'_n}{b'_n} \to 0 \quad and \quad \frac{b_n}{b'_n} \to 1.$$

Replace in the theorem X_n by $a'_n + b'_nX_n$.

14.3 Extensions. The results and terminology of this chapter extend at once to families of r.v.'s, and we shall content ourselves with a few generalities.

The *law of a random vector* $X = \{X_1, \cdots, X_N\}$ with d.f. F_X on R^N is represented by the ch.f. f_X on R^N defined by the N-uple integral

$$f_X(u) = \int e^{iux} \, dF_X(x), \quad ux = u_1x_1 + \cdots + u_Nx_N$$

or, explicitly, by

$$f_X(u_1, \cdots, u_N) = \overset{N\text{-uple}}{\int \cdots \int} e^{i(u_1x_1 + \cdots + u_Nx_N)} \, d_1d_2 \cdots d_N F_X(x_1, \cdots, x_N).$$

The integral which appears in the inversion formula becomes an N-uple Riemann-Stieltjes integral $\int_{-U_1}^{+U_1} \cdots \int_{-U_N}^{+U_N}$ and the "kernel" $\dfrac{e^{-iua} - e^{-iub}}{iu}$

becomes $\displaystyle\prod_{k=1}^{N} \frac{e^{-iu_ka_k} - e^{-iu_kb_k}}{iu_k}$.

We observe that there is a one-to-one correspondence between the law of the random vector $X = \{X_1, \cdots, X_N\}$ and the laws of the r.v.'s $uX = u_1X_1 + \cdots + u_NX_N$, *where u varies over* R^N, since

$$f_X(tu) = f_{uX}(t), \quad t \in R$$

and, in particular, $f_X(u) = f_{uX}(1)$.

Finally, the *law of a random function* $X = \{X_t, t \in T\}$ is the set of joint laws of all its finite subfamilies.

§ 15.   NONNEGATIVE-DEFINITENESS; REGULARITY

15.1 Ch.f.'s and nonnegative-definiteness. The class of ch.f.'s has been defined to be the class of Fourier-Stieltjes transforms of d.f.'s. Conversely, given a continuous function g on R, we can recognize whether or not it is a ch.f. by applying the inversion formula: if the right-hand side of the inversion formula exists and is nonnegative for all pairs $a < b$ of finite numbers, then g is a ch.f. up to a multiplicative constant. If g is absolutely integrable on R, then it suffices to apply Corollary 1 of the inversion formula and verify that the function F' is nonnegative. A very important criterion of a different type is that of nonnegative-definiteness that we investigate now.

Let g be a real or complex-valued function on a set $D_S \subset R$ obtained by forming all differences of the elements of a set $S \neq \theta$; for example, $S = [0, U)$ and $D_S = (-U, +U)$, $S = $ set of all positive integers and $D_S = $ set of all integers. Sets D_S are necessarily symmetric with respect to the origin $u = 0$ and contain it. We say that g on D_S is *nonnegative-definite* if for every finite set $S_n \subset S$ and every real or complex-valued function h on S_n

$$\sum_{u,v \in S_n} g(u - v)h(u)\bar{h}(v) \geqq 0;$$

we shall omit mention of D_S when $D_S = R$.

a. *If g on D_S is nonnegative-definite, then, for every $u \in D_S$,*

$$g(0) \geqq 0, \quad g(-u) = \bar{g}(u), \quad |g(u)| \leqq g(0).$$

If, moreover, $D_S \supset (-U, +U)$ and g is continuous at the origin, then g is uniformly continuous on the set of limit points of D_S.

Proof. We apply the defining relation with

$$S_1 = \{0\}, \quad S_2 = \{0, u\}, \quad S_3 = \{0, u, u'\}.$$

With S_1 we obtain $g(0) \geqq 0$. It follows with S_2 that $g(u)h(u) + g(-u)\bar{h}(u)$ is real and hence $g(-u) = \bar{g}(u)$ (take $h(u) = 1$ and $h(u) = i$). We use these two properties below.

The discriminant of a nonnegative quadratic form being nonnegative, elementary computations with S_2 yield $|g(u)| \leqq g(0)$. For the last assertion we exclude the trivial case $g(0) = 0$ which implies $g = 0$, and, to simplify the writing, assume that $g(0) = 1$ (it suffices to replace g by $g/g(0)$). The same discriminant property but with S_3 yields, by elementary computations,

$$|g(u) - g(u')|^2 \leqq 1 - |g(u - u')|^2 - 2\Re\{\bar{g}(u)g(u')(1 - g(u - u'))\}$$

Therefore, if g is continuous at the origin, that is, if $g(u - u') \to g(0) = 1$ as $u' \to u$, then $g(u') \to g(u)$. The proof is complete.

The foregoing proposition shows that a nonnegative-definite function g on R continuous at the origin has properties similar to those of ch.f.'s. In fact, g coincides on R with a ch.f.—up to a multiplicative constant; and this is what we intend to prove now. According to **a**, if $g(0) = 0$, then $g = 0$ so that, by excluding this trivial case and dividing by $g(0)$ *we can and will assume from now on that $g(0) = 1$.*

b. HERGLOTZ LEMMA. *A function g on the set $D_S = \{\cdots -2c, -c, 0, +c, +2c, \cdots\}$ is nonnegative-definite if, and only if, it coincides on this set with a ch.f. $f(u) = \displaystyle\int_{-\pi/c}^{+\pi/c} e^{iux}\, dF(x)$.*

Proof. We can assume that $c > 0$. If g on D_S is nonnegative-definite, then, for every integer n and every finite number x,

$$G'_n(x) = \frac{1}{2\pi} \sum_{k=-n+1}^{n-1} \left(1 - \frac{|k|}{n}\right) g(kc) e^{-ikx}$$

$$= \frac{1}{2\pi n} \sum_{j=1}^{n} \sum_{h=1}^{n} g((j-h)c) e^{-i(j-h)x} \geq 0.$$

Upon multiplying by e^{ikx} with some fixed value of k and integrating over $[-\pi, +\pi)$, we obtain

$$\left(1 - \frac{|k|}{n}\right) g(kc) = \int_{-\pi}^{+\pi} e^{ikx} G'_n(x)\, dx = \int_{-\pi/c}^{+\pi/c} e^{i(kc)x}\, dF_n(x)$$

where F_n is a d.f. with $F_n(-\pi/c) = 0$, $F_n(+\pi/c) = g(0) = 1$. The "only if" assertion follows, on account of the weak compactness and Helly-Bray lemma, by letting $n \to \infty$ along a suitable subsequence of integers. The "if" assertion is immediate (as below).

A. BOCHNER'S THEOREM. *A function g on R is nonnegative-definite and continuous if, and only if, it is a ch.f.*

Proof. The "if" assertion (Mathias) is immediate, since, if g is a ch.f. with d.f. G, then, letting u and v range over an arbitrary but finite set in R,

$$\sum_{u,v} g(u-v) h(u) \bar{h}(v) = \int \left\{ \sum_{u,v} e^{i(u-v)x} h(u) \bar{h}(v) \right\} dG(x)$$

$$= \int \left| \sum_{u} e^{iux} h(u) \right|^2 dG(x) \geq 0.$$

Conversely, let g on R be nonnegative-definite and continuous. It co-incides on R with a ch.f. if it does so on the set S_r (dense in R) of all rationals of the form $k/2^n$, $k = 0, \pm1, \pm2, \cdots, n = 1, 2, \cdots$. For every integer n, let S_n be the corresponding subset of all rationals of the form $k/2^n$ so that $S_n \uparrow S_r$. Since g is nonnegative-definite on R, it is nonnegative-definite on every S_n. Therefore, by **b**, there exist ch.f.'s f_n such that $g(k/2^n) = f_n(k/2^n)$ whatever be k and n. Since $S_n \uparrow S_r$, it follows that $f_n \to g$ on S_r. Let $0 \leqq \theta, \theta_n \leqq 1$, so that, by **b**,

$$1 - \Re f_n(\theta/2^n) = \int_{-\pi}^{+\pi} (1 - \cos \theta x) \, dF_n(2^n x)$$

$$\leqq \int_{-\pi}^{+\pi} (1 - \cos x) \, dF_n(2^n x) = 1 - \Re g(1/2^n).$$

Therefore, by the elementary inequality $|a + b|^2 \leqq 2|a|^2 + 2|b|^2$ and the increments inequality, for every fixed $h = (k_n + \theta_n)/2^n$,

$$|1 - f_n(h)|^2 \leqq 2|1 - f_n(k_n/2^n)|^2 + 4(1 - \Re f_n(\theta_n/2^n))$$

$$\leqq 2|1 - g(k_n/2^n)| + 4(1 - \Re g(1/2^n)).$$

Since g is continuous at the origin, it follows by 13.4, 2°, that the se-quence f_n of ch.f.'s is equicontinuous. Hence, by Ascoli's theorem, it contains a subsequence converging to a continuous function f, so that $g = f$ on S_r and hence on R. Since by the continuity theorem f is a ch.f., the proof is complete.

The "only if" assertion can be proved directly, and this direct proof will extend to a more general case: For every $T > 0$ and $x \in R$

$$p_T(x) = \frac{1}{T} \int_0^T \int_0^T g(u - v)e^{-i(u-v)x} \, du \, dv \geqq 0,$$

since, g on R being nonnegative-definite and continuous, the integral can be written as a limit of nonnegative Riemann sums. Let $u = v + t$, integrate first with respect to v and set $g_T(t) = \left(1 - \dfrac{|t|}{T}\right) g(t)$ or 0 ac-cording as $|t| \leqq T$ or $|t| \geqq T$. The above relation becomes

$$p_T(x) = \int e^{-itx} g_T(t) \, dt \geqq 0.$$

Now multiply both sides by $\dfrac{1}{2\pi}\left(1 - \dfrac{|x|}{X}\right) e^{iux}$ and integrate with respect to x on $(-X, +X)$. The relation becomes

$$\frac{1}{2\pi}\int_{-X}^{+X}\left(1 - \frac{|x|}{X}\right)p_T(x)e^{iux}\,dx = \frac{1}{2\pi}\int \frac{\sin^2\frac{1}{2}X(t-u)}{\frac{1}{4}X(t-u)^2}\,g_T(t)\,dt.$$

The left-hand side is a ch.f. (since its integrand is a product of e^{iux} by a nonnegative function) and the right-hand side converges to $g_T(u)$ as $X \to \infty$. Therefore, g_T is the limit of a sequence of ch.f.'s. Since it is continuous at the origin, the continuity theorem applies and g_T is a ch.f. Since $g_T \to g$ as $T \to \infty$, the same theorem applies, and the assertion is proved.

Extension 1. The question arises whether in **A** continuity at the origin is necessary. Let g on R be nonnegative-definite and Lebesgue-measurable.

By integrating

$$\sum_{u_j, u_k \in S_n} g(u_j - u_k)e^{i(u_j - u_k)x} \geqq 0, \quad x \in R$$

with respect to every $u \in S_n$ over $(0, T)$, we obtain

$$nT^n + n(n-1)T^{n-2}\int_0^T\int_0^T g(u-v)e^{i(u-v)x}\,du\,dv \geqq 0.$$

Dividing by $n(n-1)T^{n-2}$ and letting $n \to \infty$, it follows that

$$\int_0^T\int_0^T g(u-v)e^{i(u-v)x}\,du\,dv \geqq 0.$$

Therefore, the direct proof of the "only if" assertion in **A** continues to apply, but instead of the continuity theorem use 12.2**A** Corollary 2, and we obtain $g = f$ ch.f. almost everywhere (in Lebesgue measure). The "if" assertion is modified accordingly. Thus (F. Riesz)

A′. *A function g on R is nonnegative-definite and Lebesgue-measurable if, and only if, it coincides a.e. with a ch.f.*

Extension 2. It can be shown that Herglotz lemma remains valid with $D_S = \{-Nc, -(N-1)c, \cdots, 0, \cdots (N-1)c, Nc\}$ whatever be the fixed integer N. Then, replacing S_r and S_n by their intersections with $(-U, +U)$ whatever be the fixed U, the proof of **A** remains valid. Thus (Krein)

A″. *A function g on $(-U, +U)$ is nonnegative-definite and continuous if, and only if, it coincides on $(-U, +U)$ with a ch.f.*

REMARK 1.   The proofs of **A** and **A″** use only the fact that g is continuous at the origin, so that these theorems imply the last assertion in **a**.

REMARK 2.   The foregoing proofs show that in the definition of a nonnegative-definite g it suffices to take $h(u) = e^{iux}$ where x runs over R.   Also if g is Lebesgue-measurable, then the definition can be taken to be

$$\int_0^{T_n} \int_0^{T_n} g(u - v)e^{i(u-v)x} \, du \, dv \geq 0$$

for every $x \in R$ and a sequence $T_n \to \infty$.

According to the second extension, a function which coincides with a ch.f. on $(-U, +U)$ can be extended to a ch.f. on R.   The problem which arises is under what conditions this extension is unique.   This is part of the problem we investigate in the following subsection.

*15.2   Regularity and extension of ch.f.'s.   According to 14.1a, if $f = 1$ on an interval $(-U, +U)$, then $f = 1$ on R.   Also according to 13.4, 3°, if $f_n \to 1$ on $(-U, +U)$ then $f_n \to 1$ on R.   Thus, in these cases a ch.f. is determined by its values on an interval, and convergence of a sequence of ch.f.'s on R follows from its convergence on an interval. We intend to investigate more general conditions under which these properties hold.   To simplify the writing, we assume that the ch.f.'s are those of r.v.'s, that is, take the value 1 at $u = 0$.

a. *If $\hat{f}$ is the integral ch.f. corresponding to the ch.f. f, then*

$$\left| \frac{\hat{f}(u + h) - \hat{f}(u - h)}{2h} \right|^2 \leq \frac{1}{2}\{1 + \Re f(h)\}.$$

For, from

$$\frac{\sin^2 x}{x^2} = \frac{\sin^2 2\dfrac{x}{2}}{4 \sin^2 \dfrac{x}{2}} \cdot \frac{\sin^2 \dfrac{x}{2}}{\left(\dfrac{x}{2}\right)^2} \leq \cos^2 \frac{x}{2} = \frac{1 + \cos x}{2},$$

it follows, upon applying the Schwarz inequality, that

$$\left| \frac{\hat{f}(u + h) - \hat{f}(u - h)}{2h} \right|^2 = \left| \int e^{iux} \frac{\sin hx}{hx} \, dF(x) \right|^2$$

$$\leq \int \frac{1 + \cos hx}{2} \, dF(x) = \frac{1}{2}\{1 + \Re f(h)\}.$$

We extend now the uniform convergence theorem 13.2C. Let f_n be ch.f.'s.

b. *If $f_n \to g$ on $(-U, +U)$ and g is continuous at $u = 0$, then the f_n are equicontinuous and the convergence is uniform.*

Proof. Because of 13.4 $(1°, 2°)$ and Ascoli's theorem, it suffices to prove that the f_n are equicontinuous at $u = 0$. If this conclusion is not true, then there exist an $\epsilon > 0$, a sequence $n' \to \infty$, and a sequence $u_{n'} \to 0$, such that $|f_{n'}(u_{n'})| < 1 - \epsilon$ for all n'; given a positive $h \in (-U, +U)$, we take $m_{n'} = \left[\dfrac{h}{u_{n'}}\right]$, so that $m_{n'}u_{n'} \to h$. Upon applying **a** with $u = kh$ and summing over $k = -m + 1, -m + 3, \cdots, m - 1$, we obtain by the elementary inequality $|a_1 + \cdots + a_m|^2 \leqq m|a_1|^2 + \cdots + m|a_m|^2$

$$\left|\frac{\hat{f}(mh) - \hat{f}(-mh)}{2mh}\right|^2 \leqq \frac{1}{2}\{1 + \Re f(h)\}.$$

It follows that

$$\left|\frac{1}{2m_{n'}u_{n'}}\int_{-m_{n'}u_{n'}}^{+m_{n'}u_{n'}} f_{n'}(v)\, dv\right|^2 \leqq \frac{1}{2}\{1 + \Re f_{n'}(u_{n'})\} < 1 - \frac{\epsilon}{2}$$

and, letting $n' \to \infty$, we have

$$\left|\frac{1}{2h}\int_{-h}^{+h} g(v)\, dv\right|^2 \leqq 1 - \frac{\epsilon}{2}.$$

Since $1 = f_n(0) \to g(0)$ and g is continuous at $u = 0$, it follows, letting $h \to 0$, that $1 \leqq 1 - \dfrac{\epsilon}{2}$. Therefore, *ab contrario*, the f_n are equicontinuous at $u = 0$, and the assertion is proved.

A. CONTINUITY THEOREM ON AN INTERVAL. *If $f_n \to f_U$ on $(-U, +U)$ and f_U is continuous at $u = 0$, then f_U extends to a ch.f. f on R; if the extension f is unique, then $f_n \to f$ on R.*

Proof. According to **b**, the f_n are equicontinuous. Therefore, by Ascoli's theorem, the sequence f_n is compact in the sense of uniform convergence and, since $f_n \to f_U$ on $(-U, +U)$, all its limit ch.f.'s coincide with f_U on $(-U, +U)$. It follows that, if there is only one ch.f. f which coincides with f_U on $(-U, +U)$, then $f_n \to f$ on R.

The second part of the problem raised above is reduced to its first part: find ch.f.'s determined by their values on an interval $(-U, +U)$. A partial answer is given by the following theorem (Marcinkiewicz).

B. Extension theorem for ch.f.'s. *If the restriction f_U of a ch.f. f to an interval $(-U, +U)$ is regular or is the boundary function of a regular function, then f_U determines f.*

This theorem follows, by the unicity of analytic continuation, from the three propositions below of independent interest. Let $f(z) = \int e^{izx} \, dF(x)$, where $z = u + iv$ is a point of the complex plane $R_u \times R_v$.

a. *$f(z)$ is regular in a circle $|z| < R$ if, and only if, for every positive $r < R, \int e^{r|x|} \, dF(x)$ is finite.*

Proof. The "if" assertion is immediate and it suffices to prove the "only if" assertion.

Let

$$ m^{(n)} = \int x^n \, dF(x) \quad \text{and} \quad \mu^{(n)} = \int |x|^n \, dF(x). $$

If $f(z)$ is regular for $|z| < R$, then, for every positive $r < R$,

$$ \sum \frac{1}{n!} |m^{(n)}| r^n < \infty, $$

and, in particular,

$$ \sum \frac{1}{(2n)!} \mu^{(2n)} r^{2n} < \infty. $$

Since

$$ (\mu^{(2n-1)})^{\frac{1}{2n-1}} \leq (\mu^{(2n)})^{\frac{1}{2n}}, $$

it follows that

$$ \sum \frac{1}{(2n-1)!} \mu^{(2n-1)} r^{2n-1} < \infty $$

and, hence,

$$ \int e^{r|x|} \, dF(x) = \sum \frac{1}{n!} \mu^{(n)} r^n < \infty. $$

This proves the assertion.

b. *If $f(z)$ is regular in the circle $|z| < R$ or in the rectangle $|\Re z| < U$, $|\Im z| < R$, then $f(z)$ is regular in the strip $|\Im z| < R$.*

Proof. The first assertion follows at once from **a**. As for the second assertion, let V be the largest number such that $f(z)$ is regular in the circle $|z| < V$ and assume that $V < R$. According to **a**, $f(z)$ is regular in the strip $|\Im z| < V$. But it is also regular in the rectangle $|\Re z| < U$, $|\Im z| < R$ and, hence, in the circle whose radius equals min $(R, \sqrt{U^2 + V^2})$. Therefore V cannot be less than R and the proof is concluded.

For every ch.f. f, we have $f(z) = f^+(z) + f^-(z)$ where

$$f^+(z) = \int_0^\infty e^{izx}\, dF(x) \quad \text{and} \quad f^-(z) = \int_{-\infty}^0 e^{izx}\, dF(x)$$

are regular for $\Im z > 0$ and $\Im z < 0$, respectively. Therefore, if, say, $f^+(z)$ is regular for $0 > \Im z > -R$, then $f(z)$ is regular for $0 > \Im z > -R$, so that the ch.f. with values $f(x)$ is the boundary function of a regular function. Thus, the following proposition completes the proof of the foregoing extension theorem.

c. $f^+(z)$ *is regular for* $0 > \Im z > -R$ *if, and only if, for every positive*
$r < R, \int_0^\infty e^{rx}\, dF(x)$ *is finite.*

Proof. The "if" assertion is immediate. As for the "only if" assertion, we observe that, since $f^+(z)$ is regular for $\Im z > 0$ and continuous for $\Im z \geq 0$, regularity for $0 > \Im z > -R$ implies, by a well-known symmetry property, regularity for $|\Im z| < R$ and, hence, according to **a**, $\int_0^\infty e^{rx}\, dF(x)$ is finite for $0 < r < R$.

PARTICULAR CASES.   Upon applying what precedes, we have

1° *If* $f_n(u) \to e^{iua}$ *on* $(-U, +U)$, *then* $f_n(u) \to e^{iua}$ *for every* $u \in R$.

2° *If* $f_n(u) \to e^{-\frac{u^2}{2}}$ *on* $(-U, +U)$, *then* $f_n(u) \to e^{-\frac{u^2}{2}}$ *for every* $u \in R$.

3° *If* $f_n \to f$ *on* $(-U, +U)$ *and* f *is ch.f. of a r.v. bounded either above or below, then* $f_n \to f$ *on* R.

d. UNICITY LEMMA.   *Let* $g(z)$ *be regular for* $\Im z > 0$ *and continuous for* $\Im z \geq 0$.
If $g(z) = f^+(z)$ *for* $z = 0$ *then* $g(z) = f^+(z)$ *for* $z \geq 0$.

For, $h(z) = g(z) - f^+(z)$ being regular for $\Im z > 0$ and continuous for $\Im z \geq 0$ with $h(z) = 0$ for $z = 0$ extends, by analytic continuation to an entire function vanishing for $z = 0$ hence vanishing everywhere.

*15.3   Composition and decomposition of regular ch.f.'s.   Let F denote the composed $F_1 * F_2$ of d.f.'s F_1 and F_2. In the case of f or f_1, f_2 regular, the composition theorem 13.4A can be completed as follows:

A. COMPOSITION THEOREM FOR REGULAR CH.F.'S. *$f(z)$ is regular in the strip $|\Im z| < R$ if, and only if, $f_1(z)$ and $f_2(z)$ are regular in $|\Im z| < R$.* This theorem follows at once, by 15.2a and **b**, from the

COMPOSITION LEMMA. *If $F = F_1 * F_2$ then, for every v,*

$$\int e^{vx}\,dF(x) = \int e^{vx}\,dF_1(x)\int e^{vx}\,dF_2(x),$$

and there exist finite numbers $\alpha_j > 0$, $\beta_j \geqq 0$ such that

$$\int e^{vx}\,dF(x) \geqq \alpha_j e^{-\beta_j|v|}\int e^{vx}\,dF_j(x), \quad j = 1, 2.$$

Proof. We exclude the trivial case of degenerate F_1 or F_2. The first assertion follows, using Fatou's lemma, in a way similar to that of the proof of the composition theorem 13.3A, whether the integrals are finite or not.

As for the second assertion, for every b, either

$$\int e^{vx}\,dF_1(x) \geqq \int_b^\infty e^{vx}\,dF_1(x) \geqq e^{bv}F_1[b, +\infty)$$

or

$$\int e^{vx}\,dF_1(x) \geqq \int_{-\infty}^b e^{vx}\,dF_1(x) \geqq e^{bv}F_1(b),$$

according as $v \geqq 0$ or $v < 0$. Let β_2 be the larger of two finite numbers $|b_1|$ and $|b_2|$ such that

$$a_1 = F_1[b_1, +\infty) > 0 \quad \text{and} \quad a_2 = F_1(b_2) > 0$$

and let α_2 be the smaller of a_1 and a_2. Then the inequalities above and the first assertion yield

$$\int e^{vx}\,dF(x) \geqq \alpha_2 e^{-\beta_2|v|}\int e^{vx}\,dF_2(x)$$

and the proof is complete.

COMPLEMENTS AND DETAILS

Unless otherwise stated, functions F, with or without affixes, are d.f.'s of r.v.'s: $F(-\infty) = 0$, $F(+\infty) = 1$, and functions f, with same affixes if any, are corresponding ch.f.'s.

1. If F is purely discontinuous and the discontinuity set is dense in R, then the nondecreasing inverse function is singular.

2. If $F_{X_n} \xrightarrow{c} F_X$ and μ is any limit point of the sequence $\mu(X_n)$ of medians of the X_n, then μ is a median of X. In particular, if $\mu(X)$ is the unique median of X, then $\mu(X_n) \to \mu(X)$. (Take $x' < \mu < x''$ to be continuity points of F, then $F(x') \leqq \frac{1}{2}$.)

3. *P. Lévy's space.* Let $\mathfrak{F}$ be the space of all d.f.'s F of r.v.'s. Set $d(F, F')$ to be the infimum of all those h for which $F(x - h) - h \leqq F'(x) \leqq F(x + h) + h$ whatever be $x \in R$.

(a) Draw a graph and interpret $d(F, F')$ geometrically by considering lengths of segments intercepted by the graphs of F and F' on parallels to the second bisector.

(b) The function d so defined is a distance, and $(\mathfrak{F}, d)$ is a complete metric space.

(c) The following three assertions are equivalent:

$$F_n \xrightarrow{c} F, \quad d(F_n, F) \to 0, \quad \int g \, dF_n \to \int g \, dF$$

for every function g continuous and bounded on R.

(d) A set S in $\mathfrak{F}$ is compact if, and only if, $F(x) \to 0$ as $x \to -\infty$ and $F(x) \to 1$ as $x \to +\infty$, uniformly on S.

4. Establish the following correspondences for laws.

Binomial: $p_k = C_n^k p^k q^{n-k}, \; k \leqq n, \; f(u) = (pe^{iu} + q)^n.$

Poissonian: $p_k = \dfrac{\lambda^k}{k!} e^{-\lambda}, \; k = 0, 1, \cdots, \; f(u) = e^{\lambda(e^{iu} - 1)}.$

Uniform: $F'(x) = \dfrac{1}{b - a}$ in (a, b), and 0 outside, $f(u) = \dfrac{e^{ibu} - e^{iau}}{i(b - a)u}.$

Cauchy: $F'(x) = \dfrac{1}{\pi} \dfrac{a}{a^2 + (x - b)^2}, \; a > 0, \; f(u) = e^{-a|u| + ibu}.$

Laplace: $F'(x) = \dfrac{1}{2a} e^{-|x - b|/a}, \; a > 0, \; f(u) = (1 + a^2u^2)^{-1} e^{ibu}.$

Normal: $F'(x) = \dfrac{1}{\sigma\sqrt{2\pi}} e^{-(x - m)^2/2\sigma^2}, \sigma > 0, \; f(u) = e^{imu - \frac{\sigma^2 u^2}{2}}.$

Squared Normal: $(m = 0, \sigma = 1)$: $F'(x) = \dfrac{1}{\sqrt{2\pi x}} e^{-x/2}$ for $x > 0$ and $= 0$ for $x \leqq 0, \; f(u) = (1 - 2iu)^{-\frac{1}{2}}.$

Γ-type: $F'(x) = \dfrac{c^\gamma}{\Gamma(\gamma)} x^{\gamma - 1} e^{-cx}$ for $x > 0, \; c > 0, \; \gamma > 0, \; 0$ for $x \leqq 0, \; f(u) = \left(1 - \dfrac{iu}{c}\right)^{-\gamma}.$

5. The composed $\overline{F}$ of F with the uniform distribution on $(-h, +h)$ is given by

$$\overline{F}(x) = \frac{1}{2h} \int_{x-h}^{x+h} F(y) \, dy, \quad \overline{f}(u) = \frac{\sin hu}{hu} f(u).$$

An absolutely convergent inversion integral follows:

$$\frac{1}{2h} \int_x^{x+2h} F(y) \, dy - \frac{1}{2h} \int_{x-2h}^x F(y) \, dy = \frac{1}{\pi} \int_{-\infty}^\infty \left(\frac{\sin u}{u}\right)^2 e^{-iux/h} f\left(\frac{u}{h}\right) du.$$

Deduce the continuity theorem.

6. Let $M_h f = \dfrac{2}{\pi h} \displaystyle\int_0^\infty |f(u)|^2 \dfrac{\sin^2 hu}{u^2}\, du,\ h > 0$, and let.

$$Mf = \lim_{u \to \infty} \frac{1}{2u} \int_{-u}^{+u} |f(v)|^2\, dv.$$

(a) $M_h f$ is nondecreasing in h and converges to 1 or Mf according as $h \to \infty$ or $h \to 0$. $\displaystyle\lim_{m\to\infty} \lim_{n\to\infty} M_h(\prod_{k=m+1}^{n} f_k)$ is either 0 or 1 (identically in h).

(b) $Mf = \sum p_k^2$ where the p_k are jumps of F; $Mf_1 f_2 \geqq Mf_1 \cdot Mf_2$; $M_h f = 2\displaystyle\int_0^{2h}\left(1 - \frac{x}{2h}\right) dF^s(x)$ where F^s is d.f. with ch.f. $f^s = |f|^2$. (The sum is the jump at 0 of $\mathfrak{L}(X) * \mathfrak{L}(-X)$ where X is a r.v. with d.f. F.)

(c) If $f_n \to f$ with $Mf = 0$, then $Mf_n \to 0$; the converse is not necessarily true. If $\prod_{k=1}^{n} f_k \to f$, then $M(\prod_{k=1}^{n} f_k) \to Mf$.

7. A law is a "lattice" law if the only possible values are of form $a + ns$ only, $s > 0;\ n = 0, \pm 1, \cdots$; if s is the largest possible, then s is the "step" of the law. The step is well determined.

(a) A law is a lattice law if, and only if, $|f(u_0)| = 1$ for an $u_0 \neq 0$. The step s is given by the property that $|f(u)| < 1$ in $0 < |u| < 2\pi/s$ and $f(2\pi/s) = 1$.

(b) Let $p_n = P[X = a + ns]$ where X has a lattice law with step s. Then

$$p_n = \frac{s}{2\pi}\int_{-\pi/s}^{+\pi/s} e^{-iau - insu} f(u)\, du,$$

$$F(x_2) - F(x_1) = \frac{s}{2\pi}\int_{-\pi/s}^{|\pi/s|} \frac{e^{iux_1} - e^{-iux_2}}{2i\sin\dfrac{su}{2}} f(u)\, du$$

where $x_1 = a + ms - \frac{1}{2}s,\ x_2 = a + ns + \frac{1}{2}s,\ n \geqq m$.

8. If the moment m_k exists and is finite, then

$$\log f(u) = \sum_{k=1}^{n} \frac{a_k}{k!} (iu)^k + o(u^k).$$

The a_k are called semi-invariants; formally

$$\sum_{n=1}^{\infty} \frac{a_n}{n!} z^n = \log \sum_{n=0}^{\infty} \frac{m_n}{n!} z^n.$$

Deduce the expression of a few first semi-invariants in terms of moments, and conversely. Prove that

$$|a_k| \leqq k^k \mu_k.$$

$(\log \sum_{k=1}^{n} \dfrac{m_k}{k!} z^k$ is majorized by $\sum_{k=1}^{\infty} \dfrac{1}{k}(e^{\mu k^{1/k_z}} - 1)^k.)$

9. If the derivative F' on R exists and is finite, then $f(u) \to 0$ as $|u| \to \infty$. (Use Riemann-Lebesgue lemma.)

If the nth derivative $F^{(n)}$ on R exists, is finite, and is absolutely integrable, then $f(u) = o(|u|^{1-n})$ for $|u| \to \infty$. (Integrate by parts.)

10. Let X be a r.v. with d.f. F.

(a) If $P[|X| \geq x] \to 0$ as $x \to \infty$ faster than any power of x^{-1}, then all moments exist and are finite. (Integrate by parts $\int |x|^n \, dF(x)$.)

A pr. law is determined by the sequence of moments assumed finite if the series $\sum_{n=0}^{\infty} \dfrac{m_n}{n!} u^n$ has a nonnull radius of convergence ρ. (Use Schwarz's inequality to show that the series with the m_n replaced by μ_n majorizes the expansion of f about any value of u, and then use analytic continuation.)

(b) Formally, by integration by parts,

$$f(z) = 1 - iz\int_{-\infty}^{0} e^{izx}F(x)\,dx + iz\int_{0}^{\infty} e^{izx}(1 - F(x))\,dx.$$

If $P[|X| \geq x] \to 0$ as $x \to \infty$ faster than e^{-rx} for every positive $r < \rho$, then $f(z)$ is analytic in the strip $|\Im z| < \rho$. If $\rho = \infty$, then $f(z)$ is an entire function.

(c) If $e^{|x|^r}F'(x) \geq c > 0$ on R for an $r < \frac{1}{2}$, then the pr. law is not determined by its moments.

11. If f' exists and is finite on R, $\int |x| \, dF(x)$ may be infinite: take

$$f(u) = c \sum_{n=2}^{\infty} \frac{\cos nu}{n^2 \log n}.$$

(The differentiated series converges uniformly but $\sum 1/n \log n = \infty$.) Let $m' = \lim_{a \to +\infty} \int_{-a}^{+a} x\, dF(x)$ be the "symmetric" first moment. If m' exists and is finite, $f'(0)$ may not exist: take a Weierstrass non-differentiable function $c \sum a^n \cos b^n u$.

If the derivative at $u = 0$ of $\Re f$ exists, then

$$\frac{f(h) - 1}{h} = o(1) + i\int_{-1/h}^{+1/h} x\, dF(x), \quad 0 < h \to 0.$$

(Set $G(x) = F(x) - G(x)$, $H(x) = F(x) + F(-x)$, so that $|\Delta H| \leq \Delta G$. Show that $\int \dfrac{\sin^2 (hx/2)}{h}\, dG(x) \to 0$ as $h \to 0$, $\int_{1/h}^{\infty} \dfrac{\sin (hx)}{x}\, dH(x) = o(1)$.)

Under the foregoing condition, f' exists and is finite if, and only if, $m' = \lim_{a \to \infty} \int_{-a}^{+a} x\, dF(x)$ exists and is finite, and then $f'(0) = im'$. Extend to any derivative of odd order. What about those of even order?

12. If g on R is not constant and $g(u) = 1 + o(u) + o(u^2)$ near $u = 0$ with $o(u)$ an odd function, then g is not a ch.f. (Observe that $g(u)g(-u) = 1 + o(u^2)$.)
Examples: e^{-u^4}, $e^{-|u|^r}$ for $r > 2$, $e^{-u^4 - u^6}$, $1/(1 + u^4)$.

13. Let g on R be real, even and continuous, with $g(0) = 1$, $g(u) \to 0$ as $u \to \infty$.

If g is convex from below, on $[0, +\infty)$, then g is a ch.f. (To prove $\int_{0}^{\infty} g(u) \cos xu \, du \geq 0$ for $x > 0$; observe that on $[0, \infty)$, say, the left-hand side derivative g' exists and is nondecreasing, with $g'(u) \leq 0$ and $g'(u) \to 0$ as $u \to \infty$. Set $h = -g'$ so that, by integration by parts,

$$x \int_0^\infty g(u) \cos xu \, du = \int_0^\infty h(u) \sin xu \, du.$$

For $x > 0$, the last integral is

$$\int_0^{\pi/x} \left\{ h(u) - h\left(u + \frac{\pi}{x}\right) + h\left(u + \frac{2\pi}{x}\right) - h\left(u + \frac{3\pi}{x}\right) + \cdots \right\} \sin xu \, du \geq 0.)$$

Examples: $e^{-|u|}$, $1/(1 + |u|)$, $1 - |u|$ for $|u| \leq 1$ and 0 for $|u| > 1$.

14. (a) Two ch.f.'s may coincide on intervals without being identical.

Take $F'_1(x) = \dfrac{1 - \cos x}{\pi x^2}$ hence $f_1(u) = 1 - |u|$ for $|u| \leq 1$ and 0 for $|u| > 1$, and take F_2 defined by $p_0 = \frac{1}{2}$, $p_{\pm \pi(2k+1)} = \dfrac{2}{\pi^2(2k + 1)^2}$; $f_2(u)$ is periodic of period two and coincides with f_1 on $[-1, +1]$. Or, take f to be a ch.f. of the type described in 13 with f' continuous and strictly increasing on $[0, \infty)$. Replace two arbitrarily small arcs of the graph of f which are symmetric with respect to the y-axis by their chords, and compare the function so defined with f.

(b) The compositions of a law with either one of two distinct laws may coincide ($f_1f_1 = f_1f_2$).

(c) If $f_n \to f$ on $[-U, +U]$, the same may not be true on R.

15. f on R is a ch.f. if, and only if, there exists a sequence g_n such that

$$\int |g_n(v)|^2 \, dv \to 1 \quad \text{and} \quad \int g_n(u + v)\bar{g}_n(v) \, dv \to f(u) \quad \text{uniformly in every finite}$$

interval.

(For the "if" assertion, observe that every integral is positive-definite. For the "only if" assertion, divide $[-n, +n]$ into n^2 equal subintervals, set $F_n(-n) = 0$, $F_n(n) = 1$, $F_n = F$ at the subdivision points, and linear inside every subinterval; set $c_n g_n(u) = \displaystyle\int_{-n}^{+n} \sqrt{F'_n(x)} \, e^{iux} \, dx$ with $g_n(0) = 1$. Compute f_n and observe that $f_n \to f$.)

16. Let g and h be bounded and continuous on R, with $\bar{g}(u) = g(-u)$, and let $\lambda(u)$ be an arbitrary finite function on R.

If for every finite set A of values of u

$$\left| \sum_{u \in A} \sum_{v \in A} g(u - v)\lambda(u)\bar{\lambda}(v) \right| \leq \sum_{u \in A} \sum_{v \in A} h(u - v)\lambda(u)\bar{\lambda}(v),$$

then

$$h(u) = \int e^{iux} \, dH(x)$$

where H is a d.f. up to a multiplicative constant.

The foregoing inequalities represent a necessary and sufficient condition for g to be of the form

$$g(u) = \int e^{iux} \, dG(x)$$

with $|\Delta G| \leq \Delta H$. Find the relation between discontinuity and continuity points of G and H.

17. The uniqueness and composition properties determine "essentially" the form of ch.f.'s. Let K on $R \times R$ be bounded and continuous. If the functions g on R are defined by $g(u) = \int K(u, x) \, dF(x)$ for every d.f. F, and the uniqueness and composition properties hold, then $K(u, x) = e^{ixh(u)}$ and $f(h(u)) = g(u)$.

18. Normal vectors. A normal vector $X = (X_k, k \leq n)$ is so defined that all r.v.'s of the form $\sum_k u_k X_k$ are normal. Let the X_k be centered at expectations.

A ch.f. f on R^n is that of a normal vector (centered at its expectation) if, and only if,

$$\log f(u_1, \cdots, u_n) = Q(u_1, \cdots, u_n) = -\tfrac{1}{2} \sum_{jk} m_{jk} u_j u_k \geq 0$$

where $m_{jk} = EX_j X_k$.

If the inequality is strict, then the normal d.f. is defined by

$$\frac{\partial^n}{\partial_{x_1} \cdots \partial_{x_n}} F(x_1, \cdots, x_n) = \frac{1}{(2\pi)^{n/2} D^{\frac{1}{2}}} e^{-\frac{1}{2}g(x_1, \cdots, x_n)}$$

where $D = \| m_{jk} \| > 0$ and $g(x_1, \cdots, x_n) = \frac{1}{D} \sum_{jk} D_{jk} x_j x_k$ is the reciprocal form of $Q(u_1, \cdots, u_n)$ with the variables x_k. What if $Q \geq 0$?

19. If (X, Y) is a normal pair centered at expectations, then $EXY/\sigma X \sigma Y = \cos p\pi$ where $p = P[XY < 0]$. (Compute $P[XY < 0]$ using the d.f.)

Part Three

INDEPENDENCE

Until very recently, probability theory could have been defined to be the investigation of the concept of independence. This concept continues to provide new problems. Also it has originated and continues to originate most of the problems where independence is not assumed.

The main model is that of sequences of sums of independent random variables. The main problems are the Strong Central Limit Problem and the (Laws) Central Limit Problem. The first is concerned with almost sure convergence and stability properties. The second one is concerned with convergence of laws. All general results were obtained since 1900.

Chapter V

SUMS OF INDEPENDENT RANDOM VARIABLES

Two properties play a basic role in the study of independent r.v.'s: the Borel zero-one law and the multiplication theorem for expectations. Two general a.s. limit problems for sums of independent r.v.'s have been investigated: the a.s. convergence problem and the a.s. stability problem. Both of them took their present form in the second quarter of this century.

§ 16. CONCEPT OF INDEPENDENCE

CONVENTION. To avoid endless repetitions, we make the convention that, unless otherwise stated,

— r.v.'s, random vectors and, in general, random functions are defined on a fixed but otherwise arbitrary pr. space $(\Omega, \mathcal{C}, P)$.
— indices t vary on a fixed but otherwise arbitrary index set T, and events of a class have the index of the class.

16.1 Independent classes and independent functions. Events A_t are said to be *independent* if, for every finite subset $(t_1, \cdots, t_n)$,

(I)
$$P \bigcap_{k=1}^{n} A_{t_k} = \prod_{k=1}^{n} P A_{t_k}.$$

In fact, the concept of independence is relative to families of classes (see Application 1° below).

Classes $\mathcal{C}_t$ of events are said to be *independent* if their events are independent; in other words, if events selected arbitrarily one from each class are independent. Clearly, if the $\mathcal{C}_t$ are independent so are the $\mathcal{C}'_{t'} \subset \mathcal{C}_{t'}$, $t' \in T' \subset T$. Because of its constant use, we state this fact as a theorem.

A. *Subclasses of independent classes are independent.*

Let X_t be r.v.'s or random vectors or, in general, random functions. Let $\mathfrak{B}(X_t)$ be the sub σ-field of events induced by X_t, that is, the inverse image under X_t of the Borel field in the range space of X_t.

The X_t are said to be *independent* if they induce independent σ-fields $\mathfrak{B}(X_t)$. Then classes $\mathfrak{B}_t \subset \mathfrak{B}(X_t)$ are independent. Since a Borel function of X_t induces a sub σ-field $\mathfrak{B}_t$ of events contained in $\mathfrak{B}(X_t)$, it follows that

A'. BOREL FUNCTIONS THEOREM. *Borel functions of independent random functions are independent.*

Independent classes can be enlarged, to some extent, without destroying independence. More precisely

Let $\mathfrak{C}_t$ be independent classes. Independence is preserved if to every $\mathfrak{C}_t$ we adjoin

1° *the null and the a.s. events,* for (I) is trivially true—both sides reducing to 0—when at least one of the events which figure in it is null, while (I) with n indices reduces to (I) with fewer indices when at least one of the events which figure in it is a.s.;

2° *the proper differences of its elements and, in particular, their complements* (because of 1°), for if $A_{t_1} \supset A'_{t_1}$, then

$$P(A_{t_1} - A'_{t_1})A_{t_2} \cdots A_{t_n} = PA_{t_1}A_{t_2} \cdots A_{t_n} - PA'_{t_1}A_{t_2} \cdots A_{t_n}$$

$$= (PA_{t_1} - PA'_{t_1})PA_{t_2} \cdots PA_{t_n}$$

$$= P(A_{t_1} - A'_{t_1})PA_{t_2} \cdots PA_{t_n};$$

3° *the countable sums of its elements,* for

$$P(\sum_j A_{t_1}{}^j)A_{t_2} \cdots A_{t_n} = \sum_j PA_{t_1}{}^j A_{t_2} \cdots A_{t_n}$$

$$= (\sum_j PA_{t_1}{}^j)PA_{t_2} \cdots PA_{t_n}$$

$$= P(\sum_j A_{t_1}{}^j)PA_{t_2} \cdots PA_{t_n};$$

4° *the limits of sequences of its elements,* for if $A_{t_1}{}^m \to A_{t_1}$ as $m \to \infty$, then

$$PA_{t_1}A_{t_2} \cdots A_{t_n} \leftarrow PA_{t_1}{}^m A_{t_2} \cdots A_{t_n}$$

$$= PA_{t_1}{}^m PA_{t_2} \cdots PA_{t_n} \to PA_{t_1}PA_{t_2} \cdots PA_{t_n}.$$

It follows easily that

B. EXTENSION THEOREM. *Minimal σ-fields over independent classes $\mathcal{C}_t$ closed under finite intersections are independent.*

Applications. 1° If the events A_t are independent, so are the σ-fields $(A_t, A_t{}^c, \emptyset, \Omega)$.

2° If the inverse images $\mathcal{C}_t$ of the classes of all intervals $(-\infty, x_t)$ in Borel spaces R_t are independent, so are the inverse images $\mathcal{B}_t$ of the Borel fields in the R_t. For, every $\mathcal{C}_t$ is closed under finite intersections and $\mathcal{B}_t$ is the minimal σ-field over $\mathcal{C}_t$.

*3° Let $\mathcal{B}_t$ be σ-fields (or fields) of events and let T_s be a subset of the index set T. The compound σ-field $\mathcal{B}_{T_s}$ with components $\mathcal{B}_t$, $t \in T_s$, is the minimal σ-field over the class $\mathcal{C}_{T_s}$ of all finite intersections of events A_t, $t \in T_s$, and contains all its components; since the $\mathcal{B}_t$ are closed under finite intersections so is $\mathcal{C}_{T_s}$. $\mathcal{B}_{T_s}$ is a compound sub σ-field of $\mathcal{B}_T$ and, if T_s is finite, then $\mathcal{B}_{T_s}$ is a "finitely compound" sub σ-field.

If compound σ-fields are independent, then, by **A**, their finitely compound sub σ-fields are independent. Conversely, if the finitely compound sub σ-fields are independent, then, by the extension theorem, the compound σ-fields are independent. We state these facts as a theorem.

C. COMPOUNDS THEOREM. *Compound σ-fields are independent if, and only if, their finitely compound sub σ-fields are independent.*

In particular, if the $\mathcal{B}_t$ are independent, so are the $\mathcal{B}_{T_s}$ for every partition of T into set T_s.

Families $X_{T_s} = \{X_t, t \in T_s\}$ of r.v.'s induce sub σ-fields $\mathcal{B}(X_{T_s})$ of events. Every $\mathcal{B}(X_{T_s})$ is the minimal σ-field over the class $\mathcal{C}(X_{T_s})$ of inverse images of all intervals in the range space R_{T_s} of X_{T_s}. But the intervals in the Borel space R_{T_s} are products of intervals in the factor spaces R_t, with only a finite number of factor intervals different from the whole factor spaces, and the inverse image of any factor space is Ω. Therefore the elements of $\mathcal{C}(X_{T_s})$ are all the finite intersections of elements of the $\mathcal{B}(X_t)$. It follows that the σ-field $\mathcal{B}(X_{T_s})$ is a compound of the σ-fields $\mathcal{B}(X_t)$, and theorem **C** becomes

C′. FAMILIES THEOREM. *Families of random variables are independent if, and only if, their finite subfamilies are mutually independent.*

Thus, in the last analysis, independence of random functions reduces to independence of random vectors.

To conclude this investigation of the definition of independence, let us observe that all which precedes applies to complex r.v.'s, to complex random vectors, and, in general, to complex random functions $X_t = X'_t + iX''_t$ considered as vector random functions (X'_t, X''_t), $t \in T$.

16.2 Multiplication properties. The direct definition of independent r.v.'s is as follows:

Random variables X_t, $t \in T$, are *independent* if, for every finite class $(S_{t_1}, \cdots, S_{t_n})$ of Borel sets in R,

$$P \bigcap_{k=1}^{n} [X_{t_k} \in S_{t_k}] = \prod_{k=1}^{n} P[X_{t_k} \in S_{t_k}].$$

The basic expectation property of independent r.v.'s is expressed by

a. MULTIPLICATION LEMMA. *If* $X_1, \cdots, X_n$ *are independent non-negative r.v.'s, then* $E \prod_{k=1}^{n} X_k = \prod_{k=1}^{n} EX_k$.

Proof. It suffices to prove the assertion for two independent r.v.'s X and Y, for then the general case follows by induction. First, let $X = \sum_j x_j I_{A_j}$ and $Y = \sum_k y_k I_{B_k}$ be nonnegative simple (or elementary) r.v.'s; we can always take the x_j, and, similarly, the y_k, to be all distinct, so that $A_j = [X = x_j]$, $B_k = [Y = y_k]$. Since X and Y are independent, $PA_j B_k = PA_j PB_k$ and, hence,

$$EXY = \sum_{j,k} x_j y_k PA_j PB_k = \sum_j x_j PA_j \cdot \sum_k y_k PB_k = EXEY.$$

Now, let X and Y be nonnegative r.v.'s and set

$$A_{nj} = \left[\frac{j-1}{2^n} \leq X < \frac{j}{2^n} \right], \quad B_{nk} = \left[\frac{k-1}{2^n} \leq Y < \frac{k}{2^n} \right].$$

Since X and Y are independent so are these events and, hence, so are the simple r.v.'s

$$X_n = \sum_{j=1}^{n2^n} \frac{j-1}{2^n} I_{A_{nj}}, \quad Y_n = \sum_{k=1}^{n2^n} \frac{k-1}{2^n} I_{B_{nk}}.$$

But $0 \leq X_n \uparrow X$, $0 \leq Y_n \uparrow Y$, so that $0 \leq X_n Y_n \uparrow XY$ and, by what precedes, $EX_n Y_n = EX_n EY_n$. Therefore, by the monotone convergence theorem, $EXY = EXEY$, and the lemma is proved.

A. MULTIPLICATION THEOREM. *Let* $X_1, \cdots, X_n$ *be independent r.v.'s. If these r.v.'s are integrable so is their product, and* $E \prod_{k=1}^{n} X_k = \prod_{k=1}^{n} EX_k$. *Conversely, if their product is integrable and none is degenerate at* 0, *then they are integrable.*

Proof. It suffices to prove the assertion for two independent r.v.'s X and Y. We observe that independence of X and Y implies that the nonnegative r.v.'s $X' = X^+$ or X^- or $|X|$ and $Y' = Y^+$ or Y^- or $|Y|$ are independent, so that, by **a**, $EX'Y' = EX'EY'$. Now, if X and Y are integrable so are X' and Y' and, by the foregoing equality, so is $X'Y'$. Therefore $|XY|$ and hence XY are integrable and, by the same equality,

$$EXY = E(X^+ - X^-)(Y^+ - Y^-)$$

$$= EX^+EY^+ - EX^+EY^- - EX^-EY^+ + EX^-EY^-$$

$$= EXEY.$$

Conversely, if XY is integrable so that $E|X|E|Y| = E|XY| < \infty$, and neither X nor Y degenerates at 0 so that $E|X|$ and $E|Y|$ do not vanish, then $E|X|$ and $E|Y|$ are finite, and the proof is concluded.

Extension. The multiplication theorem remains valid for independent complex r.v.'s $X_k = X'_k + iX''_k$, since it applies to every term of the expansion of $\prod_{k=1}^n (X'_k + iX''_k)$. In particular, according to the Borel functions theorem, if the X_k are independent so are the e^{iuX_k} and, hence,

$$Ee^{iu \sum_{k=1}^n X_k} = E \prod_{k=1}^n e^{iuX_k} = \prod_{k=1}^n Ee^{iuX_k}.$$

In other words,

COROLLARY. *Ch.f.'s of sums of independent r.v.'s are products of ch.f.'s of the summands.*

This proposition, to be used extensively in the following chapter, is but a special case of a property which can serve as an equivalent definition of independent r.v.'s, as follows:

Let F_t and f_t, $F_{t_1 \cdots t_n}$ and $f_{t_1 \cdots t_n}$ be the d.f.'s and ch.f.'s of the r.v. X_t and of the random vector $(X_{t_1} \ldots X_{t_n})$, respectively.

B. EQUIVALENCE THEOREM. *The three following definitions of independence of the r.v.'s X_t are equivalent.*

For every finite class of Borel sets S_t and of points $x_t, u_t \in R$

(I₁) $$P \bigcap_{k=1}^n [X_{t_k} \in S_{t_k}] = \prod_{k=1}^n P[X_{t_k} \in S_{t_k}],$$

(I₂) $$F_{t_1 \cdots t_n}(x_{t_1}, \cdots, x_{t_n}) = F_{t_1}(x_{t_1}) \cdots F_{t_n}(x_{t_n}),$$

(I₃) $$f_{t_1 \cdots t_n}(u_{t_1}, \cdots, u_{t_n}) = f_{t_1}(u_{t_1}) \cdots f_{t_n}(u_{t_n}).$$

Proof. (I_1) implies (I_2) by taking $S_t = (-\infty, x_t)$. Conversely, (I_2) implies (I_1) with $S_t = (-\infty, x_t)$ and, on account of 16.1, Application 2°, this implies (I_1) for all S_t.

(I_2) implies (I_3), for (I_2) implies (I_1) which implies (I_3) exactly as the multiplication theorem implies its corollary. Conversely, (I_3) implies (I_2), for the inversion formula for one- and multi-dimensional ch.f.'s shows at once that if (I_3) is true, then, for all continuity intervals,

$$F_{t_1 \cdots t_n}[a_{t_1}, \cdots, a_{t_n}; b_{t_1}, \cdots, b_{t_n}] = F_{t_1}[a_{t_1}, b_{t_1}] \cdots F_{t_n}[a_{t_n}, b_{t_n}],$$

and (I_2) follows by letting the $a_t \to -\infty$ and $b_t \uparrow x_t$. This completes the proof.

Extension. The equivalence theorem is valid when the X_t are random vectors, for the proof applies word by word, provided R is replaced by the range space R_t of X_t.

16.3 Sequences of independent r.v.'s. At the root of known a.s. limit properties of sequences of independent r.v.'s lies the celebrated

A. Borel zero-one criterion. *If the events A_n are independent, then $P(\limsup A_n) = 0$ or 1 according as $\sum PA_n < \infty$ or $= \infty$.*

Proof. Since

$$P(\limsup A_n) = \lim_m \lim_n P \bigcup_{k=m}^{n} A_k = \lim_m \lim_n (1 - P \bigcap_{k=m}^{n} A_k^c)$$

and the events A_n and hence A_n^c are independent, the assertion follows by passing to the limit in the elementary inequality

$$1 - \exp\left[-\sum_{k=m}^{n} PA_k\right] \leqq 1 - \prod_{k=m}^{n} (1 - PA_k) \leqq \sum_{k=m}^{n} PA_k.$$

Since, whatever be the events A_n, $\sum PA_n < \infty$ implies that

$$\lim_m \lim_n P \bigcup_{k=m}^{n} A_k \leqq \lim_m \lim_n \sum_{k=m}^{n} PA_k = 0,$$

the "zero" part of this criterion is valid *with no assumption of independence:*

a. Borel-Cantelli lemma. *If $\sum PA_n < \infty$, then $P(\limsup A_n) = 0$.*

Corollary 1. *If the events A_n are independent and $A_n \to A$, then $PA = 0$ or 1.*

Corollary 2. *If the r.v.'s X_n are independent and $X_n \xrightarrow{\text{a.s.}} 0$, then $\sum P[|X_n| \geqq c] < \infty$ whatever be the finite number $c > 0$.*

For $X_n \xrightarrow{\text{a.s.}} 0$ implies that, if $A_n = [|X_n| \geq c]$, then $P(\limsup A_n) = 0$, and independence of the X_n implies that of the A_n.

Because of its intuitive appeal, instead of "$\limsup A_n$" we shall sometimes write "A_n i.o."; to be read "A_n's occur infinitely often" or "infinitely many A_n occur." This terminology corresponds to the fact that $\limsup A_n$ is the set of all those elementary events which belong to infinitely many A_n or, equivalently, to some of "the A_n, A_{n+1}, $\cdots$ however large be n"—the "tail" of the sequence A_n. To the "tail" of the sequence A_n of events corresponds the "tail" of the sequence I_{A_n} of their indicators. More generally, the "tail" of a sequence X_n of r.v.'s is "the sequence X_n, X_{n+1}, $\cdots$ however large be n."

To be precise, let X_1, X_2, $\cdots$ be a sequence of r.v.'s and let $\mathcal{B}(X_n)$, $\mathcal{B}(X_n, X_{n+1})$, $\cdots$, $\mathcal{B}(X_n, X_{n+1}, \cdots)$, $\mathcal{B}(X_{n+1}, X_{n+2}, \cdots)$, $\cdots$ be sub σ-fields of events induced by the random functions within the brackets. We give a precise meaning to $\limsup \mathcal{B}(X_n)$, as follows: The sequence $\mathcal{B}(X_n)$, $\mathcal{B}(X_n, X_{n+1})$, $\cdots$ is a nondecreasing sequence of σ-fields, its supremum or union is a field, and the minimal σ-field over this field is $\mathcal{B}(X_n, X_{n+1}, \cdots)$ or, writing loosely, "$\sup_{m \geq n} \mathcal{B}(X_m)$." In turn, the sequence $\mathcal{B}(X_n, X_{n+1}, \cdots)$, $\mathcal{B}(X_{n+1}, X_{n+2}, \cdots)$, $\cdots$ is a nonincreasing sequence of σ-fields and its limit or intersection is a σ-field $\mathcal{C}$ contained in $\mathcal{B}(X_n, X_{n+1}, \cdots)$ however large be n or, writing loosely, "$\limsup \mathcal{B}(X_n)$." The σ-field $\mathcal{C}$ will be called the *tail σ-field* of the sequence X_n or "the sub σ-field of events induced by the tail of the sequence X_n." Let us observe that all the foregoing σ-fields and, in particular, the tail σ-field, are contained in the σ-field $\mathcal{B}(X_1, X_2, \cdots)$ induced by the whole sequence X_n. The elements of the tail σ-field $\mathcal{C}$ are *tail events* and the numerical (finite or not) $\mathcal{C}$-measurable functions, that is, those functions which induce sub σ-fields of events contained in $\mathcal{C}$ are *tail functions*—they are defined on the "tail" of the sequence. For example, the limits inferior and superior of the sequence X_n and of the sequence $(X_1 + X_2 + \cdots + X_n)/b_n$, where $b_n \to \infty$, are tail functions (not necessarily finite), while the sets of convergence of these sequences, as well as the set of convergence of the series $\sum X_n$, are tail events.

To Borel's result corresponds the basic Kolmogorov's

B. ZERO-ONE LAW. *On a sequence of independent r.v.'s, the tail events have for pr. either 0 or 1 and the tail functions are degenerate.*

In other words, the tail σ-field of a sequence of independent r.v.'s is equivalent to $\{\emptyset, \Omega\}$.

Proof. We observe that an event A is independent of itself if, and only if, $PAA = PA \cdot PA$, that is, if $PA = 0$ or 1—and such events are mutually independent. Thus, the first assertion means that the tail σ-field $\mathcal{C}$ of the sequence X_n of independent r.v.'s is independent of itself. Since $\mathcal{C} \subset \mathcal{B}(X_{n+1}, X_{n+2}, \cdots)$ whatever be n and, because of the independence assumption, $\mathcal{B}(X_1, \cdots, X_n)$ is independent of $\mathcal{B}(X_{n+1}, X_{n+2}, \cdots)$, it follows that $\mathcal{C}$ is independent of $\mathcal{B}(X_1, X_2, \cdots, X_n)$ whatever be n. Therefore, $\mathcal{C}$ is independent of $\mathcal{B}(X_1, X_2, \cdots)$ and, being contained in $\mathcal{B}(X_1, X_2, \cdots)$, it is independent of itself. This proves the first assertion and the second follows, since, if X is a tail function, then it is a.s. $\{\emptyset, \Omega\}$-measurable hence degenerates.

CoROLLARY. *If X_n are independent r.v.'s, then the sequence X_n either converges a.s. or diverges a.s.; and similarly for the series $\sum X_n$. Moreover, the limits of the sequences X_n and $(X_1 + \cdots + X_n)/b_n$ where $b_n \uparrow \infty$, are degenerate.*

*16.4 Independent r.v.'s and product spaces.

Let X_t, where t runs over an index set T, be independent r.v.'s with d.f.'s F_{X_t} on R_t. Because of the correspondence theorem, every F_{X_t} determines a pr. P_{X_t} on the Borel field $\mathcal{B}_t$ in R_t. On account of the product-measure theorem, the P_{X_t} determine a product-measure $\prod P_{X_t}$ on the product Borel field $\prod \mathcal{B}_t$ in the product space $\prod R_t$. On the other hand, the law of the family $X = \{X_t, t \in T\}$, represented by the family of d.f.'s $\{F_{X_{t_1}, \cdots, X_{t_N}}\}$ of all finite subfamilies of X determines, by the correspondence theorem, a family $\{P_{X_{t_1}, \cdots, X_{t_N}}\}$ of consistent measures on the product Borel fields $\prod_{k=1}^{N} \mathcal{B}_{t_k}$. Owing to the consistent measures theorem, this family of pr.'s determines a pr. P_X on $\prod \mathcal{B}_t$.

Since the X_t are independent,

$$F_{X_{t_1} \cdots X_{t_N}} = F_{X_{t_1}} \times \cdots \times F_{X_{t_N}}$$

so that

$$P_{X_{t_1} \cdots X_{t_N}} = P_{X_{t_1}} \times \cdots \times P_{X_{t_N}}$$

and, therefore, P_X coincides with $\prod P_{X_t}$. In other words,

A. *The pr. space induced on its range space by a family of independent r.v.'s is the product of pr. spaces induced on their respective range spaces by the r.v.'s of the family.*

Let us observe that this reduces the multiplication theorem to the Fubini theorem.

The question arises whether the converse is true: Given a product pr. space $(\prod R_t, \prod \mathfrak{B}_t, \prod P_t)$, is there a family $\{X_t, t \in T\}$ of independent r.v.'s on some pr. space $(\Omega, \mathcal{Q}, P)$ which induces this product pr. space? Equivalently, given a family $\{F_t, t \in T\}$ of d.f.'s with variation 1, is there a family $\{X_t, t \in T\}$ of independent r.v.'s with $F_{X_t} = F_t$?

If the pr. space on which the r.v.'s have to be defined is fixed, then, in general, the answer is in the negative, since on a fixed pr. space even one r.v. with a given d.f. might not exist. However, if we are at liberty to select the pr. space on which to define r.v.'s, and *we shall always do so*, then the answer is in the affirmative, as follows:

Let the pr. space be the product pr. space $(\prod R_t, \prod \mathfrak{B}_t, \prod P_t)$ where, if the F_t are given, the P_t are determined upon applying the correspondence theorem. The r.v.'s X_t, defined on this pr. space by $X_t(x) = x_t$, $x = \{x_t, t \in T\}$, are then independent, since their pr.d.'s are P_t and their d.f.'s are F_t. Thus

B. *The relation* $X_t(x) = x_t, x = \{x_t, t \in T\}$ *establishes a one-to-one correspondence between families* $\{X_t\}$ *of independent r.v.'s and product pr. spaces on* $\prod R_t$.

REMARK. There exist pr. spaces on which can be defined all possible families of independent r.v.'s with a given index set T. For example, take the pr. space $(\Omega, \mathcal{Q}, P)$ where $\Omega = \prod \Omega_t$ with $\Omega_t = (0, 1)$ and $P = \prod P_t$ on the Borel field $\mathcal{Q}$ in Ω, with P_t being the Lebesgue measure on the Borel field in Ω_t (class of Borel sets in Ω_t). Then the r.v.'s X_t— inverse functions of arbitrarily given d.f.'s F_t—are independent and $F_{X_t} = F_t$.

Extension. The preceding considerations apply, word for word, to random vectors. They also apply to arbitrary random functions, provided we consider that the d.f. of a random function is defined in terms of its "finite sections," that is, the family of d.f.'s of projections of the random function on finite subspaces.

§ 17. CONVERGENCE AND STABILITY OF SUMS; CENTERING AT EXPECTATIONS AND TRUNCATION

This section and the following one are devoted to the investigation of sums $S_n = \sum_{k=1}^{n} X_k$ of independent r.v.'s $X_1, X_2, \cdots$ and, especially, of their limit properties—convergence to r.v.'s and stability.

Given two numerical sequences a_n and $b_n \uparrow \infty$, we say that the sequence S_n is *stable* in pr. or a.s. if $\dfrac{S_n}{b_n} - a_n \xrightarrow{\text{P}} 0$ or $\dfrac{S_n}{b_n} - a_n \xrightarrow{\text{a.s.}} 0$. In fact, a stability property is at the root of the whole development of pr. theory. If $X_1, X_2, \cdots$ are independent and identically distributed indicators with $P[X_n = 1] = p$ and $P[X_n = 0] = q = 1 - p$, we have the *Bernoulli case*. The first stability property is the

BERNOULLI LAW OF LARGE NUMBERS: *In the Bernoulli case* $\dfrac{S_n}{n} - p$
$\xrightarrow{\text{P}} 0$.

The Central Limit Problem, to which the following chapter is devoted, is the direct descendant of its sharpening by de Moivre and by Laplace. On the other hand, the following strengthening

BOREL STRONG LAW OF LARGE NUMBERS: *In the Bernoulli case*

$$\frac{S_n}{n} - p \xrightarrow{\text{a.s.}} 0,$$

is at the origin of the results given in this chapter. Perhaps the importance of the methods overwhelms that of the results and emphasis will be laid upon the methods. These methods are (1) centering at expectations and truncation and (2) centering at medians and symmetrization.

17.1   Centering at expectations and truncation.   We say that we *center* X at c if we replace X by $X - c$. If X is integrable, then we can center it at its expectation EX and, thus, X is replaced by $X - EX$. In other words, a r.v. is *centered at its expectation* if, and only if, its expectation exists and equals 0.

Let X be integrable. The second moment of $X - EX$ is called *variance* of X; it exists but may be infinite and will be denoted by $\sigma^2 X$. Thus

$$\sigma^2 X = E(X - EX)^2 = EX^2 - (EX)^2.$$

Since, for every finite c, we have

$$\sigma^2(X - c) = E(X - c - E(X - c))^2 = E(X - EX)^2,$$

centerings do not modify variances.

The importance of variances is due to the fact that we have at our disposal bounds, in terms of variances of summands, of pr.'s of events defined in terms of sums S_n of independent r.v.'s; we shall find and use such bounds in this section. However, variances can be introduced only when the summands are integrable. Moreover, the bounds mentioned above are nontrivial only when the variances are finite. This seems to limit the use of such bounds to square-integrable summands. Yet this obstacle can be overcome by means of the *truncation method*.

We *truncate* X at $c > 0$ (finite) when we replace X by $X^c = X$ or 0 according as $|X| < c$ or $|X| \geq c$, and X^c is X *truncated at* c. It follows that, if F is the d.f. of X, then all moments of X^c

$$EX^c = \int_{|x|<c} x\, dF, \quad E(X^c)^2 = \int_{|x|<c} x^2\, dF, \text{ etc.,}$$

exist and are finite. We can always select c sufficiently large so as to make $P[X \neq X^c] = P[|X| \geq c]$ arbitrarily small. Furthermore, we can always select the c_j sufficiently large so as to make $P \bigcup [X_j \neq X_j^{c_j}]$ arbitrarily small, since, given $\epsilon > 0$, we have

$$P \bigcup [X_j \neq X_j^{c_j}] \leq \sum P[|X_j| \geq c_j] < \epsilon$$

if, say, the c_j are selected so as to make $P[|X_j| \geq c_j] < \dfrac{\epsilon}{2^j}$. Thus, to every countable family of r.v.'s we can make correspond a family of bounded r.v.'s which differs from the first on an event of arbitrarily small pr. Moreover, if we are interested primarily in limit properties there is no need for arbitrarily small pr., for the following reasons.

Let two sequences X_n and X'_n of r.v.'s be called *tail-equivalent* if they differ a.s. only by a finite number of terms; in other words, if for a.e. $\omega \in \Omega$ there exists a finite number $n(\omega)$ such that for $n \geq n(\omega)$ the two sequences $X_n(\omega)$ and $X'_n(\omega)$ are the same; in symbols $P[X_n \neq X'_n \text{ i.o.}] = 0$. If the sequences X_n and X'_n only converge on the same event up to some null subset, then we say that they are *convergence equivalent*. Let $S_n = \sum_{k=1}^{n} X_k$ and $S'_n = \sum_{k=1}^{n} X'_k$. Since

$$P[X_n \neq X'_n \text{ i.o.}] = \lim_n P \bigcup_{k=n}^{\infty} [X_k \neq X'_k] \leq \lim_n \sum_{k=n}^{\infty} P[X_k \neq X'_k]$$

it follows that

a. Equivalence lemma. *If the series $\sum P[X_n \neq X'_n]$ converges, then the sequences X_n and X'_n are tail-equivalent and, hence, the series $\sum X_n$ and $\sum X'_n$ are convergence-equivalent and the sequences $\dfrac{S_n}{b_n}$ and $\dfrac{S'_n}{b_n}$, where $b_n \uparrow \infty$, converge on the same event and to the same limit, excluding a null event.*

17.2 Bounds in terms of variances. To avoid repetitions, we make the convention that, unless otherwise stated, $S_0 = 0$, $S_n = \sum\limits_{k=1}^{n} X_k$, $n = 1, 2, \cdots$, and the summands $X_1, X_2, \cdots$ are independent r.v.'s.

Let $X_1, X_2, \cdots$, be integrable. Since centerings do not modify the variances, we can assume, when computing variances, that these r.v.'s are centered at expectations. Then

$$\sigma^2 S_n = E S_n{}^2 = \sum_{k=1}^{n} E X_k{}^2 + \sum_{\substack{j, k=1 \\ j \neq k}}^{n} E X_j X_k = \sum_{k=1}^{n} \sigma^2 X_k,$$

since independence of X_j and X_k entails, by 15.2,

$$E X_j X_k = E X_j \cdot E X_k = 0.$$

Thus, we obtain the classical

BIENAYMÉ EQUALITY. *If the r.v.'s X_n are independent and integrable, then*

$$\sigma^2 S_n = \sum_{k=1}^{n} \sigma^2 X_k.$$

The basic inequalities 9.3A become

a. $\dfrac{\sum\limits_{k=1}^{n} \sigma^2 X_k - \epsilon^2}{\text{a.s. sup } (S_n - E S_n)^2} \leqq P[|S_n - E S_n| \geqq \epsilon] \leqq \dfrac{1}{\epsilon^2} \sum\limits_{k=1}^{n} \sigma^2 X_k.$

The right-hand side inequality is the celebrated BIENAYMÉ-TCHEBICHEV INEQUALITY. Applied to $(S_{n+k} - E S_{n+k}) - (S_n - E S_n)$ and to $S_n - E S_n$ with ϵ replaced by ϵb_n, it yields, by passage to the limit,

b. *If the series $\sum \sigma^2 X_n$ converges, then the series $\sum (X_n - E X_n)$ converges in pr. If $\dfrac{1}{b_n{}^2} \sum\limits_{k=1}^{n} \sigma^2 X_k \to 0$, then $\dfrac{S_n - E S_n}{b_n} \xrightarrow{\text{P}} 0$.*

This last property is due to Tchebichev (when $b_n = n$). In the Bernoulli case, where $b_n = n$, $E X_n = p$, $\sigma^2 X_n = pq$, it reduces to the Bernoulli law of large numbers. It is of some interest to observe that Borel's strengthening can also be obtained by means of the Bienaymé-Tchebichev inequality (see Introductory part).

So far, the assumption of independence was used only to establish that the summands were *orthogonal*, that is, $E X_j X_k = 0 (j \neq k)$ when

X_j and X_k are centered at expectations. In fact, the foregoing results remain valid under the even less restrictive assumption of orthogonality of S_{n-1} and X_n, $n = 1, 2, \cdots$, since, then,

$$\sigma^2 S_n = \sigma^2 S_{n-1} + \sigma^2 X_n,$$

and the Bienaymé equality follows by induction.

But, in the case of independence, the r.v.'s $S_{n-1} I_{A_{n-1}}$ and X_n are orthogonal, not only for $A_{n-1} = \Omega$ but also for *every* event A_{n-1} defined in terms of $X_1, X_2, \cdots, X_{n-1}$. Therefore, it is to be expected that the foregoing results can be strengthened by using more completely the properties of independence, in particular the orthogonality property just mentioned.

A. Kolmogorov inequalities. *If the independent r.v.'s X_k are integrable and the $|X_k| \leqq c$ finite or not, then, for every $\epsilon > 0$,*

$$1 - \frac{(\epsilon + 2c)^2}{\sum\limits_{k=1}^{n} \sigma^2 X_k} \leqq P[\max_{k \leqq n} |S_k - ES_k| \geqq \epsilon] \leqq \frac{1}{\epsilon^2} \sum_{k=1}^{n} \sigma^2 X_k.$$

If one of the variances is infinite, then the right-hand side inequality is trivial and the left-hand side inequality has no content (for, then, $c = \infty$), so that we assume that all variances are finite. In that case, the left-hand side inequality is trivial when c is infinite and therefore we assume, in proving this inequality, that, moreover, c is finite.

Proof. We can assume, without restricting the generality, that the X_n and hence the S_n are centered at expectations, provided we note that $|X| \leqq c$ implies $|EX| \leqq c$ and, hence, $|X - EX| \leqq 2c$.

Let

$$A_k = [\max_{j \leqq k} |S_j| < \epsilon],$$

$$B_k = A_{k-1} - A_k = [|S_1| < \epsilon, \cdots, |S_{k-1}| < \epsilon, |S_k| \geqq \epsilon]$$

so that

$$A_0 = \Omega, \quad A_n^c = \sum_{k=1}^{n} B_k, \quad B_k \subset [|S_{k-1}| < \epsilon, |S_k| \geqq \epsilon].$$

1° Since $S_k I_{B_k}$ and $S_n - S_k$ are orthogonal, it follows that

$$\int_{B_k} S_n^2 = E(S_n I_{B_k})^2$$

$$= E(S_k I_{B_k})^2 + E((S_n - S_k) I_{B_k})^2 \geqq E(S_k I_{B_k})^2 \geqq \epsilon^2 P B_k.$$

Summing over $k = 1, \cdots, n$, we obtain

$$\sum_{k=1}^{n} \sigma^2 X_k = ES_n{}^2 \geq \int_{A_n{}^c} S_n{}^2 = \sum_{k=1}^{n} \int_{B_k} S_n{}^2 \geq \epsilon^2 \sum_{k=1}^{n} PB_k = \epsilon^2 PA_n{}^c,$$

and the right-hand side inequality is proved.

2° Since

$$S_{k-1} I_{A_{k-1}} + X_k I_{A_{k-1}} = S_k I_{A_{k-1}} = S_k I_{A_k} + S_k I_{B_k}$$

and $S_{k-1} I_{A_{k-1}}$ and X_k are orthogonal while $I_{A_k} I_{B_k} = 0$, it follows that

$$E(S_{k-1} I_{A_{k-1}})^2 + \sigma^2 X_k \cdot PA_{k-1} = E(S_k I_{A_k})^2 + E(S_k I_{B_k})^2.$$

Since $PA_{k-1} \geq PA_n$ and $|X_k| \leq 2c$, and hence

$$|S_k I_{B_k}| \leq |S_{k-1} I_{B_k}| + |X_k I_{B_k}| \leq (\epsilon + 2c) I_{B_k},$$

it follows that

$$E(S_{k-1} I_{A_{k-1}})^2 + \sigma^2 X_k \cdot PA_n \leq E(S_k I_{A_k})^2 + (\epsilon + 2c)^2 PB_k.$$

Summing over $k = 1, \cdots, n$, we obtain

$$(\sum_{k=1}^{n} \sigma^2 X_k) PA_n \leq E(S_n I_{A_n})^2 + (\epsilon + 2c)^2 \sum_{k=1}^{n} PB_k$$

$$\leq \epsilon^2 PA_n + (\epsilon + 2c)^2 PA_n{}^c \leq (\epsilon + 2c)^2,$$

and the left-hand side inequality follows.

17.3 Convergence and stability. We apply now Kolmogorov inequalities and the truncation method to convergence and stability problems for consecutive sums S_n of independent r.v.'s $X_1, X_2, \cdots$.

I. CONVERGENCE. In this Chapter, convergence means convergence to a *finite* number or to a *finite* function (r.v.).

a. *If $\sum \sigma^2 X_n$ converges, then $\sum (X_n - EX_n)$ converges a.s. If $\sum \sigma^2 X_n$ diverges and the X_n are uniformly bounded, then $\sum (X_n - EX_n)$ diverges a.s. Thus, if the X_n are uniformly bounded, then $\sum (X_n - EX_n)$ converges a.s. if, and only if, $\sum \sigma^2 X_n$ converges.*

This follows, by letting $m, n \to \infty$ in Kolmogorov's inequalities with S_k replaced by $S_{m+k} - S_m$.

b. *If the X_n are uniformly bounded and $\sum X_n$ converges a.s., then $\sum \sigma^2 X_n$ and $\sum EX_n$ converge.*

Proof. To the r.v.'s X_n we associate r.v.'s X'_n such that X_n and X'_n are identically distributed for every n and X_1, X'_1, X_2, X'_2, $\cdots$ is a sequence of independent r.v.'s. We form the "symmetrized" sequence $X_n{}^s = X_n - X'_n$ of independent r.v.'s, and have

$$| X_n{}^s | \leq | X_n | + | X'_n | \leq 2c, \quad EX_n{}^s = EX_n - EX'_n = 0,$$

$$\sigma^2 X_n{}^s = \sigma^2 X_n + \sigma^2 X'_n = 2\sigma^2 X_n.$$

Since $\sum X_n$ converges a.s., so does $\sum X'_n$ and hence $\sum X_n{}^s$ ($= \sum X_n - \sum X'_n$). It follows, by **a**, that $\sum \sigma^2 X_n{}^s$ and hence $\sum \sigma^2 X_n$ converge and, again by **a**, $\sum (X_n - EX_n)$ converges a.s., so that $\sum EX_n = \sum X_n - \sum (X_n - EX_n)$ converges. The assertion is proved.

Let X^c be X truncated at (a finite) $c > 0$. We have Kolmogorov's

A. THREE-SERIES CRITERION. *The series $\sum X_n$ of independent summands converges a.s. to a r.v. if, and only if, for a fixed $c > 0$, the three series*

$$\text{(i)} \ \sum P[| X_n | \geq c], \quad \text{(ii)} \ \sum \sigma^2 X_n{}^c, \quad \text{(iii)} \ \sum EX_n{}^c,$$

converge.

Proof. Convergence of (i) entails, by the equivalence lemma, convergence-equivalence of $\sum X_n$ and $\sum X_n{}^c$, and convergence of (ii) and (iii) entails, by **a**, a.s. convergence of $\sum X_n{}^c$. This proves the "if" assertion.

Conversely, let $\sum X_n$ converge a.s. so that $X_n \overset{\text{a.s.}}{\to} 0$. By 16.3A, Cor. 2, (i) converges, so that, by the equivalence lemma, $\sum X_n{}^c$ converges a.s. and, by **b**, (ii) and (iii) converge. This proves the "only if" assertion.

COROLLARY. *If at least one of the three series in* **A** *does not converge, then $\sum X_n$ diverges a.s.*

For, by 16.3B (Corollary), $\sum X_n$ either converges a.s. or diverges a.s.

REMARK. In the proof of **b** we introduced a "symmetrized" sequence. This is an application of the "symmetrization method," to be expounded in the next section.

II. A.S. STABILITY. We seek conditions under which $\dfrac{S_n}{b_n} - a_n \overset{\text{a.s.}}{\to} 0$ when $b_n \uparrow \infty$, and require the following elementary proposition.

TOEPLITZ LEMMA. *Let a_{nk}, $k = 1, 2, \cdots, k_n$, be numbers such that, for every fixed k, $a_{nk} \to 0$ and, for all n, $\sum_k |a_{nk}| \leq c < \infty$; let $x'_n = \sum_k a_{nk} x_k$.*

Then, $x_n \to 0$ entails $x'_n \to 0$ and, if $\sum_k a_{nk} \to 1$, then $x_n \to x$ finite entails $x'_n \to x$. In particular, if $b_n = \sum_{k=1}^{n} a_k \uparrow \infty$, then $x_n \to x$ finite entails $\dfrac{1}{b_n} \sum_{k=1}^{n} a_k x_k \to x$.

The proof is immediate. If $x_n \to 0$ then, for a given $\epsilon > 0$ and $n \geq n_\epsilon$ sufficiently large, $|x_n| < \dfrac{\epsilon}{c}$ so that

$$| x'_n | \leq \sum_{k < n_\epsilon} | a_{nk} x_k | + \epsilon.$$

Letting $n \to \infty$ and then $\epsilon \to 0$, it follows that $x'_n \to 0$. The second assertion follows, since then

$$x'_n = \sum_k (a_{nk}) x + \sum_k a_{nk}(x_k - x) \to x.$$

And setting $a_{nk} = \dfrac{a_k}{b_n}$, $k \leq n$, the particular case is proved.

The particular case yields the powerful

KRONECKER LEMMA. *If $\sum x_n$ converges to s finite and $b_n \uparrow \infty$, then $\dfrac{1}{b_n} \sum_{k=1}^{n} b_k x_k \to 0$.*

For, setting $b_0 = 0$, $a_k = b_k - b_{k-1}$, $s_{n+1} = \sum_{k=1}^{n} x_k$, we have

$$\frac{1}{b_n} \sum_{k=1}^{n} b_k x_k = \frac{1}{b_n} \sum_{k=1}^{n} b_k(s_{k+1} - s_k) = s_{n+1} - \frac{1}{b_n} \sum_{k=1}^{n} a_k s_k \to s - s = 0.$$

We are now in a position to prove Kolmogorov's proposition below.

A. *If the integrable r.v.'s X_n are independent, then $\sum \dfrac{\sigma^2 X_n}{b_n{}^2} < \infty$, $b_n \uparrow \infty$, entails $\dfrac{S_n - ES_n}{b_n} \xrightarrow{\text{a.s.}} 0$.*

For, by **Ia**, convergence of $\sum \dfrac{\sigma^2 X_n}{b_n{}^2}$ entails a.s. convergence of $\sum \dfrac{X_n - EX_n}{b_n}$, and the Kronecker lemma applies.

We can now prove an extension of Borel's strong law of large numbers.

B. Kolmogorov strong law of large numbers. *If the independent r.v.'s X_n are identically distributed with a common law $\mathcal{L}(X)$, then*
$$\frac{X_1 + \cdots + X_n}{n} \xrightarrow{\text{a.s.}} c \text{ finite if, and only if, } E|X| < \infty; \text{ and then } c = EX.$$

Proof. We set $A_n = [|X| \geq n]$, $A_0 = \Omega$, and observe that, for every n, $PA_n = P[|X_n| \geq n]$, while

$$\sum PA_n = \sum (n-1)(PA_{n-1} - PA_n) \leq \sum E|X| I_{A_{n-1}-A_n}$$
$$\leq \sum n(PA_{n-1} - PA_n) \leq 1 + \sum PA_n$$

or
$$\sum PA_n \leq E|X| \leq 1 + \sum PA_n.$$

If $\dfrac{S_n}{n} \xrightarrow{\text{a.s.}} c$ finite, then $\dfrac{X_n}{n} = \dfrac{S_n}{n} - \dfrac{n-1}{n} \dfrac{S_{n-1}}{n-1} \xrightarrow{\text{a.s.}} 0$ and, hence, by 16.3a, Cor. 2, $\sum PA_n < \infty$. This proves the "only if" assertion and it remains to prove that, if $E|X| < \infty$, then $\dfrac{S_n}{n} \xrightarrow{\text{a.s.}} EX$.

Let $E|X| < \infty$ and set $\bar{S}_n = \sum\limits_{k=1}^{n} \bar{X}_k$, where $\bar{X}_k$ represents X_k truncated at k. Since
$$\sum P[|X_n| \geq n] = \sum PA_n \leq E|X| < \infty,$$
it follows that the sequences S_n/n and $\bar{S}_n/n$ have same limit, and it suffices to prove that $\dfrac{\bar{S}_n}{n} \xrightarrow{\text{a.s.}} EX$. Since, by the dominated convergence theorem,
$$E\bar{X}_n = EXI_{A_n^c} \to EX$$
and, hence, by the Toeplitz lemma, $\dfrac{E\bar{S}_n}{n} \to EX$, it suffices to prove that $\dfrac{\bar{S}_n - E\bar{S}_n}{n} \xrightarrow{\text{a.s.}} 0$. But
$$\sum \frac{\sigma^2 \bar{X}_n}{n^2} \leq \sum \frac{E\bar{X}_n{}^2}{n^2} = E \sum \frac{X^2}{n^2} I_{A_n^c} \leq 2 + E|X| < \infty,$$

since, setting $B_m = [\, m - 1 \leqq |X| < \ m \,]$, we have $A_n{}^c B_m = \emptyset$ or B_m according as $n < m$ or $n \geqq m$ and, hence,

$$\sum_n \frac{X^2}{n^2} I_{A_n{}^c B_m} \leqq m^2 \left(\frac{1}{m^2} + \frac{1}{(m+1)^2} + \cdots \right) I_{B_m}$$

$$\leqq \left(1 + m^2 \int_m^\infty \frac{dx}{x^2} \right) I_{B_m} \leqq (2 + |X|) I_{B_m},$$

so that, by summing over m, we obtain the bound $2 + E|X|$. Thus, theorem **A** applies, and the proof is complete.

*17.4 **Generalization.** Let c, with or without affixes, be finite positive numbers and let g_n be continuous and nondecreasing functions on $[0, +\infty]$ such that $g_n(0) = 0$ *and* $g_n(x) \geqq cx^2$ *or* $\geqq c'$ *according as* $0 < x < c_n$ *or* $x \geqq c_n$.

a. *If the series* (i) $\sum P[|X_n| \geqq c_n]$ *and* (ii) $\sum E g_n(|X_n{}^{c_n}|)$ *converge, then* $\sum (X_n - EX_n{}^{c_n})$ *converges a.s.*

For convergence of (i) entails, by the equivalence lemma, convergence-equivalence of $\sum (X_n - EX_n{}^{c_n})$ and $\sum (X_n{}^{c_n} - EX_n{}^{c_n})$ and, by **Ia**, this last series converges a.s., since convergence of (ii) entails

$$\sum \sigma^2 X_n{}^{c_n} \leqq \sum E|X_n{}^{c_n}|^2 \leqq \frac{1}{c} \sum E g_n(|X_n{}^{c_n}|) < \infty.$$

b. *If the series* (i) $\sum E g_n(|X_n|)$ *or* (ii) $\sum \int_0^{c_n} P[|X_n| \geqq x] \, dg_n(x)$ *converges, then* $\sum (X_n - EX_n{}^{c_n})$ *converges a.s.*

For convergence of (i) entails

$$\sum P[|X_n| \geqq c_n] \leqq \frac{1}{c'} \sum E g_n(|X_n|) < \infty$$

and

$$\sum E g_n(|X_n{}^{c_n}|) \leqq \sum E g_n(|X_n|) < \infty,$$

so that **a** applies.

Similarly, convergence of (ii) entails, by integration by parts,

$$\infty > \sum \int_0^{c_n} P[|X_n| \geqq x] \, dg_n(x) = \sum g_n(c_n) P[|X_n| \geqq c_n]$$

$$+ \int_0^{c_n} g_n(x) \, dP[|X_n| < x]$$

$$\geqq c' \sum P[|X_n| \geqq c_n] + \sum E g_n(|X_n{}^{c_n}|),$$

so that **a** applies.

A. *If the series* (i) $\sum Eg_n\left(\dfrac{|X_n|}{b_n}\right)$ *or* (ii) $\sum \displaystyle\int_0^{c_n} P[|X_n| \geqq b_n x]\, dg_n(x)$ *converges, then*

$$\sum \frac{X_n - EX_n^{b_n c_n}}{b_n} \ \text{converges a.s. and} \ \frac{1}{b_n}\sum_{k=1}^n (X_k - EX_k^{b_n c_n}) \xrightarrow{\text{a.s.}} 0.$$

Moreover, if (i) *converges and* $g_n(x) \geqq c''x$ *for* $0 < x \leqq c_n$ *or for* $x \geqq c_n$, *then* $EX_n^{b_n c_n}$ *can be replaced by* 0 *or by* EX_n, *respectively.*

Proof. The first assertion follows from **b** and the Kronecker lemma. As for the second assertion, if $\sum'$ and $\sum''$ denote summations over those values of n for which the first, respectively, the second, assumption about g_n holds, then

$$\sum' \frac{1}{b_n} E|X_n^{b_n c_n}| = \sum' \int_0^{b_n c_n} \frac{x}{b_n}\, dP[|X_n| < x]$$

$$\leqq \frac{1}{c''}\sum' \int_0^{b_n c_n} g_n\left(\frac{x}{b_n}\right) dP[|X_n| < x]$$

$$\leqq \frac{1}{c''}\sum Eg_n\left(\frac{|X_n|}{b_n}\right) < \infty,$$

and

$$\left|\sum'' \frac{1}{b_n} EX_n - \sum'' \frac{1}{b_n} EX_n^{b_n c_n}\right| \leqq \sum'' \frac{1}{b_n} E|X_n - X_n^{b_n c_n}|$$

$$= \sum'' \int_{b_n c_n}^{\infty} \frac{x}{b_n}\, dP[|X_n| < x]$$

$$\leqq \frac{1}{c''}\sum'' \int_{b_n c_n}^{\infty} g_n\left(\frac{x}{b_n}\right) dP[|X_n| < x]$$

$$\leqq \frac{1}{c''}\sum Eg_n\left(\frac{|X_n|}{b_n}\right) < \infty.$$

This completes the proof.

Particular cases. 1° Let $g_n(x) = |x|^{r_n}$ with $0 < r_n \leqq 2$. Theorem **A** yields

If $b_n \uparrow \infty$ *and* $\sum \dfrac{E|X_n|^{r_n}}{b_n^{r_n}} < \infty$, *then* $\dfrac{1}{b_n}\sum_{k=1}^n (X_k - a_k) \xrightarrow{\text{a.s.}} 0$ *where* $a_k = 0$ *or* EX_k *according as* $0 < r_n < 1$ *or* $1 \leqq r_n \leqq 2$.

For $r_n \equiv 2$, we find 17.3IIA.

2°   Let $g_n(x) = x^2$ for $0 \leq x \leq 1$ and, to simplify the writing, set $q(x) = P[|X| \geq x]$, $q_n(x) = P[|X_n| \geq x]$.   Theorem **A** yields

$$\textit{If} \int_0^1 x\{\sum q_n(b_n x)\}\, dx < \infty, \textit{ then } \sum \frac{1}{b_n}(X_n - EX_n^{b_n}) \textit{ converges a.s.}$$

$$\textit{and } \frac{1}{b_n} \sum_{k-1}^n (X_k - EX_k^{b_k}) \overset{\text{a.s.}}{\longrightarrow} 0.$$

We require the following

MOMENTS LEMMA.   *For every $r > 0$ and $x > 0$*

$$x^r \sum q(n^{\frac{1}{r}}x) \leq E|X|^r \leq 1 + x^r \sum q(n^{\frac{1}{r}}x).$$

This follows from

$$E|X|^r = -\int_0^\infty t^r\, dq(t) = -\sum \int_{(n-1)^{\frac{1}{r}}x}^{n^{\frac{1}{r}}x} t^r\, dq(t)$$

and

$$(n-1)x^r\{q((n-1)^{\frac{1}{r}}x) - q(n^{\frac{1}{r}}x)\}$$

$$\leq -\int_{(n-1)^{\frac{1}{r}}x}^{n^{\frac{1}{r}}x} t^r\, dq(t) \leq nx^r\{q((n-1)^{\frac{1}{r}}x) - q(n^{\frac{1}{r}}x)\},$$

by summing the inequalities over $n = 1, 2, \cdots$ and rearranging the terms.

3°   If $b_n = n^{\frac{1}{r}}$ and the laws of the r.v.'s X_n are uniformly bounded by the law of a r.v. X, that is, $q_n \leq q$, then $E|X|^r < \infty$ entails

$$\int_0^1 x(\sum q_n(n^{\frac{1}{r}}x))\, dx \leq \int_0^1 x(\sum q(n^{\frac{1}{r}}x))\, dx \leq E|X|^r \int_0^1 \frac{dx}{x^{r-1}},$$

so that the right-hand side is finite for $r < 2$.   Therefore, on account of 2°,

If $q_n \leq q$ and $E|X|^r < \infty$ with $r < 2$, then $\dfrac{1}{n^{\frac{1}{r}}} \sum_{k=1}^n (X_k - a_k) \overset{\text{a.s.}}{\longrightarrow} 0$

where $a_k = 0$ or EX_k according as $r < 1$, or ≥ 1.

4°   If $F_n = F$, then the converse is also true. More precisely (Kolmogorov: $r = 1$; Marcinkiewicz: $r \neq 1$),

Let the independent r.v.'s X_n be identically distributed with common law $\mathcal{L}(X)$, and let $0 < r < 2$.

If $E|X|^r < \infty$, then $\dfrac{1}{n^{1/r}} \displaystyle\sum_{k=1}^{n} (X_k - a_k) \xrightarrow{\text{a.s.}} 0$ with $a_k = 0$ or EX according as $r < 1$ or $r \geqq 1$.

Conversely, if $\dfrac{1}{n^{1/r}} \displaystyle\sum_{k=1}^{n} (X_k - a_k) \xrightarrow{\text{a.s.}} 0$, then $E|X|^r < \infty$.

Proof. The first assertion is a particular case of the preceding proposition. As for the converse proposition, we use the symmetrization method expounded in the following section.

Let X'_n be a sequence independent of the sequence X_n and with same distribution, and let X' be independent of X and with same distribution; set $X_n{}^s = X_n - X'_n$ and $X^s = X - X'$. Then, on account of the assumption,

$$Y_n = \frac{1}{n^{1/r}} \sum_{k=1}^{n} X_k{}^s = \frac{1}{n^{1/r}} \sum_{k=1}^{n} (X_k - a_k) - \frac{1}{n^{1/r}} \sum_{k=1}^{n} (X'_k - a_k) \xrightarrow{\text{a.s.}} 0$$

and, hence,

$$\frac{X_n{}^s}{n^{1/r}} = Y_n - \left(\frac{n-1}{n}\right)^{1/r} Y_{n-1} \xrightarrow{\text{a.s.}} 0.$$

Since the $X_n{}^s$ are independent r.v.'s, it follows that, for every $x > 0$,

$$\sum q^s(n^{1/r}x) = \sum P[|X_n{}^s| \geqq n^{1/r}x] < \infty.$$

Therefore, by the moments lemma, $E|X^s|^r < \infty$ so that, by 17.1A, Corollary 2,

$$E|X - \mu X|^r \leqq 2E|X^s|^r < \infty$$

and, hence, by the c_r-inequality,

$$E|X|^r \leqq c_r E|X - \mu X|^r + c_r|\mu X|^r < \infty.$$

The proof is complete.

*§ 18. CONVERGENCE AND STABILITY OF SUMS; CENTERING AT MEDIANS AND SYMMETRIZATION

While centering at expectations goes back to Bernoulli and use of bounds in terms of variances goes back to Tchebichev, centering at medians and symmetrization are relatively recent. Yet, not only do they complete the first ones, but they also tend to replace them altogether. Moreover, medians always exist and the ch.f.'s of symmetrized r.v.'s, being real-valued, are much easier to handle than complex-valued ones.

*18.1   Centering at medians and symmetrization. Let F be the d.f. of a r.v. X. There exists at least one finite number μX called a *median* of X, such that

$$P[X \geq \mu X] \geq \tfrac{1}{2} \leq P[X \leq \mu X]$$

or, equivalently,

$$F(\mu X) \leq \tfrac{1}{2}, \quad F(\mu X + 0) \geq \tfrac{1}{2}.$$

For, F being nondecreasing on R with $F(-\infty) = 0$, $F(+\infty) = 1$, the graph of $y = F(x)$ completed at its discontinuity points by the segments $(x, F(x))$ to $(x, F(x + 0))$ has either a point or a segment parallel to the x-axis, in common with the line $y = \tfrac{1}{2}$. According to the foregoing definition, the abscissae of the common point or of the common segment are medians of X so that either X has a unique median or it has for medians all points of a closed interval on R—the *median segment* of X.

It follows from the definition of medians that, for every finite number c, we can set $\mu(cX) = c\mu X$. Furthermore, there is a relation between μX, EX, and $\sigma^2 X$, namely,

a. *If X is integrable, then* $|\mu X - EX| \leq \sqrt{2\sigma^2 X}$.

For, by Tchebichev's inequality,

$$P[|X - EX| \geq \sqrt{2\sigma^2 X}] \leq \tfrac{1}{2},$$

so that

$$EX - \sqrt{2\sigma^2 X} \leq \mu X \leq EX + \sqrt{2\sigma^2 X}.$$

A r.v. X and its law as well as its d.f. F and ch.f. f are said to be *symmetric* if, for every x,

(1) $P[X \leq x] = P[X \geq -x];$

equivalently,

(2) $F(-x + 0) = 1 - F(x),$

or, for every pair $a < b$ of continuity points of F,

(3) $F[a, b) = F[-b, -a),$

or

(4) $f = \bar{f}$ is real.

The *symmetrization* procedure consists in assigning to a r.v. X the *symmetrized* r.v. $X^s = X - X'$, where X' is independent of X and has the same distribution. More generally, if $X = \{X_t, t \in T\}$ is a family of r.v.'s, then the *symmetrized family* is $X^s = \{X_t - X'_t, t \in T\}$ where the family X' is independent of X and has same distribution. If X has affixes we affix them to X^s as well as to its d.f. and ch.f. Clearly

b. *To a r.v. X with ch.f. f, there corresponds a symmetric r.v. $X^s = X - X'$ where X and X' are independent and identically distributed, and $f^s = |f|^2$ is the ch.f. of X^s.*

We arrive now at inequalities which are the basic reason for centering at medians.

A. Weak symmetrization inequalities. *For every ϵ and every a,*

(i) $$\tfrac{1}{2}P[X - \mu X \geq \epsilon] \leq P[X^s \geq \epsilon]$$

and

(ii) $$\tfrac{1}{2}P[|X - \mu X| \geq \epsilon] \leq P[|X^s| \geq \epsilon] \leq 2P\left[|X - a| \geq \frac{\epsilon}{2}\right].$$

Proof. Since $X^s = X - X'$ where X and X' are independent and identically distributed, it follows that to a median $\mu = \mu X$ corresponds an equal median $\mu = \mu X'$ and

$$P[X^s \geq \epsilon] = P[(X - \mu) - (X' - \mu) \geq \epsilon] \geq P[X - \mu \geq \epsilon, X' - \mu \leq 0]$$

$$= P[X - \mu \geq \epsilon] \cdot P[X' - \mu \leq 0] \geq \tfrac{1}{2}P[X - \mu \geq \epsilon].$$

This proves inequality (i) which, together with the inequality obtained by changing in (i) X into $-X$, entails the left-hand side inequality in (ii). The right-hand side inequality in (ii) follows from the identical distribution of X and X' only, by

$$P[|X^s| \geq \epsilon] = P[|(X - a) - (X' - a)| \geq \epsilon]$$

$$\leq P\left[|X - a| \geq \frac{\epsilon}{2}\right] + P\left[|X' - a| \geq \frac{\epsilon}{2}\right]$$

$$= 2P\left[|X - a| \geq \frac{\epsilon}{2}\right].$$

Corollary 1. *If $X_n - a_n \xrightarrow{P} 0$, then $X_n{}^s \xrightarrow{P} 0$ and $a_n - \mu X_n \to 0$, and conversely.*

This follows by letting $n \to \infty$ in (ii) where X is replaced by X_n.

COROLLARY 2. *For $r > 0$ and every a,*

$$\tfrac{1}{2}E\,|\,X - \mu X\,|^r \le E\,|\,X^s\,|^r \le 2c_r E\,|\,X - a\,|^r$$

where $c_r = 1$ or 2^{r-1} according as $r \le 1$ or $r \ge 1$.

Proof. The right-hand side inequality follows, by the c_r-inequality, from

$$E|\,X^s\,|^r = E\,|\,(X - a) - (X' - a)\,|^r \le c_r E\,|\,X - a\,|^r + c_r\,E\,|\,X' - a\,|^r$$
$$= 2c_r E\,|\,X - a\,|^r.$$

As for the left-hand side inequality, it is trivial when $E\,|\,X^s\,|^r = \infty$ and then, according to the inequality just proved (with $a = \mu X$), $E\,|\,X - \mu X\,|^r = \infty$; thus, we can assume that $E|\,X^s\,|^r$ is finite. Let

$$q(t) = P[|\,X - \mu X\,| \ge t] \quad \text{and} \quad q^s(t) = P[|\,X^s\,| \ge t]$$

so that, by $\mathbf{A}$(ii),

$$q(t) \le 2q^s(t).$$

It follows, upon integrating by parts, that

$$E|\,X - \mu X\,|^r = -\int_0^\infty t^r\,dq(t) = \int_0^\infty q(t)\,d(t^r) \le 2\int_0^\infty q^s(t)\,d(t^r)$$

$$= -2\int_0^\infty t^r\,dq^s(t) = 2E|\,X^s\,|^r,$$

and the proof is concluded.

This corollary was used at the end of the preceding section.

We pass now to symmetrized families and recall that, if two families $\{X_t,\ t \in T\}$ and $\{X'_t,\ t \in T\}$ are independent, then events defined in terms of the X_t and in terms of the X'_t, respectively, are independent. We require the following

c. LEMMA FOR EVENTS. *Let events with subscript 0 be empty. If, for every integer $j \ge 1$, $A_j A_{j-1}{}^c \cdots A_0{}^c$ and B_j are independent, then*

$$P \bigcup A_j B_j \ge \alpha P \bigcup A_j, \quad \alpha = \inf PB_j.$$

More generally, if $(A_j + A_j')(A_{j-1} + A_{j-1}')^c \cdots (A_0 + A_0')^c$ are independent of B_j and of B'_j, then

$$P \bigcup (A_j B_j + A'_j B'_j) \ge \alpha P \bigcup (A_j + A'_j), \quad \alpha = \inf (PB_j, PB'_j).$$

Proof. The same method applies to both cases. For instance

$$P \bigcup A_j B_j = PA_1B_1 + P(A_1B_1)^c A_2 B_2 + P(A_1B_1)^c(A_2B_2)^c A_3 B_3 + \cdots$$
$$\geq PA_1B_1 + PA_1^c A_2 B_2 + PA_1^c A_2^c A_3 B_3 + \cdots$$
$$\geq PA_1 \cdot PB_1 + PA_1^c A_2 \cdot PB_2 + PA_1^c A_2^c A_3 \cdot PB_3 + \cdots$$
$$\geq \alpha(PA_1 + PA_1^c A_2 + PA_1^c A_2^c A_3 + \cdots) = \alpha P \bigcup A_j.$$

B. SYMMETRIZATION INEQUALITIES. *For every ϵ and every $a_j, j \leq n$,*

(i) $$\tfrac{1}{2} P[\sup_j (X_j - \mu X_j) \geq \epsilon] \leq P[\sup_j X_j^s \geq \epsilon]$$

and

(ii) $$\tfrac{1}{2} P[\sup_j |X_j - \mu X_j| \geq \epsilon] \leq P[\sup_j |X_j^s| \geq \epsilon]$$

$$\leq 2P\left[\sup_j |X_j - a_j| \geq \frac{\epsilon}{2}\right].$$

Proof. Since $X_j^s = X_j - X'_j$ and the families $\{X_j\}$ and $\{X'_j\}$ are independent and identically distributed, it follows that to medians $\mu_j = \mu X_j$ correspond equal medians $\mu_j = \mu X'_j$; setting

$$A_j = [X_j - \mu_j \geq \epsilon'], \quad B_j = [X'_j - \mu_j \leq 0], \quad C_j = [X_j^s \geq \epsilon'],$$

so that $A_j B_j \subset C_j$, the lemma for events applies, with $\alpha = \tfrac{1}{2}$, and

$$\tfrac{1}{2} P \bigcup A_j \leq P \bigcup A_j B_j \leq P \bigcup C_j.$$

This proves (i) by letting $\epsilon' \uparrow \epsilon$, and (ii) follows by arguments similar to those used in the proof of **A** and by the lemma for events.

COROLLARY. *If $X_n - a_n \xrightarrow{\text{a.s.}} 0$, then $X_n^s \xrightarrow{\text{a.s.}} 0$ and $a_n - \mu X_n \to 0$; and conversely.*

By centering sums of independent r.v.'s at suitable medians, we obtain inequalities which can play the role of Kolmogorov's inequalities.

C. P. LÉVY INEQUALITIES. *If $X_1, \cdots, X_n$ are independent r.v.'s and $S_k = \sum_{j=1}^{k} X_j$, then, for every ϵ,*

(i) $$P[\max_{k \leq n} (S_k - \mu(S_k - S_n)) \geq \epsilon] \leq 2P[S_n \geq \epsilon]$$

and

(ii) $$P[\max_{k \leq n} |S_k - \mu(S_k - S_n)| \geq \epsilon] \leq 2P[|S_n| \geq \epsilon].$$

Proof. Let $S_0 = 0$, $S^*_k = \max_{j \le k} (S_j - \mu(S_j - S_n))$ and set

$$A_k = [S^*_{k-1} < \epsilon, \quad S_k - \mu(S_k - S_n) \ge \epsilon],$$

$$B_k = [S_n - S_k - \mu(S_n - S_k) \ge 0]$$

where $\mu(S_n - S_k) = -\mu(S_k - S_n)$. Since

$$[S^*_n \ge \epsilon] = \sum_{k=1}^{n} A_k, \quad [S_n \ge \epsilon] \supset \sum_{k=1}^{n} A_k B_k, \quad PB_k \ge \tfrac{1}{2},$$

(i) follows upon applying the lemma for events or, directly, by

$$P[S_n \ge \epsilon] \ge \sum_{k=1}^{n} PA_k PB_k \ge \tfrac{1}{2} \sum_{k=1}^{n} PA_k = \tfrac{1}{2} P[S^*_n \ge \epsilon].$$

By changing the signs of all r.v.'s which figure in (i) and combining with (i), inequality (ii) follows, and the proof is complete.

REMARK. Let $X_1, \cdots, X_n$ be independent, square-integrable, and centered at expectations. Since, by **a**,

$$\left| \mu(S_k - S_n) \right| \le \sqrt{2\sigma^2(S_n - S_k)} \le \sqrt{2\sigma^2 S_n}$$

inequality (i) remains valid if $\mu(S_k - S_n)$ is replaced by $-\sqrt{2\sigma^2 S_n}$ and, hence, changing ϵ into $\epsilon - \sqrt{2\sigma^2 S_n}$,

$$P[\max S_k \ge \epsilon] \le 2P[S_n \ge \epsilon - \sqrt{2\sigma^2 S_n}].$$

*18.2 Convergence and stability. We are now in possession of the basic tools and shall apply them to the investigation of convergence and stability of sums $S_n = \sum_{k=1}^{n} X_k$ of independent r.v.'s. We recall that here we say that a sequence of r.v.'s converges a.s. if it converges a.s. to a r.v., and their sequence of laws converges if it converges to the law of a r.v., that is, converges completely.

I. CONVERGENCE. Whatever be the sequence of r.v.'s, we have the comparison table of convergences below:

convergence a.s. $\Rightarrow$ convergence in pr. $\Rightarrow$ convergence of laws

$$\Uparrow$$

convergence in q.m.

("in q.m." means "in the 2nd mean" and reads "in quadratic mean").

For series of independent r.v.'s, reverse implications are also true, either with no restriction or under a uniform boundedness restriction. More precisely

a. IMPROVED CONVERGENCE LEMMA. *For series of independent r.v.'s:*

(i) *Convergence a.s. and convergence in pr. are equivalent.*

(ii) *If the summands are uniformly bounded and centered at expectations, then convergence a.s., convergence in pr., convergence in q.m., and convergence of laws, are equivalent.*

Proof. 1° Let $S_n \xrightarrow{\text{P}} S$, so that, by 6.3A, there exists a subsequence $S_{n_k} \xrightarrow{\text{a.s.}} S$ with $\sum_k P\left[|S_{n_{k+1}} - S_{n_k}| \geq \dfrac{1}{2^k} \right] < \infty$. Let $n_k < n \leq n_{k+1}$ and set $T_k = \max_n | S_n - S_{n_k} - \mu(S_n - S_{n_{k+1}}) |$, so that, by P. Lévy's inequality (ii),

$$\sum_k P\left[T_k \geq \frac{1}{2^k} \right] \leq 2 \sum_{k} P\left[|S_{n_{k+1}} - S_{n_k}| \geq \frac{1}{2^k} \right] < \infty$$

and, hence, $T_k \xrightarrow{\text{a.s.}} 0$ as $k \to \infty$. Therefore,

$$| S_n - S - \mu(S_n - S_{n_{k+1}}) | \leq | S_n - S_{n_k} - \mu(S_n - S_{n_{k+1}}) | + | S_{n_k} - S |$$

$$\leq T_k + | S_{n_k} - S | \xrightarrow{\text{a.s.}} 0,$$

that is, $S_n - \mu(S_n - S_{n_{k+1}}) \xrightarrow{\text{a.s.}} S$ and, *a fortiori*, $S_n - \mu(S_n - S_{n_{k+1}}) \xrightarrow{\text{P}} S$. Since $S_n \xrightarrow{\text{P}} S$, it follows that $\mu(S_n - S_{n_{k+1}}) \to 0$ and, hence, $S_n \xrightarrow{\text{a.s.}} S$. Thus, convergence in pr. of the series $\sum X_n$ entails its convergence a.s. and, the converse being always true, the first assertion is proved.

2° Let $| X_n | \leq c < \infty$ and $EX_n = 0$. The series $\sum X_n$ converges in q.m. if, and only if, as $m, n \to \infty$

$$E(S_m - S_n)^2 = \sum_{m+1}^{n} \sigma^2 X_k \to 0$$

or, equivalently, $\sum \sigma^2 X_n < \infty$; then it converges in pr. and, hence, by the first assertion, it converges a.s. But if $\mathscr{L}(S_n) \xrightarrow{c} \mathscr{L}(S)$, so that for all u in some neighborhood of the origin

$$-\sum \log |f_n| = - \log |f_S| < \infty,$$

then, by 12.4$\mathbf{B}'$, for u belonging to the intersection of this neighborhood with $(-1/c, +1/c)$,

$$2 \sum \sigma^2 X_n = \sum \sigma^2 X_n{}^s \leqq - \frac{3}{u^2} \sum \log |f_n(u)|^2 < \infty,$$

and the second assertion follows.

The three-series criterion follows from this improved convergence lemma exactly as it followed from the convergence lemma in section 16.

REMARK. A better insight into the behavior of the series is provided by the Liapounov theorem for the bounded case, according to which, if $s_n{}^2 = \sum_{k=1}^{n} \sigma^2 X_k \to \infty$ and $ES_n = 0$, then, for any fixed $a > 0$ and $\epsilon > 0$ and n large enough to have $\epsilon s_n > a$, we have

$$(1) \qquad P[|S_n| \geqq a] \geqq P[|S_n| \geqq \epsilon s_n] \to \frac{1}{\sqrt{2\pi}} \int_{|x| \geqq \epsilon} e^{-x^2/2}\, dx.$$

Thus, as $\epsilon \to 0$, $P[|S_n| \geqq a] \to 1$ for any fixed but arbitrarily large a, and the sequence $\mathcal{L}(S_n)$ of laws diverges to a law degenerate at infinity. The second assertion follows *ab contrario*, and we see that when the sequence of laws does not converge, then, as $n \to \infty$, the distribution of S_n escapes to infinity in the fashion described by (1).

So far we have been concerned with convergence of a given series. Yet various auxiliary centering constants appeared during the investigation, and the problem arises whether, given the series $\sum X_n$ of independent r.v.'s, there exist centering constants a_n such that the series $\sum (X_n - a_n)$ converges. If $\sum (X_n - a_n)$ converges a.s. for some numerical constants a_n, we say that the series $\sum X_n$ is *essentially convergent*; otherwise, we say that it is *essentially divergent*, since, then, by the corollary of the zero-one law, $\sum (X_n - a_n)$ diverges a.s. whatever be the a_n. As above, our problem is to find criteria for this dichotomy and to find the suitable centering constants when the series is essentially convergent; at the same time, we shall be able to improve the preceding results (see also 37.1).

b. ESSENTIAL CONVERGENCE LEMMA. *The series $\sum X_n$ is essentially convergent if, and only if, the symmetrized series $\sum X_n{}^s$ converges a.s.*

Proof. If $\sum X_n{}^s$ converges a.s., then, for every finite $c > 0$, using 17.1A, by the three series criterion,

$$\sum P[|X_n - \mu X_n| \geqq c] \leqq \sum 2P[|X_n{}^s| \geqq c] < \infty$$

and, upon integrating by parts,

$$\tfrac{1}{2} \sum \sigma^2 (X_n - \mu X_n)^c \leqq \sum \sigma^2 (X_n{}^s)^c + c^2 \sum P[|X_n{}^s| \geqq c] < \infty.$$

Therefore, the series $\sum \{X_n - \mu X_n - E(X_n - \mu X_n)^c\}$ converges a.s. and the "if" assertion is proved while the "only if" assertion is immediate.

From this proof follows the

A. TWO-SERIES CRITERION. *The series $\sum X_n$ is essentially convergent if, and only if, for some arbitrarily fixed $c > 0$, the two series $\sum P[|X_n - \mu X_n| \geq c]$ and $\sum \sigma^2(X_n - \mu X_n)^c$ converge; then the centered series $\sum \{X_n - \mu X_n - E(X_n - \mu X_n)^c\}$ converges a.s.*

The essential convergence lemma permits us to improve further the convergence lemma.

B. EQUIVALENCE THEOREM. *For series of independent r.v.'s, convergence of laws, convergence in pr. and a.s. convergence are equivalent.*

Proof. It suffices to prove that convergence of laws implies a.s. convergence. Let f_n be the ch.f. of X_n so that $|f_n|^2$ is ch.f. of X_n^s. If $\prod_{k=1}^{n} f_k \to f$ ch.f., then $\prod_{k=1}^{n} |f_k|^2 \to |f|^2$ and, by 13.4 B', the two series $\sum P[|X_n^s| \geq c]$ and $\sum \sigma^2(X_n^s)^c$ converge. Since $E(X_n^s)^c = 0$, it follows, by the three series criterion, that the symmetrized series $\sum X_n^s$ converges a.s. Therefore, by the essential convergence lemma, there exist constants a_n such that the series $\sum (X_n - a_n)$ converges a.s. to a r.v. and *a fortiori* its law converges completely, so that, for every u, $\prod_{k=1}^{n} e^{-ia_k u} f_k(u) \to f'(u)$, where f' is a ch.f. By taking u close enough to 0 so that $f(u)f'(u) \neq 0$, it follows that the series $\sum a_n$ converges and, hence, the series $\sum X_n$ converges a.s. This completes the proof.

COROLLARY 1. *A series $\sum X_n$ of independent r.v.'s converges a.s. if, and only if, $\prod_{k=1}^{n} f_k \to f$ and f is continuous at the origin or $f \neq 0$ on a set of positive Lebesgue measure.*

This follows by the continuity theorem or 12.4, 4°.

COROLLARY 2. *A series $\sum X_n$ of independent r.v.'s is essentially convergent or divergent according as*

$$\lim \prod_{k=1}^{n} |f_k| \neq 0 \quad \text{on a set of positive Lebesgue measure or}$$

$$\lim \prod_{k=1}^{n} |f_k| = 0 \quad a.e.$$

This follows by 13.4, 4° and **b**.

II. STABILITY. Given sequences a_n and $b_n \uparrow \infty$, we seek conditions for a.s. stability of sequences S_n of sums of independent r.v.'s. On account of the corollary to the symmetrization lemma, a first condition is that $a_n = \mu\left(\dfrac{S_n}{b_n}\right) + o(1)$. Thus, it suffices to take $a_n = \mu\left(\dfrac{S_n}{b_n}\right)$ and investigate conditions under which $\dfrac{S_n}{b_n} - \mu\left(\dfrac{S_n}{b_n}\right) \xrightarrow{\text{a.s.}} 0$.

We have $b_n \uparrow \infty$ and, moreover, assume that there exists a subsequence b_{n_k} and finite numbers c, c' such that, for all k sufficiently large, $1 < c' \leqq \dfrac{b_{n_{k+1}}}{b_{n_k}} \leqq c < \infty$. Roughly speaking, this assumption means that the sequence b_n does not increase too fast, and it is always satisfied (with an arbitrary $c > 1$) when $\dfrac{b_{n+1}}{b_n} \to 1$. Let $S_{n_0} = 0$ and $T_k = \dfrac{S_{n_k} - S_{n_{k-1}}}{b_{n_k}}$.

A. A.S. STABILITY CRITERION. (i) $\dfrac{S_n}{b_n} - \mu\left(\dfrac{S_n}{b_n}\right) \xrightarrow{\text{a.s.}} 0$ if, and only if,
(ii) $T_k - \mu T_k \xrightarrow{\text{a.s.}} 0$ as $k \to \infty$ or, equivalently, (ii') for every $\epsilon > 0$, $\sum P[|T_k - \mu T_k| \geqq \epsilon] < \infty$.

Proof. Since the T_k are nonoverlapping sums of independent r.v.'s, it follows, by 16.3A, that conditions (ii) and (ii') are equivalent. And, on account of the symmetrization lemma, it suffices to prove equivalence of (i) and (ii) for symmetric summands; then the medians which figure in these conditions vanish.

If $\dfrac{S_n}{b_n} \xrightarrow{\text{a.s.}} 0$, then

$$\frac{S_{n_k}}{b_{n_k}} \xrightarrow{\text{a.s.}} 0 \text{ as } k \to \infty, \quad T_k = \frac{S_{n_k} - S_{n_{k-1}}}{b_{n_k}} = \frac{S_{n_k}}{b_{n_k}} - \frac{b_{n_{k-1}}}{b_{n_k}} \frac{S_{n_{k-1}}}{b_{n_{k-1}}} \xrightarrow{\text{a.s.}} 0,$$

and the "only if" assertion is proved.

Conversely, if $T_k \xrightarrow{\text{a.s.}} 0$, then, by the Toeplitz lemma,

$$\frac{S_{n_k}}{b_{n_k}} = \frac{1}{b_{n_k}} \sum_{j=1}^{k} b_{n_j} T_j \xrightarrow{\text{a.s.}} 0.$$

Furthermore, upon setting $U_k = \max_{n_{k-1} < n \leqq n_k} \dfrac{|S_n - S_{n_{k-1}}|}{b_{n_k}}$ and applying P. Lévy's inequality we obtain, for every $\epsilon > 0$,

$$\sum P[U_k \geqq \epsilon] \leqq 2 \sum P[|T_k| \geqq \epsilon] < \infty,$$

so that $U_k \xrightarrow{\text{a.s.}} 0$.   Therefore, for $n_{k-1} < n \leq n_k$,

$$\left| \frac{S_n}{b_n} \right| = \left| \frac{S_n - S_{n_{k-1}}}{b_n} + \frac{S_{n_{k-1}}}{b_n} \right| \leq \frac{b_{n_k}}{b_n} U_k + \frac{b_{n_{k-1}}}{b_n} \left| \frac{S_{n_{k-1}}}{b_{n_{k-1}}} \right|$$

$$\leq cU_k + \frac{S_{n_k}}{b_{n_k}} \xrightarrow{\text{a.s.}} 0,$$

and the "if" assertion is proved.

COROLLARY 1.   *If* $|X_n| < b_n$, *then* $\dfrac{S_n - ES_n}{b_n} \xrightarrow{\text{a.s.}} 0$ *if, and only if,*
$T_k - ET_k \xrightarrow{\text{a.s.}} 0$ *as* $k \to \infty$ *or, equivalently, for every* $\epsilon > 0$,

$$\sum_k P[|T_k - ET_k| \geq \epsilon] < \infty.$$

Proof.   The "only if" assertion is proved as that of the foregoing criterion.   As for the "if" assertion, set $X_{nk} = (X_n - EX_n)/b_{n_k}$, $n_{k-1} < n \leq n_k$, so that $\sum_n X_{nk} = T_k - ET_k \xrightarrow{\text{a.s.}} 0$.   Note that $|X_{nk}| < 2$ and apply 13.5, 3° and 18.1 a.   It follows that

$$|\mu T_k - ET_k| \leq \sqrt{2t_k^2} \to 0,$$

so that $T_k - \mu T_k \xrightarrow{\text{a.s.}} 0$ and, by the foregoing criterion, $\dfrac{S_n}{b_n} - \mu\left(\dfrac{S_n}{b_n}\right)$
$\xrightarrow{\text{a.s.}} 0$.   But

$$\left| \mu\left(\frac{S_n}{b_n}\right) - E\left(\frac{S_n}{b_n}\right) \right| \leq \sqrt{2\frac{s_n^2}{b_n^2}} \to 0, \quad (s_n^2 = \sigma^2 S_n),$$

since, for $n_{k-1} < n \leq n_k$,

$$\frac{1}{c^2}\frac{s_n^2}{b_n^2} \leq \frac{s_{n_k}^2}{b_{n_k}^2} = \frac{1}{b_{n_k}^2}\sum_{j=1}^{k} b_{n_j}^2 t_j^2 \to 0.$$

Therefore, $\dfrac{S_n - ES_n}{b_n} \xrightarrow{\text{a.s.}} 0$, and the proof is concluded.

COROLLARY 2.   *If the* X_n *are centered at expectations and* $\sum \dfrac{\sigma^2 X_n}{b_n^2} < \infty$,
then $\dfrac{S_n}{b_n} \xrightarrow{\text{a.s.}} 0$.

Let $\bar{X}_n$ be X_n truncated at b_n, and set $\bar{S}_n = \sum_{k=1}^{n} \bar{X}_k$, $\bar{T}_k = \dfrac{S_{n_k} - S_{n_{k-1}}}{b_{n_k}}$.
Since, by Tchebichev's inequality,

$$\sum P[|X_n \neq \bar{X}_n] = \sum P[|X_n| \geq b_n] \leq \sum \frac{\sigma^2 X_n}{b_n^2} < \infty,$$

it follows, by the equivalence lemma, that the sequences $\dfrac{S_n}{b_n}$ and $\dfrac{\bar{S}_n}{b_n}$ are
tail-equivalent. But $E\bar{S}_n/b_n \to 0$ since $|EX_k - E\bar{X}_k| \leq \sigma^2 X_k/b_k$ while

$$\epsilon^2 P[|\bar{T}_k| \geq \epsilon] \leq \sum_{n > n_{k-1}}^{n_k} \frac{\sigma^2 \bar{X}_n}{b_{n_k}^2} \leq \sum_{n > n_{k-1}}^{n_k} \frac{\sigma^2 \bar{X}_n}{b_n^2} \leq \sum_{n > n_{k-1}}^{n_k} \frac{\sigma^2 X_n}{b_n^2},$$

so that

$$\epsilon^2 \sum P[|\bar{T}_k| \geq \epsilon] \leq \sum \frac{\sigma^2 X_n}{b_n^2} < \infty,$$

Corollary 1 applies, $\dfrac{\bar{S}_n}{b_n} \xrightarrow{\text{a.s.}} 0$ and, therefore, $\dfrac{S_n}{b_n} \xrightarrow{\text{a.s.}} 0$.

*§ 19. EXPONENTIAL BOUNDS AND NORMED SUMS

In this section, the r.v.'s X_n, $n = 1, 2, \cdots$, are independent and centered at expectations with variance $\sigma_n^2 = \sigma^2 X_n = EX_n^2$; and $S_n = \sum_{k=1}^{n} X_k$ are their consecutive sums, so that $ES_n = 0$, $s_n^2 = \sigma^2 S_n = \sum_{k=1}^{n} \sigma_k^2$. We exclude the trivial case of degenerate summands.

19.1 Exponential bounds. Kolmogorov's inequalities led, in Section 17, to asymptotic properties of sums S_n. His inequalities below, where to simplify the writing we drop the subscript n, will lead to deeper results but under more restrictive assumptions.

A. EXPONENTIAL BOUNDS. *Let $c = \max\limits_{k \leq n} \left| \dfrac{X_k}{s} \right|$ and let $\epsilon > 0$.*

(i) *If $\epsilon c \leq 1$, then $P\left[\dfrac{S}{s} > \epsilon \right] < \exp\left[-\dfrac{\epsilon^2}{2}\left(1 - \dfrac{\epsilon c}{2} \right) \right]$ and, if $\epsilon c \geq 1$,*
then $P\left[\dfrac{S}{s} > \epsilon \right] < \exp\left[-\dfrac{\epsilon}{4c} \right].$

(ii) *Given $\gamma > 0$, if $c = c(\gamma)$ is sufficiently small and $\epsilon = \epsilon(\gamma)$ is sufficiently large, then $P\left[\dfrac{S}{s} > \epsilon \right] > \exp\left[-\dfrac{\epsilon^2}{2}(1 + \gamma) \right].$*

Proof. 1° Let $t > 0, |X| \leq c < \infty, EX = 0$ and $\sigma^2 = \sigma^2 X$. Since

$$|EX^n| \leq c^n, \quad Ee^{tX} = 1 + \frac{t^2}{2!} EX^2 + \frac{t^3}{3!} EX^3 + \cdots,$$

$$e^{t(1-t)} < 1 + t < e^t,$$

it follows that, for $tc \leq 1$,

$$Ee^{tX} < 1 + \frac{t^2\sigma^2}{2} \left(1 + \frac{tc}{3} + \frac{t^2c^2}{3.4} + \cdots \right) < 1 + \frac{t^2\sigma^2}{2} \left(1 + \frac{tc}{2} \right)$$

$$< \exp \left[\frac{t^2\sigma^2}{2} \left(1 + \frac{tc}{2} \right) \right]$$

and

$$Ee^{tX} > 1 + \frac{t^2\sigma^2}{2} \left(1 - \frac{tc}{3} - \frac{t^2c^2}{3.4} - \cdots \right) > 1 + \frac{t^2\sigma^2}{2} \left(1 - \frac{tc}{2} \right)$$

$$> \exp \left[\frac{t^2\sigma^2}{2} (1 - tc) \right].$$

Replacing X by $\dfrac{X_k}{s}$, setting $S' = \dfrac{S}{s}$, and taking into account that

$$Ee^{tS'} = \prod_{k=1}^{n} E \exp \left[\frac{tX_k}{s} \right]$$

we obtain

(1) $$\exp \left[\frac{t^2}{2} (1 - tc) \right] < Ee^{tS'} < \exp \left[\frac{t^2}{2} \left(1 + \frac{tc}{2} \right) \right], \quad tc \leq 1.$$

Inequalities (i) follow then from

$$P[S' > \epsilon] \leq e^{-t\epsilon} Ee^{tS'} < \exp \left[-t\epsilon + \frac{t^2}{2} \left(1 + \frac{tc}{2} \right) \right]$$

where t is replaced by ϵ or $\dfrac{1}{c}$ according as $\epsilon c \leq 1$ or ≥ 1.

2° The proof of inequality (ii) is much more involved. Let α and β be two positive numbers less than 1; they will be selected later in terms of the given number γ. According to (1), we can take c sufficiently small $(\leq \alpha/t)$ so as to have

(2) $$Ee^{tS'} > \exp \left[\frac{t^2}{2} (1 - \alpha) \right].$$

On the other hand, setting $q(x) = P[S' > x]$ and integrating by parts, we have

$$Ee^{tS'} = -\int e^{tx}\, dq(x) = t\int e^{tx}q(x)\, dx.$$

We decompose the interval $(-\infty, +\infty)$ of integration into the five intervals $I_1 = (-\infty, 0]$, $I_2 = (0, t(1-\beta)]$, $I_3 = (t(1-\beta), t(1+\beta)]$, $I_4 = (t(1+\beta), 8t]$ and $I_5 = (8t, +\infty)$ and search for upper bounds of the integral over I_1 and I_5 and over I_2 and I_4. We have

$$J_1 = t\int_{-\infty}^{0} e^{tx}q(x)\, dx < t\int_{-\infty}^{0} e^{tx}\, dx = 1.$$

On account of (i), we have on I_5, for $8tc < 1$,

$$q(x) < \exp\left[-\frac{x}{4c}\right] < \exp[-2tx] \quad \text{for} \quad x \geq \frac{1}{c}$$

$$q(x) < \exp\left[-\frac{x^2}{2}\left(1 - \frac{xc}{2}\right)\right] \leq \exp\left[-\frac{x^2}{4}\right] < \exp[-2tx] \quad \text{for} \quad x < \frac{1}{c}.$$

Therefore, for c sufficiently small ($<1/8t$)

$$J_5 = t\int_{8t}^{\infty} e^{tx}q(x)dx < t\int_{8t}^{\infty} e^{-tx}\, dx < 1$$

and

(3) $$J_1 + J_5 < 2.$$

On the intervals I_2 and I_4 we have $x < \dfrac{1}{c}$ for c sufficiently small and, by (i),

$$e^{tx}q(x) < \exp\left[tx - \frac{x^2}{2}\left(1 - \frac{xc}{2}\right)\right] \leq \exp\left[tx - \frac{x^2}{2}(1 - 4tc)\right] = e^{g(x)}.$$

The quadratic expression $g(x)$ attains its maximum for $x = \dfrac{t}{1-4tc}$ which, for $c < \beta/4t(1+\beta)$, lies in I_3. Therefore, for c sufficiently small and $x \in I_2$,

$$g(x) \leq g(t(1-\beta)) = \frac{t^2}{2}(1-\beta)(1+\beta+4tc-4tc\beta) < \frac{t^2}{2}\left(1 - \frac{\beta^2}{2}\right)$$

and, then,

$$J_2 = t\int_{0}^{t(1-\beta)} e^{tx}q(x)\, dx < t\int_{0}^{t(1-\beta)} e^{g(x)}\, dx < t^2 \exp\left[\frac{t^2}{2}\left(1 - \frac{1}{2}\beta^2\right)\right];$$

similarly,

$$J_4 = t \int_{t(1+\beta)}^{8t} e^{tx} q(x) \, dx < t \int_{t(1+\beta)}^{8t} e^{g(x)} \, dx < 8t^2 \exp\left[\frac{t^2}{2}\left(1 - \frac{1}{2}\beta^2\right)\right].$$

We set now $\alpha = \dfrac{\beta^2}{4}$ and $t = \dfrac{\epsilon}{1-\beta}$ so that, by (2),

$$(4) \quad J_2 + J_4 < 9t^2 \exp\left[\frac{t^2}{2}\left(1 - \frac{1}{2}\beta^2\right)\right]$$

$$< \frac{9\epsilon^2}{(1-\beta)^2} \exp\left[-\frac{\epsilon^2\beta^2}{8(1-\beta)^2}\right] E \exp\left[\frac{\epsilon}{1-\beta} S'\right].$$

Since the last expectation and the inverse of its coefficient increase indefinitely as $\epsilon \to \infty$, it follows, by (3) and (4), that for ϵ sufficiently large

$$J_1 + J_5 < 2 < \tfrac{1}{4} E e^{tS'}, \quad J_2 + J_4 < \tfrac{1}{4} E e^{tS'}.$$

Then

$$J_3 - t \int_{t(1-\beta)}^{t(1+\beta)} e^{tx} q(x) \, dx > \tfrac{1}{2} E e^{tS'},$$

a fortiori,

$$2t^2 \beta e^{t^2(1+\beta)} q(\epsilon) > \frac{1}{2} \exp\left[\frac{t^2}{2}(1 - \alpha)\right]$$

and, since as $\epsilon \to \infty, \dfrac{1}{4t^2} \exp\left[\dfrac{t^2}{2}\alpha\right] \to \infty$, replacing t by its value, it

follows that, for ϵ sufficiently large,

$$q(\epsilon) > \frac{1}{4t^2\beta} \exp\left[\frac{t^2}{2}\alpha\right] \exp\left[-\frac{t^2}{2}(1 + 2\alpha + 2\beta)\right]$$

$$> \exp\left[-\frac{\epsilon^2}{2}\frac{1 + 2\alpha + 2\beta}{(1-\beta)^2}\right].$$

But, given $\gamma > 0$, we can select $\beta > 0$ so as to have

$$\frac{1 + 2\beta + \dfrac{\beta^2}{2}}{(1-\beta)^2} \leq 1 + \gamma.$$

Therefore, for $c = c(\gamma)$ sufficiently small and $\epsilon = \epsilon(\gamma)$ sufficiently large,

$$q(\epsilon) > \exp\left[-\frac{\epsilon^2}{2}(1 + \gamma)\right],$$

and (ii) is proved.

*19.2 Stability. The a.s. stability criterion (which is due to Prokhorov for $b_n = n$) is a criterion in the sense that it is both necessary and sufficient. Yet, it is not satisfactory, since, because of the independence of the summands, it has to be expected that a satisfactory criterion ought to be expressed in terms of individual summands and not in terms of nonoverlapping sums. The nearest to this requirement is a criterion in terms of variances (due also to Prokhorov for $b_n = n$), valid when the summands are suitably bounded, and whose proof is based upon the exponential bounds.

Let $b_n \uparrow \infty, 0 < \delta' \leq \dfrac{b_{n_{k+1}}}{b_{n_k}} \leq c < \infty$ and set $T_k = \dfrac{S_{n_k} - S_{n_{k-1}}}{b_{n_k}},\ t_k{}^2 =$ $\sigma^2 T_k = \dfrac{1}{b_{n_k}{}^2} \sum\limits_{n_{k-1} < n \leq n_k} \sigma^2 X_k$. We write $\log_2$ for loglog.

A. *If* $\dfrac{|X_n|}{b_n} = o(\log_2{}^{-1} b_n)$ *then* $\dfrac{S_n}{b_n} \xrightarrow{\text{a.s.}} 0$ *if, and only if, for every* $\epsilon > 0$, *the series* (i) $\sum \exp\left[-\dfrac{\epsilon^2}{t_k{}^2} \right]$ *converges.*

Proof. For n sufficiently large $\dfrac{|X_n|}{b_n} < 1$, so that corollary 1 of the a.s. stability criterion applies: for every $\epsilon > 0$

(ii) $\sum P[|T_k| > \epsilon] < \infty$.

We have to prove that convergence of series (i) for some ϵ implies that of series (ii) for the same or distinct ϵ; and conversely. On the other hand, elementary computations show that, setting $c_k = \max\limits_{n_{k-1} < n \leq n_k} \dfrac{|X_n|}{b_n}$, the assumption made implies that $c_k = \dfrac{a_k}{\log k}$ with $a_k \to 0$ as $k \to \infty$.

We use now the upper exponential bounds and observe that for $c_k \dfrac{\epsilon}{t_k{}^2} \geq 1$ and k sufficiently large

$$P[|T_k| > \epsilon] < 2 \exp\left[-\dfrac{\epsilon}{4a_k} \log k \right] = 2\left(\dfrac{1}{k}\right)^{\frac{\epsilon}{4a_k}} < \dfrac{2}{k^2}$$

and

$$\exp\left[-\dfrac{\epsilon^2}{4t_k{}^2} \right] \leq \exp\left[-\dfrac{\epsilon^2}{4\epsilon c_k} \right] = \exp\left[-\dfrac{\epsilon}{4a_k} \log k \right] < \dfrac{1}{k^2}$$

so that the corresponding sums in (i) and (ii) converge and we can

neglect all those terms for which $c_k \dfrac{\epsilon}{t_k{}^2} \geqq 1$.

Since for $c_k \dfrac{\epsilon}{t_k{}^2} < 1$

$$P[|\, T_k\,| > \epsilon] < 2 \exp\left[-\frac{\epsilon^2}{t_k{}^2}\left(1 - \frac{c_k \epsilon}{2 t_k{}^2}\right)\right] < 2 \exp\left[-\frac{\epsilon^2}{4 t_k{}^2}\right],$$

it follows that convergence of series (i) for every $\epsilon > 0$ entails that of series (ii) for every $\epsilon > 0$. Conversely, if series (ii) converges, then

$T_k \xrightarrow{\text{a.s.}} 0$ and $t_k{}^2 \to 0$, so that, for k sufficiently large, $\dfrac{\epsilon}{t_k}$ is as large as we

please and $\dfrac{c_k}{t_k} < \dfrac{t_k}{\epsilon}$ is as small as we please. Therefore, the exponential

bound is valid with, say, $\gamma = 1$, and

$$P[|\, T_k\,| > \epsilon] > 2 \exp\left[-\frac{\epsilon^2}{t_k{}^2}\right],$$

so that convergence of series (ii) for every $\epsilon > 0$ entails that of series (i) for every $\epsilon > 0$, and the proof is concluded.

COROLLARY. *If, for an $r \geqq 1$, $\sum \dfrac{E|\, X_n\,|^{2r}}{n^{r+1}} < \infty$, then $\dfrac{S_n}{n} \xrightarrow{\text{a s}} 0$.*

For $r = 1$, this proposition coincides with Corollary 2 of the a.s. stability criterion, so that it suffices to consider the case $r > 1$ (due to Brunk).

Proof. Let $\overline{X}_n = X_n$ or 0 according as $|\, X_n\,| < n^{\frac{r+1}{2r}}$ or $\geqq n^{\frac{r+1}{2r}}$, so that

$$\frac{|\, \overline{X}_n\,|}{n} = O(\log_2{}^{-1} n), \quad \sum \frac{E|\, \overline{X}_n\,|^{2r}}{n^{r+1}} \leqq \sum \frac{E|\, X_n\,|^{2r}}{n^{r+1}} < \infty$$

and, by Tchebichev's inequality,

$$\sum P[X_n \neq \overline{X}_n] = \sum P[|\, X_n\,| \geqq n^{\frac{r+1}{2r}}] \leqq \sum \frac{E|\, X_n\,|^{2r}}{n^{r+1}} < \infty.$$

Therefore, on account of the equivalence lemma, it suffices to prove that the assertion holds for r.v.'s X_n which satisfy the assumption made in **A**.

But, upon applying for $r > 1$ the inequality $E^r |X| \leq E |X|^r$, setting $n_k = 2^k$, and applying the c_r-inequality with $n_k - n_{k-1}$ summands, we have, summing over $n = n_{k-1} + 1, \cdots, n_k$,

$$t_k^{2r} = \frac{1}{n_k^{2r}} (E \sum X_n^2)^r \leq \frac{1}{n_k^{2r}} E(\sum X_n^2)^r$$

$$\leqq \frac{1}{n_k^{r+1}} \sum E |X_n|^{2r} \leq \sum \frac{E |X_n|^{2r}}{n^{r+1}}.$$

Therefore,

$$\sum_{k=1}^{\infty} t_k^{2r} \leqq \sum_{n=1}^{\infty} \frac{E |X_n|^{2r}}{n^{r+1}}$$

and, since we have $\exp\left[-\dfrac{\epsilon}{t_k^2}\right] < t_k^{2r}$ for k sufficiently large, criterion **A** is satisfied, and the proof is concluded.

*19.3   **Law of the iterated logarithm.**   We say that a numerical sequence b_n belongs to the *upper class* or to the *lower class* of a sequence S_n of r.v.'s, according as $P[S_n > b_n \text{ i.o.}] = 0$ or 1. *A priori*, there may be sequences b_n which belong to neither of these two classes. However, if S_n is an essentially divergent sequence of consecutive sums of independent r.v.'s, then every sequence b_n belongs to one of the foregoing two classes. The problem which arises is that of corresponding criteria. Relatively little is known about its general solution (in the case of unbounded summands), and the proofs of what is known are quite involved; the best results are due to Feller. The basic known result was first obtained by Khintchine (also P. Lévy) in the Bernoulli case as a strengthening of consecutive improvements of Borel's strong law of large numbers and, then, was extended by Kolmogoroff (also Cantelli) to more general cases, as follows:

A. Law of the iterated logarithm.   *If*

$$s_n^2 \to \infty \quad and \quad \frac{|X_n|}{s_n} = o(\log_2^{-\frac{1}{2}} s_n^2), \quad t_n = (2 \log_2 s_n^2)^{\frac{1}{2}},$$

then

$$P\left[\limsup \frac{S_n}{s_n t_n} = 1\right] = 1.$$

In other words, for every $\delta > 0$, the sequence $(1 + \delta)s_n t_n$ belongs to the upper class of the sequence S_n while the sequence $(1 - \delta)s_n t_n$ belongs to the lower class; clearly, it suffices to prove these assertions for δ arbitrarily small.

We observe that, since the assumptions remain valid if every X_n is replaced by $-X_n$, the conclusion yields

$$P\left[\lim\inf\frac{S_n}{s_n t_n} = -1\right]$$

and, therefore, it holds for both sequences S_n and $|S_n|$ if it holds for the first one.

Proof. Since $s_n{}^2 \to \infty$ and $\dfrac{s_{n+1}{}^2}{s_n{}^2} = 1 + o(\log_2{}^{-1} s_n{}^2) \to 1$, it follows that, for every $c > 1$, there exists a sequence $n_k = n_k(c) \uparrow \infty$ as $k \to \infty$, such that $s_{n_k} \sim c^k$. Let $\delta, \delta', \delta''$ be positive numbers.

1° We prove that the sequences $(1 + \delta)s_n t_n$ belong to the upper class of the sequence S_n by proving the same for the sequence $S^*_{n_k} = \max_{n \leq n_k} S_n$. For

$$P[S_n > (1 + \delta)s_n t_n \text{ i.o.}] \leqq P[S^*_{n_k} > (1 + \delta)s_{n_{k-1}} t_{n_{k-1}} \text{ i.o.}]$$

where

$$(1 + \delta)s_{n_{k-1}}t_{n_{k-1}} \sim \frac{1 + \delta}{c}\, s_{n_k}t_{n_k},$$

hence, taking $\delta' < \delta$, we can select $c > 1$ so that $\dfrac{1 + \delta}{c} > 1 + \delta'$ and

$$P[S^*_{n_k} > (1 + \delta)s_{n_{k-1}}\, t_{n_{k-1}} \text{ i.o.}] \leqq P[S^*_{n_k} > (1 + \delta')s_{n_k}t_{n_k} \text{ i.o.}].$$

Thus, the assertion will follow from the Cantelli lemma if we prove that

$$\sum P[S^*_{n_k} > (1 + \delta')s_{n_k}t_{n_k}] < \infty.$$

But, by the remark at the end of 18.1, the general term of this series is bounded by $2P\left[S_{n_k} > \left(1 + \delta' - \dfrac{\sqrt{2}}{t_{n_k}}\right)s_{n_k}t_{n_k}\right]$, where $1 + \delta' - \dfrac{\sqrt{2}}{t_{n_k}} \to 1 + \delta'$ Therefore, for $\delta'' < \delta'$ and k sufficiently large,

$$P\left[S_{n_k} > \left(1 + \delta' - \frac{\sqrt{2}}{t_{n_k}}\right)s_{n_k}t_{n_k}\right] \leqq P[S_{n_k} > (1 + \delta'')s_{n_k}t_{n_k}],$$

and it suffices to prove that the right-hand side is general term of a convergent series. This follows by applying the first upper exponential bound with $\epsilon_k = (1 + \delta'')t_{n_k}$ and $c_k = \max |X_j|/s_{n_k}$, valid for k sufficiently large since $c_k t_{n_k} \to 0$, so that

$$P[S_{n_k} > (1 + \delta'')s_{n_k}t_{n_k}] \leq \exp\left[-\tfrac{1}{2}(1 - \epsilon_k c_k/2)(1 + \delta'')^2 t_{n_k}^2\right]$$

$$\leq \exp\left[-(1 + \delta'') \log_2 s_{n_k}^2\right] \sim \frac{1}{(2k \log c)^{1+\delta''}},$$

and the assertion is proved. Furthermore, according to the considerations which follow the statement of the theorem, this assertion entails that $P[|S_n| > (1 + \delta)s_n t_n$ i.o.$] = 0$.

$2°$ It remains to prove that the sequences $(1 - \delta')s_n t_n$ belong to the lower class of the sequence S_n where we will take $1 > \delta' > \delta$. This assertion will be *a fortiori* true if we prove that it holds for a sequence S_{n_k}. Let

$$u_k^2 = s_{n_k}^2 - s_{n_{k-1}}^2 \sim s_{n_k}^2\left(1 - \frac{1}{c^2}\right),$$

$$v_k = (2 \log_2 u_k^2)^{1/2} \sim (2 \log_2 s_{n_k}^2) = t_{n_k}$$

and set

$$A_k = [S_{n_k} - S_{n_{k-1}} > (1 - \delta)u_k v_k].$$

We prove first that $P[A_k$ i.o.$] = 1$, as follows: The sums $S_{n_k} - S_{n_{k-1}}$, being nonoverlapping sums of independent r.v.'s, are independent and, by the Borel criterion, it suffices to prove that $\sum PA_k = \infty$. But, $\epsilon_k = (1 - \delta)v_k \to \infty$ while $c_k = \max_{n_{k-1} < n \leq n_k} (|X_n|/u_k) \to 0$ as $k \to \infty$; hence the lower exponential bound for PA_k applies with $1 + \gamma = \dfrac{1}{1 - \delta}$.

Therefore,

$$PA_k > \exp\left[-\tfrac{1}{2}(1 + \gamma)(1 - \delta)^2 v_k^2\right] = \exp\left[-(1 - \delta) \log_2 u_k^2\right]$$

$$\sim \frac{1}{(2k \log c)^{1-\delta}},$$

the series $\sum PA_k$ diverges, and $P[A_k$ i.o.$] = 1$.

On the other hand, if $B_k = [|S_{n_{k-1}}| \leq 2s_{n_{k-1}}t_{n_{k-1}}]$, then, according to the end of $1°$, $P[B_k^c$ i.o.$] = 0$; thus, from some value $n = n(\omega)$ on $|S_{n_{k-1}}(\omega)| \leq 2s_{n_{k-1}}t_{n_{k-1}}$ except for ω belonging to the null event $[B_k^c$ i.o.$]$. Therefore, $P[A_k B_k$ i.o.$] = 1$, and this entails the assertion. For,

$$A_k B_k \subseteq [S_{n_k} > (1 - \delta)u_k v_k - 2s_{n_{k-1}}t_{n_{k-1}}],$$

$$(1 - \delta)u_k v_k - 2s_{n_{k-1}}t_{n_{k-1}} \sim \left\{ (1 - \delta)\left(1 - \frac{1}{c^2}\right)^{\frac{1}{2}} - \frac{2}{c} \right\} s_{n_k}t_{n_k}$$

and, if we take c sufficiently large so that for $\delta' > \delta$

$$(1 - \delta)\left(1 - \frac{1}{c^2}\right)^{\frac{1}{2}} - \frac{2}{c} > 1 - \delta',$$

then

$$1 = P[A_k B_k \text{ i.o.}] \leqq P[S_{n_k} > (1 - \delta')s_{n_k}t_{n_k} \text{ i.o.}].$$

The proof is terminated.

COMPLEMENTS AND DETAILS

As throughout this chapter, $S_n = \sum_{k=1}^{n} X_k$ and a.s. convergence is to a r.v.

1. If the ch.f. of a sum of two r.v.'s is the product of the ch.f.'s of the summands, the summands may not be independent. Construct examples. Here is one: X is a Cauchy r.v.—with ch.f. $e^{-|u|}$; consider $X + Y$ where $Y = cX$, $c > 0$.

2. Let X, Y be independent r.v.'s and let $r \geqq 1$.
If X and Y are centered at expectations, then $E|X + Y|^r$ majorizes $E|X|^r$ and $E|Y|^r$. More generally, if, say, A is an event defined on X, then $E|X + Y|^r I_A \geqq E|X|^r I_A$.
If $E|X + Y|^r$ is finite, so are $E|X|^r$ and $E|Y|^r$. (Since $|x|^r = |E(x + Y)|^r \leqq E|x + Y|^r$, it follows that

$$E|X + Y|^r I_A = \int_A dF_X(x) \left\{ \int |x + y|^r dF_Y(y) \right\} \geqq \int_A |x|^r dF_X(x) = E|X|^r I_A.$$

For $r > 1$ the first assertion implies the second one. For $r = 1$, set $A = [|X| < a]$ and observe that $E|X + Y| \geqq E(|Y| - a)I_A = (E|Y| - a)PA$.)

3. *Generalized Kolmogorov inequality.* Let X_1, X_2, $\cdots$ be independent r.v.'s centered at expectations, and let $r \geqq 1$. Set $C = [\sup_{k \leqq n} |S_k| \geqq c]$ and prove that

$$c^r PC \leqq E|S_n|^r I_C \leqq E|S_n|^r.$$

Apply to the same problems to which Kolmogorov's inequality was applied. For example, if $S_n \xrightarrow{r} S$, then $S_n \xrightarrow{\text{a.s.}} S$. (Set $C_k = [\sup_{j < k} |S_j| < c, |S_k| \geqq c]$, $S_0 = 0$. By 2, $E|S_n|^r I_C = \sum_{k=1}^{n} E|S_n|^r I_{C_k} \geqq \sum_{k=1}^{n} E|S_k|^r I_{C_k} \geqq c^r PC$.)

4. Let X_1, X_2, $\cdots$ be independent r.v.'s, and let $T_n{}^r = \sup_{k \leqq n} |S_k|^r$, $r \geqq 1$.
If the X_k are symmetric, then $ET_n{}^r \leqq 2E|S_n|^r$.
If the X_k are centered at expectations, then $ET_n{}^r \leqq 2^{2r+1}E|S_n|^r$.

Extend to $n = \infty$ when $S_n \xrightarrow{\text{a.s.}} S_\infty$. (If symmetric, then

$$ET_n^r = \int_0^\infty P[T_n^r \geq t] \, dt \leq 2\int_0^\infty P[|S_n|^r \geq t] \, dt = 2E|S_n|^r.$$

If centered at expectations symmetrize; then

$$|S_k|^r \leq 2^{r-1} \sup_{k \leq n} |S_k - S'_k|^r + 2^{r-1}|S'_k|^r.$$

Integrate over $X'_1, \cdots, X'_n$, take sup, integrate over $X_1, \cdots, X_n$, and apply the first assertion.)

5. Let $X_1, X_2, \cdots$ be independent r.v.'s centered at expectations, and let $r \geq 1$. If $\sum E|X_n|^{2r}/n^{r+1} < \infty$, then $\dfrac{S_n}{n} \xrightarrow{\text{a.s.}} 0$. (Apply 4 and the elementary inequality $(\sum_{k=1}^n a_k^2)^r \leq n^{r-1} \sum_{k=1}^n |a_k|^{2r}$ to obtain $E|S_n|^{2r} \leq cn^{r-1} \sum_{k=1}^n E|X_k|^{2r}$. By Tchebichev's inequality,

$$P[|S_{2^{k+1}} - S_{2^k}| \geq 2^k\epsilon] \leq c2^{r+1}\epsilon^{-2r} \sum_{j=2^k+1}^{2^{k+1}} E|X_j|^{2r}/j^{r+1}.$$

Apply the a.s. stability criterion with $n_k = 2^k$.)

6. The series $\sum c_n e^{i\theta_n}$ where the θ_n are independent r.v.'s with $Ee^{i\theta_n} = 0$, converges or diverges a.s., according as the series $\sum c_n^2$ converges or diverges.

7. If a series $\sum X_n$ of independent r.v.'s converges a.s., then by centering the summands at the terms of some convergent series, the a.s. convergence and the limit are preserved under all changes of the order of the summands. (Start with a series which converges in q.m. Use the centering in the two series criterion.)

8. A series $\sum X_n$ of independent r.v.'s with ch.f.'s f_n converges a.s. whatever be the order of summands if, and only if, $\sum |f_n - 1| < \infty$.

9. If a series $\sum X_n$ of independent r.v.'s is essentially divergent, then it degenerates at infinity: $P[|S_n| < c] \to 0$ however large be $c > 0$. State the dual form for essential convergence. (This is true for the symmetrized series. Prove and apply: if X and X' are independent and identically distributed, then $P^2[|X| < c] \leq P[|X - X'| < 2c]$.)

10. Let $\sum X_n$ be a series of independent r.v.'s with ch.f.'s f_n.

If for a subsequence of integers $m \to \infty$ there exist r.v.'s Y_m with ch.f. g_m such that S_m and $Y_m - S_m$ are independent and $|g_m|^2 \to |g|^2$ continuous at the origin, then $\sum X_n$ is essentially convergent. (This follows from $\prod_{k=1}^m |f_k| \geq |g_m| \to |g| > \epsilon > 0$ in a neighborhood of the origin.)

11. *Smoothing by addition.* Loosely speaking, a sum of independent r.v.'s is at least as "smooth" as any of its summands. More precisely, continuity or analyticity properties of the law of one of the summands continue to hold for the law of the sum. Examples:

(a) If one of the summands has a continuous law so does the sum. (Introduce the "concentration" C_X defined by $C_X(l) = \max_{x \in R} P[x \leq X \leq x + l]$, $l \geq 0$. Observe that $C_X(0) = 0$ if, and only if, F_X is continuous. By the composition theorem for independent r.v.'s X and Y, $C_{X+Y} \leq C_X$, $C_{X+Y} \leq C_Y$.)

(b) If one of the summands has an absolutely continuous law, so does the sum. (In defining the concentration replace translates of segments of length l by translates of Lebesgue sets of measure l.)

(c) If one of the summands has a strictly increasing d.f., so does the sum. What about unicity of medians?

12. The symmetrization method reduces medians to zero and transforms essentially convergent series into a.s. convergent ones. However, only centering at medians does not yield a.s. convergence. In fact, let $\sum_{n=0}^{\infty} X_n$ be an a.s. convergent series of independent summands. The sequence $\mu(S_n)$ of medians may not converge. However, if $\sum_{n=0}^{\infty} X_n$ is essentially convergent and the r.v. Y is independent of all the X_n and has a strictly increasing d.f., then, after centering the $S_n + Y$ at medians, the series converges a.s.

(For the counterexample, take $X_0 = -1$ or $+1$ with same pr. $1/2$; let $0 < p_n < 1$ with $\sum p_n < \infty$ and, for $n \geq 1$, take X_{2n-1} and X_{2n} with values $2(-1)^n$ of pr. p_n and 0 of pr. $1 - p_n$. The sequence S_n converges a.s., yet the S_n are odd integers with $\mu(S_{4n-1}) \geq 1$ and $\mu(S_{4n+1}) \leq 1$. For the last assertion use 11(c).)

13. The X_n are not assumed to be independent. If $\dfrac{S_n^2}{n^2} \xrightarrow{\text{a.s.}} U$ and the X_n are uniformly bounded, then $\dfrac{S_n}{n} \xrightarrow{\text{a.s.}} U$. What if n^2 is replaced by n^k where k is a fixed integer? What if n^2 is replaced by $[q^n]$ with $q > 1$ arbitrarily close to 1?

More generally, let $\sum P[|U_n - U| > \epsilon]/n^\alpha < \infty$ for every $\epsilon > 0$, $\sum P[|X_n| > cn^\beta] < \infty$ for some $c > 0$, $0 < \alpha \leq 1$, $\beta > 0$. If $\gamma \geq \alpha + \beta$, then $U_n \xrightarrow{\text{a.s.}} U$, where $U_n = S_n/n^\gamma$

(For the first assertion, the second part of the proof of Borel's strong law of large numbers (see Introductory Part) applies. For the second assertion, use the following property of series: if $\sum |p_n|/n^\alpha < \infty$ with $0 < \alpha \leq 1$, then $\sum_k |p_{n_k}| < \infty$ for $n_{k+1} - n_k = o(n_k^\alpha)$.)

In what follows, *the r.v.'s $X_1, X_2, \cdots$, are independent and identically distributed* with common d.f. F, and ch.f. f of a r.v. X; the trivial case of $X = 0$ a.s. is excluded. In other words, repeated trials are performed on X.

14. Random selection. Let $\nu_1 < \nu_2 < \cdots$ be integer-valued r.v.'s such that every $[\nu_j = n]$ is defined on $X_1, \cdots, X_{n-1}$. The r.v.'s $X_{\nu_1}, X_{\nu_2}, \cdots$, are independent and identically distributed—as X. (Proceed as in

$$P[X_{\nu_1} < x_1, X_{\nu_2} < x_2] = \sum_{1 \leq n_1 < n_2 < \infty} P[\nu_1 = n_1, X_{n_1} < x_1; \nu_2 = n_2, X_{n_2} < x_2]$$

$$= \sum_{1 \leq n_1 < n_2 < \infty} P[\nu_1 = n_1, X_{n_1} < x_1; \nu_2 = n_2]P[X_{n_2} < x_2]$$

$$= P[X_1 < x_1]P[X_2 < x_2].)$$

15. Deviations from the median. If X is centered at a median, then

$$E\Big|\sum_{k=1}^{n} X_k\Big| \geqq \frac{g(n)}{n} \sum_{k=1}^{n} E|X_k|, \quad g(2n+1) = g(2n+2) = \frac{(2n+1)!}{(2^n n!)^2}.$$

This inequality is not necessarily true when X is centered at its expectation. Extend to nonidentically distributed X_k's. (Divide R^n into its 2^n "octants" and consider the corresponding parts of the left-hand side. For a counterexample, take $n = 3$, $X = 1$ with pr. 2/3 and -2 with pr. 1/3.)

16. Equidistribution of sums. If X is a lattice r.v. with step h—only possible values kh, $k = 0, \pm 1, \cdots$—set $M(g) = \lim_{n \to \infty} \frac{1}{2n+1} \sum_{k=-n}^{+n} g(kh)$ and otherwise set $M(g) = \lim_{h \to \infty} \frac{1}{2h} \int_{-h}^{+h} g(x)\, dx$ for those functions g on R for which either of the foregoing limits exists and is finite.

(a) In the first case $M(e^{iux}) = 1$ or 0, according as $u = 0 \left(\operatorname{mod} \dfrac{2\pi}{h}\right)$ or $u \neq 0$ $\left(\operatorname{mod} \dfrac{2\pi}{h}\right)$. In the second case $M(e^{iux}) = 1$ or 0 according as $u = 0$ or $u \neq 0$.

(b) For every $u \in R$,

$$Y_n = \frac{1}{n} \sum_{k=1}^{n} e^{iuS_k} \xrightarrow{\text{a.s.}} M(e^{iux}).$$

(This is immediate in the lattice case and if $u = 0$. Otherwise $f(u) \neq 1$ and

$$E|Y_n|^2 = \frac{1}{n} + \frac{2}{n^2} \Re \sum_{j>k} f^{j-k}(u) \leqq \frac{c}{n}$$

where c is finite. Use *13*.)

(c) The family G of functions g on R such that $\dfrac{1}{n} \sum_{k=1}^{n} g(S_k) \xrightarrow{\text{a.s.}} M(g)$ contains all almost periodic functions and functions with period p Riemann-integrable on $[o, p]$. (G contains all functions $g(x) = e^{iux}$. It is closed under additions, multiplications by complex numbers, conjugations, and uniform passages to the limit. M is a linear monotone operation on G.)

If $g_n \in G$ and $g_n \xrightarrow{u} g$, then $M(g_n) \to M(g)$. If $g'_n, g''_n \in G$ and $M(g'_n) - M(g''_n) \to 0$, then for every g such that $g'_n \leqq g \leqq g''_n$ whatever be n, $g \in G$ and $M(g) = \lim M(g'_n) = \lim M(g''_n)$.

(d) For X degenerate at an irrational a, the classical equidistribution (modulo 1) of the fractional parts of na follows: for g bounded with $g(x) \to c$ finite as $x \to \pm\infty$,

$$\frac{1}{n} \sum_{k=1}^{n} g(S_k) \xrightarrow{\text{a.s.}} c.$$

For every finite segment I, (no. of $S_1, \cdots, S_n$ in I)$/n \xrightarrow{\text{a.s.}} 0$.

17. Normal r.v.'s. Let X be normal with $EX = 0$, $EX^2 = 1$, let g on R^n be a finite Borel function, and set $\overline{X} = S_n/n$.

(a) If $g(x_1 + c, \cdots, x_n + c) = g(x_1, \cdots, x_n)$ for all x_k, $c \in R$, then the ch.f. of the pair $\overline{X}$, $g(X_1, \cdots, X_n)$ is $f(u, v) = f_1(u) f_2(u, v)$ where $f_1(u) = e^{-u^2/2}$ is ch.f. of $\overline{X}$ and $f_2(u, v) = (2\pi)^{-n/2} \int h(x_1, \cdots, x_n) \, dx_1, \cdots, dx_n$ with

$$\log h(x_1, \cdots, x_n) = -\frac{1}{2} \sum_k \left(x_k - \frac{iu}{n} \right)^2 + ivg(x_1, \cdots, x_n).$$

(b) If f_2 is analytic in u, then $\overline{X}$ and $g(X_1, \cdots, X_n)$ are independent. In particular, $\overline{X}$ is independent of $\max_{j, k} |X_j - X_k|$ and of $\sum_{k=1}^{n} |X_k - \overline{X}|^r$, $r > 0$. (f_2 is independent of u: set $u = inc$ and use the translation property of g.)

(c) Let p with or without affixes denote a pr. density with respect to the Lebesgue measure. Let $p(x) = \dfrac{1}{b\sqrt{2\pi}} \exp\left[-(x - a)^2/2b^2\right]$ be the pr. density of X_k, and set

$$S^2 = \frac{1}{n} \sum_{k=1}^{n} (X_k - \overline{X})^2, \quad \overline{S} = S/\sqrt{n}, \quad Y = \frac{\sqrt{n}}{b}(\overline{X} - a), \quad Z = \frac{\sqrt{2n}}{b}(\overline{S} - b).$$

Then the pr. density of Y is $\dfrac{1}{\sqrt{2\pi}} e^{-x^2/2}$, the pr. density of Z converges to $\dfrac{1}{\sqrt{2\pi}} e^{-x^2/2}$, the pr. density of (Y, Z) converges to $\dfrac{1}{\sqrt{2\pi}} e^{-(x^2+v^2)/2}$, and

$$E\overline{S} = b\left(1 + O\left(\frac{1}{n}\right)\right), \quad \sigma^2 \overline{S} = \frac{\sigma^2}{2n}\left(1 + O\left(\frac{1}{n}\right)\right).$$

Chapter VI

CENTRAL LIMIT PROBLEM

The Central Limit Problem of probability theory is the problem of convergence of laws of sequences of sums of r.v.'s.

For more than two centuries a particular case—the Classical Limit Problem—has been the limit problem of probability theory. The precise formulation of this case and its solution were obtained in the second quarter of this century. At the very time that this particular problem was receiving its definite answer, the much more general Central Limit Problem appeared, and was solved almost at once, thanks to the powerful ch.f.'s tool and to the truncation and symmetrization methods.

§ 20. DEGENERATE, NORMAL, AND POISSON TYPES

20.1 First limit theorems and limit laws. Three limit theorems and corresponding limit laws are at the origin of the classical limit problem. Let S_n be the number of occurrences of an event of pr. p in n independent and identical trials; to avoid trivialities we assume that $pq \neq 0$, where $q = 1 - p$. If X_k denotes the indicator of the event in the kth trial, then $S_n = \sum_{k=1}^{n} X_k$, $n = 1, 2, \cdots$, where the summands are independent and identically distributed indicators—this is the Bernoulli case. Since $EX_k = p$, $EX_k^2 = p$ and, hence, $\sigma^2 X_k = p - p^2 = pq$, it follows that

$$ES_n = \sum_{k=1}^{n} EX_k = np, \quad \sigma^2 S_n = \sum_{k=1}^{n} \sigma^2 X_k = npq.$$

The first limit theorem of pr. theory, published in 1713, says that $\dfrac{S_n}{n} \xrightarrow{P} p$. Bernoulli found it by a direct but cumbersome analysis of the asymptotic behavior of the "binomial pr.'s" $P[S_n = k] = C_n{}^k p^k q^{n-k}$, $k = 0, 1, 2, \cdots, n$.

Sharpening this analysis, de Moivre obtained the second limit theorem which, in its integral form due to Laplace, says that

$$P\left[\frac{S_n - np}{\sqrt{npq}} < x\right] \rightarrow \frac{1}{\sqrt{2\pi}} \int_{-\infty}^{x} \exp\left[-\frac{1}{2}y^2\right] dy, \quad -\infty \leqq x \leqq \infty.$$

The third limit theorem was obtained by Poisson, who modified the Bernoulli case by assuming that the pr. $p = p_n$ depends upon the total number n of trials in such a manner that $np_n \rightarrow \lambda > 0$. Thus, writing now X_{nk} and S_{nn} instead of X_k and S_n, the *Poisson case* corresponds to sequences of sums $S_{nn}{}^{\boldsymbol{\cdot}} = \sum_{k=1}^{n} X_{nk}$, $n = 1, 2, \cdots$, where, for every fixed n, the summands X_{nk} are independent and identically distributed indicators with $P[X_{nk} = 1] = \frac{\lambda}{n} + o\left(\frac{1}{n}\right)$. By a direct analysis of the asymptotic behavior of the binomial pr.'s, much easier to carry than the preceding ones, Poisson proved that

$$P[S_{nn} = k] \rightarrow \frac{\lambda^k}{k!} e^{-\lambda}, \quad k = 0, 1, 2, \cdots.$$

Thus are born the three basic laws of pr. theory.

1°   *The degenerate law* $\mathfrak{L}(0)$ of a r.v. degenerate at 0 with d.f. having one point of increase only at $x = 0$ and ch.f. reduced to 1.

2°   *The normal law* $\mathfrak{N}(0, 1)$ of a *normal r.v.* with d.f. defined by

$$F(x) = \frac{1}{\sqrt{2\pi}} \int_{-\infty}^{x} \exp\left[-\frac{1}{2}y^2\right] dy$$

and ch.f. given by

$$f(u) = \frac{1}{\sqrt{2\pi}} \int \exp\left[iux - \frac{x^2}{2}\right] dx$$

$$= \exp\left[-\frac{u^2}{2}\right] \cdot \frac{1}{\sqrt{2\pi}} \int_{-\infty-iu}^{+\infty-iu} \exp\left[-\frac{z^2}{2}\right] dz = \exp\left[-\frac{u^2}{2}\right].$$

The well-known value of the last integral is obtained by using Cauchy contour integration theorem.

3° *The Poisson law* $\mathcal{P}(\lambda)$ of a *Poisson r.v.* with d.f. defined by

$$F(x) = e^{-\lambda} \sum_{k=0}^{[x]} \frac{\lambda^k}{k!},$$

and ch.f. given by

$$f(u) = e^{-\lambda} \sum_{k=0}^{\infty} e^{iuk} \frac{\lambda^k}{k!} = e^{-\lambda} \sum_{k=0}^{\infty} \frac{(\lambda e^{iu})^k}{k!} = e^{\lambda(e^{iu}-1)}.$$

While the first two limit laws played a central role in the development of pr. theory, Poisson's law long stood isolated and ignored. We shall see later that there was a deep reason for this isolation and also that, unexpectedly enough, Poisson's law is, in a sense to be made precise, more fundamental for the central limit problem than the two others. With the notation introduced above, the three first limit theorems can be summarized as follows:

A. FIRST LIMIT THEOREMS. *In the Bernoulli case* $\mathcal{L}\left(\dfrac{S_n - ES_n}{n}\right) \to$ $\mathcal{L}(0)$ *and* $\mathcal{L}\left(\dfrac{S_n - ES_n}{\sigma S_n}\right) \to \mathcal{N}(0, 1)$, *while in the Poisson case* $\mathcal{L}(S_{nn}) \to$ $\mathcal{P}(\lambda)$.

The proof by means of ch.f.'s reduces to elementary computations. We have, taking limited expansions of exponentials,

$$E \exp\left[iu \frac{S_n - np}{n}\right] = \prod_{k=1}^{n} E \exp\left[iu \frac{X_k - p}{n}\right]$$

$$= \left(p \exp\left[\frac{iuq}{n}\right] + q \exp\left[-\frac{iup}{n}\right]\right)^n$$

$$= \left(1 + o\left(\frac{u}{n}\right)\right)^n \to 1;$$

$$E \exp\left[iu \frac{S_n - np}{\sqrt{npq}}\right] = \prod_{k=1}^{n} E \exp\left[iu \frac{X_k - p}{\sqrt{npq}}\right]$$

$$= \left(p \exp\left[\frac{iuq}{\sqrt{npq}}\right] + q \exp\left[\frac{-iup}{\sqrt{npq}}\right]\right)^n$$

$$= \left(1 - \frac{u^2}{2n} + o\left(\frac{u^2}{n}\right)\right)^n \to \exp\left[-\frac{u^2}{2}\right];$$

$$E \exp [iuS_{nn}] = \prod_{k=1}^{n} E \exp [iuX_{nk}] = (p_n \exp [iu] + q_n)^n$$

$$= \left(1 + \frac{\lambda}{n} (\exp [iu] - 1) + o\left(\frac{1}{n}\right)\right)^n$$

$$\to \exp [\lambda(e^{iu} - 1)].$$

The three first limit laws give rise to the three first limit types:

the *degenerate type* of degenerate laws $\mathcal{L}(a)$ with $f(u) = e^{iua}$;

the *normal type* of normal laws $\mathfrak{N}(a, b^2)$ with $f(u) = \exp\left[iua - \frac{b^2}{2} u^2\right]$;

the *Poisson type* of Poisson laws $\mathcal{P}(\lambda; a, b)$ with

$$f(u) = \exp [iua + \lambda(e^{iub} - 1)].$$

The three first limit theorems extend at once by means of the convergence of types theorem; we leave the corresponding statements to the reader.

*20.2 **Composition and decomposition.** The three first limit types possess an important closure property. Its deep parts are the normal and the Poisson "decompositions" discovered between 1935 and 1937. P. Lévy surmised and Cramer proved the first one and, then, Raikov proved the second one.

Let $\mathcal{L}(X)$, $\mathcal{L}(X_1)$, $\mathcal{L}(X_2)$ be laws of r.v.'s with corresponding ch.f.'s f, f_1, f_2. We say that $\mathcal{L}(X)$ is *composed* of $\mathcal{L}(X_1)$ and $\mathcal{L}(X_2)$ or that $\mathcal{L}(X_1)$ and $\mathcal{L}(X_2)$ are *components* of $\mathcal{L}(X)$ if, X_1 and X_2 being independent, $\mathcal{L}(X) = \mathcal{L}(X_1 + X_2)$ or, equivalently, if $f = f_1 f_2$.

A. COMPOSITION AND DECOMPOSITION THEOREM. *The degenerate and the normal types are closed under compositions and under decompositions. The same is true of every family of Poisson laws $\mathcal{P}(\lambda; a, b)$ with the same b.*

To avoid exceptions we consider degenerate laws as degenerate normal and as degenerate Poisson ones.

Proof. 1° Closure under compositions

$$\mathcal{L}(a_1) * \mathcal{L}(a_2) = \mathcal{L}(a_1 + a_2)$$

$$\mathfrak{N}(a_1, b_1^2) * \mathfrak{N}(a_2, b_2^2) = \mathfrak{N}(a_1 + a_2, b_1^2 + b_2^2)$$

$$\mathcal{P}(\lambda_1; a_1, b) * \mathcal{P}(\lambda_2; a_2, b) = \mathcal{P}(\lambda_1 + \lambda_2; a_1 + a_2, b)$$

follows at once by means of ch.f.'s, for

$$e^{iua_1} \cdot e^{iua_2} = e^{iu(a_1+a_2)}$$

$$\exp\left[iua_1 - \frac{b_1^2}{2}u^2\right] \cdot \exp\left[iua_2 - \frac{b_2^2}{2}u^2\right]$$

$$= \exp\left[iu(a_1 + a_2) - \frac{b_1^2 + b_2^2}{2}u^2\right]$$

$$\exp\left[iua_1 + \lambda_1(e^{iub} - 1)\right] \cdot \exp\left[iua_2 + \lambda_2(e^{iub} - 1)\right]$$

$$= \exp\left[iu(a_1 + a_2) + (\lambda_1 + \lambda_2)(e^{iub} - 1)\right].$$

The decomposition property of the degenerate type is immediate. For, if for every $u \in R, f_1(u)f_2(u) = e^{iua}$, then $|f_1||f_2| = 1$ and, since $|f_1| \leqq 1, |f_2| \leqq 1$, it follows that $|f_1| = |f_2| = 1$, so that by 14.1a

$$f_1(u) = e^{iua_1}, \quad f_2(u) = e^{iua_2}, \quad u \in R.$$

The proof in the normal and Poisson cases is much more involved. To begin with, we can, by a linear change of variable, make $a = 0$ and $b = 1$ in the laws to be decomposed. Thus, we have to seek ch.f.'s f_1 and f_2 such that, for every $u \in R$,

$$f_1(u)f_2(u) = e^{-\frac{u^2}{2}}$$

or

$$f_1(u)f_2(u) = e^{\lambda(e^{iu}-1)}.$$

2° We consider first the normal decomposition and apply 15.3A.

Since $e^{-\frac{z^2}{2}}$ is an entire nonvanishing function in the complex plane, the same is true of $f_1(z)$ and $f_2(z)$, and there exists a constant $c > 0$ such that $|f_1(z)| \leqq e^{c|z|^2}$. Therefore, upon taking the principal branch of $\log f_1(z)$ (vanishing at $u = 0$), it follows from the Hadamard factorization theorem that $\log f_1(z)$ is a polynomial in z of, at most, second degree. Since $f_1(u)$ being a ch.f., reduces to 1 at $u = 0$, equals $\bar{f}_1(-u)$, and is bounded on R, it follows that

$$\log f_1(u) = iua_1 - \frac{b_1^2}{2}u^2, \quad u \in R,$$

where a and b are real numbers. Similarly for $f_2(u)$, and the normal decomposition is proved.

3° It remains for us to consider the Poisson decomposition. Let X_1 and X_2 be two independent r.v.'s with d.f.'s F_1 and F_2, and let F be the d.f. of their sum. Since

$$[a_1 \leqq X_1 < b_1][a_2 \leqq X_2 < b_2] \subset [a_1 + a_2 \leqq X_1 + X_2 < b_1 + b_2]$$

and X_1, X_2 are independent, we have

(1) $F_1[a_1, b_1]F_2[a_2, b_2] \leqq F[a_1 + a_2, b_1 + b_2)$

and, letting $b_1, b_2 \to \infty$, it follows that

(2) $F(a_1 + a_2) \leqq F_1(a_1) + F_2(a_2)$.

Let now α_1 and α_2 be points of increase of F_1 and F_2, respectively. If $\alpha_1 \in (a_1, b_1)$ and $\alpha_2 \in (a_2, \nu_2)$ whence $\alpha_1 + \alpha_2 \in (a_1 + a_2, b_1 + b_2)$, then the left-hand side in (1) is positive and, hence, $\alpha_1 + \alpha_2$ is point of increase of F. Moreover, if α_1 and α_2 are first points of increase, then, taking $a_1 < \alpha_1$ and $a_2 < \alpha_2$ in (2), we have $F(a_1 + a_2) = 0$, and, hence, $\alpha_1 + \alpha_2$ is the first point of increase of F.

Now let F be the Poisson d.f. corresponding to $\mathcal{P}(\lambda)$; its only points of increase are $k = 0, 1, 2, \cdots$. Therefore, on account of what precedes, all points of increase α_1 and α_2 of its components F_1 and F_2 are such that $\alpha_1 + \alpha_2 = $ some k and the first points of increase are α and $-\alpha$ where α is some finite number. It follows, replacing $F_1(x)$ by $F_1(x - \alpha)$ and $F_2(x)$ by $F_2(x + \alpha)$ (this does not change F), that the new d.f.'s have $k = 0, 1, 2, \cdots$ as the only possible points of increase. Thus, we can set for the corresponding ch.f.'s

$$f_1(u) = \sum_{k=0}^{\infty} a_k e^{iuk}, \quad f_2(u) = \sum_{k=0}^{\infty} b_k e^{iuk}$$

with

$$a_0, b_0 > 0,\ a_k, b_k \geqq 0 \quad \text{for} \quad k > 0,\ \sum_{k=0}^{\infty} a_k = \sum_{k=0}^{\infty} b_k = 1.$$

Upon setting $z = e^{iu}$, $\varphi_1(z) = f_1(u)$, $\varphi_2(z) = f_2(u)$, we have to find nonvanishing functions φ_1 and φ_2 such that

$$\varphi_1(z)\varphi_2(z) = \sum_{k,l=0}^{\infty} a_k b_l z^{k+l} = \sum_{k=0}^{\infty} \frac{\lambda^k e^{-\lambda}}{k!} z^k.$$

Therefore,

$$a_0 b_k + a_1 b_{k-1} + \cdots + a_k b_0 = \frac{\lambda^k e^{-\lambda}}{k!}, \quad k = 0, 1, 2, \cdots,$$

and it follows that

$$a_k \leq \frac{1}{b_0} \frac{\lambda^k e^{-\lambda}}{k!}, \quad | \varphi_1(z) | \leq \frac{1}{b_0} e^{\lambda(|z|-1)}.$$

Thus, $\varphi_1(z)$ and similarly $\varphi_2(z)$ are nonvanishing entire functions at most of first order. It follows from the Hadamard factorization theorem that they are of the form $e^{cz+c'}$. Since $f_1(u)$ reduces to 1 at $u = 0$ and is bounded by 1, we have

$$\log f_1(u) = \lambda_1(e^{iu} - 1), \quad \lambda_1 \geq 0.$$

Similarly for $f_2(u)$, and the Poisson decomposition is proved. This terminates the proof of the theorem.

§ 21. EVOLUTION OF THE PROBLEM

21.1 The problem and preliminary solutions. From the time of Laplace and until 1935, the limit problem aims at weakenings of the assumptions under which the *law of large numbers* (convergence to $\mathcal{L}(0)$) and the *normal convergence* (convergence to $\mathfrak{N}(0, 1)$) hold. This classical problem can be stated as follows:

Let $S_n = \sum_{k=1}^{n} X_k$ be consecutive sums of independent r.v.'s. Find conditions under which

$$\mathcal{L}\left(\frac{S_n - ES_n}{n}\right) \to \mathcal{L}(0), \quad \mathcal{L}\left(\frac{S_n - ES_n}{\sigma S_n}\right) \to \mathfrak{N}(0, 1).$$

It is implicitly assumed, in the first case, that the summands are integrable, and in the second case that their squares also are integrable. To simplify the writing, we shall center the summands at expectations, so that, *in this section*, $EX_k = 0$, $ES_n = 0$. We also set $f_k(u) = Ee^{iuX_k}$, $\sigma_k = \sigma X_k$ and $s_n = \sigma S_n$, and exclude the trivial case of all summands degenerate.

Although not the first historically, the solution of the extension of the Bernoulli case to independent and identically distributed summands (not necessarily indicators) is immediate—when ch.f.'s are used.

A. *If the summands are independent, identically distributed, and centered at expectations, then $\mathcal{L}\left(\dfrac{S_n}{n}\right) \to \mathcal{L}(0)$ and $\mathcal{L}\left(\dfrac{S_n}{s_n}\right) \to \mathfrak{N}(0, 1)$.*

For, if f is the common ch.f. of the summands, then, by using its limited expansions, we have

$$E \exp\left[iu \frac{S_n}{n} \right] = \left(f\left(\frac{u}{n}\right) \right)^n = \left(1 + o\left(\frac{u}{n}\right) \right)^n \to 1,$$

and, since $s_n^2 = n\sigma^2 > 0$,

$$E \exp\left[iu \frac{S_n}{s_n} \right] = \left(f\left(\frac{u}{s_n}\right) \right)^n = \left(1 - \frac{\sigma^2}{2s_n^2} u^2 + o\left(\frac{\sigma^2}{s_n^2} u^2\right) \right)^n$$

$$= \left(1 - \frac{u^2}{2n} + o\left(\frac{u^2}{n}\right) \right)^n \to \exp\left[-\frac{u^2}{2} \right].$$

However, the first reasonably general conditions are the following.

B. *Let $S_n = \sum\limits_{k=1}^{n} X_k$ and $s_n = \sigma S_n$, where the summands are independent r.v.'s centered at expectations.*

(i) *If $\dfrac{1}{n^{1+\delta}} \sum\limits_{k=1}^{n} E| X_k |^{1+\delta} \to 0$ for a positive $\delta \leq 1$, then*

$$\mathcal{L}\left(\frac{S_n}{n} \right) \to \mathcal{L}(0).$$

(ii) *If $\dfrac{1}{s_n^{2+\delta}} \sum\limits_{k=1}^{n} E| X_k |^{2+\delta} \to 0$ for a positive δ, then*

$$\mathcal{L}\left(\frac{S_n}{s_n} \right) \to \mathfrak{N}(0, 1).$$

The assumptions imply finiteness of moments $E| X_k |^{1+\delta}$ and $E| X_k |^{2+\delta}$, respectively.

The first assertion is slightly more general than the classical ones. For $\delta = 1$, it becomes the celebrated *Tchebichev's theorem*. It also contains *Markov's theorem*: if $E| X_k |^{1+\delta} \leq c < \infty$, then $\mathcal{L}\left(\dfrac{S_n}{n} \right) \to \mathcal{L}(0)$

(since, then, the asserted condition becomes $\dfrac{c}{n^\delta} \to 0$); since, for $\delta > 1$,

$E X_k^2 \leq (E| X_k |^{1+\delta})^{2/1+\delta}$ Markov's theorem is valid with any $\delta > 0$.

The second assertion is the celebrated *Liapounov's theorem* which has been the turning point for the entire Central Limit theorem. Moreover, while the ch.f.'s were known to and used by Laplace, the first continuity theorem for ch.f.'s:

$$\text{if } f_n(u) \to e^{-\frac{u^2}{2}}, \quad \text{then } \mathcal{L}(X_n) \to \mathfrak{N}(0, 1),$$

is to be found, proved but not stated, in Liapounov's proof of his theorem. We observe that (ii) has content only when at least one of the r.v.'s is not degenerate at zero and, then, the hypothesis implies that $s_n \to \infty$.

Proof. 1° To begin with, let us reduce in (ii) the case $\delta > 1$ to $\delta = 1$, so that it will suffice to assume that $0 < \delta \leq 1$.

Let Y be a r.v. whose d.f. is $\dfrac{1}{n} \sum\limits_{k=1}^{n} F_k$ and, hence,

$$E|Y|^r = \frac{1}{n} \sum_{k=1}^{n} E|X_k|^r.$$

According to 9.3b. $\log E|Y|^r$ is a convex from below function of $r > 0$. Therefore, for $2 + \delta > 3$, we have

$$\delta \cdot \log E|Y|^3 \leq (\delta - 1) \log E|Y|^2 + \log E|Y|^{2+\delta}$$

or, equivalently,

$$\frac{1}{s_n^3} \sum_{k=1}^{n} E|X_k|^3 \leq \left(\frac{1}{s_n^{2+\delta}} \sum_{k=1}^{n} E|X_k|^{2+\delta} \right)^{1/\delta}.$$

It follows that, if the condition in (ii) holds for a $\delta > 1$, then it holds for $\delta = 1$. Thus, in what follows we can limit ourselves to $0 < \delta \leq 1$.

2° We use limited expansions of ch.f.'s, the continuity theorem, and the expansion $\log (1 + z) = z + o(|z|)$ valid for $|z| < 1$. As usual, θ with or without affixes denotes quantities bounded by 1.

Condition (i) implies that

$$\max_{k \leq n} \frac{E|X_k|^{1+\delta}}{n^{1+\delta}} \leq \frac{1}{n^{1+\delta}} \sum_{k=1}^{n} E|X_k|^{1+\delta} \to 0,$$

so that, for u arbitrary but fixed,

$$f_k \left(\frac{u}{n} \right) = 1 + \frac{2^{1-\delta}}{1 + \delta} \theta_{nk} |u|^{1+\delta} \frac{E|X_k|^{1+\delta}}{n^{1+\delta}} \to 1$$

uniformly in $k \leq n$. Therefore, for n sufficiently large,

$$\sum_{k=1}^{n} \log f_k \left(\frac{u}{n} \right) = 2\theta_n |u|^{1+\delta} \cdot \frac{1}{n^{1+\delta}} \sum_{k=1}^{n} E|X_k|^{1+\delta} \to 0,$$

and the first assertion is proved.

Condition (ii) implies that

$$\max_{k \le n} \left(\frac{\sigma_k}{s_n}\right)^{2+\delta} \le \max_{k \le n} \frac{E|X_k|^{2+\delta}}{s_n^{2+\delta}} \le \frac{1}{s_n^{2+\delta}} \sum_{k=1}^n E|X_k|^{2+\delta} \to 0,$$

so that, for u arbitrary but fixed,

$$f_k\left(\frac{u}{s_n}\right) = 1 - \frac{u^2}{2} \cdot \frac{\sigma_k^2}{s_n^2} + \frac{2^{1-\delta}}{(1+\delta)(2+\delta)} \theta'_{nk}|u|^{2+\delta} \frac{E|X_k|^{2+\delta}}{s_n^{2+\delta}} \to 1$$

uniformly in $k \le n$. Therefore, for n sufficiently large,

$$\sum_{k=1}^n \log f_k\left(\frac{u}{s_n}\right) = -\frac{u^2}{2}(1 + o(1))$$

$$+ 2\theta'_n|u|^{2+\delta} \frac{1}{s_n^{2+\delta}} \sum_{k=1}^n E|X_k|^{2+\delta} \to -\frac{u^2}{2},$$

and Liapounov's theorem is proved

BOUNDED CASE. *If the summands are uniformly bounded, then* $\mathcal{L}(S_n/n) \to \mathcal{L}(0)$. *If, moreover,* $s_n \to \infty$, *then* $\mathcal{L}(S_n/s_n) \to \mathfrak{N}(0, 1)$.

For, if $|X_k| \le c < \infty$, then $E|X_k|^{1+\delta} \le c^{1+\delta}$ and $E|X_k|^{2+\delta} \le c^\delta \sigma_k^2$, and, hence,

$$\frac{1}{n^{1+\delta}} \sum_{k=1}^n E|X_k|^{1+\delta} \le \frac{c^{1+\delta}}{n^\delta} \to 0,$$

$$\frac{1}{s_n^{2+\delta}} \sum_{k=1}^n E|X_k|^{2+\delta} \le \frac{c^\delta}{s_n^\delta} \to 0 \quad \text{as} \quad s_n \to \infty.$$

Tools for solution. The preceding theorem is not satisfactory since moments of higher order than those which figure in the formulation of the problem are used. Yet a restatement of this theorem with $\delta = 1$, together with the truncation method, will provide the stepping stone towards the solution.

a. BASIC LEMMA. *If* $S_{nn} = \sum_{k=1}^n X_{nk}$, *where the summands are independent r.v.'s (centered at expectations), then*

(i)   *if* $\dfrac{1}{n^2} \sum_{k=1}^n E|X_{nk}|^2 \to 0$, *then* $\mathcal{L}\left(\dfrac{S_{nn}}{n}\right) \to \mathcal{L}(0)$

(ii)   *if* $\dfrac{1}{s_{nn}^3} \sum_{k=1}^n E|X_{nk}|^3 \to 0$, *then* $\mathcal{L}\left(\dfrac{S_{nn}}{s_{nn}}\right) \to \mathfrak{N}(0, 1)$.

It suffices to replace in the proof of 21.2B subscripts k and n by double subscripts nk and nn, respectively.

In order to use the truncation method we shall require a weak form of the equivalence lemma. We say that two sequences $\mathcal{L}(X_n)$ and $\mathcal{L}(X'_n)$ of laws are *equivalent* if, for every subsequence $\mathcal{L}(X_{n'}) \to \mathcal{L}(X)$, we have $\mathcal{L}(X'_{n'}) \to \mathcal{L}(X)$, and conversely.

b. Law-equivalence lemma. *If $X_n - X'_n \overset{P}{\to} 0$ or $P[X_n \neq X'_n] \to 0$, then the sequences $\mathcal{L}(X_n)$ and $\mathcal{L}(X'_n)$ of laws are equivalent.*

For the second condition implies the first one which, by 10.1d, implies the asserted equivalence.

21.2   Solution of the Classical Limit Problem. We are now in a position to give a complete solution of the problem.

$X_1, X_2, \cdots$ are independent r.v.'s centered at expectations, with d.f.'s $F_1, F_2, \cdots$, ch.f.'s $f_1, f_2, \cdots$, and variances $\sigma_1^2, \sigma_2^2, \cdots$; $S_n = \sum_{k=1}^{n} X_k$ are their consecutive sums with variances $s_n^2 = \sum_{k=1}^{n} \sigma_k^2$. To simplify the writing, we make the convention that all summations are over $k = 1, \cdots, n$.

A. Classical degenerate convergence criterion. $\mathcal{L}\left(\dfrac{S_n}{n}\right) \to \mathcal{L}(0)$ *if, and only if,*

(i)
$$\sum \int_{|x| \geq n} dF_k \to 0,$$

(ii)
$$\frac{1}{n} \sum \int_{|x| < n} x \, dF_k \to 0,$$

(iii)
$$\frac{1}{n^2} \sum \left\{ \int_{|x| < n} x^2 \, dF_k - \left(\int_{|x| < n} x \, dF_k \right)^2 \right\} \to 0.$$

Proof. 1° Let (i), (ii), and (iii) hold. We wish to prove that $\mathcal{L}\left(\dfrac{S_n}{n}\right) \to \mathcal{L}(0)$. In what follows we apply the law equivalence lemma and the first part of the basic lemma.

Let $S_{nn} = \sum X_{nk}$, where $X_{nk} = X_k$ or 0 according as $|X_k| < n$ or $|X_k| \geq n$. On account of (i)

$$P\left[\frac{S_{nn}}{n} \neq \frac{S_n}{n} \right] \leq \sum P[X_{nk} \neq X_k] = \sum \int_{|x| \geq n} dF_k \to 0,$$

so that it suffices to prove that $\mathcal{L}\left(\dfrac{S_{nn}}{n}\right) \to \mathcal{L}(0)$. But, on account of (ii),

$$\frac{1}{n} E S_{nn} = \frac{1}{n} \sum \int_{|x|<n} x \, dF_k \to 0,$$

so that it suffices to prove that $\mathcal{L}\left(\dfrac{S_{nn} - E S_{nn}}{n}\right) \to \mathcal{L}(0)$. But this follows, by Tchebichev inequality, from (iii) and

$$\frac{1}{n^2} \sum E |X_{nk} - EX_{nk}|^2$$

$$= \frac{1}{n^2} \sum \left\{ \int_{|x|<n} x^2 \, dF_k - \left(\int_{|x|<n} x \, dF_k \right)^2 \right\} \to 0.$$

2° Conversely, let $\mathcal{L}\left(\dfrac{S_n}{n}\right) \to \mathcal{L}(0)$; equivalently, $\dfrac{S_n}{n} \xrightarrow{P} 0$ or $g_n(u) =$

$\prod\limits_{k=1}^{n} f_k(u/n) \to 1$ uniformly on every finite interval. Let n be suffi-ciently large so that $\log |g_n(u)|$ is bounded on $[-c, +c]$. By the weak symmetrization lemma and the second truncation inequality

$$\frac{1}{2} \sum_{k=1}^{n} P\left[\left| \frac{X_k - \mu X_k}{n} \right| \geq c^{-1} \right] \leq \sum_{k=1}^{n} P\left[\left| \frac{X_k{}^s}{n} \right| \geq c^{-1} \right]$$

$$\leq 7 \int_0^c \log |g_n(v)|^2 \, dv \to 0.$$

Since

$$\frac{X_n}{n} = \frac{S_n}{n} - \frac{n-1}{n} \frac{S_{n-1}}{n-1} \xrightarrow{P} 0,$$

so that $\mu X_n/n \to 0$, it follows that the foregoing relation with $c > 1$ yields (i) and, hence, $\mathcal{L}(S_{nn}/n) \to \mathcal{L}(0)$. But, by the first truncation in-equality,

$$(1) \qquad 2 \sum_{k=1}^{n} \sigma^2(X_{nk}/n) = \sum_{k=1}^{n} \sigma^2(X_{nk}{}^s/n) \leq -3 \log |g_n(1)|^2 \to 0,$$

so that (iii) holds, and, by Tchebichev inequality, $\dfrac{S_{nn} - ES_{nn}}{n} \xrightarrow{P} 0$. Therefore,

$$\frac{ES_{nn}}{n} = \frac{S_{nn}}{n} - \frac{S_{nn} - ES_{nn}}{n} \to 0$$

and (ii) holds. The proof is completed.

Observe that centering at, and in fact existence of, expectations were not required. Also, according to the proof,

$$\mathcal{L}\left(\frac{S_n - ES_{nn}}{n}\right) \to \mathcal{L}(0) \Leftrightarrow \text{(i) } and \text{ (iii) } hold.$$

B. CLASSICAL NORMAL CONVERGENCE CRITERION. $\mathcal{L}\left(\dfrac{S_n}{s_n}\right) \to \mathfrak{N}(0, 1)$

and $\max\limits_{k \leq n} \dfrac{\sigma_k}{s_n} \to 0$ if, and only if, for every $\epsilon > 0$,

$$g_n(\epsilon) = \frac{1}{s_n{}^2} \sum \int_{|x| \geq \epsilon s_n} x^2 \, dF_k \to 0.$$

The "if" part is due to Lindeberg and the "only if" part is due to Feller.

Proof. 1° Let $g_n(\epsilon) \to 0$ for every $\epsilon > 0$. We apply the law equivalence lemma and the basic lemma.

Since $g_n(\epsilon) \to 0$ for every $\epsilon > 0$, there is a sufficiently slowly decreasing sequence $\epsilon_n \downarrow 0$ such that $\dfrac{1}{\epsilon_n{}^2} g_n(\epsilon_n) \to 0$ and, *a fortiori*, $\dfrac{1}{\epsilon_n} g_n(\epsilon_n) \to 0$, $g_n(\epsilon_n) \to 0$ (it suffices to select a sequence $n_k \uparrow \infty$ as $k \to \infty$ such that $g_n\left(\dfrac{1}{k}\right) < \dfrac{1}{k^3}$ for $n \geq n_k$ and, then, take $\epsilon_n = \dfrac{1}{k}$ for $n_k \leq n < n_{k+1}$). We have

$$\max_{k \leq n} \frac{\sigma_k{}^2}{s_n{}^2} \leq \max_{k \leq n} \frac{1}{s_n{}^2} \int_{|x| \geq \epsilon_n s_n} x^2 \, dF_k + \epsilon_n{}^2 \leq g_n(\epsilon_n) + \epsilon_n{}^2 \to 0,$$

and the "if" assertion will be proved if we show that $\mathcal{L}\left(\dfrac{S_n}{s_n}\right) \to \mathfrak{N}(0, 1)$.

Let $X_{nk} = X_k$ or 0 according as $|X_k| < \epsilon_n s_n$ or $|X_k| \geq \epsilon_n s_n$. Since

$$P\left[\frac{S_{nn}}{s_n} \neq \frac{S_n}{s_n}\right] \leq \sum P[X_{nk} \neq X_k] = \sum \int_{|x| \geq \epsilon_n s_n} dF_k \leq \frac{1}{\epsilon_n{}^2} g_n(\epsilon_n) \to 0,$$

it suffices to prove that $\mathcal{L}\left(\dfrac{S_{nn}}{s_n}\right) \to \mathfrak{N}(0, 1)$.

Since the X_k are centered at expectations, we have

$$|EX_{nk}| = \left|\int_{|x| < \epsilon_n s_n} x \, dF_k\right| = \left|\int_{|x| \geq \epsilon_n s_n} x \, dF_k\right| \leq \frac{1}{\epsilon_n s_n} \int_{|x| \geq \epsilon_n s_n} x^2 \, dF_k.$$

Therefore,

$$\frac{1}{s_n} \sum |EX_{nk}| \leq \frac{g_n(\epsilon_n)}{\epsilon_n} \to 0$$

and, setting $s_{nn}^2 = \sigma^2 S_{nn}$, we obtain

$$1 - \frac{s_{nn}^2}{s_n^2} = \frac{1}{s_n^2} \sum \int_{|x| \geq \epsilon_n s_n} x^2 \, dF_k + \frac{1}{s_n^2} (\sum |EX_{nk}|)^2 \leq g_n(\epsilon_n) + \frac{g_n^2(\epsilon_n)}{\epsilon_n^2} \to 0.$$

Thus, it suffices to prove that $\mathcal{L}\left(\dfrac{S_{nn} - ES_{nn}}{s_{nn}}\right) \to \mathfrak{N}(0, 1)$. But, this follows from

$$\frac{1}{s_{nn}^3} \sum E |X_{nk} - EX_{nk}|^3 \leq \frac{2\epsilon_n s_n}{s_{nn}^3} \sum E(X_{nk} - EX_{nk})^2 \leq 2\epsilon_n \frac{s_n}{s_{nn}} \to 0,$$

and the "if" assertion is proved.

2° It remains to prove the "only if" assertion.

Since $\max\limits_{k \leq n} \dfrac{\sigma_k}{s_n} \to 0$, it follows from

$$f_k\left(\frac{u}{s_n}\right) = 1 - \theta_k \frac{u^2}{2} \frac{\sigma_k^2}{s_n^2}$$

that

$$\max_{k \leq n} \left| f_k\left(\frac{u}{s_n}\right) - 1 \right| \to 0, \quad \sum \left| f_k\left(\frac{u}{s_n}\right) - 1 \right|^2 \to 0.$$

Therefore, for n sufficiently large, $\log f_k\left(\dfrac{u}{s_n}\right)$ exists, so that

$$E \exp\left[iu \frac{S_n}{s_n} \right] = \prod_{k=1}^{n} f_k\left(\frac{u}{s_n}\right) \to \exp\left[-\frac{u^2}{2} \right]$$

becomes

$$\sum \log f_k\left(\frac{u}{s_n}\right) \to -\frac{u^2}{2}$$

and, since $\log z = z - 1 + \theta |z - 1|^2$,

$$\left| \frac{u^2}{2} - \sum \left\{ 1 - f_k\left(\frac{u}{s_n}\right) \right\} \right| \to 0.$$

Upon taking the real parts, we obtain

$$\frac{u^2}{2} - \sum \int_{|x| < \epsilon s_n} \left(1 - \cos \frac{ux}{s_n}\right) dF_k$$

$$= \sum \int_{|x| \geq \epsilon s_n} \left(1 - \cos \frac{ux}{s_n}\right) dF_k + o(1).$$

Since

$$\sum \int_{|x| < \epsilon s_n} \left(1 - \cos \frac{ux}{s_n}\right) dF_k \leq \frac{u^2}{2s_n^2} \sum \int_{|x| < \epsilon s_n} x^2 \, dF_k$$

$$= \frac{u^2}{2s_n^2} \left(s_n^2 - \sum \int_{|x| \geq \epsilon s_n} x^2 \, dF_k\right) = \frac{u^2}{2} (1 - g_n(\epsilon))$$

and

$$\sum \int_{|x| \geq \epsilon s_n} \left(1 - \cos \frac{ux}{s_n}\right) dF_k \leq 2 \sum \int_{|x| \geq \epsilon s_n} dF_k$$

$$\leq \frac{2}{\epsilon^2 s_n^2} \sum \int_{|x| \geq \epsilon s_n} x^2 \, dF_k \leq \frac{2}{\epsilon^2},$$

it follows that

$$\frac{u^2}{2} g_n(\epsilon) \leq \frac{2}{\epsilon^2} + o(1).$$

Therefore, letting $n \to \infty$ and then $u \to \infty$ in

$$0 \leq g_n(\epsilon) \leq \frac{2}{u^2} \left(\frac{2}{\epsilon^2} + o(1)\right),$$

we obtain $g_n(\epsilon) \to 0$. This concludes the proof.

*21.3 Normal approximation. In his celebrated investigation of normal convergence, Liapounov examined not only conditions for, but also the speed of, this convergence. His results were greatly improved by Berry (and, independently, by Esseen) and to present the basic one we shall proceed in steps.

Let F and G be d.f.'s of r.v.'s with corresponding ch.f.'s f and g, and let $H = F - G$, $h = f - g$. We exclude the trivial case of $\alpha = \sup |H| = 0$, that is, $H = h = 0$.

a. *If G is continuous on R, then there exists a finite number s such that either $H(s) = \mp\alpha$ or $H(s + 0) = \alpha$.*

Proof. Let x_n be a sequence such that $\left| H(x_n) \right| \rightarrow \alpha$. It contains a subsequence $x_{n'} \rightarrow s$ finite or infinite. Since $H(x) \rightarrow 0$ as $x \rightarrow \mp\infty$ and $\alpha > 0$, s must be finite.

The sequence $x_{n'}$ contains a subsequence $x_{n''}$ such that either $H(x_{n''}) \rightarrow -\alpha$ or $H(x_{n''}) \rightarrow +\alpha$. It suffices to consider one case only, say the first, for the same argument is valid for the other. Thus, let $x_{n''} \rightarrow s$, $H(x_{n''}) \rightarrow -\alpha$; we know that H is continuous from the left.

If the sequence $x_{n''}$ contains a subsequence converging to s from the left, then $-\alpha = \lim H(x_{n''}) = H(s)$, and the assertion is proved. Otherwise, this sequence contains a subsequence converging to s from the right, $-\alpha = H(s + 0)$ and, G being continuous on R,

$$-\alpha \leqq H(s) \leqq F(s + 0) - G(s) = F(s + 0) - G(s + 0) = -\alpha,$$

so that $-\alpha = H(s)$. The assertion is proved.

Let p be the derivative of a symmetric d.f. (of a r.v.) differentiable on R, so that $p(x) = p(-x)$, $x \in R$.

b. *If G has a derivative G' on R, then there exists a finite number "a" such that*

$$\left| \int H(x + a)p(x)\,dx \right| \geqq \frac{\alpha}{2}\left(1 - 6\int_{\frac{\alpha}{2\beta}}^{\infty} p(x)\,dx\right), \quad \beta = \sup \left| G' \right|.$$

Proof. If $\beta = \infty$, then $\dfrac{\alpha}{2\beta} = 0$, and the inequality is trivially true whatever be a. Thus, it suffices to prove it when $\beta < \infty$. Let $\gamma = \dfrac{\alpha}{2\beta} > 0$.

We have, for an arbitrary a,

(1) $$\left| \int H(x + a)p(x)\,dx \right|$$

$$\geqq \left| \int_{|x|<\gamma} H(x + a)p(x)\,dx \right| - \left| \int_{|x|\geqq\gamma} H(x + a)p(x)\,dx \right|$$

and

(2) $$\left| \int_{|x|\geqq\gamma} H(x + a)p(x)\,dx \right| \leqq \alpha\int_{|x|\geqq\gamma} p(x)\,dx.$$

On the other hand, according to **a**, there exists a finite number s such that, say, $-\alpha = H(s)$. For $|x| < \gamma$, we have, setting $a = s - \gamma$ so that

$$s - 2\gamma < x + a < s, \quad x - \gamma < 0,$$

the relation

$$G(x + a) = G(s) + \theta(x - \gamma)G'(x'), \quad |\theta| \leq 1, \quad s - 2\gamma < x' < s.$$

Thus, for $|x| < \gamma$,

$$H(x + a) = F(x + a) - G(s) - \theta(x - \gamma)G'(x')$$

$$\leq F(s) - G(s) - \beta(x - \gamma)$$

$$= -\alpha - \beta(x - \gamma) = -\beta(x + \gamma),$$

and it follows that

$$(3) \qquad \int_{|x|<\gamma} H(x + a)p(x)\, dx \leq -\beta \int_{|x|<\gamma} (x + \gamma)p(x)\, dx$$

$$= -\beta\gamma \int_{|x|<\gamma} p(x)\, dx$$

$$= -\frac{\alpha}{2}\left(1 - \int_{|x|\geq\gamma} p(x)\, dx\right).$$

Upon substituting in (1) the bounds given by (2) and (3), we obtain

$$\left| \int H(x + a)p(x)\, dx \right| \geq \frac{\alpha}{2}\left(1 - 3\int_{|x|\geq\gamma} p(x)\, dx\right)$$

and the assertion follows. In the case $\alpha = H(s + 0)$, the argument is similar.

Let $\bar\omega$ be a real ch.f. with $\int |\bar\omega(u)|\, du < \infty$, so that the corresponding d.f. has a symmetric derivative continuous on R, given by

$$p(x) = \frac{1}{2\pi}\int e^{-iux}\bar\omega(u)\, du = \frac{1}{2\pi}\int \cos ux\cdot\bar\omega(u)\, du.$$

    c. *For every* $a \in R$

$$\frac{1}{2\pi}\int \left| \frac{h(u)\bar\omega(u)}{u} \right| du \geq \left| \int H(x + a)p(x)\, dx \right|.$$

Proof. We can assume $\dfrac{h(u)\bar{\omega}(u)}{u}$ to be integrable, for otherwise the

inequality is trivially true. According to the composition theorem, $h\bar{\omega}$ is the Fourier-Stieltjes transform of $\overline{H}$ defined by

$$\overline{H}(x) = \int H(x - y)p(y)\, dy.$$

Since $\dfrac{h(u)\bar{\omega}(u)}{u}$ is integrable, the inversion formula yields

$$\overline{H}(x) - \overline{H}(x') = \frac{1}{2\pi}\int \frac{e^{-iux} - e^{-iux'}}{-iu}\, h(u)\bar{\omega}(u)\, du.$$

But, as $x' \to -\infty$, $H(x') \to 0$ and, by the Riemann-Lebesgue theorem,

$\displaystyle\int e^{-iux'}\frac{h(u)\bar{\omega}(u)}{-iu}\, du \to 0$. Therefore,

$$\int H(x - y)p(y)\, dy = \frac{1}{2\pi}\int e^{-iux}\frac{h(u)\bar{\omega}(u)}{-iu}\, du$$

and, hence, replacing x by a, y by $-x$, and taking into account that p is symmetric, we obtain

$$\int H(x + a)p(x)\, dx = \frac{1}{2\pi}\int e^{-iua}\frac{h(u)\bar{\omega}(u)}{-iu}\, du.$$

The asserted inequality follows.

We are now in a position to establish the basic inequality below, of independent interest. We shall require a real integrable function $\bar{\omega}_0$ defined by $\bar{\omega}_0(u) = 1 - \dfrac{|u|}{U}$ or 0 according as $|u| < U$ or $|u| \geqq U$.

Its Fourier-Stieltjes transform p_0 is given by

$$p_0(x) = \frac{1}{2\pi}\int_{-U}^{+U}\left(1 - \frac{|u|}{U}\right)\cos ux\, du = \frac{1 - \cos Ux}{\pi x^2 U}$$

and we have $p_0 \geqq 0$, $\int p_0(x)\, dx = 1$, so that $\bar{\omega}_0$ is a ch.f.

A. Basic inequality. *If G has a derivative G' on R, then, for every $U > 0$,*

$$\sup |H| \leqq \frac{2}{\pi}\int_0^U \left|\frac{h(u)}{u}\right|\, du + \frac{24}{\pi U}\sup |G'|.$$

Proof. Upon replacing $\bar{\omega}$ and p by $\bar{\omega}_0$ and p_0, the propositions **b** and **c** yield the inequality

$$\frac{1}{2\pi}\int_{-U}^{U}\left|\frac{h(u)}{u}\right|\,du \geq \frac{1}{2\pi}\int\left|\frac{h(u)\bar{\omega}_0(u)}{u}\right|\,du \geq \frac{\alpha}{2}\left(1 - 6\int_{\gamma}^{\infty}p_0(x)\,dx\right)$$

where

$$\int_{\gamma}^{\infty}p_0(x)\,dx = \frac{1}{\pi}\int_{\gamma}^{\infty}\frac{1 - \cos Ux}{x^2 U}\,dx \leq \frac{2}{\pi}\int_{\gamma U}^{\infty}\frac{dx}{x^2} = \frac{2}{\pi\gamma U} = \frac{4\beta}{\pi\alpha U}.$$

Therefore,

$$\frac{1}{\pi}\int_{0}^{U}\left|\frac{h(u)}{u}\right|\,du \geq \frac{\alpha}{2} - \frac{12\beta}{\pi U}$$

and the asserted inequality follows.

In order to apply the basic inequality to the normal approximation problem, we have to bound the corresponding h. Let F^*_n and G^* be the d.f.'s of $\mathcal{L}\left(\dfrac{S_n}{s_n}\right)$ and $\mathfrak{N}(0, 1)$ and let $h^*_n = f^*_n - e^{-\frac{u^2}{2}}$ denote the difference of the corresponding ch.f.'s. The summands X_n are independent r.v.'s centered at expectations, and we set $\gamma_n{}^3 = E|X_n|^3$, $g_n{}^3 = \dfrac{2^3}{s_n{}^3}\sum_{k=1}^{n}\gamma_k{}^3$. We exclude the case of one of the γ_n infinite, for then the normal approximation theorem below is trivially true.

d. *If* $|u| < \dfrac{2}{g_n{}^3}$, *then* $|h^*_n(u)| \leq 2g_n{}^3|u|^3\exp\left[-\dfrac{u^2}{3}\right]$.

Proof. 1° First, we prove the assertion under the supplementary condition $|u| \geq \dfrac{1}{g_n}$. Then $g_n{}^3|u|^3 \geq 1$ and it suffices to prove that $|h^*_n(u)| \leq 2\exp\left[-\dfrac{u^2}{3}\right]$. But, since

$$|h^*_n(u)| \leq |f^*_n(u)| + \exp\left[-\frac{u^2}{2}\right] \leq |f^*_n(u)| + \exp\left[-\frac{u^2}{3}\right],$$

it will suffice to prove that $|f^*_n(u)|^2 \leq \exp\left[-\dfrac{2u^2}{3}\right]$.

Consider the symmetrized r.v. $X_k - X'_k$ where X_k and X'_k are independent and identically distributed, so that its ch.f. is $|f_k|^2$ and

$$E(X_k - X'_k)^2 = 2\sigma_k{}^2, \quad E|X_k - X'_k|^3 \leq 2^3\gamma_k{}^3 < \infty.$$

Therefore,

$$|f_k(u)|^2 \leq 1 - \sigma_k^2 u^2 + \frac{2^2}{3} \gamma_k^3 |u|^3 \leq \exp\left[-\sigma_k^2 u^2 + \frac{2^2}{3} \gamma_k^3 |u|^3\right]$$

and, replacing u by $\dfrac{u}{s_n}$ and summing over $k = 1, \cdots, n$, we obtain, using the fact that, by assumption, $g_n^3 |u| < 2$,

$$|f^*_n(u)|^2 \leq \exp\left[-u^2 + \frac{g_n^3 |u|^3}{3.2}\right] \leq \exp\left[-u^2 + \frac{u^2}{3}\right]$$

$$= \exp\left[-\frac{2}{3} u^2\right].$$

2° It remains to prove the assertion when $|u| < \dfrac{1}{g_n}$ and, hence,

$$\frac{\sigma_k}{s_n} |u| \leq \frac{\gamma_k}{s_n} |u| \leq \frac{g_n}{2} |u| < \frac{1}{2}.$$

Then, we have

$$f_k\left(\frac{u}{s_n}\right) = 1 - \frac{\sigma_k^2}{2s_n^2} u^2 + \theta \frac{\gamma_k^3}{6s_n^3} |u|^3 = 1 - r_k,$$

where $|r_k| < \frac{1}{2}$, so that

$$\log f_k\left(\frac{u}{s_n}\right) = -r_k + \theta'_k r_k^2.$$

On the other hand,

$$|r_k|^2 \leq 2\left(\frac{\sigma_k^2 u^2}{2s_n^2}\right)^2 + 2\left(\frac{\gamma_k^3 |u|^3}{6s_n^3}\right)^2 \leq \frac{\gamma_k^3}{3s_n^3} |u|^3.$$

so that

$$\log f_k\left(\frac{u}{s_n}\right) = -\frac{\sigma_k^2}{2s_n^2} u^2 + \theta''_k \frac{|\gamma_k|^3}{2s_n^3} |u|^3$$

and, summing over k, we obtain

$$\log f^*_n(u) = -\frac{u^2}{2} + \theta \frac{g_n^3}{2^4} |u|^3.$$

Since, for every number a, $e^a = 1 + \theta'_a e^{\theta' a}$, it follows, taking $a = \dfrac{g_n^3}{2^4} |u|^3 \le \dfrac{1}{2^4}$ so that $e^a < 2$, that

$$\left| f^*_n(u) - \exp\left[-\frac{u^2}{2}\right] \right| \le 2 \frac{g_n^3}{2^4} |u|^3 \exp\left[-\frac{u^2}{2}\right]$$

$$\le 2 g_n^3 |u|^3 \exp\left[-\frac{u^2}{3}\right],$$

and the proof is complete.

B. Normal approximation theorem. *There exists a numerical constant $c < \infty$ such that, for all x and all n, if F^*_n is d.f. of $\mathcal{L}(S_n/s_n)$ and G^* is d.f. of $\mathfrak{N}(0, 1)$, then*

$$| F^*_n(x) - G^*(x) | \le \frac{c}{s_n^3} \sum_{k=1}^{n} E| X_k |^3.$$

For, upon replacing h^*_n by its bound obtained above in the basic inequality with $U = \dfrac{2}{g_n^3}$, $F = F^*_n$, and $G = G^*$ hence $\sup |G'| = \dfrac{1}{\sqrt{2\pi}}$, we obtain

$$\alpha \le \frac{4}{\pi} g_n^3 \left(\int_0^\infty u^2 \exp\left[-\frac{u^2}{3}\right] du + \frac{3}{\sqrt{2\pi}} \right).$$

§ 22. CENTRAL LIMIT PROBLEM; THE CASE OF BOUNDED VARIANCES

22.1 Evolution of the problem. The classical limit problem deals with independent summands X_n with finite first moments and, in the normal convergence case, with finite second moments as well. Those moments are used for changing origins and scales of values of the consecutive sums $S_n = \sum_{k=1}^{n} X_k$ so as to avoid shifts of the pr. spreads towards infinite values. There is no reason for these choices of "norming" quantities except an historical one; they are a straightforward extension to more general cases of the norming quantities which appeared in the Bernoulli case. *A priori*, there is no reason to expect that these quantities will continue to play the same role in the general case. Furthermore, whether they are available (that is, exist and are finite) or not, other choices might achieve the same purpose. Thus, the problem becomes a search for conditions under which the law of large numbers and the normal convergence hold for normed sums $\frac{S_n}{b_n} - a_n$. The methods remain those of the classical problem, but the computations become more involved. However, remnants of the two first limit theorems in the Bernoulli case are still visible. For there is no other reason to expect or to look for limit laws which are either degenerate or normal.

The real liberation which gave birth to the Central Limit Problem came with a new approach due to P. Lévy. He stated and solved the following problem: Find the family of *all possible limit laws* of normed sums of independent and identically distributed r.v.'s. We saw that when these r.v.'s have a finite second moment, the limit law (with classical norming quantities) is normal. Thus, P. Lévy was concerned primarily with the novel case of infinite second moments and finite or infinite first moments.

Naturally, the question of all possible limit laws of normed sums with independent, but not necessarily identically distributed, r.v.'s arises at once. Yet, the Poisson limit theorem is still out, for it is relative to sequences of sums and not to sequences of normed consecutive sums. Moreover, as we shall find it later (end 24.4), under "natural" restrictions Poisson laws *cannot* be limit laws of sequences of normed sums—which explains their isolation. But sequences $\frac{S_n}{b_n} - a_n$ are a particular form of sequences $\sum_{k=1}^{n} X_{nk} \left(\text{set } X_{nk} = \frac{X_k}{b_n} - \frac{a_n}{n} \right)$, and this provides the final modification of the problem.

The general outline of the Central Limit Problem is now visible: Find *the* limit laws of sequences of sums of independent summands and find conditions for convergence to a specified one. Yet, so general a problem is without content. In fact, let Y_n be arbitrary r.v.'s, set $X_{n1} = Y_n$ and $X_{nk} = 0$ a.s. for $k > 1$ and every n. Then the sequence of laws becomes the sequence $\mathcal{L}(Y_n)$, so that the family of possible limit laws contains *any* law $\mathcal{L}$—take $\mathcal{L}(Y_n) \equiv \mathcal{L}$. Thus, some restriction is needed.

To find a "natural" one, let us consider the problems which led to this one. Their common feature is that the number of summands increases indefinitely and that the limit law remains the same if an arbitrary but finite number of summands is dropped. To emphasize this feature, we are led to the following "natural" restriction: the summands

X_{nk} are *uniformly asymptotically negligible (uan)*, that is, $X_{nk} \xrightarrow{P} 0$ uniformly in k or, equivalently, for every $\epsilon > 0$,

$$\max_k P[|X_{nk}| \geq \epsilon] \to 0.$$

Finally, the precise formulation of the problem is as follows:

CENTRAL LIMIT PROBLEM. *Let* $S_{nk_n} = \sum_{k=1}^{k_n} X_{nk}$ *be sums of uan independent summands* X_{nk}, *with* $k_n \to \infty$.
     1° *Find the family of all possible limit laws of these sums.*
     2° *Find conditions for convergence to any specified law of this family.*

To simplify the writing, we make the following conventions valid for the whole chapter.
     (i) $k = 1, \cdots, k_n$, $k_n \to \infty$, the summations $\sum_k$, the products $\prod_k$, the maxima $\max_k$, are over these values of k, and the limits are taken as usually for $n \to \infty$, unless otherwise stated.
     (ii) F_{nk} and f_{nk} denote the d.f. and the ch.f. of r.v.'s X_{nk}, F_n and f_n denote the d.f. and the ch.f. of $\sum_k X_{nk}$. Thus, the uan condition becomes:

$$\max_k \int_{|x| \geq \epsilon} dF_{nk} \to 0 \text{ for every } \epsilon > 0, \text{ and the assumption of independ-}$$

ence becomes $f_n = \prod_k f_{nk}$. The problem becomes

Given sequences $f_n = \prod_k f_{nk}$ *of products of ch.f.'s of uan r.v.'s:* 1° *Find all ch.f.'s* f *such that* $f_n \to f$; 2° *Find conditions under which* $f_n \to f$ *given.*
If these ch.f.'s have log's on $I = [-U, +U]$, we always select their principal branches—continuous and vanishing at $u = 0$, and then on I: $\log f_n = \sum_k \log f_{nk}$, $f_n \to f$ (uniformly) $\Leftrightarrow \log f_n \to \log f$ (uniformly).

The solution of the problem is due to the introduction, by de Finetti, of the "infinitely decomposable" family of laws and to the discovery of their explicit representation by Kolmogorov in the case of finite second moments and by P. Lévy in the general case.

It has been obtained, with the help of the preceding family of laws, by the efforts of Kolmogorov, P. Lévy, Feller, Bawly, Khintchine, Marcin-kiewicz, Gnedenko, and Doblin (1931–1938). The final form is essentially due to Gnedenko.

22.2 The case of bounded variances. As a preliminary to the investigation of the general problem, and independently of it, we examine here the particular "case of bounded variances"—a "natural" extension of the classical normal convergence problem. It is much less involved computationally than the general one, while the method of attack is essentially the same.

We consider sums $\sum_k X_{nk}$ of independent r.v.'s, centered at expectations, with d.f.'s F_{nk}, ch.f.'s f_{nk} and finite variances $\sigma_{nk}{}^2 = \sigma^2 X_{nk}$ such that

(C): $\max_k \sigma_{nk}{}^2 \to 0$ and $\sum_k \sigma_{nk}{}^2 \leq c < \infty$, where c is a constant independent of n.

Since, for every $\epsilon > 0$,

$$\max_k P[|X_{nk}| \geq \epsilon] \leq \frac{1}{\epsilon^2} \max_k \sigma_{nk}{}^2 \to 0,$$

the uan condition is satisfied and the model is a particular case of that of the Central Limit Problem. The boundedness of the sequence of variances of the sums entails finiteness of the variance of the limit law.

a. COMPARISON LEMMA. *Under* (C), $\log f_{nk}(u)$ *exists and is finite for $n \geq n_u$ sufficiently large and, for any fixed u,*

$$\sum_k \{\log f_{nk}(u) - (f_{nk}(u) - 1)\} \to 0.$$

Proof. Since $f_{nk}(u) = 1 - \theta_{nk} \dfrac{\sigma_{nk}{}^2}{2} u^2$, it follows from (C) that

$$\max_k |f_{nk}(u) - 1| \leq \frac{u^2}{2} \max_k \sigma_{nk}{}^2 \to 0, \quad \sum_k |f_{nk}(u) - 1| \leq \frac{c}{2} u^2.$$

Therefore, for $n \geq n_u$ sufficiently large, $|f_{nk}(u) - 1| \leq \frac{1}{2}$, so that the $\log f_{nk}(u)$ exist and are finite,

$$\log f_{nk}(u) = f_{nk}(u) - 1 + \theta'_{nk} |f_{nk}(u) - 1|^2,$$

and it follows that

$$\left| \sum_k \{\log f_{nk}(u) - (f_{nk}(u) - 1)\} \right|$$

$$\leq \sum_k |f_{nk}(u) - 1|^2$$

$$\leq \max_k |f_{nk}(u) - 1| \sum_k |f_{nk}(u) - 1| \to 0.$$

The comparison lemma is proved.

Let

$$\psi_n(u) = \sum_k (f_{nk}(u) - 1) = \sum_k \int (e^{iux} - 1) \, dF_{nk}.$$

Since

$$\int x \, dF_{nk} = 0, \quad \sum_k \int x^2 \, dF_{nk} \leqq c,$$

we have

$$\psi_n(u) = \sum_k \int (e^{iux} - 1 - iux) \frac{1}{x^2} \cdot x^2 \, dF_{nk}$$

or

$$\psi_n(u) = \int (e^{iux} - 1 - iux) \frac{1}{x^2} \, dK_n,$$

where K_n on R is a continuous from the left nondecreasing function with $K_n(-\infty) = 0$, Var $K_n \leqq c < \infty$, defined by

$$K_n(x) = \sum_k \int_{-\infty}^{x} y^2 \, dF_{nk},$$

and the integrand, defined by continuity at $x = 0$, takes there the value $-u^2/2$. The comparison lemma becomes

a′. *Under* (C), $\log \prod_k f_{nk} - \psi_n \to 0$.

Functions of the foregoing type will be denoted in this subsection by ψ and K, with or without affixes. Thus, unless otherwise stated, ψ is a function defined on R by

$$\psi(u) = \int (e^{iux} - 1 - iux) \frac{1}{x^2} \, dK(x),$$

and K is a d.f.—up to a multiplicative constant—with $K(-\infty) = 0$, Var $K \leqq c$; ψ and K will have same affixes if any.

b. *Every* e^{ψ} *is a ch.f. with null first moment and finite variance* $\sigma^2 =$ Var K, *and is a limit law under* (C).

Proof. The integrand is bounded in x and continuous in u (or x) for every fixed x (or u). It follows that ψ is continuous on R and is limit of Riemann-Stieltjes sums of the form $\sum_k \{iua_{nk} + \lambda_{nk}(e^{iub_{nk}} - 1)\}$ where

$$\lambda_{nk} = \frac{1}{x_{nk}^2} K[x_{nk}, x_{n,k+1}), \quad a_{nk} = -\lambda_{nk}x_{nk}, \quad b_{nk} = x_{nk};$$

we can and do take all subdivision points $x_{nk} \neq 0$. Since every summand is log of a (Poisson type) ch.f., the sums are log of ch.f.'s, and so is their limit ψ according to the continuity theorem. The second assertion follows, since, by elementary computations,

$$(e^\psi)'_{u=0} = (\psi)'_{u=0} = 0, \quad (e^\psi)''_{u=0} = (\psi'')_{u=0} = - \operatorname{Var} K.$$

Finally, let X_{nk}, $k = 1, \cdots, n$ be independent r.v.'s with common log of ch.f. being ψ/n. Since ψ/n corresponds to K/n, we have $\sigma^2 X_{nk}$ $= \operatorname{Var} K/n$ while $EX_{nk} = 0$. Since $\sum_{k=1}^{n} X_{nk}$ has for ch.f. e^ψ whatever be n and condition (C) is fulfilled, the last assertion is proved.

c. UNIQUENESS LEMMA. *ψ determines K, and conversely.*

Proof. Since

$$-\psi''(u) = \int e^{iux}\, dK(x), \quad \operatorname{Var} K < \infty, \quad K(-\infty) = 0,$$

the inversion formula applies and K is determined by ψ by means of ψ''. The converse is obvious.

d. CONVERGENCE LEMMA. *Let (C) hold. If $K_n \overset{w}{\to} K$, then $\psi_n \to \psi$. Conversely, if $\psi_n \to \log f$, then $K_n \overset{w}{\to} K$ and $\log f = \psi$ determined by K.*

Proof. The first assertion follows at once from the extended Helly-Bray lemma. As for the converse, since the variations are uniformly bounded, the weak compactness theorem applies and there exists a K (with $\operatorname{Var} K \leq c$) such that $K_{n'} \overset{w}{\to} K$ as $n' \to \infty$ along some subsequence of integers. Therefore, by the same lemma, $\psi_{n'} \to \psi = \log f$ since $\psi_n \to \log f$. But, by the uniqueness lemma, $\psi = \log f$ determines K, and it follows that $K_n \overset{w}{\to} K$. The proposition is proved.

Upon applying the foregoing lemmas, the answer to our problem follows:

A. BOUNDED VARIANCES LIMIT THEOREM. *If independent summands X_{nk} are centered at expectations and $\max_k \sigma_{nk}^2 \to 0$, $\sum_k \sigma_{nk}^2 \leq c < \infty$ for all n, then*

1° *the family of limit laws of sequences $\mathcal{L}(\sum_k X_{nk})$ coincides with the family of laws of r.v.'s centered at expectations with finite variances and ch.f.'s of the form $f = e^\psi$, where ψ is of the form*

$$\psi(u) = \int (e^{iux} - 1 - iux) \frac{1}{x^2}\, dK(x),$$

with K continuous from the left and nondecreasing on R and
Var $K \leqq c < \infty$; ψ *determines K and conversely.*

2° $\mathcal{L}(\sum_k X_{nk}) \to \mathcal{L}(X)$ *with ch.f. necessarily of the form e^{ψ} if, and*

only if, $K_n \xrightarrow{w} K$ *where K_n are defined by*

$$K_n(x) = \sum_k \int_{-\infty}^x y^2 \, dF_{nk}.$$

If $\sum_k \sigma_{nk}^2 \leqq c < \infty$ *is replaced by* $\sum_k \sigma_{nk}^2 \to \sigma^2 X < \infty$, *then* $K_n \xrightarrow{w} K$

is to be replaced by $K_n \xrightarrow{c} K$.

Proof. 1° follows from **b**, the comparison lemma and the convergence lemma.

2° follows from 1° and the convergence lemma; and the particular case follows from the fact that the assumption made becomes

$$\text{Var } K_n = \sum_k \sigma_{nk}^2 \to \sigma^2 X = \text{Var } K.$$

EXTENSION. So far the r.v.'s under consideration were all centered at expectations. If we suppress this condition and set

$$a_{nk} = EX_{nk}, \quad \overline{F}_{nk}(x) = F_{nk}(x + a_{nk}), \quad \overline{f}_{nk}(u) = e^{-iua_{nk}}f_{nk}(u),$$

then the foregoing results continue to apply, provided F_{nk} and f_{nk} are replaced everywhere by $\overline{F}_{nk}$ and $\overline{f}_{nk}$; and then we write $\overline{\psi}$ instead of ψ. Going back to the noncentered r.v.'s, we have to introduce limit laws $\mathcal{L}(X)$ with finite variances but not necessarily null expectations $a = EX$, whose log's of ch.f.'s are of the form $\psi(u) = iua + \overline{\psi}(u)$, so that

$$\left(\frac{d\psi(u)}{du}\right)_0 = ia.$$

The uniqueness lemma becomes: ψ determines a and K, and conversely.

In the convergence lemma, $K_n \xrightarrow{w} K$ is replaced by $K_n \xrightarrow{w} K$ and $a_n \to a$.

The same is to be done in the limit theorem with $a_n = \sum_k a_{nk}$ and F_{nk} replaced by $\overline{F}_{nk}$.

Thus, the convergence criterion A2° becomes

EXTENDED CONVERGENCE CRITERION. *If independent summands X_{nk} are such that* $\max_k \sigma_{nk}^2 \to 0$ *and* $\sum_k \sigma_{nk}^2 \leqq c < \infty$, *then* $\mathcal{L}(\sum_k X_{nk}) \to$

$\mathcal{L}(X)$ *with ch.f. necessarily of the form* e^ψ *if, and only if,* $K_n \overset{w}{\to} K$ *and* $\sum_k a_{nk} \to a$ *where*

$$K_n(x) = \sum_k \int_{-\infty}^x y^2 \, dF_{nk}(y + a_{nk}), \quad a_{nk} = EX_{nk}.$$

If $\sum_k \sigma_{nk}^2 \leq c < \infty$ *is replaced by* $\sum_k \sigma_{nk}^2 \to \sigma^2 X < \infty$, *then* $K_n \overset{w}{\to} K$ *is to be replaced by* $K_n \overset{c}{\to} K$.

Particular cases:

1° NORMAL CONVERGENCE. The normal law $\mathfrak{N}(0, 1)$ corresponds to $\psi(u) = -\dfrac{u^2}{2}$ and, hence, to K defined by $K(x) = 0$ or 1 according as $x < 0$ or $x > 0$ (because of the uniqueness lemma, it suffices to verify that this K gives the above ψ).

NORMAL CONVERGENCE CRITERION. *Let the independent summands* X_{nk}, *centered at expectations, be such that* $\sum_k \sigma_{nk}^2 = 1$ *for all* n: *then* $\mathcal{L}(\sum_k X_{nk}) \to \mathfrak{N}(0, 1)$ *and* $\max_k \sigma_{nk}^2 \to 0$ *if, and only if, for every* $\epsilon > 0$,

$$g_n(\epsilon) = \sum_k \int_{|x| \geq \epsilon} x^2 \, dF_{nk} \to 0.$$

Proof. Since

$$\max_k \sigma_{nk}^2 = \max_k \int x^2 \, dF_{nk}(x) \leq \epsilon^2 + \max_k \int_{|x| \geq \epsilon} x^2 \, dF_{nk} \leq \epsilon^2 + g_n(\epsilon),$$

it follows that $g_n(\epsilon) \to 0$ for every $\epsilon > 0$ implies (letting $n \to \infty$ and then $\epsilon \to 0$ in the foregoing relation) $\max_k \sigma_{nk}^2 \to 0$. Then, immediate computations show that the convergence criterion **A2°** is equivalent to $g_n(\epsilon) \to 0$ for every $\epsilon > 0$.

Upon setting $X_{nk} = \dfrac{X_k}{s_n}$, $k = 1, \cdots, n$, $EX_k = 0$, $s_n^2 = \sum_k \sigma^2 X_k$, we obtain the classical normal convergence criterion. Liapounov's theorem follows from

$$\int_{|x| \geq \epsilon s_n} x^2 \, dF_k \leq \frac{1}{\epsilon^\delta s_n^\delta} \int |x|^{2+\delta} \, dF_k.$$

2° POISSON CONVERGENCE. The Poisson law $\mathcal{P}(\lambda)$ corresponds to $\psi(u) = iu\lambda + \lambda(e^{iu} - 1 - iu) = iu\lambda + \bar\psi(u)$ and, hence, the function

K which corresponds to ψ is defined by $K(x) = 0$ or λ according as $x < 1$ or $x > 1$. The extended convergence criterion yields, by immediate transformations, the following

POISSON CONVERGENCE CRITERION. *If the independent summands X_{nk} are such that* $\max \sigma_{nk}^2 \to 0$ *and* $\sum_k \sigma_{nk}^2 \to \lambda$, *then* $\mathcal{L}(\sum_k X_{nk}) \to \mathcal{P}(\lambda)$ *if, and only if,* $\sum_k EX_{nk} \to \lambda$ *and, for every* $\epsilon > 0$,

$$\sum_k \int_{|x-1| \geq \epsilon} x^2 \, dF_{nk}(x + EX_{nk}) \to 0.$$

*§ 23. SOLUTION OF THE CENTRAL LIMIT PROBLEM

We consider now the general problem. As was pointed out, the method of attack will be essentially the same as in the case of bounded variances. The computational difficulties will arise from two facts. (1) Even existence of first moments is not assumed, and the centerings, instead of being at expectations, will have to be at truncated expectations. (2) The functions K defined previously are not necessarily of bounded variation and, even when they are, they are not assumed to be of uniformly bounded variation. They will have to be replaced by functions of the form $\Psi_n(x) = \sum_k \int_{-\infty}^x \frac{y^2}{1 + y^2} \, d\overline{F}_{nk}$ where $\overline{F}_{nk}$ will be d.f.'s of the summands centered at truncated expectations. This will lead to limit laws with log ch.f.'s of a more complicated form, which we investigate first.

23.1 A family of limit laws; the infinitely decomposable laws. A law $\mathcal{L}$ and its ch.f. f are said to be *infinitely decomposable* (*i.d.*) if, for every integer n, there exist (on some pr. space) n independent and identically distributed r.v.'s X_{nk}, such that $\mathcal{L} = \mathcal{L}(\sum_{k=1}^n X_{nk})$; in other words, for every n there exists a ch.f. f_n such that $f = f_n{}^n$. If $f \neq 0$, then $\log f$ exists and is finite and $f_n = e^{(1/n) \log f}$; unless otherwise stated, we select for log of a ch.f. its principal branch (vanishing at $u = 0$) and for the nth root of f we take the function defined by the preceding equality.

Clearly, if a law is i.d., so is its type. The degenerate, normal, and Poisson type are i.d., since if $\log f(u) = iua$ or $iua - \sigma^2 \frac{u^2}{2}$ or $iua + \lambda(e^{iub} - 1)$, then $\frac{1}{n} \log f(u)$ has the same form whatever be n. More generally, the limit laws e^ψ obtained in the case of bounded variances are i.d., since the corresponding functions ψ are such that ψ/n is log of a ch.f. of the same form (with $a/n \in R$ and K/n d.f. up to a multiplicative constant). In fact

a. *The i.d. family belongs to that of limit laws of the Central Limit Problem.*

For, on the one hand, the uan condition for independent and identically distributed r.v.'s X_{nk} which figure in the definition of i.d. laws becomes convergence of their common law to the degenerate at 0, that is, $f_n \to 1$; on the other hand,

b. *If, for every n, $f = f_n{}^n$ where f_n is a ch.f., then $f_n \to 1$; and, moreover, $f \neq 0$.*

Proof. Since $|f| \leq 1$, we have $|f_n|^2 = |f|^{2/n} \to g$ with $g(u) = 0$ or 1 according as $f(u) = 0$ or $f(u) \neq 0$. Since f is continuous and $f(0) = 1$, there exists a neighborhood of the origin where $|f(u)| > 0$ and, hence, $g(u) = 1$, so that g is continuous in this neighborhood. Thus, the sequence $|f_n|^2$ of ch.f.'s converges to a function g continuous at the origin, the continuity theorem applies, and g is a ch.f. Therefore, g is continuous on R with $g(0) = 1$ and, since it takes at most two values 0 and 1, it reduces to 1. Consequently, $f \neq 0$, $\log f$ exists and is finite, and $f_n = e^{\frac{1}{n}\log f} \to 1$. The proposition is proved.

We shall see later that the family of limit laws of the problem *coincides* with the i.d. one. This explains the property below.

A. Closure theorem. *The i.d. family is closed under compositions and passages to the limit.*

Proof. If f and f' are i.d. ch.f.'s, then, for every n, there exist ch.f.'s f_n and f'_n such that $f = f_n{}^n$, $f' = f'_n{}^n$, so that $ff' = (f_n f'_n)^n$ where $f_n f'_n$ are ch.f.'s, and the first assertion is proved.

On the other hand, if a sequence f_n of i.d. ch.f.'s converges to a ch.f. f, then, for every integer m, $|f_n|^{\frac{2}{m}} \to |f|^{\frac{2}{m}}$ and, by the continuity theorem, $|f|^{\frac{2}{m}}$ is a ch.f. Therefore, $|f|^2$ is an i.d. ch.f. and, hence, by **b**, $f \neq 0$. Since $\log f$ exists and is finite, and

$$f_n{}^{\frac{1}{m}} = e^{\frac{1}{m}\log f_n} \to e^{\frac{1}{m}\log f} = f^{\frac{1}{m}},$$

it follows that $f^{1/m}$ is a ch.f., so that f is i.d. This concludes the proof.

The basic feature of i.d. laws (hence, as we shall see, of all the limit laws of the Central Limit Problem) is that they are constructed by means of Poisson type laws. This is made precise in the theorem below and explicited by the representation theorem which will follow.

B. Structure theorem. *A ch.f. is i.d. if, and only if, it is the limit of sequences of products of Poisson type ch.f.'s.*

In other words, the class of i.d. laws coincides with the limit laws of sequences of sums of independent Poisson type r.v.'s.

Proof. Products f_n of Poisson type ch.f.'s are defined by finite sums of the form

$$\log f_n(u) = \sum_k \{iua_{nk} + \lambda_{nk}(e^{iub_{nk}} - 1)\}, \quad \lambda_{nk} \geqq 0,$$

so that the functions $\frac{1}{m} \log f_n$ are log of ch.f.'s (of the same kind) whatever be the fixed integer m and the f_n are i.d.ch.f.'s. Thus, by **A**, if $f_n \to f$ ch.f., then f is i.d. This proves the "if" assertion.

Conversely, if f is i.d., then $\log f$ exists and is finite and

$$n(f^{\frac{1}{n}} - 1) \to \log f, \; f^{\frac{1}{n}}(u) - 1 = \int (e^{iux} - 1) \, dF_n(x)$$

where F_n are d.f.'s. By taking Riemann-Stieltjes sums which approximate $f^{1/n}(u) - 1$ by less than $1/n^2$, the "only if" assertion follows, and the proof is terminated.

In what precedes, $\psi_n(u) = \int (e^{iux} - 1)ndF_n(x) \to \log f(u)$ and ψ_n is itself log of an i.d.ch.f. Since $\text{Var}\,(nF_n) = n \to \infty$, brutal interchange of integration and passage to the limit is excluded. However, the integral inequality in 13.4 yields $\text{Var}\,\Psi_n \leqq c < \infty$ with $d\Psi_n(x) = (x^2/1 + x^2)ndF_n(x)$ so that the weak compactness theorem applies. But the integrand for $d\Psi_n(x)$ is undetermined at $x = 0$, and we have to modify it. This leads to the ψ-functions below:

Unless otherwise stated, ψ, with or without affixes, will denote a function defined on R by

$$\psi(u) = iu\alpha + \int \left(e^{iux} - 1 - \frac{iux}{1 + x^2}\right)\frac{1 + x^2}{x^2} \, d\Psi(x)$$

where $\alpha \in R$ and Ψ denoting a d.f.—up to a multiplicative constant, with $\Psi(-\infty) = 0$; the corresponding ψ, α, Ψ will have same affixes if any. The value of the integrand at $x = 0$, defined by continuity, is $-u^2/2$.

c. *Every e^ψ is an i.d. ch.f.*

Proof. We use repeatedly the fact that the class of log's of ch.f.'s is closed under additions.

The integrand is bounded in x and continuous in u (or x) for every fixed x (or u). It follows that the integral is continuous in u and is limit of Riemann-Stieltjes sums of the form

$$\sum_k \{iua_{nk} + \lambda_{nk}(e^{iub_{nk}} - 1)\}$$

where

$$\lambda_{nk} = \frac{1 + x_{nk}^2}{x_{nk}^2} \Psi[x_{nk}, x_{n,k+1}), \quad a_{nk} = -\lambda_{nk} \frac{x_{nk}}{1 + x_{nk}^2}, \quad b_{nk} = x_{nk};$$

we can and do take all $x_{nk} \neq 0$. Since every nonvanishing summand is log of a (Poisson type) ch.f., the sums are log's of ch.f.'s, and so is the integral according to the continuity theorem. Since $iu\alpha$ is log of a ch.f., so is ψ and, hence, so is every ψ/n corresponding to $\alpha/n \in R$ and Ψ/n—d.f. up to a multiplicative constant. The assertion is proved.

REMARK. If $\int x^2 d\Psi(x) < \infty$, then

$$\psi(u) = iua + \int (e^{iux} - 1 - iux) \frac{1}{x^2} dK(x)$$

where

$$a = \alpha + \int x \, d\Psi(x) \in R, \quad dK(x) = (1 + x^2) \, d\Psi(x),$$

and the i.d. ch.f. e^ψ has for first moment a and for variance $\text{Var } K < \infty$ (take the first two derivatives at $u = 0$). Conversely, if an i.d. ch.f. e^ψ has second (hence first) finite moment, then $\int x^2 d\Psi(x) < \infty$ (take the second symmetric derivative at $u = 0$). Thus, the family of all limit laws in the case of bounded variances coincides with the subfamily of i.d. laws with finite second moments.

We establish now two properties of functions ψ corresponding to the unicity and continuity theorems for ch.f.'s. They will be reduced to these theorems by making correspond to functions ψ functions φ and Φ, with same affixes as ψ if any. We define φ on R by

$$\varphi(u) = \psi(u) - \int_0^1 \frac{\psi(u + h) + \psi(u - h)}{2} \, dh.$$

We have, upon replacing ψ by its defining relation and interchanging the integrations,

$$\varphi(u) = \int_0^1 \left\{ \int e^{iux}(1 - \cos hx) \frac{1 + x^2}{x^2} d\Psi \right\} dh = \int e^{iux} d\Phi$$

with

$$\Phi(x) = \int_{-\infty}^x \left(1 - \frac{\sin y}{y} \right) \frac{1 + y^2}{y^2} d\Psi.$$

Since

$$0 < c' \leqq \left(1 - \frac{\sin x}{x} \right) \frac{1 + x^2}{x^2} \leqq c'' < \infty$$

where c' and c'' are independent of $x \in R$, it follows that Φ is non-decreasing on R with

$$c' \operatorname{Var} \Psi \leqq \operatorname{Var} \Phi \leqq c'' \operatorname{Var} \Psi < \infty$$

and

$$\Psi(x) = \int_{-\infty}^x d\Phi \bigg/ \left(1 - \frac{\sin y}{y} \right) \frac{1 + y^2}{y^2}.$$

C. Unicity theorem. *There is a one-to-one correspondence between functions ψ and couples (α, Ψ).*

For this reason we shall sometimes write $\psi = (\alpha, \Psi)$.

Proof. By definition, every couple (α, Ψ) determines a function ψ. Conversely, if ψ is given, then, by the foregoing considerations, ψ determines a function φ which is a ch.f. (up to a constant factor). By the inversion formula for ch.f.'s, φ determines Φ and, in its turn, Φ determines Ψ; furthermore, ψ and Ψ determine α, which completes the proof.

D. Convergence theorem. *If $\alpha_n \to \alpha$ and $\Psi_n \xrightarrow{c} \Psi$, then $\psi_n \to \psi$. Conversely, if $\psi_n \to g$ continuous at the origin, then $\alpha_n \to \alpha$ and $\Psi_n \xrightarrow{c} \Psi$ such that $g = \psi = (\alpha, \Psi)$.*

Proof. The first assertion follows at once by the Helly-Bray theorem. As for the converse, since the sequence e^{ψ_n} of i.d. ch.f.'s converges to e^g continuous at the origin, this convergence is uniform in every finite interval and, by 23.1b and A, e^g is an i.d. ch.f. with $e^g \neq 0$. Hence, g is finite and continuous on R, the sequence ψ_n converges to g uniformly on every finite interval, and

$$\varphi_n(u) \to g(u) - \int_0^1 \frac{g(u + h) + g(u - h)}{2} dh$$

continuous on R. In particular,

$$\text{Var } \Phi_n = \varphi_n(0) \to \int_0^1 \frac{g(h) + g(-h)}{2} \, dh < \infty,$$

so that variations of the Φ_n are uniformly bounded. Thus, the continuity theorem applies to the sequence φ_n, and there exists a nondecreasing function Φ of bounded variation on R such that, upon applying the Helly-Bray theorem, at every continuity point x of Φ as well as for $x = +\infty$,

$$\Psi_n(x) = \int_{-\infty}^x d\Phi_n \Big/ \left(1 - \frac{\sin y}{y}\right) \frac{1 + y^2}{y^2}$$

$$\to \Psi(x) = \int_{-\infty}^x d\Phi \Big/ \left(1 - \frac{\sin y}{y}\right) \frac{1 + y^2}{y^2}.$$

Hence, $\Psi_n \overset{c}{\to} \Psi$ and, by the same theorem,

$$iu\alpha_n = \psi_n(u) - \int \left(e^{iux} - 1 - \frac{iux}{1 + x^2}\right) \frac{1 + x^2}{x^2} \, d\Psi_n$$

$$\to g(u) - \int \left(e^{iux} - 1 - \frac{iux}{1 + x^2}\right) \frac{1 + x^2}{x^2} \cdot d\Psi$$

$$= iu\alpha.$$

This terminates the proof.

E. REPRESENTATION THEOREM. *The family of i.d. ch.f.'s coincides with the family of ch.f.'s of the form e^ψ.*

Proof. According to 23.1c, every e^ψ is an i.d. ch.f. Conversely if, for every $n, f = f_n{}^n$ where f_n is a ch.f. corresponding to a d.f. F_n, then, upon applying the preceding convergence theorem, we obtain

$$\log f(u) = \lim n(f^{1/n}(u) - 1) = \lim n(f_n(u) - 1) = \lim \int (e^{iux} - 1)n \, dF_n$$

$$= \lim \left(iu \int \frac{nx}{1 + x^2} \, dF_n \right.$$

$$+ \int \left(e^{iux} - 1 - \frac{iux}{1 + x^2}\right) \frac{1 + x^2}{x^2} \cdot \frac{x^2}{1 + x^2} n \, dF_n\right)$$

$$= \lim \psi_n = \text{some } \psi,$$

with

$$d\Psi_n(x) = n \frac{x^2}{1 + x^2} \, dF_n(x) \quad \text{and} \quad \Psi_n \overset{c}{\to} \Psi.$$

The theorem is proved.

23.2 The uan condition. The main computational difficulties arise in connection with the uan condition, and we have to investigate it in detail. We recall that given a sequence of sums $\sum\limits_{k} X_{nk}$ of independent r.v.'s, the uan condition is that, for every $\epsilon > 0$,

$$\max_{k} P[|X_{nk}| \geq \epsilon] = \max_{k} \int_{|x| \geq \epsilon} dF_{nk} \to 0.$$

a. *The uan condition implies that*

$$\max_{k}|\mu X_{nk}| \to 0, \quad \max_{k} \int_{|x| < \tau} |x|^r dF_{nk} \to 0, \quad r > 0, \quad \tau > 0 \; finite.$$

Proof. The medians of a r.v. belong to any interval such that the pr. for the r.v. to be in the interval is greater than $1/2$. Since under the uan condition $\min\limits_{k} P[|X_{nk}| < \epsilon] > 1/2$ whatever be $\epsilon > 0$, provided $n \geq n_\epsilon$ sufficiently large, it follows that $\max\limits_{k} |\mu X_{nk}| < \epsilon$ for $n \geq n_\epsilon$, and the first assertion is proved.

Under the same condition, by letting $n \to \infty$ and then $\epsilon \to 0$, we have

$$\max_{k} \int_{|x| < \tau} |x|^r dF_{nk} \leq \epsilon^r + \max_{k} \int_{\epsilon \leq |x| < \tau} |x|^r dF_{nk}$$

$$\leq \epsilon^r + \tau^r \max_{k} \int_{|x| \geq \epsilon} dF_{nk} \to 0,$$

and the second assertion is proved.

A. UAN CRITERIA. *The uan condition is equivalent to*

$$\max_{k} \int \frac{x^2}{1 + x^2} dF_{nk} \to 0 \quad or \quad \max_{k} |f_{nk} - 1| \to 0$$

uniformly on every finite interval.

Proof. Under the uan condition, by letting $n \to \infty$ and then $\epsilon \to 0$, we have

$$\max_{k} \int \frac{x^2}{1 + x^2} dF_{nk} \leq \epsilon^2 + \max_{k} \int_{|x| \geq \epsilon} dF_{nk} \to 0$$

and, for $|u| \leq b < \infty$,

$$\max_k \left| f_{nk}(u) - 1 \right|$$

$$\leq \max_k \left| \int_{|x|<\epsilon} (e^{iux} - 1)\, dF_{nk} \right| + \max_k \left| \int_{|x|\geq \epsilon} (e^{iux} - 1)\, dF_{nk} \right|$$

$$\leq b\epsilon + 2 \max_k \int_{|x|\geq \epsilon} dF_{nk} \to 0.$$

Conversely, if $\max_k \int \dfrac{x^2}{1+x^2}\, dF_{nk} \to 0$, then, for every $\epsilon > 0$,

$$\max_k \int_{|x|\geq \epsilon} dF_{nk} \leq \frac{1+\epsilon^2}{\epsilon^2} \max_k \int_{|x|\geq \epsilon} \frac{x^2}{1+x^2}\, dF_{nk} \to 0,$$

and the uan condition holds.

Since, upon replacing $f_{nk}(u)$ by $\int e^{iux}\, dF_{nk}$ and interchanging the integrations, we have

$$\max_k \int \frac{x^2}{1+x^2}\, dF_{nk} = \max_k \int_0^\infty e^{-u}(1 - \Re f_{nk}(u))\, du$$

$$\leq \int_0^\infty e^{-u} \max_k \left| f_{nk}(u) - 1 \right|\, du,$$

it follows, by the dominated convergence theorem, that $\max_k \left| f_{nk} - 1 \right|$ $\to 0$ implies the uan condition, and the proof is complete.

From now on, we fix a finite $\tau > 0$ and, for every d.f. F, with or without affixes, we set

$$a = \int_{|x|<\tau} x\, dF, \quad \overline{F}(x) = F(x + a), \quad \overline{f}(u) = \int e^{iux}\, d\overline{F}$$

with same affixes if any.

We observe that $\left| a \right| < \tau$ and that the "bar" does not mean "complex-conjugate."

Corollary 1. *Under the uan condition, $\max_k \left| \overline{f}_{nk} - 1 \right| \to 0$ uniformly on every finite interval.*

Since, by **a**, $\max_k \left| a_{nk} \right| \leq \max_k \int_{|x|<\tau} \left| x \right| dF_{nk} \to 0$, the r.v.'s $\overline{X}_{nk} = X_{nk} - a_{nk}$ obey the uan condition, and the assertion follows by **A**.

COROLLARY 2. *Under the uan condition, given $b < \infty$, all $\log f_{nk}(u)$ exist and are finite for $|u| \leq b$ and $n \geq n_b$ sufficiently large, and*

$$\log f_{nk}(u) = f_{nk}(u) - 1 + \theta_{nk}|f_{nk}(u) - 1|^2, \quad |\theta_{nk}| \leq 1;$$

similarly for the $\overline{f}_{nk}(u)$.

This follows from **A** and $\log z = (z - 1) + |\theta| z - 1|^2$ for $|z - 1| < \dfrac{1}{2}$.

From now on, if $b > 0$ is given, then we take $n \geq n_b$ so that the foregoing relations hold.

We are now in a position to establish the inequalities which will lead almost at once to the solution of the Central Limit Problem.

B. CENTRAL INEQUALITIES. *Under the uan condition, for $n \geq n_b$ sufficiently large, there exist two finite positive constants $c_1 = c_1(b, \tau)$ and $c_2 = c_2(b, \tau)$ such that*

$$c_1 \max_{|u| \leq b} |\overline{f}_{nk}(u) - 1| \leq \int \frac{x^2}{1 + x^2} d\overline{F}_{nk} \leq c_2 \int_0^b |\log|f_{nk}(u)|| \, du.$$

The inequalities follow at once, upon applying **a**, from two inequalities, valid for arbitrary r.v.'s, that we establish now. We shall use repeatedly the two relations

$$\int g(x) \, dF(x + c) = \int g(x - c) \, dF(x),$$

and

$$\int_{|x| < \tau} (x - a) \, dF = a - a \int_{|x| < \tau} dF = a \int_{|x| \geq \tau} dF.$$

B₁. LOWER BOUND. *There exists a finite positive number $c_1 = c_1(a, b, \tau)$ such that*

$$c_1 \max_{|u| \leq b} |\overline{f}(u) - 1| \leq \int \frac{x^2}{1 + x^2} dF.$$

Proof. Since, for $|u| \leq b < \infty$,

$|\overline{f}(u) - 1|$

$$= \left| \int (e^{iu(x-a)} - 1) \, dF \right| \leq 2 \int_{|x| \geq \tau} dF + b \left| \int_{|x| < \tau} (x - a) \, dF \right|$$

$$+ \frac{b^2}{2} \int_{|x| < \tau} (x - a)^2 \, dF$$

$$= (2 + |a|b) \int_{|x| \geq \tau} dF + \frac{b^2}{2} \int_{|x| < \tau} (x - a)^2 \, dF$$

where

$$\int_{|x|\geq\tau} dF \leq \frac{1 + (\tau + |a|)^2}{(\tau - |a|)^2} \int_{|x|\geq\tau} \frac{(x - a)^2}{1 + (x - a)^2} dF$$

and

$$\int_{|x|<\tau} (x - a)^2 \, dF \leq \{1 + (\tau + |a|)^2\} \int_{|x|<\tau} \frac{(x - a)^2}{1 + (x - a)^2} dF,$$

it follows that

$$|\bar{f}(u) - 1| \leq \frac{1}{c_1} \int \frac{x^2}{1 + x^2} d\bar{F}$$

where

$$\frac{1}{c_1} = \{1 + (\tau + |a|)^2\} \left\{ \frac{2 + |a|b}{(\tau - |a|)^2} + \frac{b^2}{2} \right\},$$

and the asserted inequality is proved.

Under the uan assertion, for n sufficiently large, we have, according to **a**, $|a| < \frac{\tau}{2}$, and we can take for $c_1 = c_1(b, \tau)$ the value of c_1 obtained upon replacing $|a|$ by $\frac{\tau}{2}$. This proves the left-hand side central inequality.

B$_2$. UPPER BOUND. *For $\tau > |\mu|$, μ a median of F, there exists a finite positive number $c_2 = c(\mu, b, \tau)$ such that*

$$\int \frac{x^2}{1 + x^2} d\bar{F} \leq c_2 \int_0^b (1 - |f(u)|^2) \, du.$$

If $f(u) \neq 0$ for $|u| \leq b$, then $1 - |f(u)|^2$ can be replaced by $2|\log|f(u)||$.

Proof. On account of the elementary inequality

$$1 - |f|^2 \leq - \log|f|^2 = 2|\log|f||,$$

the second assertion follows from the first one. To prove the first assertion, we shall use the symmetrization method and denote by F^s the d.f. of the symmetrized r.v. $X - X'$ where X and X' are independent and identically distributed, so that the corresponding ch.f. $f^s = |f|^2$.

From the elementary inequality

$$\left(1 - \frac{\sin bx}{bx}\right)\frac{1 + x^2}{x^2} \geq c(b) > 0, \quad x \in R,$$

and the relation (obtained upon interchanging the integrations)

$$\int_0^b (1 - |f(u)|^2)\, du = \int \left\{ \int_0^b (1 - \cos ux)\, du \right\} dF^s$$

$$= b\int\left(1 - \frac{\sin bx}{bx}\right)\frac{1 + x^2}{x^2} \cdot \frac{x^2}{1 + x^2}\, dF^s,$$

it follows that

$$(1) \qquad \int_0^b (1 - |f(u)|^2)\, du \geq bc(b)\int \frac{x^2}{1 + x^2}\, dF^s.$$

We pass now from F^s to F^μ, the d.f. of $X - \mu$, and set

$$q^\mu(t) = P[|X - \mu| \geq t], \quad q^s(t) = P[|X^s| \geq t], \quad t \in [0, +\infty),$$

so that, upon applying the weak symmetrization lemma (which says that $q^\mu \leq 2q^s$) and integrating by parts, we obtain

$$(2) \qquad \int \frac{x^2}{1 + x^2}\, dF^\mu = -\int_0^\infty \frac{t^2}{1 + t^2}\, dq^\mu = \int_0^\infty q^\mu(t)\, d\left(\frac{t^2}{1 + t^2}\right)$$

$$\leq 2\int_0^\infty q^s(t)\, d\left(\frac{t^2}{1 + t^2}\right) = 2\int \frac{x^2}{1 + x^2}\, dF^s.$$

Now, we pass from F^μ to $\overline{F}$. From the elementary inequality

$$(x - a)^2 \leq (x - \mu)^2 + 2(\mu - a)(x - a),$$

it follows that

$$\int_{|x| < \tau} (x - a)^2\, dF \leq \int_{|x| < \tau} (x - \mu)^2\, dF + 2(\tau + |\mu|)\left|\int_{|x| < \tau} (x - a)\, dF\right|$$

$$\leq \int_{|x| < \tau} (x - \mu)^2\, dF + 2\tau(\tau + |\mu|)\int_{|x| \geq \tau} dF$$

and, hence,

$$\int \frac{x^2}{1 + x^2}\, d\overline{F} = \int \frac{(x - a)^2}{1 + (x - a)^2}\, dF \leq \int_{|x| < \tau} (x - a)^2\, dF + \int_{|x| \geq \tau} dF$$

$$\leq \int_{|x| < \tau} (x - \mu)^2\, dF + \{1 + 2\tau(\tau + |\mu|)\}\int_{|x| \geq \tau} dF.$$

Since

$$\int_{|x|<\tau} (x - \mu)^2 \, dF \le \{1 + (\tau + |\mu|)^2\} \int_{|x|<\tau} \frac{(x - \mu)^2}{1 + (x - \mu)^2} \, dF$$

$$\le \{1 + (\tau + |\mu|)^2\} \int \frac{x^2}{1 + x^2} \, dF^\mu$$

and

$$\int_{|x|\ge\tau} dF \le \frac{1 + (\tau + |\mu|)^2}{(\tau - |\mu|)^2} \int_{|x|\ge\tau} \frac{(x - \mu)^2}{1 + (x - \mu)^2} \, dF$$

$$\le \frac{1 + (\tau + |\mu|)^2}{(\tau - |\mu|)^2} \int \frac{x^2}{1 + x^2} \, dF^\mu,$$

it follows that

(3) $$\int \frac{x^2}{1 + x^2} \, d\overline{F} \le c' \int \frac{x^2}{1 + x^2} \, dF^\mu$$

where

$$c' = c'(\mu, \tau) = \{1 + (\tau + |\mu|)^2\} \left\{ 1 + \frac{1 + 2\tau(\tau + |\mu|)}{(\tau - |\mu|)^2} \right\}.$$

Together, the inequalities (1), (2), and (3) yield the inequality

$$\int \frac{x^2}{1 + x^2} \, d\overline{F} \le c_2 \int_0^b (1 - |f(u)|^2) \, du$$

with $c_2 = \dfrac{2c'}{bc(b)}$, and the proof is concluded.

Under the uan condition, for $n \ge n_\tau$ sufficiently large, $|\mu| < \dfrac{\tau}{2}$ and we can take for $c_2 = c_2(b, \mu, \tau)$ the value of c_2 obtained upon replacing $|\mu|$ by $\dfrac{\tau}{2}$. This proves the right-hand side central inequality.

23.3 Central Limit Theorem. We are ready for the solution of the Central Limit Problem and can follow the same approach as in the case of bounded variances, since

a. BOUNDEDNESS LEMMA. *Under the uan condition, if* $\prod_k |f_{nk}| \to |f|$ *continuous, then there exists a finite constant* $c > 0$ *such that*

$$\sum_k \int \frac{x^2}{1 + x^2} \, d\overline{F}_{nk} \le c < \infty.$$

Proof. It suffices to prove the assertion for n sufficiently large so that, by 23.2**A**, Corollary 2, all $\log f_{nk}$ exist and are finite. Let $b > 0$ be sufficiently small so that, for $|u| \leq b$, $|f(u)| > 0$, and $\log |f(u)|$ exists and is finite. Since $|f|^2$ is a ch.f., $\sum_k \log |f_{nk}| \to \log |f|$ uniformly on $[-b, +b]$, and, by the right-hand side central inequality,

$$\sum_k \int \frac{x^2}{1 + x^2} \, d\overline{F}_{nk} \leq -c_2 \sum_k \int_0^b \log |f_{nk}(u)| \, du$$

$$\to -c_2 \int_0^b \log |f(u)| \, du < \infty.$$

The assertion follows.

b. COMPARISON LEMMA. *Under the uan condition, if there exists a constant c such that whatever be n*

$$\sum_k \int \frac{x^2}{1 + x^2} \, d\overline{F}_{nk} \leq c < \infty,$$

then

$$\sum_k \{\log \overline{f}_{nk}(u) - (\overline{f}_{nk}(u) - 1)\} \to 0, \quad u \in R.$$

Proof. By 23.2**A**, Corollaries 1 and 2, $\max_k |\overline{f}_{nk} - 1| \to 0$ and, given $b > 0$, for $|u| \leq b$ and n sufficiently large,

$$\log \overline{f}_{nk} = \overline{f}_{nk} - 1 + \theta_{nk} |\overline{f}_{nk} - 1|^2, \quad |\theta_{nk}| \leq 1.$$

By the left-hand side central inequality

$$\sum_k |\overline{f}_{nk}(u) - 1| \leq \frac{1}{c_1} \sum_k \int \frac{x^2}{1 + x^2} \, d\overline{F}_{nk} \leq \frac{c}{c_1} < \infty.$$

It follows that by taking $b > |u|$, where $u \in R$ is arbitrarily fixed,

$$\left| \sum_k \{\log \overline{f}_{nk}(u) - (\overline{f}_{nk}(u) - 1)\} \right| \leq \sum |\overline{f}_{nk}(u) - 1|^2$$

$$\leq \frac{c}{c_1} \max_k |\overline{f}_{nk}(u) - 1| \to 0,$$

and the theorem is proved.

Since (omitting the subscripts)

$$\log \overline{f}(u) - (\overline{f}(u) - 1) = \log f(u) - \left\{iua + \int (e^{iux} - 1) \, d\overline{F}\right\}$$

and

$$\int (e^{iux} - 1)\, d\overline{F} = iu \int \frac{x}{1 + x^2}\, d\overline{F}$$

$$+ \int \left(e^{iux} - 1 - \frac{iux}{1 + x^2} \right) \frac{1 + x^2}{x^2} \cdot \frac{x^2}{1 + x^2}\, d\overline{F},$$

the sums which figure in the comparison lemma are

$$\log \prod_k f_{nk}(u) - \psi_n(u)$$

where

$$\psi_n(u) = iu\alpha_n + \int \left(e^{iux} - 1 - \frac{iux}{1 + x^2} \right) \frac{1 + x^2}{x^2}\, d\Psi_n(x)$$

with

$$\alpha_n = \sum_k \left\{ a_{nk} + \int \frac{x}{1 + x^2}\, d\overline{F}_{nk} \right\}, \quad d\Psi_n(x) = \sum_k \frac{x^2}{1 + x^2}\, d\overline{F}_{nk}(x).$$

A. Central limit theorem. *Let X_{nk} be uan independent summands.*
1° *The family of limit laws of sequences $\mathcal{L}(\sum_k X_{nk})$ coincides with the family of i.d. laws or, equivalently, with the family of laws with log of ch.f. $\psi = (\alpha, \Psi)$ defined by*

$$\psi(u) = iu\alpha + \int \left(e^{iux} - 1 - \frac{iux}{1 + x^2} \right) \frac{1 + x^2}{x^2}\, d\Psi(x)$$

where $\alpha \in R$, and Ψ is a d.f. up to a multiplicative constant.
2° *$\mathcal{L}(\sum_k X_{nk}) \to \mathcal{L}(X)$ with log of ch.f. necessarily of the form $\psi = (\alpha, \Psi)$ if, and only if,*

$$\Psi_n \xrightarrow{c} \Psi, \quad \alpha_n \to \alpha,$$

where

$$\alpha_n = \sum_k \left\{ a_{nk} + \int \frac{x}{1 + x^2}\, d\overline{F}_{nk} \right\}, \quad \Psi_n(x) = \sum_k \int_{-\infty}^x \frac{y^2}{1 + y^2}\, d\overline{F}_{nk}$$

and

$$a_{nk} = \int_{|x| < \tau} x\, dF_{nk}, \quad \overline{F}_{nk}(x) = F_{nk}(x + a_{nk}),$$

with $\tau > 0$ finite and arbitrarily fixed.

Proof. Every i.d. law is a limit law of the Central Limit Problem. Conversely, if, under the uan condition, $\prod_k f_{nk} \to f$ ch.f., then, on account of the boundedness lemma, the comparison lemma applies and $e^{\psi_n} \to f$. Thus, on the one hand, by the closure theorem for i.d. laws $f = e^\psi$ is i.d. and 1° is proved. On the other hand, $\psi_n \to \psi$ and hence, by the convergence theorem for i.d. laws, $\Psi_n \overset{c}{\to} \Psi$, $\alpha_n \to \alpha$, and the "only if" part of 2° is proved.

Conversely, if $\alpha_n \to \alpha$ and $\Psi_n \overset{c}{\to} \Psi$, so that

$$\operatorname{Var} \Psi_n = \sum_k \int \frac{x^2}{1 + x^2} \, d\bar{F}_{nk} \to \operatorname{Var} \Psi < \infty$$

and the comparison lemma applies, then $\psi_n \to \psi$ hence $\prod_k f_{nk} \to e^\psi$, and the "if" part of 2° is proved. This terminates the proof.

Extension. It may happen that under the uan condition, the sequence $\mathcal{L}(\sum X_{nk})$ does not converge, yet the sequence $\mathcal{L}(\sum_k X_{nk} - a_n)$ converges for suitably chosen constants a_n; this is the situation in the Bernoulli case and, more generally, in the classical limit problem where $X_{nk} = X_k/b_n$ with $b_n = n$ or s_n. Then $\prod_k f_{nk}(u)$ is replaced by $e^{-iua_n} \prod_k f_{nk}(u)$ and the boundedness lemma can still be used, since it refers only to the moduli of products. On the other hand, the sums in the comparison lemma can be written $\log \{e^{-iua_n} \prod_k f_{nk}(u)\} - \{-iua_n + \psi_n(u)\}$. Since $-iua_n + \psi_n(u)$ is still a ψ-function, the Central Limit theorem remains valid, provided α_n is replaced by $\alpha_n - a_n$, and the theorem can be stated as follows:

B. EXTENDED CENTRAL LIMIT THEOREM. *Let X_{nk} be uan independent summands.*

1° *The family of limit laws of sequences $\mathcal{L}(\sum_k X_{nk} - a_n)$ coincides with the family of i.d. laws.*

2° *There exist constants a_n such that the sequence $\mathcal{L}(\sum X_{nk} - a_n)$ converges if, and only if, $\Psi_n \overset{c}{\to}$ some Ψ, where*

$$\Psi_n(x) = \sum_k \int_{-\infty}^x \frac{y^2}{1 + y^2} \, d\bar{F}_{nk}.$$

Then all admissible a_n are of the form $a_n = \alpha_n - \alpha + o(1)$ where α is an arbitrary finite number and $\alpha_n = \sum_k \left\{ a_{nk} + \int \frac{x}{1 + x^2} \, d\bar{F}_{nk} \right\}$, and all possible limit laws have for lóg of ch.f. $\psi = (\alpha, \Psi)$.

23.4   Central convergence criterion.   The convergence criterion 22.3A 2° is expressed in terms of expressions twice removed from the primary datum—the d.f.'s of the summands, and the probabilistic meaning of these expressions is somewhat hidden. We transform it by unpleasant but elementary computations as follows:

A. CENTRAL CONVERGENCE CRITERION.   *If X_{nk} are uan independent summands, then*

$$\prod_k f_{nk} \to f = e^\psi, \quad \psi = (\alpha, \Psi),$$

if, and only if,

(i) *at every continuity point $x \neq 0$ of Ψ*

$$\sum_k F_{nk}(x) \to \int_{-\infty}^x \frac{1 + y^2}{y^2} \, d\Psi \quad \text{for} \quad x < 0,$$

$$\sum_{k} \{1 - F_{nk}(x)\} \to \int_x^\infty \frac{1 + y^2}{y^2} \, d\Psi \quad \text{for} \quad x > 0$$

(ii) *as $n \to \infty$ and then $\epsilon \to 0$*

$$\sum_k \left\{ \int_{|x|<\epsilon} x^2 \, dF_{nk} - \left(\int_{|x|<\epsilon} x \, dF_{nk} \right)^2 \right\} \to \Psi(+0) - \Psi(-0)$$

(iii) *for a fixed $\tau > 0$ such that $\pm\tau$ are continuity points of Ψ*

$$\sum_k \int_{|x|<\tau} x \, dF_{nk} \to \alpha + \int_{|x|<\tau} x \, d\Psi - \int_{|x|\geq\tau} \frac{1}{x} \, d\Psi.$$

The iterated limit in (ii) is the generalized iterated limit $\varliminf_{\epsilon \to 0} \varlimsup_n$.

Proof.   We have to prove that the three stated conditions are equivalent to

(C) $\qquad\qquad \Psi_n \overset{c}{\to} \Psi \quad \text{with} \quad d\Psi_n(x) = \dfrac{x^2}{1 + x^2} \sum_k d\overline{F}_{nk}$

and

(C') $\sum_k \left\{ a_{nk} + \int \dfrac{x}{1 + x^2} \, d\overline{F}_{nk} \right\} \to \alpha, \quad \text{with} \quad a_{nk} = \int_{|x|<\tau} x \, dF_{nk},$

$\overline{F}_{nk}(x) = F_{nk}(x + a_{nk}).$

1° Let x be continuity points of Ψ. It is readily seen that condition (C) can be written as follows:

$$\Psi_n(x) \to \Psi(x) \quad \text{for} \quad x < 0,$$

$$\Psi_n(+\infty) - \Psi_n(x) \to \Psi(+\infty) - \Psi(x) \quad \text{for} \quad x > 0$$

and, as $n \to \infty$ and then $\epsilon \to 0$,

$$\Psi_n(+\epsilon) - \Psi_n(-\epsilon) \to \Psi(+0) - \Psi(-0).$$

It follows, upon replacing Ψ_n by its defining expression and applying the Helly-Bray theorem, that (C) is equivalent to

$$(\text{C}_1) \qquad \sum_k \overline{F}_{nk}(x) \to \int_{-\infty}^{x} \frac{1+y^2}{y^2} \, d\Psi \quad \text{for} \quad x < 0,$$

$$\sum_k \{1 - \overline{F}_{nk}(x)\} \to \int_{x}^{\infty} \frac{1+y^2}{y^2} \, d\Psi \quad \text{for} \quad x > 0$$

and

$$(\text{C}_2) \qquad \sum_k \int_{|x|<\epsilon} \frac{x^2}{1+x^2} \, d\overline{F}_{nk} \to \Psi(+0) - \Psi(-0)$$

$$\text{as} \quad n \to \infty \quad \text{and then} \quad \epsilon \to 0.$$

Let $a_n = \max_k \int_{|x|<\tau} |x| \, d\overline{F}_{nk}$ so that $|a_{nk}| \leq a_n \to 0$. Since

$$\sum_k F_{nk}(x - a_n) \leq \sum_k \overline{F}_{nk}(x) \leq \sum_k F_{nk}(x + a_n),$$

and the continuity points x of Ψ are continuity points of the integrals in (C_1), it follows at once that the first parts of (i) and (C_1) are equivalent; similarly for the second parts. Thus (C_1) is equivalent to (i).

2° Since

$$\frac{1}{1+\epsilon^2} \sum_k \int_{|x|<\epsilon} x^2 \, d\overline{F}_{nk} \leq \sum_k \int_{|x|<\epsilon} \frac{x^2}{1+x^2} \, d\overline{F}_{nk} \leq \sum_k \int_{|x|<\epsilon} x^2 \, d\overline{F}_{nk},$$

condition (C_2) is equivalent to

$$\sum_k \int_{|x|<\epsilon} x^2 \, d\overline{F}_{nk} \to \Psi(+0) - \Psi(-0) \quad \text{as} \quad n \to \infty \quad \text{and then} \quad \epsilon \to 0.$$

But, on account of (i), as $n \to \infty$ and then $\epsilon \to 0$,

$$\left| \sum_k \int_{|x|<\epsilon} x^2 \, dF_{nk} - \sum_k \int_{|x|<\epsilon} (x - a_{nk})^2 \, dF_{nk} \right|$$

$$\sum_k \left(\int_{\substack{|x|>\epsilon \\ |x-a_{nk}|<\epsilon}} + \int_{\substack{|x|<\epsilon \\ |x-a_{nk}|\geq\epsilon}} \right) (x - a_{nk})^2 \, dF_{nk} \leq 2 \sum_k \int_{\frac{\epsilon}{2} \leq |x| \leq 2\epsilon} (x - a_{nk})^2 \, dF_{nk} \to 0$$

and, since $a_n \to 0$, we have, for $\epsilon < \tau$,

$$\left| \sum_k \int_{|x|<\epsilon} (x - a_{nk})^2 \, dF_{nk} - \sum_k \left\{ \int_{|x|<\epsilon} x^2 \, dF_{nk} - \left(\int_{|x|<\epsilon} x \, dF_{nk} \right)^2 \right\} \right|$$

$$= \left| \sum_k \left(\int_{\epsilon \leq |x| < \tau} x \, dF_{nk} \right)^2 - \sum_k a_{nk}{}^2 \int_{|x|\geq\epsilon} dF_{nk} \right|$$

$$\leq (\tau a_n + a_n{}^2) \sum_k \int_{|x|\geq\epsilon} dF_{nk} \to 0.$$

Therefore, under (i) or its equivalent (C_1), condition (C_2) is equivalent to (ii). Thus, condition (C) is equivalent to (i) and (ii).

3° It remains to prove that, under (C) or its equivalent (i) and (ii), condition (C′) is equivalent to (iii). Since

$$\sum_k \int \frac{x}{1 + x^2} \, dF_{nk}$$

$$\doteq \sum_k \int_{|x|<\tau} x \, dF_{nk} - \sum_k \int_{|x|<\tau} \frac{x^3}{1 + x^2} \, dF_{nk} + \sum_k \int_{|x|\geq\tau} \frac{x}{1 + x^2} \, dF_{nk}$$

and, $\pm\tau$ being continuity points of Ψ, we have, by the Helly-Bray theorem,

$$\sum_k \int_{|x|<\tau} \frac{x^3}{1 + x^2} \, dF_{nk} = \int_{|x|<\tau} x \, d\Psi_n \to \int_{|x|<\tau} x \, d\Psi$$

$$\sum_k \int_{|x|\geq\tau} \frac{x}{1 + x^2} \, dF_{nk} = \int_{|x|\geq\tau} \frac{1}{x} \, d\Psi_n \to \int_{|x|\geq\tau} \frac{1}{x} \, d\Psi,$$

it suffices to prove that $\sum_k \int_{|x|<\tau} x \, dF_{nk} \to 0$. This assertion follows from the fact that $a_n \to 0$ and $\pm\tau$, being continuity points of Ψ, are continuity points of integrals in (i), so that, by (i),

$$
\left| \sum_k \int_{|x|<\tau} x \, d\overline{F}_{nk} \right| \leqq \left| \sum_k \int_{|x|<\tau} (x - a_{nk}) \, dF_{nk} \right|
$$

$$
+ \left| \sum_k \left\{ \int_{|x-a_{nk}|<\tau} (x - a_{nk}) \, dF_{nk} \right. \right.
$$

$$
\left. \left. - \int_{|x|<\tau} (x - a_{nk}) \, dF_{nk} \right\} \right|
$$

$$
\leqq a_n \sum_k \int_{|x|\geqq\tau} dF_{nk} + (\tau + a_n) \int_{\tau \leqq |x| < \tau + a_n} dF_{nk} \to 0.
$$

This terminates the proof.

REMARK 1. In the course of the proof, it was found that condition (i) can be written with $\overline{F}_{nk}$ instead of F_{nk} and condition (ii) is equivalent to

(ii′)
$$
\sum_k \int_{|x|<\epsilon} x^2 \, d\overline{F}_{nk} \to \Psi(+0) - \Psi(-0)
$$

as $n \to \infty$ and then $\epsilon \to 0$.

REMARK 2. In conditions (ii) or (ii′), the passages to the limit can be taken indifferently to be $\lim\limits_{\epsilon \to 0} \lim\limits_n \sup$ or $\lim\limits_{\epsilon \to 0} \lim\limits_n \inf$, instead of the generalized iterated limit; we leave the verification to the reader.

Upon using the extended Central Limit theorem, the central convergence criterion extends at once to sums with variable origin, as follows:

B. EXTENDED CENTRAL CONVERGENCE CRITERION. *If X_{nk} are uan independent summands, then there exist constants a_n such that $e^{-iua_n} \prod_k f_{nk}(u)$ $\to e^{\psi(u)}$ where $\psi = (\alpha, \Psi)$ if, and only if, conditions (i) and (ii) of the central convergence criterion hold. Then the admissible a_n are of the form*

$$
a_n = \sum_k \int_{|x|<\tau} x \, dF_{nk} - \alpha - \int_{|x|<\tau} x \, d\Psi + \int_{|x|\geqq\tau} \frac{1}{x} \, d\Psi + o(1)
$$

where $\pm\tau$ are fixed continuity points of Ψ.

This criterion implies properties of $\min\limits_k X_{nk}$ and $\max\limits_k X_{nk}$. In fact, it takes then a more intuitive form, as follows:

C. EXTREMA CRITERION. *Let X_{nk} be uan independent summands, and let $X_{nk}^{\epsilon} = X_{nk}$ or 0 according as $|X_{nk}| < \epsilon$ or $|X_{nk}| \geqq \epsilon$.*

The sequence $\mathcal{L}(\sum_{k} X_{nk} - a_n)$ converges for suitable constants a_n if, and only if, the sequences $\mathcal{L}(\min_{k} X_{nk})$, $\mathcal{L}(\max_{k} X_{nk})$ and $\sum_{k} \sigma^2 X_{nk}^{\epsilon}$ converge as $n \to \infty$ and then $\epsilon \to 0$.

More precisely, $\mathcal{L}(\sum_{k} X_{nk} - a_n) \to \mathcal{L}(X)$ with $\mathcal{L}(X)$ necessarily an i.d. law (α, Ψ) if, and only if, as $n \to \infty$ and then $\epsilon \to 0$,

$$\sum \sigma^2 X_{nk}^{\epsilon} \to \Psi(+0) - \Psi(-0)$$

and

$$\mathcal{L}(\min_{k} X_{nk}) \to \mathcal{L}(Y), \quad \mathcal{L}(\max_{k} X_{nk}) \to \mathcal{L}(Z)$$

with

$F_Y(x) = 1 - e^{-L(x)}$ or 1 and $F_Z(x) = 0$ or $e^{L(x)}$, according as $x < 0$ or $x > 0$,

where

$$L(x) = \int_{-\infty}^{x} \frac{1+y^2}{y^2} d\Psi(y), \ x < 0; \quad L(x) = -\int_{x}^{\infty} \frac{1+y^2}{y^2} d\Psi(y), \ x > 0.$$

Proof. Let G_n be the d.f. of $\min X_{nk}$, so that $1 - G_n = \prod_{k=1}^{k_n} (1 - F_{nk})$. For every fixed $x > 0$, $F_{nk}(x) \overset{k \leqq k_n}{\to} 1$ uniformly in k and, hence, $G_n(x) \to 1$. For every fixed $x < 0$, $F_{nk}(x) \to 0$ uniformly in k and, hence, for n sufficiently large,

$$\log (1 - G_n(x)) = \sum_{k} \log (1 - F_{nk}(x)) = -(1 + o(1)) \sum_{k} F_{nk}(x).$$

Therefore, the assertion relative to F_Y is equivalent to the first part of condition (i) of the central convergence criterion; similarly for the assertion relative to F_Z. The theorem follows.

23.5 Normal, Poisson, and degenerate convergence. We apply now the central convergence criterion to the three first-discovered limit types. We set

$$a_{nk}(\tau) = \int_{|x|<\tau} x \, dF_{nk}, \quad \sigma_{nk}^2(\tau) = \int_{|x|<\tau} x^2 \, dF_{nk} - \left(\int_{|x|<\tau} x \, dF_{nk} \right)^2$$

1° A normal law $\mathfrak{N}(\alpha, \sigma^2)$ corresponds to $\psi(u) = iu\alpha - \dfrac{\sigma^2}{2} u^2$, that is, $\psi = (\alpha, \Psi)$ where $\Psi(x) = 0$ or σ^2 according as $x < 0$ or $x > 0$.

NORMAL CONVERGENCE CRITERION. *If X_{nk} are independent summands, then, for every $\epsilon > 0$,*

$$\mathcal{L}(\sum_k X_{nk}) \to \mathfrak{N}(\alpha, \sigma^2) \quad and \quad \max_k P[|X_{nk}| \geq \epsilon] \to 0$$

if, and only if, for every $\epsilon > 0$ and a $\tau > 0$,

(i) $$\sum_k P[|X_{nk}| \geq \epsilon] \to 0$$

(ii) $$\sum_k \sigma_{nk}^2(\tau) \to \sigma^2, \quad \sum_k a_{nk}(\tau) \to \alpha.$$

Proof. We have, under (i),

$$\max_k P[|X_{nk}| \geq \epsilon] \leq \sum_k P[|X_{nk}| \geq \epsilon] \to 0.$$

Furthermore, always under (i), if $\epsilon < \tau$, then

$$\left| \sum_k \sigma_{nk}^2(\tau) - \sum_k \sigma_{nk}^2(\epsilon) \right| \leq \sum_k \int_{\epsilon \leq |x| < \tau} x^2 \, dF_{nk} + 2\tau \sum_k \left| \int_{\epsilon \leq |x| < \tau} x \, dF_{nk} \right|$$

$$\leq 3\tau^2 \sum_k \int_{\epsilon \leq |x| < \tau} dF_{nk} \to 0$$

and the same is true of $\epsilon > \tau$; it suffices to interchange ϵ and τ in the foregoing chain of inequalities. Upon taking into account these consequences of (i), the foregoing criterion follows from the central convergence criterion applied to the limit law $\mathfrak{N}(\alpha, \sigma^2)$.

COROLLARY. *If X_{nk} are independent summands and the sequence $\mathcal{L}(\sum_k X_{nk})$ converges, then the limit law is normal and the uan condition is satisfied if, and only if, $\max_k |X_{nk}| \xrightarrow{P} 0$.*

Upon setting $p_{nk} = P[|X_{nk}| \geq \epsilon]$, it suffices to observe that, because of the independence of the summands,

$$P[\max_k |X_{nk}| \geq \epsilon] = 1 - \prod_k (1 - p_{nk}).$$

For, upon applying the elementary inequality

$$1 - \exp[-\sum_k p_{nk}] \leq 1 - \prod_k (1 - p_{nk}) \leq \sum_k p_{nk},$$

it follows that the asserted condition is equivalent to condition (i) of the above criterion.

2°   The Poisson law $\mathcal{P}(\lambda)$ corresponds to $\psi(u) = \lambda(e^{iu} - 1)$ and, consequently, to $\psi = \left(\dfrac{\lambda}{2}, \Psi\right)$ with $\Psi(x) = 0$ or $\dfrac{\lambda}{2}$ according as $x < 1$ or $x > 1$. Upon applying the central convergence criterion and observing that the condition relative to the $\sigma_{nk}{}^2(\epsilon)$ reduces exactly as in the normal case, we obtain the

POISSON CONVERGENCE CRITERION. *If X_{nk} are uan independent summands, then $\mathcal{L}(\sum\limits_{k} X_{nk}) \to \mathcal{P}(\lambda)$ if, and only if, for every $\epsilon \in (0, 1)$ and a $\tau \in (0, 1)$,*

(i) $\sum\limits_{k} \int_{|x| \geq \epsilon,\, |x-1| \geq \epsilon} dF_{nk} \to 0$ *and* $\sum\limits_{k} \int_{|x-1| < \epsilon} dF_{nk} \to \lambda$

(ii) $\sum\limits_{k} \sigma_{nk}{}^2(\tau) \to 0$ *and* $\sum\limits_{k} a_{nk}(\tau) \to 0.$

3°   The degenerate law $\mathcal{L}(0)$ can be considered as a degenerate normal $\mathfrak{N}(0, 0)$ so that the normal convergence criterion reduces to the

DEGENERATE CONVERGENCE CRITERION. *If X_{nk} are independent summands, then $\mathcal{L}(\sum\limits_{k} X_{nk}) \to \mathcal{L}(0)$ and the uan condition is satisfied if, and only if, for every $\epsilon > 0$ and a $\tau > 0$*

(i) $\sum\limits_{k} \int_{|x| \geq \epsilon} dF_{nk} \to 0$

(ii) $\sum\limits_{k} \sigma_{nk}{}^2(\tau) \to 0, \quad \sum\limits_{k} a_{nk}(\tau) \to 0.$

COROLLARY 1. *If X_k are independent summands and $b_n \uparrow \infty$, then $\mathcal{L}\left(\dfrac{S_n}{b_n}\right) \to 0$ if, and only if, for every $\epsilon > 0$*

(i) $\sum\limits_{k} \int_{|x| \geq \epsilon b_n} dF_k \to 0$

(ii) $\dfrac{1}{b_n{}^2} \sum\limits_{k} \left\{ \int_{|x| < b_n} x^2\, dF_k - \left(\int_{|x| < b_n} x\, dF_k \right)^2 \right\} \to 0,$

        $\dfrac{1}{b_n} \sum\limits_{k} \int_{|x| < b_n} x\, dF_k \to 0.$

Because of the above criterion, taking $\tau = 1$ and observing that for $X_{nk} = \dfrac{X_k}{b_n}$, $F_{nk}(x) = F_k(b_n x)$, it remains only to prove that $\mathcal{L}\left(\dfrac{S_n}{b_n}\right) \to$ $\mathcal{L}(0)$ implies the uan condition. This follows from the fact that $P\left[\left|\dfrac{S_n}{b_n}\right| < \epsilon\right] > 1 - \delta$, for $n \geqq n_{\epsilon,\delta}$ sufficiently large, implies that, for $n > n_{\epsilon,\delta}$,

$$P\left[\left|\frac{X_n}{b_n}\right| < 2\epsilon\right] = P\left[\left|\frac{S_n}{b_n} - \frac{b_{n-1}}{b_n}\frac{S_{n-1}}{b_{n-1}}\right| < 2\epsilon\right]$$

$$\geqq P\left[\left|\frac{S_n}{b_n}\right| < \epsilon\right]\left[\left|\frac{S_{n-1}}{b_{n-1}}\right| < \epsilon\right] \geqq 1 - 2\delta.$$

REMARK. For the degenerate convergence criterion, (ii) and (i) with $\epsilon = \tau$ imply that $\mathcal{L}(\sum_k X_{nk}) \to \mathcal{L}(0)$. For, as in 21.2A, by Tchebichev inequality, (ii) implies that $\mathcal{L}(\sum_k X_{nk}{}^\tau) \to \mathcal{L}(0)$ and then, by 21.1b, (i) implies that $\mathcal{L}(\sum_k X_{nk}) \to \mathcal{L}(0)$.

In particular, in Corollary 1, we may take $\epsilon = 1$. Thus, for $b_n = n$, we have

COROLLARY 2. *If X_k are independent summands, then* $\mathcal{L}\left(\dfrac{S_n}{n}\right) \to \mathcal{L}(0)$ *if, and only if,*

(i)
$$\sum_k \int_{|x| \geqq n} dF_k \to 0,$$

(ii)
$$\frac{1}{n^2}\sum_k \left\{ \int_{|x|<n} x^2\, dF_k - \left(\int_{|x|<n} x\, dF_k\right)^2 \right\} \to 0,$$

(iii)
$$\frac{1}{n}\sum_k \int_{|x|<n} x\, dF_k \to 0.$$

This is the classical degenerate convergence criterion.

The reader is invited to specialize 23.4C to the three foregoing cases.

In particular, it implies the corollary to the normal convergence criterion. As for the Poisson case, $dL(x) = 0$ or λ according as $x \neq 1$ or $x = 1$ so that

If $\mathcal{L}(\sum_k X_{nk}) \to \mathcal{L}(X)$, *then* $\mathcal{L}(X) = \mathcal{P}(\lambda)$ *if and only if* $\mathcal{L}\min_k(X_{nk}) \to$ $\mathcal{L}(0)$ *and* $\mathcal{L}(\max_k X_{nk}) \to \mathcal{L}(0, 1)$ *with two values 0 and 1 only of pr.* $e^{-\lambda}$ *and* $1 - e^{-\lambda}$, *respectively.*

*§ 24.   NORMED SUMS

***24.1   The problem.** Let $\dfrac{S_n}{b_n} - a_n$ be normed sums with d.f. G_n and ch.f. g_n, where $S_n = \sum\limits_{k=1} X_k$ are consecutive sums of independent r.v.'s X_k with d.f.'s F_k and ch.f.'s f_k, and where a_n, $b_n > 0$ are finite numbers; thus

$$g_n(u) = e^{-iua_n} \prod_{k=1}^{n} f_k\left(\frac{u}{b_n}\right).$$

In what follows k runs over $1, \cdots, n$; $n = 1, \cdots$
   If the $X_{nk} = X_k/b_n$ obey the uan condition:

$$\max_k P[|X_k| \geq \epsilon b_n] \to 0 \quad \text{or} \quad \max_k \int \frac{x^2}{b_n{}^2 + x^2} dF_k(x) \to 0$$

$$\text{or} \quad \max_k \left| f_k\left(\frac{u}{b_n}\right) - 1 \right| \to 0,$$

then, according to the extended Central Limit Theorem, all possible limit laws of sequences $\dfrac{S_n}{b_n} - a_n$ of normed sums form a family $\mathfrak{N}$ of i.d. laws, and the extended central convergence criterion applies with $F_{nk}(x) = F_k(b_n x)$.

However, in the case of normed sums, new problems arise.
   1°   Given a sequence X_n of independent r.v.'s, find whether there exist sequences a_n and $b_n > 0$ such that the uan condition (for the X_k/b_n) is satisfied and $g_n \to f$ ch.f., necessarily of the form e^ψ with $\psi = (\alpha, \Psi)$; and if such sequences exist, then characterize them.
   2°   Characterize the family $\mathfrak{N}$; in other words, characterize those i.d. ch.f.'s e^ψ and the corresponding functions Ψ which represent limit laws of normed sums obeying the uan condition.
   But on the one hand, according to the convergence of types theorem, there always exist sequences a_n and $b_n > 0$ such that the limit laws of $\dfrac{S_n}{b_n} - a_n$ are degenerate and, on the other hand, all degenerate laws belong to $\mathfrak{N}$: $e^{iua} = (e^{iua/n})^n$. Thus, whenever convenient, *we can and do exclude degenerate limit laws from our considerations.*

   a. *If $g_n \to f$ nondegenerate ch.f., then the uan condition for the X_k/b_n implies that $b_n \to \infty$ and $b_{n+1}/b_n \to 1$.*

Proof. We have

$$g_n(u) = e^{-iua_n} \prod_k f_k\left(\frac{u}{b_n}\right) \to f(u) \quad \text{nondegenerate.}$$

If $b_n \not\to \infty$, then the sequence b_n contains a bounded subsequence and, by the Bolzano-Weierstrass lemma, this subsequence contains another sequence $b_{n'} \to b$ finite as $n' \to \infty$. Setting $u_{n'} = b_{n'}u$, the uan condition implies that for every k, $f_k(u) = f_k(u_{n'}/b_{n'}) \to 1$; hence, $f_k = 1$ and $f = 1$. This contradicts the nondegeneracy assumption so that, *ab contrario*, $b_n \to \infty$.

Since $X_{n+1}/b_{n+1} \overset{P}{\to} 0$, it follows by the law-equivalence lemma that the limit laws of the sequences $\dfrac{S_n}{b_n} - a_n$ and $\dfrac{S_n}{b_{n+1}} - a_{n+1} = \dfrac{S_{n+1}}{b_{n+1}} - a_{n+1} - \dfrac{X_{n+1}}{b_{n+1}}$ are the same. Thus $e^{-iua_n'} g_n(b_n'u) \to f(u)$ as $n' \to \infty$, with $b_n' = b_n/b_{n+1}$ and f nondegenerate. It follows, by the corollary to the convergence of types theorem, that $b_{n+1}/b_n \to 1$. The proof is complete.

*24.2 Norming sequences. We have at our disposal the necessary tools to solve the problem of existence and determination of norming sequences a_n and $b_n > 0$. Given the summands, we know, according to the convergence of types theorem, that 1° all the limit laws belong— if they exist—to the positive type of *one* i.d. law and 2° it suffices to find one pair of such sequences. Furthermore, on account of the extended convergence criterion (with $X_{nk} = X_k/b_n$), 3° if there exists a limit i.d. positive type, then the a_n are determined by the expression given there, 4° the uan condition is satisfied and $g_n \to e^\psi$ if, and only if,

$$\max_k \int \frac{x^2}{b_n^2 + x^2} dF_k(x) \to 0 \quad \text{and} \quad \Psi_n \overset{c}{\to} \Psi$$

where Ψ_n are defined on R by

$$\text{(D)} \quad \Psi_n(x) = \sum_k \int_{-\infty}^{b_n x} \frac{y^2}{b_n^2 + y^2} dF_k(y + b_{nk}), \quad b_{nk} = \int_{|x| < b_n \tau} x \, dF_k(x)$$

with $\pm\tau \neq 0$ fixed continuity points of Ψ (we shall see later that any τ is admissible, so that we may set, say, $\tau = 1$). The theorem below completes the answer. As usual, the superscript "s" will denote the operation of symmetrization.

A. Norming theorem. *There exist sequences b_n such that* $\mathcal{L}\left(\dfrac{S_n}{b_n} - a_n\right) \to \mathcal{L}(X)$ *for suitable a_n if, and only if, there exists a Ψ such that, upon setting in (D), $b_n = b'_n > 0$ determined by*

$$\frac{1}{2}\sum_{k=1}^{n} \int \frac{x^2}{b'^2_n + x^2}\, dF_k{}^s(x) = \Psi(+\infty).$$

we have

(i) $\max_k \int \dfrac{x^2}{b'^2_n + x^2}\, dF_k(x) \to 0,$

(ii) $\Psi_n \overset{c}{\to} \Psi.$

Proof. The "if" assertion follows by taking normed sums $\dfrac{S_n}{b'_n} - a_n$. Because of the corollary to the convergence of types theorem and of the extended central convergence criterion, the "only if" assertion will follow by proving that if $\mathcal{L}\left(\dfrac{S_n}{b_n} - a_n\right) \to \mathcal{L}(X)$ with ch.f. e^ψ, $\psi = (\alpha, \Psi)$, then $b'_n/b_n \to 1$.

Upon symmetrizing, the hypothesis becomes $\mathcal{L}(S_n{}^s/b_n) \to \mathcal{L}(X^s)$ and the corresponding Ψ^s is defined by

$$\Psi^s(x) = \Psi(x) + \Psi(+\infty) - \Psi(-x + 0).$$

Thus $\Psi_n{}^s \overset{c}{\to} \Psi^s$ where $\Psi_n{}^s$ are defined by

$$\Psi_n{}^s(x) = \sum_{k=1}^{n} \int_{-\infty}^{b_n x} \frac{x^2}{b_n{}^2 + x^2}\, dF_k{}^s(x).$$

Upon using $\Psi^s(+\infty) = 2\Psi(+\infty)$, and (D) with b_n replaced by b'_n, it follows that

$$\Delta_n = \sum_{k=1}^{n} \left\{ \int \frac{x^2}{b_n{}^2 + x^2}\, dF_k{}^s(x) - \int \frac{x^2}{b'^2_n + x^2}\, dF_k{}^s(x) \right\} \to 0.$$

On the other hand, since degenerate limit laws are excluded, Ψ^s does not reduce to a constant. Therefore, there exists an $a > 0$ such that $2\delta = \Psi^s(a) - \Psi^s(-a + 0) > 0$ and, hence, for $n \geq n_a$ sufficiently large,

$$\sum_{k=1}^{n} \int_{-ab_n}^{+ab_n} \frac{x^2}{b_n{}^2 + x^2}\, dF_k{}^s(x) > \delta > 0.$$

It follows that

$$0 \leftarrow |\Delta_n| = |b_n^2 - b'_n^2| \sum_{k=1}^{n} \int \frac{x^2}{(b_n^2 + x^2)(b'_n^2 + x^2)} \, dF_k^s(x)$$

$$\geq \frac{|b_n^2 - b'_n^2|}{b_n^2 + a^2 b'_n^2} \cdot \sum_{k=1}^{n} \int_{-ab_n}^{+ab_n} \frac{x^2}{(b'_n^2 + x^2)} \, dF_k^s(x)$$

$$\geq \frac{|(b_n/b'_n)^2 - 1|}{1 + a^2 b_n^2/b'_n^2} \cdot \delta \geq 0,$$

so that $b_n/b'_n \to 1$, and the proof is complete.

*24.3 Characterization of $\mathfrak{N}$. We characterize $\mathfrak{N}$ by a decomposability property and, then, we characterize the corresponding functions Ψ.

In order to define the decomposability property we prove

a. *If to a ch.f. f there corresponds a number $c > 0$ and a nondegenerate ch.f. f_c such that, for every $u, f(u) = f(cu)f_c(u)$, then $c < 1$.*

Proof. If $c = 1$, then $f_c = 1$. If $c > 1$, then, replacing repeatedly in the assumed relation u by $\frac{u}{c}$ and $|f_c|$ by 1, we have

$$1 \geq |f(u)| \geq \left|f\left(\frac{u}{c}\right)\right| \geq \left|f\left(\frac{u}{c^2}\right)\right| \geq \cdots \geq \lim \left|f\left(\frac{u}{c^n}\right)\right| = f(0) = 1$$

and f is degenerate, so that f_c is degenerate. The assertion follows *ab contrario*.

We say that a law and its ch.f. f are *self-decomposable* if, for every $c \in (0, 1)$, there exists a ch.f. f_c such that, for every $u, f(u) = f(cu)f_c(u)$. Clearly, a degenerate ch.f. is self-decomposable and all its components f_c are also degenerate.

b. *If f is self-decomposable, then $f \neq 0$.*

Proof. If $f(2a) = 0$ and $f(u) \neq 0$ for $0 \leq u < 2a$, then $f_c(2a) = 0$. Upon replacing t and h by a in

$$|f_c(t + h) - f_c(t)|^2 \leq 2\{1 - \Re f_c(h)\},$$

we obtain

$$|f_c(a)|^2 \leq 2\{1 - \Re f_c(a)\}.$$

This leads to a contradiction since, by letting $c \to 1$, we obtain

$f_c(a) = \dfrac{f(a)}{f(ca)} \to 1$ and the inequality becomes $1 \leqq 0$. The assertion follows *ab contrario*.

A. SELF-DECOMPOSABILITY CRITERION. *A law belongs to $\mathfrak{N}$ if, and only if, it is self-decomposable.*

Proof. A degenerate law certainly belongs to $\mathfrak{N}$, so that it suffices to consider nondegenerate laws with ch.f. f.

1° If f is self-decomposable, then let $X_k(k = 1, \cdots, n)$ be independent r.v.'s, with ch.f. f_k defined by

$$f_k(u) = f_{\frac{k-1}{k}}(ku) = \frac{f(ku)}{f((k-1)u)}.$$

Since $f_k\left(\dfrac{u}{n}\right) \to 1$ uniformly in k and the ch.f. of $\dfrac{S_n}{n}$ is given by

$$\prod_k f_k\left(\frac{u}{n}\right) = f(u),$$

the "if" assertion follows.

2° Conversely, let f belong to $\mathfrak{N}$. There exist normed sums $\dfrac{S_n}{b_n} - a_n$ with ch.f. g_n such that, denoting by f_k the ch.f. of summands X_k,

$$g_n(u) = e^{-iua_n} \prod_k f_k\left(\frac{u}{b_n}\right) \to f(u)$$

and, by 24.1b, $b_n \to \infty$, $\dfrac{b_{n+1}}{b_n} \to 1$. Then, given $c \in (0, 1)$, we can make correspond to every integer n an integer $m < n$ such that $\dfrac{b_m}{b_n} \to c$ and $m, n - m \to \infty$ as $n \to \infty$. Since

$$(1) \quad g_n(u) = \left\{ e^{-iuca_m} \prod_{k=1}^m f_k\left(\frac{b_m}{b_n} \cdot \frac{u}{b_m}\right) \right\} \left\{ e^{-iu(a_n - ca_m)} \prod_{k=m+1}^n f_k\left(\frac{u}{b_n}\right) \right\}$$

where $g_n(u) \to f(u)$, and the first bracket converges to $f(cu)$, it follows that the ch.f. $g_{m,n}$, whose values figure within the second bracket, converges to the continuous function f_c defined by $f_c(u) = \dfrac{f(u)}{f(cu)}$. Therefore, by the continuity theorem, f_c is a ch.f., and the proof is concluded.

COROLLARY. *A self-decomposable ch.f. f and its components f_c are i.d.*

Proof. Since f belongs to $\mathfrak{N}$, f is i.d. On the other hand, upon taking for f_k the ch.f. of r.v.'s X_k defined in 1° and $m < n$ such that $\dfrac{m}{n} \to c$, we have

$$f(u) = \prod_{k=1}^{m} f_k\left(\frac{m}{n} \cdot \frac{u}{m}\right) \prod_{k=m+1}^{n} f_k\left(\frac{u}{n}\right).$$

The first product converges to $f(cu)$; the second one converges to $f_c(u)$. Thus, f_c is ch.f. of the limit law of sums $\sum\limits_{k=m+1}^{n} X_{nk}$ where the summands $X_{nk} = \dfrac{X_k}{n}$ obey the uan condition. Therefore, f_c is an id. ch.f., and the proof is concluded.

We express now the self-decomposability criterion in terms of functions Ψ which figure in the representation of the i.d. self-decomposable ch.f.'s.

B. Ψ-criterion. *Self-decomposable laws coincide with i.d. laws with functions Ψ such that on $(-\infty, 0)$ and on $(0, +\infty)$, their left and right derivatives, denoted indifferently by $\Psi'(x)$, exist and $\dfrac{1+x^2}{x}\Psi'(x)$ do not increase.*

Proof. Because of the preceding corollary, the self-decomposability property of a ch.f. f, necessarily of the form e^ψ, is as follows: for every $c \in (0, 1)$ the difference $\psi_c(u) = \psi(u) - \psi(cu)$ defines a ψ-function (a log of an i.d. ch.f.).

Upon replacing x by $c^{-1}x$, we can write

$$(1) \quad \psi(cu) = iu\left\{ c\alpha + (1 - c^2)\int \frac{x}{1+x^2}\, d\Psi(c^{-1}x) \right\}$$

$$+ \int \left(e^{iux} - 1 - \frac{iux}{1+x^2} \right) \frac{1 + c^{-2}x^2}{c^{-2}x^2}\, d\Psi(c^{-1}x).$$

Thus

$$\psi_c(u) = iu\alpha_c + \int \left(e^{iux} - 1 - \frac{iux}{1+x^2} \right) \frac{1 + x^2}{x^2}\, d\Psi_c(x),$$

where α_c is a finite number and

$$(2) \quad d\Psi_c(x) = d\Psi(x) - \frac{1 + c^{-2}x^2}{c^{-2}(1+x^2)}\, d\Psi(c^{-1}x), \quad \Psi_c(-\infty) = 0.$$

Since Ψ_c is a difference of two Ψ-functions, its variation on R is bounded. It follows readily that ψ_c is a ψ-function if, and only if, Ψ_c is nondecreasing on R. Since

$$(3) \qquad \Psi_c(+0) - \Psi_c(-0) = (1 - c^2)\{\Psi(+0) - \Psi(-0)\} \geq 0,$$

the self-decomposability property becomes $d\Psi_c(x) \geq 0$ for every $c \in (0, 1)$ and $x \neq 0$ or, equivalently, on account of (2), for every $c \in (0, 1)$ and arbitrary $x' < x''$, $x'x'' > 0$,

$$(4) \quad \int_{x'}^{x''} \frac{1 + y^2}{y^2} \, d\Psi_c(y)$$

$$= \int_{x'}^{x''} \frac{1 + y^2}{y^2} \, d\Psi(y) - \int_{x'}^{x''} \frac{1 + c^{-2}y^2}{c^{-2}y^2} \, d\Psi(c^{-1}y) \geq 0.$$

It remains to show that this last inequality implies and is implied by the one asserted in the theorem.

If

$$J(x) = \int_{+\infty}^{e^x} \frac{1 + y^2}{y^2} \, d\Psi(y), \quad x \in R,$$

then, by setting in (4) $x' = e^{x-h}$, $x'' = e^x$, $c = e^{-h}$, we obtain

$$J(x) - J(x - h) \geq J(x + h) - J(x) \quad \text{or} \quad J(x) \geq \frac{J(x + h) + J(x - h)}{2}.$$

Therefore, the nondecreasing finite function J on R is convex (from above) and, consequently, J is continuous and its left and right derivatives $J'(x)$ exist and do not increase on R. Since

$$\frac{J(x + h) - J(x)}{e^h - 1} = \frac{1 + e^{2(x+\theta h)}}{e^{x+2\theta h}} \cdot \frac{\Psi(e^{x+h}) - \Psi(e^x)}{e^{x+h} - e^x}, \quad 0 \leq \theta \leq 1,$$

it follows, letting $h \to 0$ and setting $e^x = y$, that the left and right derivatives $\Psi'(y)$ exist and that $\dfrac{1 + y^2}{y} \Psi'(y)$ do not increase on $(0, \infty)$.

Similarly, introducing $J^-(x) = \displaystyle\int_{-\infty}^{-e^x} \frac{1 + y^2}{y^2} \, d\Psi(y)$, we find that the same is true on $(-\infty, 0)$. Thus (4) implies the asserted property of Ψ.

Conversely, if this asserted property is true then, for every $c \in (0, 1)$ and $x' < x''$, $x'x'' > 0$

$$\int_{x'}^{x''} \frac{1 + y^2}{y^2} d\Psi(y) = \int_{x'}^{x''} \frac{1 + y^2}{y} \Psi'(y) \frac{dy}{y}$$

$$\geq \int_{x'}^{x''} \frac{1 + c^{-2}y^2}{c^{-1}y} \Psi'(c^{-1}y) \frac{dy}{y}$$

$$= \int_{x'}^{x''} \frac{1 + c^{-2}y^2}{c^{-2}y^2} d\Psi(c^{-1}y),$$

so that the inequality in (4) holds and the conclusion is reached.

REMARK. Since Poisson laws correspond to functions Ψ discontinuous at some $x \neq 0$, they do not belong to the family $\mathfrak{N}$. This explains the isolation in which they remained as long as only limit laws of normed sums were considered.

*24.4   Identically distributed summands and stable laws. The first *family* $\mathfrak{N}_I$ of limit laws to be investigated by P. Lévy, was that of limit laws of normed sums $\dfrac{S_n}{b_n} - a_n$ of independent and identically distributed summands X_k with an arbitrary common ch.f. f_0. In other words, $\mathfrak{N}_I$ is defined as the family of laws whose ch.f.'s f are such that

$$g_n(u) = e^{-iua_n} f_0{}^n \left(\frac{u}{b_n} \right) \to f(u), \quad u \in R.$$

Clearly, the uan condition is satisfied, so that $\mathfrak{N}_I \subset \mathfrak{N}$. The self-decomposability concept and the criteria for $\mathfrak{N}$ are easily particularized for $\mathfrak{N}_I$, as follows; we exclude degenerate limit laws which, clearly, belong to $\mathfrak{N}_I$. Let a law and its ch.f. f be called *stable* if, for arbitrary $b > 0$, $b' > 0$, there exist finite numbers a and $b'' > 0$ such that

$$f(b''u) = e^{iua} f(bu) f(b'u), \quad u \in R.$$

Upon replacing $b''u$ by u and setting $c = \dfrac{b}{b''}$, $c' = \dfrac{b'}{b''}$, we obtain

$$f(u) = e^{iu\frac{a}{b''}} f(cu) f(c'u) = f(cu) f_c(u)$$

where

$$f_c(u) = e^{iu\frac{a}{b''}} f(c'u).$$

The self-decomposability criterion for $\mathfrak{N}$ becomes

A. Stability criterion. *A law belongs to $\mathfrak{N}_I$ if, and only if, it is stable.*

Proof. The "if" assertion follows from the fact that stability of f implies, taking $f_0 = f$, that the ch.f. of S_n is of the form $f^n(u) = e^{iua_n}f(b_n u)$ so that, norming S_n with these quantities a_n and b_n, we have $g_n = f$.

Conversely, leaving out—to simplify the writing—factors of the form e^{iua}, which does not restrict the generality, we have to prove that

$$f_0{}^n \left(\frac{u}{b_n} \right) \to f(u), \; u \in R, \text{ implies that to arbitrary } b > 0, \, b' > 0, \text{ there}$$

corresponds $b'' > 0$ such that $f(b''u) = f(bu)f(b'u)$. Since $b_n \to \infty$ and $\dfrac{b_{n+1}}{b_n} \to 1$, we can assign to every integer n integers m and m' such that $\dfrac{b_m}{b_n} \to b, \dfrac{b_{m'}}{b_n} \to b'$. Then

$$f_0{}^{m+m'} \left(\frac{b_{m+m'}}{b_n} \cdot \frac{u}{b_{m+m'}} \right) = f_0{}^m \left(\frac{b_m}{b_n} \cdot \frac{u}{b_m} \right) f_0{}^{m'} \left(\frac{b_{m'}}{b_n} \cdot \frac{u}{b_{m'}} \right)$$

and the right-hand side converges to $f(bu)f(b'u)$, while, according to the convergence of types theorem, there exists $b'' > 0$ such that the left-hand side converges to $f(b''u)$. The conclusion is reached.

Thus, a stable law is self-decomposable and, moreover, f_c belongs to the positive type of f; in particular f is an i.d. ch.f.

The Ψ-criterion for $\mathfrak{N}$ is easily transformed and, furthermore, the stable ch.f.'s are obtained in terms of elementary functions of analysis, as follows.

B. *A function f is a stable ch.f. if, and only if, either*

(i) $$\log f(u) = i\alpha u - b|u|^\gamma \left\{ 1 + ic \frac{u}{|u|} \tan \frac{\pi}{2} \gamma \right\}$$

or

(ii) $$\log f(u) = i\alpha u - b|u| \left\{ 1 + ic \frac{u}{|u|} \cdot \frac{2}{\pi} \log |u| \right\}$$

with

$$\alpha \gtrless 0, \quad b \geqq 0, \quad |c| \leqq 1, \quad \gamma \in (0, 1) \cup (1, 2].$$

We observe that $\gamma = 2$ gives the normal laws and that real stable ch.f.'s are of the form $e^{-b|u|^{\gamma}}$, $0 < \gamma \leq 2$.

Proof. If the asserted forms of f are ch.f.'s, then they are clearly stable. Thus, we have to prove that these forms are ch.f.'s and that stable ones are of this form. The first assertion will follow if we can determine functions Ψ such that $\log f = (\alpha, \Psi)$.

Let $f = e^{\psi}$ be a stable ch.f., that is, for arbitrary $b > 0$ and $b' > 0$, there exist a and $b'' > 0$ such that

$$iua + \psi(bu) + \psi(b'u) = \psi(b''u).$$

1° We follow the pattern of Ψ-criterion's proof $\left(\text{with } c = \dfrac{b}{b'}\right)$. Upon replacing ψ by its representation in terms of α and Ψ, the foregoing requirement reduces to

$$\frac{1 + b^2 x^2}{b^2} d\Psi(bx) + \frac{1 + b'^2 x^2}{b'^2} d\Psi(b'x) = \frac{1 + b''^2 x^2}{b''^2} d\Psi(b''x).$$

Upon introducing the functions J and J^- defined on R by

$$J(x) = \int_{\infty}^{e^x} \frac{1 + y^2}{y^2} d\Psi(y), \quad J^-(x) = \int_{-\infty}^{-e^x} \frac{1 + y^2}{y^2} d\Psi(y), \quad x \in R,$$

and setting $e^h = b$, $e^{h'} = b'$, $e^{h''} = b''$, this requirement becomes

(1) $\quad \{\Psi(+0) - \Psi(-0)\}(b^2 + b'^2 - b''^2) = 0$

and

(2) $\qquad\qquad J(x + h) + J(x + h') = J(x + h''),$

$$J^-(x + h) + J^-(x + h') = J^-(x + h''), \quad x \in R,$$

where h, h' are arbitrary numbers and h'' is a function of h and h'.

Let $\Psi(+\infty) - \Psi(+0) > 0$ so that J does not vanish. If, in the foregoing relation in J, we set repeatedly $h' = h$, it follows that, for arbitrary positive integers n and sn,

$$nJ(x + h) = J(x + h''_n), \quad snJ(x + h) = J(x + h''_{sn}).$$

Therefore, to every rational $s > 0(s' > 0)$ there corresponds a number $t(t')$, such that, for every x,

(3) $\qquad\qquad sJ(x) = J(x + t).$

Since J is continuous from the left and nondecreasing, with $J \leqq 0$,

$J(+\infty) = 0$, it follows that $t' \downarrow t$ as $s' \uparrow s$, so that J is continuous, (3) holds for irrational s as well, and

$$t \downarrow {}^{\infty}_{-\infty} \quad \text{as} \quad s \uparrow {}^{\infty}_{0}.$$

Since J does not vanish, we can assume—by changing the origin if necessary—that $J(0) \neq 0$. Then, setting $J_0 = J/J(0)$, it follows, by

$$sJ(0) = J(t), \quad s'J(0) = J(t'), \quad s'J(t) = J(t + t'),$$

that

$$J_0(t)J_0(t') = J_0(t + t'), \quad t, t' \in R.$$

The only nonvanishing continuous solution of this functional equation, with $J_0(\infty) = 0$, is proportional to $e^{-\gamma t}$ with $\gamma > 0$. Therefore, setting $y = e^t$ and going back to Ψ, the derivative $\Psi'(y)$ exists for $y > 0$ and

$$\frac{1 + y^2}{y} \Psi'(y) = \beta' y^{-\gamma}, \quad \beta' \geq 0,$$

taking into account the vanishing case. Since Ψ is of bounded variation on $(0, +\infty)$, it follows that $\int_\epsilon^0 y^{1-\gamma}\, dy$ is finite for $\epsilon > 0$ and, hence, $\gamma < 2$. Furthermore, replacing J in (2) by its above-found expression, we have

$$b^\gamma + b'^\gamma = b''^\gamma, \quad 0 < \gamma < 2.$$

Similarly, with J^-: for $y < 0$

$$\frac{1 + y^2}{y} \Psi'(y) = -\beta |y|^{-\gamma'}, \quad \beta \geq 0,$$

with $b^{\gamma'} + b'^{\gamma'} = b''^{\gamma'}$, hence $\gamma = \gamma'$ (set $b = b' = 1$).

Therefore, on account of (1), either $b^2 + b'^2 = b''^2$ so that J and J^- vanish and f is a normal ch.f., or $\Psi(+0) - \Psi(-0) = 0$ and, for $y \neq 0$, $\Psi'(y)$ is given by the foregoing relations.

2° According to what precedes, a stable ch.f. f is either normal or of the form

$$(1) \quad \log f(u) = iu\alpha + \beta \int_{-\infty}^{-0} \left(e^{iux} - 1 - \frac{iux}{1 + x^2} \right) \frac{dx}{|x|^{1+\gamma}}$$

$$+ \beta' \int_0^{+\infty} \left(e^{iux} - 1 - \frac{iux}{1 + x^2} \right) \frac{dx}{x^{1+\gamma}}.$$

If $0 < \gamma < 1$, then it is possible to take out of the bracket the term $-\dfrac{iux}{1 + x^2}$ and, by modifying α, we obtain

$$(2) \quad \log f(u) = iu\alpha' + \beta \int_{-\infty}^0 (e^{iux} - 1) \frac{dx}{|x|^{1+\gamma}} + \beta' \int_0^\infty (e^{iux} - 1) \frac{dx}{x^{1+\gamma}}.$$

Let $u > 0$. Setting $ux = v$ and integrating along the closed contour formed by the positive halves of the real and imaginary axes and a circumference centered at the origin of radius $r \to \infty$, it follows, by the Cauchy theorem, that

$$(3) \qquad \int_0^\infty (e^{iux} - 1) \frac{dx}{x^{1+\gamma}} = |u|^\gamma e^{-\frac{\pi}{2}\gamma i} \Gamma(-\gamma),$$

where

$$\Gamma(-\gamma) = \int_0^\infty (e^{-v} - 1) \frac{dv}{v^{1+\gamma}} < 0.$$

The first integral in (1) follows by taking the complex-conjugate of (3) and, for $u < 0$, $\log f(u)$ is obtained by taking the complex-conjugate of $\log f(|u|)$. Upon substituting in (2) and setting

$$b = -\Gamma(-\gamma)(\beta + \beta') \cos \frac{\pi}{2} \gamma, \quad c = \frac{\beta - \beta'}{\beta + \beta'},$$

so that $b \geq 0, |c| \leq 1$, we obtain the asserted form (i) of $\log f(u)$.

If $1 < \gamma < 2$, then we can take out of the bracket in (1) the term $-\dfrac{iux}{1 + x^2} + iux$, and (2) is replaced by

$$(4) \quad \log f(u) = iu\alpha'' + \beta \int_{-\infty}^0 (e^{iux} - 1 - iux) \frac{dx}{|x|^{1+\gamma}}$$

$$+ \beta' \int_0^\infty (e^{iux} - 1 - iux) \frac{dx}{|x|^{1+\gamma}}.$$

Proceeding as above we obtain the same form (i) of $\log f(u)$.

If $\gamma = 1$, the foregoing modifications of the third term in the bracket in (1) are no more possible. But, for $u > 0$,

$$\int_{+0}^{+\infty} \left(e^{iux} - 1 - \frac{iux}{1 + x^2} \right) \frac{dx}{x^2}$$

$$= \int_0^\infty \frac{\cos ux - 1}{x^2} dx + i \int_{+0}^\infty \left(\sin ux - \frac{ux}{1 + x^2} \right) \frac{dx}{x^2}$$

$$= -\frac{\pi}{2} u + iu \lim_{\epsilon \downarrow 0} \left\{ \int_{\epsilon u}^{+\infty} \frac{\sin v}{v^2} dv - \int_\epsilon^\infty \frac{dv}{v(1 + v^2)} \right\}$$

$$= -\frac{\pi}{2} u - iu \lim_{\epsilon \downarrow 0} \int_\epsilon^{\epsilon u} \frac{\sin v}{v^2} dv + iu \lim_{\epsilon \downarrow 0} \int_\epsilon^\infty \left(\frac{\sin v}{v^2} - \frac{1}{v(1 + v^2)} \right) dv.$$

The limit of the second integral exists and is finite, and that of the first one is log u. The asserted form (ii) of log $f(u)$ readily follows, and the conclusion is reached.

24.5. Lévy representation. This subsection may read immediately after 24.1 except for reformulations of results in the intervening subsections.

So far we used systematically *Khintchine representation* of i.d. ch.f.'s e^ψ with $\psi = (\alpha, \Psi)$ representing

$$\psi(u) = i\alpha u + \int\left(e^{iux} - 1 - \frac{iux}{1 + x^2}\right)\frac{1 + x^2}{x^2}\, d\Psi(x)$$

where $\alpha \in R$ and the Khintchine function Ψ is bounded nondecreasing with $\Psi(-\infty) = 0, \Psi(+\infty) < \infty$ or, in terms of the measure which corresponds biunivoquely to it and is also denoted by Ψ, the *Khintchine measure* Ψ on R (that is, on the Borel field of R), is bounded. Ψ has no direct probabilistic meaning but presents definite technical advantages: It permits a simple description of the i.d. family with $\psi = (\alpha, \Psi), \alpha \in R$, Ψ bounded measure on R, as well as a simple description of convergence of i.d. laws: $\psi_n = (\alpha_n, \Psi_n) \to \psi = (\alpha, \Psi)$ if and only if $\alpha_n \to \alpha$, $\Psi_n \xrightarrow{c} \Psi$.

In fact, "Lévy representation" below was the initial one and is central to and born from P. Lévy probabilistic analysis of decomposable processes (§41).

Let barred integral sign mean that the origin is excluded from the interval of integration and, as usual, we omit its endpoints when they are $-\infty$ and $+\infty$.

P. Lévy representation of i.d. ch.f.'s e^ψ with $\psi = (\alpha, \beta^2, L)$ is given by

$$\psi(u) = i\alpha u - \frac{\beta}{2} u^2 + \int\left(e^{iux} - 1 - \frac{iux}{1 + x^2}\right) dL(x)$$

where $\alpha, \beta \in R$ and the Lévy function L defined on $R - \{0\}$ is nondecreasing on $(-\infty, 0)$ and on $(0, +\infty)$ with $L(\pm\infty) = 0$ and $\int_{-x}^{x} y^2\, dL(y) < \infty$ for some hence every finite $x > 0$. The corresponding *Lévy measure* L on $R - \{0\}$ is bounded outside every neighborhood of the origin but may be infinite on $R - \{0\}$.

The somewhat involved characterization of Lévy function explains why Khintchine representation is frequently favored despite its lack of direct probabilistic meaning.

The following correspondence is immediate:

a. CORRESPONDENCE LEMMA. *There is a one-to-one correspondence between Lévy and Khintchine representations.*

It is given by $\beta^2 = \Psi(+0) - \Psi(-0)$ *and*

$$(1) \qquad\qquad dL(x) = \frac{1+x^2}{x^2}\, d\Psi(x), \quad x \neq 0,$$

or, more precisely, with $x > 0$,

$$(1') \quad L(-x) = \int_{-\infty}^{-x} \frac{1+y^2}{y^2}\, d\Psi(y), \quad L(x) = \int_{-\infty}^{x} \frac{1+y^2}{y^2}\, d\Psi(y)$$

and, conversely,

$$(1'') \quad \Psi(-x) = \int_{-\infty}^{-x} \frac{y^2}{1+y^2}\, dL(y), \quad \Psi(x) = \int_{+\infty}^{x} \frac{y^2}{1+y^2}\, dL(y) + \beta^2.$$

The continuity sets $C(L)$ *and* $C(\Psi)$ *are the same on* $R - \{0\}$.

A. I.D. CONVERGENCE CRITERION.

$$\psi_n = (\alpha_n, \beta_n{}^2, L_n) \to \psi = (\alpha, \beta^2, L)$$

if and only if

(i) $$L_n \xrightarrow{w} L \text{ on } R - \{0\}$$

(ii) $$\int_{-x}^{x} y^2\, dL_n(y) + \beta_n{}^2 \to \beta^2 \text{ as } n \to \infty \text{ then } 0 < x \to 0$$

(iii) $$\alpha_n \to \alpha$$

Proof. Since $\psi_n = (\alpha_n, \Psi_n) \to \psi = (\alpha, \Psi)$ if and only if $\alpha_n \to \alpha$ and $\Psi_n \xrightarrow{c} \Psi$, it suffices to prove that $\Psi_n \xrightarrow{c} \Psi \Leftrightarrow$ (i) and (ii) hold.

We use **a** and Helly–Bray lemma and theorem without further comment. Let $x > 0$.

Let $\Psi_n \xrightarrow{c} \Psi$. Clearly (i) follows. Since for $\pm x \in C(\Psi)$

$$\int_{-x}^{x} \frac{y^2}{1+y^2}\, dL_n(x) + \beta_n{}^2 = \Psi_n(x) - \Psi_n(-x)$$

(ii) follows as $n \to \infty$ then $0 < x \to 0$ hence without the above restriction on $\pm x$ since $\Psi(x) - \Psi(-x)$ is monotone in x.

Conversely, let (i) and (ii) hold. Clearly $\Psi_n(-x) \to \Psi(-x)$ for $-x \in C(L)$. For $0 < \epsilon < x \in C(L)$, from

$$\Psi_n(x) = \int_{-\infty}^{-\epsilon} \frac{y^2}{1+y^2}\, dL_n(y) + \int_{-\epsilon}^{+\epsilon} \frac{y^2}{1+y^2}\, dL_n(y)$$

$$+ \beta_n{}^2 + \int_{\epsilon}^{x} \frac{y^2}{1+y^2}\, dL_n(y)$$

it follows that, as $n \to \infty$ then $\epsilon \to 0$,

$$\Psi_n(x) \to \Psi(-0) + \beta^2 + \Psi(x) - \Psi(0) = \Psi(x).$$

The same is true for $x = +\infty$ so that $\Psi_n(+\infty) \to \Psi(+\infty)$. Thus $\Psi_n \overset{c}{\to} \Psi$ and the proof is terminated.

Reformulations. Lévy representation is visible in the main results and also in the proofs in the preceding subsections:

1. EXTREMA CRITERION. Its statement in 23.4C is already in terms of Lévy function L and of $\beta^2 = \Psi(+0) - \Psi(-0)$ of the i.d. limit law.

2. EXTENDED CENTRAL CONVERGENCE CRITERION. This most important result of the section 23.4B is to be reformulated as follows.
Let $x > 0$ and set

$$L_n(-x) = \sum_k F_{nk}(-x), \quad L_n(x) = \sum_k (F_{nk}(x) - 1).$$

Then, in terms of L and β^2 of the limit i.d. law, the criterion conditions are

$$L_n \overset{w}{\to} L \text{ and } \sum_k \sigma^2 X^\epsilon \to \beta^2 \text{ as } n \to \infty \text{ then } 0 < \epsilon \to 0$$

Furthermore, Lévy functions L_n have a direct probabilistic meaning in terms of the summands $X_n,\ k = 1,\ \cdots,\ k_n$:

$$L_n(-x) = E(\text{number of the } X_{nk} \text{ in } (-\infty, x))$$
$$- L_n(x) = E\ (\text{number of the } X_{nk} \text{ in } [x, \infty)).$$

3. The proof of the Ψ-criterion 24.3B is, in fact, in terms of L. For, the functions $\mathcal{J}$ and $\mathcal{J}^-$ therein are given by $\mathcal{J}(x) = -L(e^x)$ and $\mathcal{J}^-(x) = L(-e^x)$.

Lévy functions of stable laws within the proof of 24.4B are:

$$\gamma = 2: \quad L = 0\text{---normal law}$$
$$0 < \gamma < 2: \quad dL(x) = \beta/|x|^{1+\gamma}\, dx \text{ for } x < 0,$$
$$dL(x) = \beta'/|x|^{1+\gamma}\, dx \text{ for } x > 0$$

CLP for iid summands. In what follows, f_n and f_n are ch.f.'s.

We intend to solve directly the Central Limit Problem (CLP for short) for independent identically distributed (*iid* for short) summands. We shall use **A** and generalize results in 24.1 replacing $f_n{}^n = f$ for every n by $f_n{}^n \to f$.

b. *If $f_n \to 1$ in some neighborhood $[-U, +U]$ of the origin then on $[-U, U]$, from some $n = n(U)$ on, $\log f_n$ exist and are bounded and*

$$\log f_n = -\sum_{m=1}^{\infty} \frac{1}{m}(1 - f_n)^m = (f_n - 1)(1 + o(1)).$$

For, on $[-U, +U]$, $f_n \to 1$ uniformly so that, from some $n = n(U)$ on, $|1 - f_n| < 1/2$ hence $\log f_n$ exists and is continuous and thus is bounded, and

$$\log f_n = \log(1 - (1 - f_n)) = -(1 - f_n) - \frac{1}{2}(1 - f_n)^2 - \cdots$$
$$= (f_n - 1)(1 + o(1)).$$

We generalize 24.1b:

c. *If $f_n{}^n \to f$ then f has no zeros and the same is true when $e^{-iua_n} f_n{}^n(u) \to f(u)$ for every $u \in R$.*

Proof. It suffices to prove that ch.f.'s $(|f_n|^2)^n \to |f|^2$ implies $|f|^2 > 0$. Suppose this "symmetrization" already took place so that $f_n{}^n \to f$ with f_n and $f \geqq 0$.

Since f is continuous with $f(0) = 1$, there is a finite interval $[-U, +U]$ on which $f > 0$ hence $\log f$ exists and is bounded. On this interval, from some n on, $\log f_n$ exist and are bounded, so that $n \log f_n \to \log f$ hence $\log f_n \to 0$, that is, $f_n \to 1$, **a** applies

$$n(f_n - 1)(1 + o(1)) = n \log f_n \to \log f$$

and $n(f_n - 1)$ remain bounded. Since, by 13.4**A**

$$n(1 - f_n(2u)) \leqq 4n(1 - f_n(u)),$$

it follows that on $[-2U, +2U]$, from some n on, $n(1 - f_n) \geqq 0$ remain bounded, so that $f_n \to 1$, **a** applies and $e^{n(f_n-1)} \to f > 0$.

Upon continuing this doubling of the intervals, any given $u \in R$ belongs to an interval on which $f > 0$ hence $f > 0$ on R, and the proposition is proved.

B. IID CONVERGENCE CRITERION. *Let ψ be continuous*

$$f_n{}^n \to f \Leftrightarrow n(f_n - 1) \to \psi, \text{ and then } f = e^\psi \text{ is i.d.}$$

More generally, if $f_n \to 1$ or $a_n/n \to 0$ then, for every $u \in R$,

$$e^{-iua_n} f_n{}^n(u) \to f(u) \Leftrightarrow -iua_n + n(f_n(u) - 1) \to \psi,$$

and then $f = e^\psi$ is i.d.

Proof. 1°. Let $n(f_n - 1) \to \psi$, so that $f_n \to 1$, **b** applies, and $f_n{}^n \to e^\psi = f$.

Conversely, let $f_n{}^n \to f$ so that, by **c**, f has no zeroes and $\log f$ exists and is continuous. Given any finite interval, it follows that on it, from some n on, $\log f_n$ exist and are bounded and, by **b**, $n(f_n - 1) \to \log f = \psi$.

2°. Let $iua_n + n(f_n(u) - 1) \to \psi(u)$ for every $u \in R$ so that $-iua_n/n + f_n(u) - 1 \to 0$ hence $a_n/n \to 0 \Leftrightarrow f_n \to 1$. With either of these equivalent conditions **b** applies and, for every $u \in R$, from some $n = n(u)$ on,

$$e^{-iua_n} f_n{}^n(u) = (e^{-iua_n/n} f_n(u))^n \to e^{\psi(u)} = f(u).$$

Conversely, let for every $u \in R$,

$$(e^{-iua_n/n} f_n(u))^n = e^{-iua} f_n{}^n(u) \to f(u)$$

so that, by **c**, $f(u) \neq 0$ hence $e^{-iua_n/n} f_n(u) \to 1$. Thus, once more, $a_n/n \Leftrightarrow f_n \to 1$ and, with either of these equivalent conditions, **b** applies and $-iua_n + n(f_n(u) - 1) \to \log f(u) = \psi(u)$.

It remains to show that the limit ch.f. f is *i.d.* This will follow from the "structure" proposition below. In fact, this proposition provides a widening of the definition in 23.1 of i.d. laws since $f_n{}^n = f$ for every n implies $f_n{}^n \to f$ but, in general, the converse is not true. It also provides a direct probabilistic proof of the structure theorem in 23.1:

Let $S_0 = 0$, $S_n = X_1 + \cdots + X_n$, $n = 1, 2, \cdots$, where the summands are iid with common ch.f. f. Let $\lambda \geq 0$. We say that a r.v. S is (λ, f)-*compound Poisson* if its d.f. is

$$F_S = e^{-\lambda} \sum_{n=0}^{\infty} \frac{\lambda^n}{n!} F_{S_n}.$$

Clearly F_S is a d.f.: It is nondecreasing with $F_S(-\infty) = 0$, $F_S(+\infty) = e^{-\lambda} \sum_{n=0}^{\infty} \frac{\lambda^n}{n!} = 1$. The corresponding ch.f. is immediate:

$$f_S = e^{\lambda(f-1)}.$$

It is an i.d. ch.f., since $e^{\frac{\lambda}{m}(f-1)}$ is the ch.f. of a $(\lambda/m, f)$-compound Poisson for $m = 1, 2, \cdots$. Also the centered ch.f. $e^{-iua+\lambda(f-1)}$ is i.d. and the i.d. assertion in **B** follows at once. And **B** yields

STRUCTURE COROLLARY. *f is i.d. if and only if there are compound Poisson f_n with $f_n{}^n \to f$.*

C. IID CENTRAL CONVERGENCE CRITERION. *Let X_{nk}, $k = 1, \cdots n$, be iid summands with common d.f. F_n and ch.f. $f_n \to 1$. Let $x > 0$.*

$$\mathcal{L}\left(\sum_k X_{nk} - a_n \right) \to \mathcal{L}(X) \text{ necessarily i.d. with } \Psi = (a, \beta^2, L)$$

if and only if

$$(C_L): \quad L_n \xrightarrow{w} L \text{ with } L_n \text{ defined by}$$

$$L_n(-x) = nF_n(-x), \quad L_n(x) = \sum_k n(F_n(x) - 1), \quad x > 0.$$

$$(C_{\beta^2}): \quad n \int_{-x}^{x} y^2 \, dF_n(y) \to \beta^2 \quad as \quad n \to \infty \quad then \quad x \to 0.$$

$$(C_a): \quad a_n = \alpha_n - \alpha + o(1) \quad \text{with} \quad \alpha_n = n \int \frac{x}{1 + x^2} \, dF_n(x).$$

Note that (C_a) characterizes all admissible a_n.

Proof. According to **B**, the required convergence is equivalent to

$$\psi_n(u) = -iua_n + n \int (e^{iux} - 1) \, dF_n(x) \to \psi(u), \, u \in R \text{ where, setting}$$

$$\alpha_n = n \int \frac{1}{1 + x^2} \, dF_n(x),$$

$$\psi_n(u) = iu(\alpha_n - a_n) + n \int \left(e^{iux} - 1 - \frac{iux}{1 + x^2} \right) dF_n(x)$$

$$= (\alpha_n - a_n, \beta_n{}^2, L_n),$$

with L_n defined by

$$L_n(-x) = nF_n(-x), \quad L_n(x) = n(F_n(x) - 1), \quad x > 0,$$

corresponding Ψ_n defined by

$$\Psi_n(z) = n \int_{-\infty}^{z} \frac{y^2}{1 + y^2} \, dF_n(y), \quad z \in R,$$

and $\beta_n{}^2$ determined by

$$n \int_{-x}^{x} y^2 \, dF_n(y) = \int_{-x}^{x} (1 + y^2) d\Psi_n(y) = \int_{-x}^{x} y^2 \, dL_n(y) + \beta_n{}^2.$$

The asserted criterion follows at once from **A**.

COMPLEMENTS AND DETAILS

1. Prove Lindeberg's theorem without using Liapounov's bounded case theorem. Then deduce Liapounov's theorem.

For Lindeberg's theorem use the expansion

$$f_k\left(\frac{u}{s_n}\right) = 1 - \frac{\sigma_k^2}{2s_n^2} u^2 + \theta_{nk} \frac{u^2}{s_n^2}\left(\epsilon\sigma_k^2|u| + \int_{|x|\geq\epsilon s_n} x^2 \, dF_k(x)\right)$$

To deduce Liapounov's theorem observe that

$$\frac{1}{s_n^2}\int_{|x|\geq\epsilon s_n} x^2 \, dF_k \leq \frac{1}{\epsilon^\delta} \cdot \frac{E|X_k|^{2+\delta}}{s_n^{2+\delta}}.$$

2. Prove directly the sufficiency of Kolmogorov's conditions for degenerate convergence. Then deduce the condition in $(1 + \delta)$.

3. Deduce the Kolmogorov and Lindeberg-Feller theorems from the degenerate and normal convergence criteria—where existence of moments is not assumed.

4. Deduce the bounded variances limit theorem from the Central Limit theorem.

5. Let $\sum_k X_{nk}$ be sums of independent uan summands centered at expectations with $\sum_k \sigma^2 X_{nk} = 1$ whatever be n. Then

$$\mathcal{L}(\sum_k X_{nk}) \to \mathfrak{N}(0, 1) \Leftrightarrow \sum_k X_{nk}^2 \xrightarrow{P} 1.$$

(Observe that the last convergence is equivalent to $\sum \int_{|x|\geq\epsilon} x^2 \, dF_{nk} \longrightarrow 0$ whatever be $\epsilon > 0$.)

6. Let $\zeta(t + iu)$, $t > 1$, be the Riemann function defined by

$$\zeta(t + iu) = \sum_n n^{-t-iu} = \prod_p (1 - p^{-t-iu})$$

where p varies over all primes. $f_t(u) = \zeta(t + iu)/\zeta(t)$ is an i.d. ch.f.

$(\log f_t(u) = \sum_p \sum_n p^{-nt}(e^{-inu \log p} - 1)/n.)$

7. An i.d. law may be composed of two non i.d. laws. In fact, there exists a non i.d. ch.f. f such that $|f|^2$ is i.d.: form the ch.f. f of X with $P[X = -1] = p(1 - p)/(1 + p)$, $P[X = k] = (1 - p)(1 + p^2)p^k/(1 + p)$, $k = 0, 1, \cdots, 0 < p < 1$.

(Put f in the form (α, Ψ); observe that Ψ so found does not satisfy the necessary requirements. Put $|f|^2$ in the form (α, Ψ).)

8. An i.d. law may be composed of an i.d. law and an indecomposable one: let $X = 0$ or 1 with pr.'s 2/3 and 1/3, respectively; the ch.f. f is indecomposable

$$\log f(u) = \log \frac{2 + e^{iu}}{3} = \sum a_n(e^{inu} - 1), \quad \sum |a_n| < \infty.$$

Set

$$\log f^+(u) = \sum\nolimits^+ a_n(e^{inu} - 1), \quad \log f^-(u) = -\sum\nolimits^- a_n(e^{inu} - 1)$$

where $\sum^+ (\sum^-)$ denotes summation over positive (negative) a_n. Then f^+ and f^- are i.d. and $f^+ = ff^-$.

Also an i.d. law may be the product of an i.d. law and two indecomposable ones: proceed as above but with f defined by $\dfrac{5 + 4 \cos u}{9} = \left| \dfrac{2 + e^{iu}}{3} \right|^2$.

9. *P. Lévy centering function.* The family of i.d. laws coincides with laws defined by

$$\log f(u) = i\alpha u - \frac{\beta^2 u^2}{2} + \int \left(e^{iux} - 1 - \frac{iux}{1 + x^2} \right) dL(x)$$

where L is defined on R, except at the origin, is nondecreasing on $(-\infty, -0)$ and on $(+0, +\infty)$, with $L(\mp\infty) = 0$ and $\int_{-\tau}^{+\tau} x^2 \, dL(x) < \infty$ for some $\tau > 0$; the barred integral sign means that the origin is excluded.

Also

$$\log f(u) = i\alpha(\tau)u - \frac{\beta^2}{2} u^2 + \int_{-\tau}^{+\tau} (e^{iux} - 1 - iux) \, dL(x)$$
$$+ \left(\int_{-\infty}^{-\tau} + \int_{+\tau}^{+\infty} \right) (e^{iux} - 1) \, dL(x).$$

This splitting of the domain of integration replaces the P. Lévy centering function $g(x) = x/(1 + x^2)$ by much simpler ones $(g(x) = x$ and $g(x) = 0)$ within the partial domains of integration.

Why was the centering function needed? Then, what are the conditions to impose upon it? Show that Feller's centering function $g(x) = \sin x$ is acceptable. Is the following one acceptable: $g(x) = x$ for $|x| < c$ for some finite positive constant c, $g(x) = c$ for $x \geqq c$ and $g(x) = -c$ for $x \leqq c$?

10. Let r.v.'s $X_{n,k}$ with d.f.'s $F_{n,k}$, $k = 1, \cdots, k_n \to \infty$, $n = 1, 2, \cdots$, be independent in k and uniformly asymptotically distributed in k, that is, there exist d.f.'s F_n such that $F_{n,k} - F_n \to 0$ uniformly in k. The nondecreasingly ranked numbers $X_{n,k}(\omega)$ into $X^*_{n,1}(\omega) \leqq \cdots \leqq X^*_{n,k_n}(\omega)$ determine "ranked" $X^*_{n,r}$ of "rank" r; the $^*X_{n,s} = X_{n,k_n+1-s}$ are of "end rank" s. Set

$$L_n = \sum_k F_{n,k}, \quad M_n = \sum_k (F_{n,k} - 1),$$

$$g_{n,r_n} = (r_n - \sum F_{n,k}) / \sqrt{\sum_k F_{n,k}(1 - F_{n,k})},$$

$$I_n = \sum_k I_{n,k}, \quad \bar{I}_n = (I_n - EI_n)/\sigma I_n, \quad I_{n,k}(x) = I_{[X_{n,k} < x]}.$$

Use throughout the fundamental relation

$$[X^*_{n,r} < x] = [I_n(x) \geqq r].$$

a) The $X^*_{n,r}$ are r.v.'s.
b) For fixed ranks r, the class of limit laws of ranked r.v.'s $X_{n,r}$ is that of laws

$\mathcal{L}(X^*_r)$ with d.f.'s $F_r{}^L = \int_0^L \frac{t^{r-1}}{(r-1)!} e^{-t}\, dt$, where the functions L on R are nondecreasing, nonnegative, and not necessarily finite.

These limit laws are laws of r.v.'s if and only if $L(-\infty) = 0$, $L(+\infty) = +\infty$. And

$$F^*_{n,r} \xrightarrow[c]{w} F_r{}^L \Leftrightarrow L_n \xrightarrow[c]{w} L.$$

c) For fixed endranks s, the class of limit laws of ranked r.v.'s $*X_{ns}$ is that of laws $\mathcal{L}(*X_s)$ with d.f.'s $^M F_s = \int_{-M}^{+\infty} \frac{t^{s-1}}{(s-1)!} e^{-t}\, dt$ where the functions M on R are nondecreasing, nonpositive, and not necessarily finite.

These limit laws are laws of r.v.'s if and only if $M(-\infty) = -\infty$, $M(+\infty) = 0$. And

$$*F_{ns} \xrightarrow[c]{w} {}^M F_s \Leftrightarrow M_n \xrightarrow[c]{w} M.$$

d) For variable ranks $r_n \to \infty$ with $s_n = k_n + 1 - r_n \to \infty$, the class of limit laws of ranked r.v.'s X^*_{n,r_n} is that of laws with d.f.'s $F^g = \frac{1}{\sqrt{2\pi}} \int_g^\infty e^{-t^2/2}\, dt$, where the functions g on R are nonincreasing, and not necessarily finite.

These limit laws are those of r.v.'s if and only if $g(-\infty) = +\infty$, $g(+\infty) = -\infty$. And

$$F^*_{n,r_n} \xrightarrow[c]{w} F^g \Leftrightarrow g_{n,r_n} \xrightarrow[c]{w} g.$$

e) What if the X_{nk} are uniformly asymptotically negligible? What if, moreover, $\mathcal{L}(\sum_k X_{nk}) \to \mathcal{L}(X)$?

f) What about joint limit laws of ranked r.v.'s?

11. Let $\mathcal{L}(X_n - a_n) \to (\alpha, \beta^2, L)$ where $X_n = \sum_k X_{nk}$ are sums of uan independent r.v.'s.

(a) The sequence $\mathcal{L}(\max_k | X_{nk} |)$ converges. Find the limit law $\mathcal{L}(X)$. Why can necessary and sufficient conditions for normality of the limit law of the sequence $\mathcal{L}(X_n - a_n)$ be expressed in terms of $\mathcal{L}(X)$? Are there other i.d. laws for which this is possible? (For n sufficiently large and $x > 0$

$$\log P[\max_k | X_{nk} | < x] = -(1 + o(1)) \sum_k P[| X_{nk} | \geq x].)$$

(b) Let $a_{nk} = \int_{|x| < \tau} x\, dF_{nk}$, $\tau > 0$ finite, $\overline{F}_{nk}(x) = F_{nk}(x + a_{nk})$ and let F'_{nk} be the d.f. of $X'_{nk} = | X_{nk} - a_{nk} |^r$ for a fixed $r > 1$.

If $\mathcal{L}(\sum_k X_{nk} - a_n) \to (\alpha, \beta^2, L)$, then there exist constants a'_n such that $\mathcal{L}(\sum_k X'_{nk} - a'_n) \to (\alpha', 0, L')$ with $L'(x) = 0$ or $L(x^{1/r}) - L(-x^{1/r})$ according as $x < 0$ or $x > 0$. (If $g \geq 0$ is even, then, for every $c > 0$,

$$\cdot \int_0^c g\, dF'_{nk} = \int_{|x| < c^{1/r}} g(| x |^r)\, d\overline{F}_{nk}, \quad \int_c^\infty g\, dF'_{nk} = \int_{|x| > c^{1/r}} g(| x |^r)\, d\overline{F}_{nk}.$$

Take $g = 1$ and $g(x) = x^2$. Observe that

$$0 \leq \int_0^\epsilon x^2 \, dF'_{nk} - \left(\int_0^\epsilon x \, dF'_{nk} \right)^2 \leq \epsilon^{2(r-1)/r} \int_{|x| < \epsilon^{1/r}} x^2 \, d\overline{F}_{nk}.)$$

(c) $\mathcal{L}(\sum_k X_{nk} - a_n) \to \mathfrak{N}(0, \beta^2)$ if, and only if, $\sum_k X'^2_{nk} \xrightarrow{P} \beta^2$. What about limit Poisson laws?

In what follows and unless otherwise stated, degenerate laws are excluded; f, with or without affixes, is a ch.f.; and, without restricting the generality, the type of f is the family of all ch.f.'s defined by $f(cu)$ for some $c > 0$.

12. f is decomposable by every f^n, $n = 2, 3, \cdots$, if, and only if, f is degenerate.

13. f is decomposable and every component belongs to its type with $f(u) = \sum f(c_j u)$, $\sum c_j^2 \geq 1$, if and only if f is normal.

14. If for an $r > 0$ and $\neq 1$, f^r belongs to the type of f, then f is i.d. If there are two such values r' and r'' of r and $\log r'/\log r''$ is irrational, then f is stable.

15. If $f_n \to f, f'_n \to f'$ and $f_n = f'_n f''_n$ for every n, then f' is a component of f.

16. f is c-decomposable if $f(u) = f(cu)f_c(u)$ for some fixed c necessarily between 0 and 1. L_c is the family of all c-decomposable laws, L_0 is the family of all laws, and L_1 is that of self-decomposable ones.

(a) $L_0 \supset L_c \supset L_1$, and if $\log c/\log c'$ is rational, then $L_c = L_{c'}$. Every L_c is closed under compositions and passages to the limit.

(b) $f \in L_c$ if, and only if, it is limit of a sequence of ch.f.'s of normed sums S_n/b_n of independent r.v.'s with $b_n/b_{n+1} \to c$.

(c) $f \in L_c$ if, and only if, it is ch.f. of $X(c) = \sum_{k=0}^{\infty} \xi_k c^k$ where the law of the series converges and the ξ_k are independent and identically distributed. Then the series converges a.s., and $f_{\xi_k} = f_c$. If ξ_k is bounded, then f is not i.d.

(d) $g(x)$ is said to be γ-convex ($\gamma > 0$ fixed) if every polygonal line inscribed in its graph with vertices projecting at distance γ on the x-axis is convex.

If ξ_k is i.d., so is $X(c)$. f i.d. with Lévy's function L belongs to L_c and f_c is i.d. only if $(-1)^j M_j$ are γ-convex for $\gamma = |\log c|$ where M_j are defined as in 9. Is the converse true?

(e) If $E\xi_k = 0$, $\sigma^2 \xi_k = 1$, then, for c, $c' \in (-1, +1)$, the covariance $EX(c)X(c') = 1/(1 - cc')$, and the random function $X(c)$ on $(-1, +1)$ exists in q.m. and is continuous and indefinitely differentiable in q.m.

Chapter VII

INDEPENDENT IDENTICALLY
DISTRIBUTED SUMMANDS

This chapter is devoted to study in some depth of consecutive sums S_1, S_2, $\cdots$ of sequences of independent identically distributed summands X_1, X_2, $\cdots$ with common law $\mathcal{L}(X)$; we shorten "independent identically distributed" to *iid*. As usual, methods are emphasized. Methods and results took their definitive form in the third quarter of this century.

In the preceding chapters some results about iid summands were obtained: Kolmogorov law of large numbers (17.3B) and its generalization 17.4, 4°, convergence of laws of normed sums to normal when the summands have finite second moments (21.1A) and the far-reaching characterization of all limit laws of normed sums (24.4), by particularizing the solution of the general central limit problem.

In this chapter, using directly 24.5, by means of Karamata theory, we obtain in §25 the above limit "stable" laws and their "domains of attraction"—those families of laws for which the laws of normed sums $S_n/b_n - a_n$ converge to any given stable one.

In §26, we study "random walks"; sequences of sums $S_1, S_2, \cdots$ *themselves* (not normed), their global and asymptotic behaviour with their dichotomy into "recurrent" and "transient" ones, and their fascinating "finite fluctuations."

§25. REGULAR VARIATION AND DOMAINS OF ATTRACTION

The domain of attraction of the normal law was found by P. Lévy, by Feller, and by Khintchine. The domains of attraction of all other stable laws were discovered by Doeblin and by Gnedenko. Much later, Feller observed that these results were in terms of Karamata regular variation theory and showed its usefulness for various limit probability problems. We follow his presentation of Karamata theory, and then apply it to the problem of stable laws and their domains of attraction. It deems advisable that at the first reading only **A** and its Corollary be covered in 25.1 and **c** in 25.2 be assumed.

25.1   Regular variation. Let U, V be positive monotone functions on $[0, \infty)$ to $[0, \infty)$ and let x, y be positive.

We say that U *varies regularly* (at $+\infty$) with exponent $a \in R$ if $U(x) = x^a V(x)$ where V *varies slowly* (at $+\infty$), that is, $V(tx)/V(t) \to 1$ as $t \to \infty$ for every x. Thus slow variation is regular variation with exponent 0. Since our only concern is with behaviour at $+\infty$, we may take $x, y > c \in R$ with $c > 0$ arbitrary but fixed, or substitute (c, ∞) for $[0, \infty)$, or assume that U, V vanish on $[0, c]$; this will be done without further comment.

A. Regular variation criterion. *Let D be a set dense in $[0, \infty)$. U varies regularly if and only if, for every $x \in D$,*

$$U(tx)/U(t) \to h(x) < \infty \quad as \quad t \to \infty,$$

and then $h(x) = x^a$ for some $a \in R$.

Proof. The "only if" assertion is trivially true. As for the "if" assertion, letting $t \to \infty$ in

$$\frac{U(tx)}{U(t)} = \frac{U(txy)}{U(ty)} \cdot \frac{U(ty)}{U(t)},$$

it follows that

$$h(xy) = h(x)h(y) \quad for \quad x, y \in D.$$

Since U is monotone, this functional equation extends to $[0, \infty)$ by taking limits from the right. But then it has a unique finite solution of the form $h(x) = x^a$ for some $a \in R$, and the proof is terminated.

Corollary. *If for every $x \in D$ dense in $[0, \infty)$,*

$$c_n U(b_n x) \to h(x) \quad finite \ positive$$

and

$$b_n \to \infty, \quad c_{n+1}/c_n \to 1,$$

then U varies regularly and $h(x) = cx^a$ for some finite a and $c > 0$.

Proof. If n is the smallest integer such that $b_n \leqq t < b_{n+1}$ then

$$\frac{U(b_n x)}{U(b_{n+1})} \leqq \frac{U(tx)}{U(t)} \leqq \frac{U(b_{n+1} x)}{U(b_n)}$$

where U is nondecreasing, while these inequalities are reversed when U is nonincreasing. By a change of scale we may assume that $1 \in D$. Then, since $c_{n+1}/c_n \to 1$ and $c_n U(b_n) \to h(1) = c > 0$, for every $x \in D$ the extreme terms converge to $h(x)/c$ hence $U(tx)/U(t) \to h(x)/c$, the above criterion applies and $h(x)/c = x^a$ for some $a \in R$.

*Let H be a positive monotone function on $[0, \infty)$ and set

$$U_a(x) = \int_0^x y^a H(y) \, dy, \quad V_a(x) = \int_x^\infty y^a H(y) \, dy$$

where $x > 0$ and a are finite.
Upon replacing if necessary 0 by $c > 0$, or assuming that H vanishes on $[0, c]$, $U_a(x)$ will be finite while $V_a(x)$ may be infinite. Since

$$U_a(x) \uparrow U_a(\infty) \quad \text{and} \quad V_a(x) \downarrow V_a(\infty) \quad \text{as} \quad x \uparrow \infty.$$

while

$$U_a(\infty) = U_a(x) + V_a(x) \quad \text{hence} \quad U_a(\infty) = U_a(\infty) + V_a(\infty),$$

it follows that

$$U_a(\infty) < \infty \Leftrightarrow V_a(\infty) = 0 \Rightarrow V_a(x) < \infty \text{ from some } x \text{ on}$$
$$U_a(\infty) = \infty \Leftrightarrow V_a(x) = \infty \text{ for every } x \Leftrightarrow V_a(\infty) = \infty.$$

a. *Let H vary slowly. Then $U_a(\infty)$ and $V_a(\infty)$ are finite for $a < -1$ and infinite for $a > -1$. Furthermore*

(i) *If $a \geqq -1$ then U_a varies regularly with exponent $a + 1$.*
(ii) *If $a < -1$ then V_a varies regularly with exponent $a + 1$, and this still holds for $a = -1$ provided V_{-1} is finite.*

Proof. Given $x > 0$ and $\epsilon > 0$, slow variation of H implies existence of $\delta > 0$ such that, for $y > \delta$,

(1) $$(1 - \epsilon)H(y) \leqq H(xy) \leqq (1 + \epsilon)H(y).$$

1°. Let $V_a(\infty) = 0$ hence $V_a(x) < \infty$ for some x on, and $U_a(\infty) < \infty$. Since

$$V_a(tx) = x^{a+1} \int_t^\infty y^a H(xy) \, dy,$$

it follows that, for $t > \delta$,

$$(1 - \epsilon)x^{a+1}V_a(t) \leq V_a(tx) \leq (1 + \epsilon)x^{a+1}V_a(t)$$

hence, letting $t \to \infty$ then $\epsilon \to 0$, $V_a(tx)/V_a(t) \to x^{a+1}$. Thus, V_a varies regularly with exponent $a + 1 \leq 0$ since V_a is nonincreasing, and $U_a(\infty) < \infty$ with $V_a(\infty) = 0$ only if $a \leq -1$.

2°.   Let $U_a(\infty) = \infty$ hence $V_a(\infty) = \infty$.   Since, for $t > \delta$,

$$U_a(tx) = U_a(\delta x) + x^{a+1}\int_\delta^t y^a H(xy)\, dy$$

hence, by (1),

$$(1 - \epsilon)x^{a+1}U_a(t) \leq U_a(tx) - U_a(\delta x) \leq (1 + \epsilon)x^{a+1}U_a(t):$$

upon dividing by $U_a(t)$ and letting $t \to \infty$ then $\epsilon \to 0$, it follows that $U_a(tx)/U_a(t) \to x^{a+1}$.   Thus, U_a varies regularly with exponent $a + 1 \geq 0$ since U_a is nondecreasing, and $U_a(\infty) = \infty$ hence $V_a(\infty) = \infty$ only if $a \geq -1$.   The assertions follow from 1° and 2°.

*B.   Main Karamata theorem.   *Let H be positive monotone on* $[0, \infty)$ *and set*

$$U_a(x) = \int_0^x y^a H(y)\, dy, \quad V_a(x) = \int_x^\infty y^a H(y)\, dy.$$

(i)   *If H varies regularly with exponent* $b \leq -a - 1$ *and* $V_a(x) < \infty$ *then, as* $t \to \infty$,

$$t^{a+1}H(t)/V_a(t) \to c = -(a + b + 1) \geq 0.$$

Conversely, if this limit exists and is positive then V_a *and H vary regularly with exponents* $-c = a + b + 1$ *and b, respectively, while if this limit is* 0 *then* V_a *varies slowly.*

(ii)   *If H varies regularly with exponent* $b \geq -a - 1$ *then, as* $t \to \infty$,

$$t^{a+1}H(t)/U_a(t) \to c = a + b + 1 \geq 0.$$

Conversely, if this limit exists and is positive then U_a *and H vary regularly with exponents* $c = a + b + 1$ *and b, respectively, while if this limit is* 0 *then* U_a *varies slowly.*

Note that when $c = 0$ the converse assertions for $c > 0$ continue to hold for V_a and for U_a, but nothing can be asserted regarding H.

Proof.   The argument for (i) and (ii) is the same, and we shall prove (i).

Set

(1) $$h(y)/y = y^a H(y)/V_a(y).$$

Since

$$y^a H(y) = -\frac{dV_a(y)}{dy},$$

upon integrating (3) over $[t, tx)$ with $x > 1$, it follows that

(2) $$-\log \frac{V_a(tx)}{V_a(t)} = \int_t^{tx} \frac{h(y)}{y}\, dy = h(t) \int_1^x \frac{h(tz)}{h(t)} \cdot \frac{1}{z}\, dz.$$

Let H vary regularly with exponent b so that, by $\mathbf{a}$, V_a varies with exponent $a + b + 1 = -c$. Thus, both sides of (1) vary regularly with exponent -1 and h varies slowly. Therefore, as $t \to \infty$, the integrand in the last integral in (2) tends to $1/z$ while the first term in (2) tends to $c \log x$ and Fatou–Lebesgue theorem implies that limsup $h(t) \le c$. Thus, h is bounded so that there is a sequence $t_n \to \infty$ with $h(t_n) \to c' \le c < \infty$. Since h varies slowly, $h(t_n y) \to c'$ for every $y > 0$ and, by the dominated convergence theorem, $c \log x = c' \log x$ hence $c' = c$ for every such sequence (t_n). Therefore, $h(t) \to c$ as $t \to \infty$ and the direct assertion is proved.

Conversely, if the limit $c \ge 0$ exists so that $h(t) \to c$ as $t \to \infty$ then, by (2), V_a varies regularly with exponent c. Moreover if $c > 0$ then this property of V_a together with (1) implies regular variation of H with exponent $-c - a - 1 = b$. This completes the proof of (i) and (ii) is proved similarly.

*C. SLOW VARIATION CRITERION.   *H varies slowly if and only if*

$$H(x) = h(x)\, exp\left\{\int_1^x \frac{g(y)}{y}\, dy\right\}$$

where $g(x) \to 0$ *and* $h(x) \to c < \infty$ *as* $x \to \infty$.

Proof. The "if" assertion is easily verified. As for the "only if" assertion, let H vary slowly. Then, by $\mathbf{B}$(ii) with $a = b = 0$,

$$H(t)/U_0(t) = (1 + g(t))/t \text{ with } g(t) \to 0 \text{ as } t \to \infty.$$

Since $H(t) = \dfrac{dU_0(t)}{dt}$, upon integrating over $[1, x)$ with $x > 1$, it follows that

$$U_0(x) = U_0(1)x\, exp\left\{\int_1^x \frac{g(y)}{y}\, dy\right\}.$$

But, by **B**(ii),

$$H(x) = h(x) U_0(x)/x,$$

and the "only if" assertion obtains.

COROLLARY.   *If H varies slowly then*, as $x \to \infty$, $H(x + y)/H(x) \to 1$ *and, given* $\delta > 0$, $x^{-\delta}H(x) \to 0$, $x^\delta H(x) \to \infty$, *and* $x^{-\delta} < H(x) < x^\delta$ *from some x on.*

*Let G be a d.f. vanishing on $(-\infty, 0)$. Let $x > 0$ be finite and set

$$\mu_\alpha(x) = \int_0^x y^\alpha dG(y), \quad \nu_\beta(x) = \int_x^\infty y^\beta dG(y).$$

Since we are concerned only with asymptotic behaviour of these integrals, whenever convenient we do take $G = 0$ in some neighborhood of the origin. *We assume that*

$$\mu_\alpha(\infty) = \lim_{x \to \infty} \mu_\alpha(x) = \infty, \quad \nu_\beta(\infty) = \lim_{x \to \infty} \nu_\beta(x) = 0$$

so that $\alpha > 0$ and $-\infty < \beta < \alpha$.

The elementary integration by parts which follows will reduce the question of regular variation of μ_α and of ν_β to the main Karamata theorem.

b. INTEGRATION BY PARTS LEMMA.   *Let x be a continuity point of G hence of μ_α and of ν_β. Then*

(i) $$\mu_\alpha(x) = -x^{\alpha-\beta}\nu_\beta(x) + (\alpha - \beta)\int_0^x y^{\beta-\alpha-1}\nu_\beta(y)\, dy$$

(ii) $$\nu_\beta(x) = -x^{\beta-\alpha}\mu_\alpha(x) + (\alpha - \beta)\int_x^\infty y^{\beta-\alpha-1}\mu_\alpha(y)\, dy.$$

Proof.   Relation (i) results at once from integration by parts of Stieltjes integrals.   Relation (ii) requires also a passage to the limit: Integration by parts on $[x, t)$ with $t > 1$ continuity point of G yields

(1) $$\nu_\beta(x) - \nu_\beta(t)$$
$$= -x^{\beta-\alpha}\mu_\alpha(x) + t^{\beta-\alpha}\mu_\alpha(t) + (\alpha - \beta)\int_x^t y^{\beta-\alpha-1}\mu_\alpha(y)\, dy.$$

Thus,

$$(\alpha - \beta)\int_x^t y^{\beta-\alpha-1}\mu_\alpha(y)\, dy \leq \nu_\beta(x) + x^{\beta-\alpha}\mu_\alpha(x)$$

so that, letting $t \to \infty$, the limit of the integral on the left is finite. Since μ_α is nondecreasing, as $t \to \infty$,

$$\frac{2^{p-\alpha}}{\beta - \alpha} t^{\beta-\alpha} \mu_\alpha(t) \leq \int_t^{2t} y^{\beta-\alpha-1} \mu_\alpha(y) \, dy \to 0$$

hence $t^{\beta-\alpha} \mu_\alpha(t) \to 0$ and, letting $t \to \infty$ in (i), (ii) obtains.

*D. VARIATION OF TRUNCATED MOMENTS. *Let* $\mu_\alpha(\infty) = \infty$ *and* $\nu_\beta(\infty) = 0$ *so that* $\alpha > 0$ *and* $-\infty < \beta < \alpha$.
(i) *If* μ_α *or* ν_β *varies regularly, then, as* $x \to \infty$,

$$x^{\alpha-\beta} \nu_\beta(x)/\mu_\alpha(x) \to c = \frac{\alpha - \gamma}{\gamma - \beta} \geq 0, \quad \beta \leq \gamma \leq \alpha.$$

(ii) *Conversely, if this limit exists then, for* $\beta < \gamma < \alpha$, μ_α *and* ν_β *vary regularly with exponents* $a = \alpha - \gamma > 0$ *and* $b = \beta - \gamma < 0$, *respectively, while* $a = 0$ *when* $\gamma = \alpha$ *and* $b = 0$ *when* $\gamma - \beta$.

Note in the boundary cases while μ_α varies slowly when $\gamma = \alpha$ and ν_β varies slowly when $\gamma = \beta$, nothing can be asserted regarding ν_β or μ_α, respectively.

Proof. 1°. Let μ_α vary regularly with exponent u. Finiteness of the integral in **b**(ii) yields $u \leq \alpha - \beta$. Since μ_α is nondecreasing $u \geq 0$. Thus, setting $u = \alpha - \gamma$, we have $\beta \leq \gamma \leq \alpha$ with $\gamma \geq 0$. Now, **b**(ii) yields

$$(1) \qquad \frac{x^{\alpha-\beta} \nu_\beta(x)}{\mu_\alpha(x)} = -1 + \frac{\alpha - \beta}{x^{\beta-\alpha} \mu_\alpha(x)} \int_x^\infty y^{\beta-\alpha-1} \mu_\alpha \, (dy)$$

so that, using **B**(i) with $H = \mu_\alpha$ and $a = \beta - \alpha - 1$, as $x \to \infty$,

$$x^{\alpha-\beta} \nu_\beta(x)/\mu_\alpha(x) \to -1 + (\alpha - \beta)/(\gamma - \beta) = (\alpha - \gamma)/(\gamma - \beta) = c$$

with $c = \infty$ when $\gamma = \beta$, and this is the asserted limit. Let ν_β vary regularly with exponent v so that $v \leq \beta$. Since ν_β is nonincreasing $v \leq 0$. Thus, setting $v = \beta - \gamma$, we have $\beta \leq \gamma \leq \alpha$ with $\gamma \geq 0$. Proceeding as above but with **b**(i) in lieu of **b**(ii) and using **B**(i) but with $H = \nu_\beta$ and $a = \alpha - \beta - 1$, once more the asserted limit obtains and (i) is proved.

2°. Conversely, let the limit $c = (\alpha - \gamma)/(\gamma - \beta)$ exist. If $0 < c < \infty$ then (1) yields, as $x \to \infty$,

$$(2) \qquad x^{\beta-\alpha} \mu_\alpha(x)/\int_x^\infty y^{\beta-\alpha-1} \mu_\alpha \, (dy) \to (\alpha - \beta)/(c + 1) = \gamma - \beta.$$

Using **B**(i), it follows that μ_α varies regularly with exponent $\alpha - \gamma > 0$ while, by (1), ν_β varies regularly with exponent $\beta - \gamma < 0$. If $c = 0$ the same argument shows that μ_α varies slowly but yields nothing about ν_β. Similarly, if $c = \infty$ then ν_β varies slowly but nothing can be asserted about μ_α.

The proof is terminated.

25.2 Domains of attraction. Throughout this subsection, $X_1, X_2,$ $\cdots$ *are iid r.v.'s with common law* $\mathcal{L}(X)$, *d.f.* F, *ch.f.* f *and* $S_n = X_1 + \cdots$ $+ X_n$, $n = 1, 2, \cdots$; *we take* $x > 0$ *and set* $\mu_2(x) = \int_{-x}^{x} y^2 dF(y)$, $q(x) = 1 - F(x) + F(-x)$.

We say that $\mathcal{L}(X)$ belongs to the domain of attraction of a law $\mathcal{L}(Y)$ or is *attracted by* $\mathcal{L}(Y)$—an attracting law, if there are a_n and $b_n > 0$ such that $\mathcal{L}(S_n/b_n - a_n) \to \mathcal{L}(Y)$. We *exclude the trivial case of degenerate attracting laws* $\mathcal{L}(Y)$ for, according to 14.2, every $\mathcal{L}(X)$ belongs to its domain of attraction with suitable a_n and b_n, and this excludes consideration of degenerate $\mathcal{L}(X)$. In fact, always according to 14.2, the above definition pertains not to individual laws but to *types of laws*.

In terms of *ch.f.'s*, $\mathcal{L}(X)$ is attracted by $\mathcal{L}(Y)$ nondegenerate means that, for every $u \in R$,

$$e^{iua_n} f^n(u/b_n) \to f_Y(u) \text{ nondegenerate.}$$

Thus, *ch.f.'s* $|f(u/b_n)|^2 \to |f_Y(u)|^2$, so that $|f_Y(u/b_n)|^2 \to 1$ with nondegenerate $|f_Y|^2$ hence $b_n \to \infty$. It follows that also $\mathcal{L}(S_n/b_{n+1} - a_n) \to \mathcal{L}(Y)$, that is, $|f(u/b_{n+1})|^2 \to |f_Y(u)|^2$ and, by the Corollary to 14.2**A**, $b_{n+1}/b_n \to 1$:

a. *If* $\mathcal{L}(S_n/b_n - a_n) \to \mathcal{L}(Y)$ *nondegenerate, then* $b_n \to \infty$ *and* $b_{n+1}/b_n \to 1$.

Since $f(u/b_n) \to 1$, 24.5**C** applies with $X_{nk} = X_k/b_n$ hence $F_n(x) = F(b_n x)$, $n = 1, \cdots, n$, and

b. $\mathcal{L}(S_n/b_n - a_n) \to \mathcal{L}(Y)$ *nondegenerate—necessarily i.d. with* $\psi = (\alpha, \beta^2, L)$, *if and only if*,

(C_L): $L_n \xrightarrow{w} L$ *where* $L_n(-x) = nF(-b_n x)$, $L_n(x) = n(F_n(x) - 1)$.

(C_{β^2}): $n\mu_2(b_n x)/b_n^2 = n\int_{-x}^{x} y^2 dF(b_n x) \to \beta^2$ *as* $n \to \infty$ *then* $x \to 0$.

(C_α): $a_n = \alpha_n - \alpha + o(1)$ *where* $\alpha_n = n\int \dfrac{x}{1 + x^2} dF(b_n x)$.

With the help of these lemmas, used without further comment, we begin by investigating condition (C_L) and its implications for Lévy functions of (nondegenerate) attracting laws $\mathcal{L}(Y)$. Clearly, the Lévy function for normal $\mathcal{L}(Y)$ is $L_2 \equiv 0$, and conversely. The others are given by

A. LEVY FUNCTIONS AND (C_L). *Let $x > 0$.*

(i) *Lévy functions L_γ of nonnormal attracting laws (Y) are given by*

$$L_\gamma(-x) = cp/x^\gamma, \quad L_\gamma(x) = -cq/x^\gamma$$

where

$$0 < \gamma < 2, \quad c > 0, \quad p, q \geqq 0 \text{ with } p + q = 1.$$

(ii) *Condition $(C_{L\gamma})$ is: as $x \to \infty$*

$$F(-x)/q(x) \to p \quad or \quad (1 - F(x))/q(x) \to 1 - p$$

and

$$q(x) = (c + o(1))h(x) \text{ where } h(x) \text{ varies slowly.}$$

The admissible b_n are characterized by $nq(b_n x) \to c$ as $n \to \infty$.

Proof. Condition (C_L) reads: for $\pm x \in C(L)$, as $n \to \infty$,

(1) $\qquad nF(-b_n x) \to L(-x) \quad and \quad$ (2) $\quad n(F(b_n x) - 1) \to L(x)$

hence

(3) $\qquad\qquad\qquad nq(b_n x) \to L(-x) - L(x).$

In fact, any two of these three relations clearly imply the remaining one.

1°. Since $L \equiv 0$ is excluded, there is an $x_0 > 0$ such that $L(-x_0) - L(x_0) > 0$ hence $L(-x) - L(x)$, being nonincreasing with increasing x, is positive for $x \in (0, x_0]$. It follows that the Corollary of 25.1A applies to (3) so that, setting $L_\gamma \equiv L$, as $n \to \infty$,

(4) $\qquad\qquad nq(b_n x) \to L_\gamma(-x) - L_\gamma(x) = c/x^\gamma$

with $c > 0$ and $\gamma > 0$.

On the other hand, upon changing in (1) and (3) the fixed x into fixed y and for every $x > 0$ selecting n to be the smallest integer such that $b_n y \leqq x \leqq b_{n+1} y$, we obtain

$$\frac{n}{n+1} \cdot \frac{(n+1)F(-b_{n+1}y)}{nq(b_n y)} \leqq \frac{F(-x)}{q(x)} \leqq \frac{n+1}{n} \cdot \frac{nF(-b_n y)}{(n+1)q(b_{n+1}y)}.$$

Upon letting $x \to \infty$ so that $n \to \infty$, the extreme sides converge to $p = L_\gamma(-y)/(L_\gamma(-y) - L_\gamma(y))$ so that

(5) $F(-x)/q(x) \to p$ with $0 \leq p \leq 1$

equivalently

(6) $(1 - F(-x))/q(x) \to 1 - p$.

Thus, replacing in (5) x by $b_n x$ with x arbitrary but fixed, as $n \to \infty$,

$$nF(-b_n x)/nq(b_n x) \to p$$

hence, by (4) and (1),

(7) $nF(-b_n x) \to L_\gamma(-x) = cp/x^\gamma$

and, similarly,

(8) $n(1 - F(b_n x)) \to L_\gamma(x) = cq/x^\gamma$.

Since the requirement for any Lévy function, $\displaystyle\int_{-x}^{x} y^2 dL_\gamma(y)$ finite, is satisfied if and only if $\gamma < 2$, we must have $0 < \gamma < 2$. Thus (i)—the asserted form of Lévy functions L_γ of nonnormal attracting laws $\mathcal{L}(Y)$—is established.

  2°. Condition (C_{L_γ}) became:

(5) $\displaystyle\lim_{x \to \infty} F(-x)/q(x) = p, \quad 0 \leq p \leq 1$,

and

(4) $\displaystyle\lim_{n} nq(b_n x) = c/x^\gamma, \quad c > 0, \quad 0 < \gamma < 2$.

According to the Corollary of 25.1A (4) implies

(9) $q(x) = (c + o(1))h(x)$ with $h(x)$ varying slowly.

On the other hand, setting $x = 1$ in (4), the scale factors b_n must satisfy

(10) $\displaystyle\lim_{n} nq(b_n) = c > 0$.

Thus, if $\mathcal{L}(S_n/b_n - a_n) \to \mathcal{L}(Y)$ nonnormal then (5), (9) and (10) hold.

  Conversely, let (5), (9) and (10) hold. From (10) it follows that $b_n = inf\{x: q(x - 0) \geq c/n \geq q(x + 0)\} \to \infty$ hence

$$\lim_{n} nq(b_n x)/c = \lim_{n} q(b_n x)/q(b_n) = x^{-\gamma} \lim_{n} h(b_n x)/h(b_n) = x^\gamma,$$

and $\lim_n nq(b_n x) = c/x^\gamma$. Thus, (C_{L_γ}) holds and admissible b_n satisfy

(10). The proof is terminated.

REMARK. Clearly, (C_{L_γ}) can also be stated in a more symmetric form:

$$F(-x) = cp(1 + o(1))h(x)/x^\gamma \quad \text{and} \quad 1 - F(x) = cq(1 + o(1))h(x)/x^\gamma.$$

In order to complete **A** we need more of Karamata theory. We write v.s. for "varies slowly."

*c. SLOW VARIATION LEMMA. *Let $x \to \infty$.*
(i) *If $\mu_2(\infty) = \infty$ then*
for $0 < \gamma < 2$:

$$x^2 q(x)/\mu_2(x) \to (2 - \gamma)/\gamma \Leftrightarrow \mu_2(x)/x^{2-\gamma} \text{ v.s.} \Leftrightarrow x^\gamma q(x) \text{ v.s.}$$

for $\gamma = 2$:

$$x^2 q(x)/\mu_2(x) \to 0 \Leftrightarrow \mu_2(x) \text{ v.s.}$$

(ii) $$0 < \mu_2(x) < \infty \begin{array}{l} \nearrow x^2 q(x)/\mu_2(x) \to 0 \\ \searrow \mu_2(x) \text{ v.s.} \end{array}$$

Proof. If $\mu_2(\infty) = \infty$ then (i) follows from 25.1D with $G(x) = F(x) - F(-x)$, $\alpha = 2$, and $\beta = 0$, so that $\nu_0(x) = q(x)$.

If $0 < \mu_2(\infty) < \infty$ then $x^2 q(x) \leq \int_{|y| \geq x} y^2 dF(x) \to 0$ consequently $x^2 q(x)/\mu_2(x) \to 0$ while, clearly, $\mu_2(tx)/\mu_2(x) \to 1$ as $t \to \infty$, that is, $\mu_2(x)$ varies slowly.

REMARK. Recall that when $0 < \mu_2(x) < \infty$ then, taking X centered at its expectation and setting $\sigma^2 = \sigma^2 X = \mu_2(\infty)$, $\mathcal{L}(S_n/\sigma \sqrt{n}) \to \mathfrak{N}(0,1)$ since

$$f^n(u/\sigma\sqrt{n})^n = \left(1 - \frac{u^2}{2}(1 + o(1))\right)^n \to e^{-u^2/2}.$$

Thus, when $0 < \mu^2(\infty) < \infty$ then $\mathcal{L}(X)$ is attracted by normal $\mathcal{L}(Y)$, and other types of attracting laws may happen *only when* $\mu_2(\infty) = \infty$.

We say that $\mathcal{L}(X)$ is *stable* if, for every n, there are a_n and $b_n > 0$ such that $\mathcal{L}(S_n/b_n - a_n) = \mathcal{L}(X)$; clearly, stable laws are attracted by themselves. Note that these are "stable" laws introduced in 24.4. We write $L_\gamma(c,p)$ for L_γ characterized by c and p as in **A**(i).

B. Stability and attraction criteria. *Let $x > 0$.*

(i) *The family of all nondegenerate attracting laws consists of all nondegenerate stable laws. They are i.d. laws with $\psi_\gamma = (\alpha, \beta_\gamma^2, L_\gamma)$, $0 < \gamma \leqq 2$ and*

or $0 < \gamma < 2$:

$$\beta_\gamma^2 = 0, \quad L_\gamma(-x) = cp/x^\gamma, \quad L_\gamma(x) = cq/x^2,$$

where $c > 0$, $p, q \geqq 0$, $p + q = 1$;

for $\gamma = 2$:

$$\beta_2^2 > 0, \quad L_2 \equiv 0.$$

(ii) *$\mathcal{L}(X)$ is attracted by some $\mathcal{L}_\gamma$ with given $\gamma \in (0,2]$ if and only if, as $x \to \infty$,*

$$x^2 q(x)/\mu_2(x) \to (2 - \gamma)/\gamma.$$

$\mathcal{L}(X)$ is attracted by $\mathcal{L}_\gamma$ with given $L_\gamma(c,p)$ if and only if, as $x \to \infty$,

for $0 < \gamma < 2$:

$$F(-x)/q(x) \to p, \quad q(x) = c(1 + o(1))h(x)/x^\gamma$$

where $h(x)$ varies slowly, and admissible b_n are characterized by $nq(b_n) \to c$ as $n \to \infty$;

for $\gamma = 2$:
$\mu_2(x)$ varies slowly and admissible b_n are characterized by $n\mu_2(b_n)/b_n^2 \to \beta_2^2 > 0$ as $n \to \infty$.

In either case, admissible a_n are characterized by

$$a_n = \alpha_n - \alpha + o(1) \text{ where } \alpha_n = n \int \frac{x}{1 + x^2} \, dF(b_n x).$$

Proof. Stability assertion is immediate. For, every stable law is attracted by itself while, conversely, the attracting laws $\mathcal{L}_\gamma$ are stable for $b_n = n^{1/\gamma}$: use the form of Lévy functions L_γ in **A**(i).

In **A**, we already found, for $0 < \gamma < 2$, (C_{L_γ}) and the L_γ as well as a characterization of admissible b_n. It remains to examine

$$(C_{\beta^2}): \quad n\mu_2(b_n x)/b_n^2 \to \beta_\gamma^2 \quad \text{as} \quad n \to \infty \quad \text{then} \quad x \to 0,$$

and to find admissible b_n for $\gamma = 2$.

1°. *Nonnormal case*: $0 < \gamma - 2$. (C_{L_γ}) is given by: as $x \to \infty$

(1) $$F(-x)/q(x) \to p, \quad 0 \leqq p \leqq 1,$$

and

(2) $q(x) = c(1 + o(1))h(x)/x^\gamma$, $h(x)$ slowly varying, $c > 0$,

or, when not specifying c,

(3) $\qquad\qquad\qquad\qquad x^\gamma q(x)$ varies slowly,

while admissible b_n are characterized by

(4) $\qquad\qquad\qquad nq(b_n) \to c > 0$ as $n \to \infty$.

We must have $\mu_2(\infty) = \infty$, for $0 < \mu_2(\infty) < \infty$ implies normality, that is, $\gamma = 2$ with $L_2 \equiv 0$. But then $\mathbf{c}(i)$ applies and (3) is equivalent to: as $x \to \infty$,

(5) $\qquad\qquad\qquad x^2 q(x)/\mu_2(x) \to (2 - \gamma)/\gamma$

and to

(6) $\qquad\qquad\qquad \mu_2(x)/x^{2-\gamma}$ varies slowly.

Upon replacing x by b_n in (5) and using (4), as $n \to \infty$, we obtain

(7) $\qquad\qquad n\mu_2(b_n)/b_n^2 \to c' = c\gamma/(\gamma - 2) > 0.$

But (6) implies that as $n \to \infty$

$$\frac{n\mu_2(b_n x)/b_n^2}{n\mu_2(b_n)/b_n^2} \to x^{2-\gamma}$$

hence, by (7),

(8) $\qquad\qquad\qquad n\mu_2(b_n x)/b_n^2 \to c'x^{2-\gamma}.$

Therefore, for $0 < \gamma < 2$, $(C_{\beta 2})$ becomes

$$0 \leftarrow n\mu_2(b_n x)/b_n^2 \to \beta_\gamma^2 \quad \text{as} \quad n \to \infty \quad \text{then} \quad x \to 0,$$

and we have the asserted $\psi_\gamma = (\alpha, 0, L_\gamma)$, and convergence.

2°. *Normal case*: $\gamma = 2$. Nondegenerate normal laws correspond to $\psi_2 = (\alpha, \beta_2^2, 0)$ with $\beta_2^2 > 0$. (C_{L_2}) and $(C_{\beta 2})$ become: as $n \to \infty$,

(1) $\qquad\qquad\qquad\qquad nq(b_n x) \to 0$

and

(2) $\qquad\qquad n\mu_2(b_n x)/b_n^2 \to \beta_2^2$, $0 < \beta_2^2 < \infty$;

setting $x = 1$, admissible b_n are characterized by

(3) $\qquad\qquad\qquad n\mu_2(b_n)/b_n^2 \to \beta_2^2.$

If $0 < \mu_2(\infty) < \infty$ so that, by c(ii), $\mu_2(\infty)$ varies slowly and $x^2q(x)/\mu_2(x) \to 0 = (2 - \gamma/\gamma$ for $\gamma = 2$, then it is easily seen that for X centered at its expectation, (1) and (2) hold with $b_n = \sigma\sqrt{n}$, $\sigma^2 = \sigma^2 X = \mu_2(\infty)$, and we have the required convergence (as we already knew). Thus, it remains to consider the case $\mu_2(\infty) = \infty$. Then, by c(i), as $x \to \infty$, $x^2q(x)/\mu_2(x) \to 0$ is equivalent to $\mu_2(x)$ varying slowly.

Let $\mu_2(x)$ vary slowly so that $\mu_2(x)/x^2 \to 0$ as $x \to \infty$. Then (3) holds for $b_n = \sup\{x: \mu_2(x)/x^2 \geq \beta_2{}^2/n\}$ and $b_n \to \infty$, so that $\lim_n \mu_2(b_nx)/\mu_2(b_n)$

$= 1$ becomes, by (3), $n\mu_2(b_nx)/b_n{}^2 \to \beta_2{}^2 > 0$, that is, (2) holds. Since $\lim_{x\to\infty} x^2q(x)/\mu_2(x) = 0$, upon replacing therein x by b_nx with $x > 0$ arbitrary but fixed, we have

$$\lim \frac{nq(b_nx)}{n\mu_2(b_nx)/b_n{}^2} = 0$$

hence, by (2), $\lim_n nq(b_nx) = 0$, that is, (1) holds. Thus, $\mu_2(x)$ varying slowly implies $\mathcal{L}(S_n/b_n - a_n) \to \mathcal{L}_2$ for admissible a_n.

Conversely, let $\mathcal{L}(S_n/b_n - a_n) \to \mathcal{L}_2$, so that (1) and (2) hence (3) hold. We prove that $\mu_2(x)$ varies slowly, that is, $(\mu_2(xt) - \mu_2(x))/\mu_2(x) \to 0$ as $x \to \infty$ for, say, $t > 1$. Let $x \to \infty$ and let n be such that $b_n \leq x < b_n$ so that $n \to \infty$. Then, since $b_n \to \infty$, (3) implies that $n\mu_2(x)/b_n{}^2 \to \beta_2{}^2 > 0$, that is, $\mu_2(x) \sim \beta_2{}^2b_n{}^2/n$. Since $b_{n+1}/b_n \to 1$, by (1),

$$\mu_2(xt) - \mu_2(t) \leq \int_{tb_{n+1}}^{b_n} x^2\,dq(x) \leq t^2b_{n+1}{}^2q(b_n)$$
$$= t^2(b_{n+1}{}^2/b_n{}^2)(b_n{}^2/n)nq(b_n) = o(b_n{}^2/n),$$

and the assertion follows.

The proof is terminated.

Consequences

1°. *For stable laws*
$$\psi_\gamma(u) = i\alpha u - c\,|u|^\gamma(1 - bh_\gamma(u)), \quad u \in R, \quad 0 < \gamma \leq 2, \quad \text{with } c > 0$$
($c = 0$ *for degenerate laws*), $b = p - q$ *hence* $|b| \leq 1$ *and*

$$h_\gamma(u) = \tan\frac{\pi}{2}\gamma \text{ or } \frac{2}{\pi}\log|u| \text{ according as } \gamma \neq 1 \text{ or } \gamma = 1.$$

Follows from **B**(i) by the computations in part 2° of the proof of 24.4**B** where β is replaced by cp, β' by cq, and b and c are interchanged.

2°. *Nondegenerate stable d.f.'s* F_γ *are infinitely differentiable and* $|F_\gamma{}^{(n)}| \leq |F_\gamma{}^{(n)}(0)|$ *positive, for every* $n = 1, 2, \cdots$.

Proof. Since, by 1°, $|f_2(u)| = \exp(-c|u|^\gamma)$ with $0 < \gamma \leq 2, c > 0, f_\gamma$ is integrable and so are the functions with values $u^n |f_\gamma(u)|$ for every n. Therefore, the inversion formula becomes

$$F_\gamma(x) - F_\gamma(a) = \frac{1}{2\pi} \int \frac{e^{-iua} - e^{-iux}}{iu} f_\gamma(u) \, du$$

and we can differentiate n times under the integral sign for the integral so obtained is absolutely convergent so that

$$F_\gamma^{(n)}(x) = \frac{(-i)^{n-1}}{2\pi} \int u^{n-1} e^{-iux} f_\gamma(u) du.$$

It follows that $|F_\gamma^{(n)}(x)| \leq |F_\gamma^{(n)}(0)| > 0$.

Let $q(x) = 1 - F_\gamma(x) + F_\gamma(-x)$ and let $\mathcal{L}_\gamma$ be nondegenerate.

3°.   *If $\mathcal{L}_\gamma$ is a stable law with $0 < \gamma < 2$, then $x^\gamma q(x) \to c > 0$ as $x \to \infty$.*

Proof. We know that $\mathcal{L}_\gamma$ attracts itself with scale factors $b_n = n^{1/\gamma}$ (also true for $\gamma = 2$); this also follows from $|f_\gamma(u)| = \exp(-c|u|^\gamma)$ since $|f_\gamma^n(n^{1/\gamma}u)| = |f_\gamma(u)|$. Therefore, by **B**(ii), replacing b_n by $n^{1/\gamma}$ in $nq(b_n) \to c > 0$, we have $(n^{1/\gamma})^\gamma q(n^{1/\gamma}) \to c$. Since $q(x)$ is nonincreasing with x increasing, taking $n^{1/\gamma} \leq x \leq (n+1)^{1/\gamma}$, we obtain

$$\frac{n+1}{n} \cdot nq((n+1)^{1/\gamma}) \leq x^\gamma q(x) \leq \frac{n}{n+1} \cdot nq(n^{1/\gamma}),$$

where the extreme terms tend to c as $x \to \infty$ hence $n \to \infty$, and $x^\gamma q(x) \to c$.

4°.   *If $\mathcal{L}(X)$ is attracted by $\mathcal{L}_\gamma$ then*
(i) $E |X|^r < \infty$ *for* $0 \leq r < \gamma \leq 2$
(ii) $E |X|^r = \infty$ *for* $r > \gamma$ *when* $0 < \gamma < 2$.
If $\mathcal{L}(X) = \mathcal{L}_\gamma$ with $0 < \gamma < 2$, then $E |X|^r$ is finite or infinite according as $0 \leq r < \gamma$ or $r \geq \gamma$.

Proof. If $\mu_2(\infty) = EX^2 < \infty$ then, by 9.3a, $E |X|^r < \infty$ for $r < \gamma = 2$, while $E |X|^r$ may be finite or infinite for $r > 2$. This shows why $\gamma = 2$ is to be excluded from (ii) and also that it suffices to prove (i) when $\mu_2(\infty) = \infty$—even for $\gamma = 2$. Then, by **c**, as $x \to \infty$,

$$x^\gamma q(x) = c(1 + o(1))h(x),$$

where $h(x)$ is slowly varying hence, by the Corollary of 25.1C, given $\delta > 0$ there is an a such that, for $x \geq a$,

$$x^{-\delta} < h(x) < x^\delta.$$

On the other hand, by integration by parts,

$$E| X |^r = \int_{-\infty}^{+\infty} | X |^r dF(x) = r \int_0^{\infty} x^{r-1} q(x) \, dx$$

so that $E| X |^r$ is finite or infinite according as $\int_a^{\infty} x^{r-1-\gamma} h(x) dx$ is finite or infinite. Since, given $\delta > 0$, for $x \geqq a$,

$$x^{r-\gamma-\delta-1} < x^{r-1-\gamma} h(x) < x^{r-\gamma-\delta-1},$$

it follows that $E \mid X \mid^r < \infty$ when $\lim_{x \to \infty} x^{r-\gamma-\delta} < \infty$ and $E \mid X \mid^r = \infty$ when $\lim_{x \to \infty} x^{r-\gamma-\delta} = \infty$.

If $0 \leqq r < \gamma \leqq 2$ then there is a positive $\delta < \gamma - r$, the first limit is finite, $E \mid X \mid^r < \infty$ and (i) is proved.

If $r > \gamma$ with $\gamma < 2$ then there is a positive $\delta < r - \gamma$, the second limit is infinite, $E \mid X \mid^r = \infty$ and (ii) is proved. It remains to show that when $\mathcal{L}(X) = \mathcal{L}_\gamma$ with $0 < \gamma < 2$ then $E \mid X \mid^\gamma = \infty$. Since, by 3°,

$$x^\gamma q(x) \to c > 0, x^{\gamma-1} q(x) \sim cx^{-1} \text{ for } x \to \infty \text{ so that } \int_a^{\infty} x^{\gamma-1} q(x) \, dx = \infty,$$

$E \mid X \mid^\gamma = \infty$ and the proof is concluded.

§ 26.   RANDOM WALK

Random walks—sequences of consecutive sums of iid summands, are present, in various guises and various degrees of generality, in an incredibly huge literature of applications of pr. theory to a very large number of concrete problems: queuing processes connected with mass service, dams, waiting times, renewal processes connected with storage and inventories, risk theory, traffic flow, particle counters, and many others. The present general random walk theory is relatively recent.

In 1921, Polya discovers "recurrence" and "nonrecurrence" phenomena in his study of some simple random walks on lattices in $R, R^2,$ and R^3. Thirty years later, in a definitive work, Chung and Fuchs settle this dichotomy problem for general random walks. Fluctuation r.v.'s defined on the n first terms of the random walk appear in the concrete problems mentioned above. But it is only in 1949 that Andersen begins his investigations into these r.v.'s for the general random walk. Since then a large number of results were obtained by many authors. They use variants of either the combinatorial or the analytic methods.

The combinatorial method initiated by Andersen threw the doors wide open. His approach was very involved. Spitzer simplified and unified the combinatorial approach and obtained some of the most important identities and limit theorems of the theory. His book, while devoted to random walk on lattices only, contains a number of deep ideas and significant examples. Feller, using ladder indices and ladder variables, first introduced by Blackwell, reduced the combinatorial approach to elementary mathematical arguments and using Feller's approach, Port, in a semi-expository paper, obtained a large number of known identities and generalized some of them.

The analytic method, as used by Pollaczec since 1930, was very involved and his work remained unnoticed until some of his results were rediscovered. Ray, Kemperman, Baxter, Wendell, . . . , simplified and unified in various ways the analytic approach and obtained further identities. Kemperman's book presents in detail the approach based on Liouville's theorem (already used by Pollaczec) and contains a large number of examples. Baxter uses a method based on Fourier-Stieltjes transforms and operators on functional Banach spaces. Wendel introduces and investigates "order statistics" of $(S_1, \cdots, S_n)$, . . .

No attempt will be made here to apply the general random walk theory to concrete problems. The interested reader will find in Feller's two volumes a large number of such problems.

26.1 Set-up and basic implications. A sequence $\bar{S} = (S_1, S_2, \cdots)$ of r.v.'s is called a *random walk* (on R) if the sequence of its *random steps* $\bar{X} = (X_1 = S_1, X_2 = S_2 - S_1, \cdots)$ at *times* $n = 1, 2, \cdots$ consists of *iid* r.v.'s $X_1, X_2, \cdots$. A random walk determines the sequence of its random steps, and conversely; similarly for the sub σ-fields of events:

$$\mathcal{B}_n = \mathcal{B}(X_1, \cdots, X_n) = \mathcal{B}(S_1, \ldots, S_n),$$
$$\mathcal{C}_n = \mathcal{B}(X_{n+1}, X_{n+2}, \cdots) = \mathcal{B}(S_{n+1} - S_n, S_{n+2} - S_n, \cdots).$$

We denote by $\mathcal{B}_\infty = \mathcal{B}(X_1, X_2, \cdots)$ the smallest σ-field generated by the field $\bigcup_{n=1}^{\infty} \mathcal{B}_n$ and $\mathcal{C} = \bigcap_{n=1}^{\infty} \mathcal{C}_n$ is the tail σ-field of the sequence $\bar{X}$; it is important to realize that, in general, $\mathcal{C}$ is not the tail σ-field $\bigcap_{n=1}^{\infty} (S_{n+1}, S_{n+2},$

$\cdots)$ of the sequence $\bar{S}$.

We shall frequently adjoin $S_0 = 0$ to the random walk so that it will become $(S_0, S_1, S_2, \cdots)$ with steps $X_n = S_n - S_{n-1}$, $n = 1, 2, \cdots$. Intuitively it means that the random walk starts at time 0 at the origin. We could also make it start at some $x \in R$ or choose S_0 to be a r.v. If the random steps obey a law $\mathcal{L}(X)$ with only values $\pm nd$, $d > 0$, $n = 0$, $1, \cdots$, then we have a very simple Markov chain with countable state space, $0, \pm d, \pm 2d, \cdots$, and initial position 0, or some $n_0 d$, or a r.v. with law $\mathcal{L}(X)$. It is strongly recommended that the reader interpret the corresponding concepts and results in III of the Introductory Part in the case of random walk theory, as found in this section.

The common law of the random steps will be denoted by $\mathcal{L}(X)$, its d.f. on R and corresponding pr. distribution on the Borel line will be denoted by the same symbol F, and its ch.f. will be f. D.f.'s and corresponding pr. distributions of their sums S_n-"positions" of the random walk at times n, will be denoted by F_n and their ch.f.'s are f^n, $n = 1$, $2, \cdots$. If $\mathcal{L}(X)$ degenerates at 0 then the random walk stays a.s. at $\{0\}$; *from now on we exclude this trivial case.* Note that if $\mathcal{L}(X)$ degenerates at $a \neq 0$ then the random walk moves a.s. by degenerate steps a from na to $(n+1)a$, $n = 1, 2, \cdots$, and $S_n \xrightarrow{\text{a.s.}} + \infty$ or $S_n \xrightarrow{\text{a.s.}} - \infty$ according as $a > 0$ or $a < 0$.

We distinguish two types of common laws $\mathcal{L}(X)$. Let $L_d = \{nd : n = 0, \pm 1, \pm 2, \cdots\}$ be a lattice of span $d > 0$. We say that X is L_d-*distributed* if $\sum_{n=-\infty}^{+\infty} P(X = nd) = 1$ and there is no lattice of larger span $d' > d$ with this property; according to the remark following 14.1a such a distribution occurs if and only if $|f(u)| = 1$ for some $u \neq 0$. If there is no $d > 0$ such that X is L_d-distributed, we set $d = 0$, $L_0 = R$, and say that X is L_0-*distributed*; thus X is L_0-distributed if and only if $|f(u)| < 1$ for all $u \neq 0$.

We now examine basic implications of the above set-up.

POSSIBLE VALUES AND STATES. We say that $x \in R$ is a *possible value* of a r.v. X if $P(X \in V_x) > 0$ for every neighborhood V_x of x. We say that x is a *possible state* of the random walk $\bar{S} = (S_1, S_2, \cdots)$ if for every given neighborhood V_x of x there is an $n = n(V_x)$ such that $P(S_n \in V_x) > 0$. In either case, it suffices to consider neighborhoods of the form $V_x = (x - \epsilon, X + \epsilon)$. Let Π_s denote the possible states of the random walk $\bar{S}$, let Π_n be the set of possible values of S_n, and set $\Pi_v = \bigcup_{n=1}^{\infty} \Pi_n$.

a. Π_s *contains* Π_v *and is closed.*

Proof. The first assertion results at once from the definitions. As for the second one, since $x_m \to x$ as $m \to \infty$ implies that, given $\epsilon > 0$, for m sufficiently large $(x_m - 1/m, x_m + 1/m) \subset (x - \epsilon, x + \epsilon)$, it follows that when the x_m are possible states there is an n such that

$$P(|S_n - x| < \epsilon) \geqq P(|S_n - x_m| < 1/m) > 0.$$

We say that x is a *discontinuity value* of X if $P(X = x) > 0$. Clearly, the set of discontinuity values of X is the set of jumps of the discontinuous part of $F_X{}^d$ of the *d.f.* F_X.

b. *If x and y are possible values of independent r.v.'s X and Y respectively, then $x + y$ is a possible value of $X + Y$.*

If x and y are discontinuity values of independent r.v.'s X and Y respectively then $x + y$ is a discontinuity value of $X + Y$, and all such values of $X + Y$ are of this form.

The first assertion obtains by

$$P(|X + Y - (x + y)| < \epsilon) \geqq P(|X - x| < \epsilon/2) \times P(|Y - y| < \epsilon/2) > 0$$

and the second one results from

$$(F_X * F_Y)^d = F_X{}^d * F_Y{}^d.$$

A. POSSIBLE VALUES THEOREM. *Let X be L_d-distributed with $d \geqq 0$.*
(i) *If neither $X \geqq 0$ a.s. nor $X \leqq 0$ a.s. then when $d > 0$, $\Pi_v = L_d$ and when $d = 0$, Π_v is dense in $L_0 = R$.*
(ii) *If either $X \geqq 0$ a.s. or $X \leqq 0$ a.s. then when $d > 0$, from some n on, nd or $-nd$, respectively, belong to Π_v and when $d = 0$, for every given $\epsilon > 0$, from some $x > 0$, Π_v intersects $(x, x + \epsilon)$ or $(-x - \epsilon, -x)$, respectively.*

Proof. We use b without further comment. We can assume that $S_1 = X_1$ has a positive value a so that $S_2 = X_1 + X_2$ has positive value $2a$; otherwise, we change X into $-X$. Thus, it suffices to prove the theorem when there are positive values $a < b$. We follow Feller.

1°. Set $J_n = [na, nb)$. For $n \geqq n_1 > a/(b - a)$, $[na, (n + 1)a) \subset J_n$ hence $\bigcup_{n \geqq n_1} J_n = [n_1 a, \infty)$ and every $x \geqq n_1 a$ belongs to some of the J_n for $n \geqq n_1$. Since the $n + 1$ points $na + k(b - a)$, $k = 0, \cdots, n$, belong to Π_v and subdivide J_n into intervals of length $b - a$, every $x \geqq n_1 a$ is at a distance at most $(b - a)/2$ from a member of Π_v.

2°. Suppose that for every given $\epsilon > 0$ there are possible values $(0 <)a < b$ with $b - a < \epsilon$. Then X is L_0-distributed for otherwise every Π_n hence Π_v is contained in L_d for some fixed $d > 0$ and we reach a contradiction.

If $X \geqq 0$ $a.s.$, the assertion in (ii) for $d = 0$ follows from 1°.

If neither $X \geqq 0$ $a.s.$ nor $X \leqq 0$ $a.s.$ then X has a possible value $c < 0$. Given $\epsilon > 0$, it follows from 1° that for arbitrary x and sufficiently large n there is a $y \in \Pi_v$ belonging to $(-nc + x, -nc + x + \epsilon)$. But $y + nc$ also belongs to Π_v. Thus, every interval of any given length $\epsilon > 0$ intersects Π_v so that Π_v is dense in $L_0 = R$ and the assertion in (i) for $d = 0$ is proved.

3°.   Suppose now that whichever be the possible values $(0 <)a < b$, there is an $\epsilon > 0$ such that $b - a \geqq \epsilon$; we may assume $b - a < 2\epsilon$ for some a and b. Then the set $J_n \Pi_v$ consists of points $na + k(b - a)$, $k = 0, \cdots , n$. Since $(n + 1)a$ is one of them, they all are multiples of $b - a$. But for any $c \in \Pi_v$, for n sufficiently large J_n has a point of the form $c + k(b - a)$ so that c is also a multiple of $b - a$. Thus X is L_d-distributed with some $d > 0$ and the proof is completed.

Corollary.   *Let X be L_d-distributed with $d \geqq 0$. If neither $X \geqq 0$ nor $X \leqq 0$ then the set of all possible states of the random walk coincides with L_d.*

Follows at once by **a**.

From now on, we take for Ω the set $\Omega = R^\infty$ of all numerical sequences $\bar{x} = (x_1, x_2, \cdots)$ and for the σ-field of events the σ-field $\mathfrak{B}$ of Borel sets in R^∞, that is, the σ-field generated by the class of all cylinders of the form $C(A_1 \times \cdots \times A_n)$, $n = 1, 2, \cdots$, where the A's are linear Borel sets. This choice does not restrict generality yet permits to avoid possible ambiguities, say, about "translations."

SLLN and 0-1 laws.

According to 17.4.4°

For $0 < r < 2$, $\dfrac{1}{n^{1/r}} \sum\limits_{k=1}^{n} (X_k - a_k) \overset{a.s.}{\longrightarrow} 0$, *with $a_k = 0$ or EX according as $r < 1$ or $r = 1$, if and only if $E|X|^r < \infty$.*

For r = 1, we have Kolmogorov strong law of large numbers, *SLLN* for short, which can be completed as follows (see also 34.4).

B.   **SLLN.**   *Let EX exist. Then $S_n/n \overset{a.s.}{\longrightarrow} EX$. Conversely, if $S_n/n \overset{a.s.}{\longrightarrow} c$ necessarily a constant (finite or infinite) then $EX = c$.*

Proof. It suffices to complete *SLLN* by considering the cases of infinite EX and c.

Let $EX = +\infty$, that is, $EX^+ = +\infty$, $EX^- < \infty$ and let $X_k(a) = X_k$ or $a \in R$ according as $X_k < a$ or $X_k \geqq a$. Set $S_n(a) = \sum_{k=1}^{n} X_k(a)$. Since $EX(a) < \infty$,

$$S_n/n \geqq S_n(a)/n \xrightarrow{a.s.} EX(a)$$

hence, letting $a \uparrow \infty$ so that, by monotone convergence theorem, $EX(a) \uparrow EX = +\infty$, we obtain $S_n/n \xrightarrow{a.s.} EX = +\infty$; similarly for $EX = -\infty$, or change X into $-X$.

For the converse, if $c = +\infty$ then, by what precedes,

$$EX^+ \leftarrow \frac{1}{n} \sum_{k=1}^{n} X_k^+ = \frac{1}{n} \sum_{k=1}^{n} X_k + \frac{1}{n} \sum_{k=1}^{n} X_k^- \rightarrow +\infty + EX^-$$

so that $EX^+ = +\infty$, hence $EX = +\infty$ since EX exists; similarly for $c = -\infty$, or change X into $-X$.

SLNN utilizes fully the *iid* property of the summands. Independence alone yields as we know (16.3B).

KOLMOGOROV ZERO-ONE LAW. *On a sequence of independent r.v.'s tail events have for pr. either 0 or 1 and tail functions are degenerate.*

This zero-one law, while applying to $\bar{X} = (X_1, X_2, \cdots)$, does not apply to the random walk $\bar{S} = (S_1, S_2, \cdots)$. Yet, the *iid* property of the summands implies "exchangeability", and a new zero-one law will apply to $\bar{S}$:

We say that a sequence $\bar{X} = (X_1, X_2, \cdots)$ of *r.v.'s is exchangeable* or that the *r.v.'s* $X_1, X_2, \cdots$ *are exchangeable* if the distribution of $\bar{X}$ is invariant under all finite exchanges of its terms or, equivalently, of their subscripts; in symbols, for every n and every one of the $n!$ permutations ω_n of $(1, \cdots, n)$ into $(k_1, \cdots, k_n)$,

$$\mathcal{L}(\bar{X}) = \mathcal{L}(\omega_n \bar{X}) = \mathcal{L}(X_{k_1}, \cdots, X_{k_n}, X_{n+1}, X_{n+2}, \cdots)$$

We say that a measurable function $g(\bar{X})$ *is exchangeable* if it is invariant under all permutations ω_n of its arguments: $g(\omega_n \bar{X}) = g(\bar{X})$, $n = 1$, $2, \cdots$; in particular, an event on $\bar{X}$ *is exchangeable* if its indicator is exchangeable. Clearly, on $\bar{X}$ every tail event and every tail function are exchangeable. In fact, by the iid property of its terms, $\bar{X}$ is exchangeable while, for every n, the sequences $(S_n, S_{n+1}, \cdots)$ are invariant under permutations ω_n of $(1, \cdots, n)$. Thus, the second assertion below follows at once, while the first one results directly from the definitions:

c. (i) *On $\overline{X}$, exchangeable events form a σ-field $\mathcal{E}$ and exchangeable functions are $\mathcal{E}$-measurable.*

(ii) *On the random walk $\overline{S}$ corresponding to the sequence $\overline{X}$ of iid steps, tail events are exchangeable (belong to $\mathcal{E}$) and tail functions are exchangeable (are $\mathcal{E}$-measurable).*

In general, tail events and tail functions on $\overline{S}$, say $[S_n \in A_n \text{ i.o.}]$ where A_n are linear Borel sets, liminf S_n, limsup S_n, while exchangeable, are not tail events on $\overline{X}$ and Kolmogorov zero-one law does not apply. Yet

B. HEWITT–SAVAGE ZERO-ONE LAW. *On a sequence of iid r.v.'s exchangeable events have for pr. either 0 or 1 and exchangeable functions are degenerate.*

To prove this theorem we require an elementary measure-theoretic proposition. Let $A \triangle B = AB^c + A^cB$.

d. APPROXIMATION LEMMA. *Let $(\Omega, \mathcal{Q}, P)$ be a pr. space. If a field $\mathcal{D}$ generates $\mathcal{Q}$ then for every given $A \in \mathcal{Q}$ and every $\epsilon > 0$ there is a $D \in \mathcal{D}$ such that $P(A \triangle D) \leq \epsilon$.*

For, clearly, the class of all sets $A \in \mathcal{Q}$ with the asserted property contains $\mathcal{D}$ and it is easily verified that this class is monotone; thus, by 1.6A, it coincides with $\mathcal{Q}$.

The approximation property can be restated as follows. Let $A \in \mathcal{Q}$ and $\epsilon_n \downarrow 0$. There are $D_n \in \mathcal{D}$ such that $P(A \triangle D_n) \leq \epsilon_n \to 0$, that is, $P(AD_n{}^c) \to 0$ and $P(A^cD_n) \to 0$. Therefore, $PD_n \to PA$ since $PA = PAD_n + PAD_n{}^c = PD_n - PA^cD_n + PAD_n{}^c$.

Proof of **B.** In our case $\mathcal{D} = \bigcup \mathcal{B}_n$ so that, given an exchangeable event A (in fact, any event) there is a sequence $B_n \in \mathcal{B}_{k_n}$ with $P(A \triangle B_n) \to 0$ hence $PB_n \to PA$; we can and do select $k_1 < k_2 < \cdots$. Let C_n be the events obtained from B_n by the permutation of $(1, \cdots, k_n, k_n + 1, \cdots, 2k_n)$ into $(k_n + 1, \cdots, 2k_n, 1, \cdots, k_n)$; thus, $\mathcal{B}_n \in \mathcal{B}_{k_n} = \mathcal{B}(X_1, \cdots, X_{k_n})$ implies $C_n \in \mathcal{C}_{k_n} = \mathcal{B}(X_{k_n+1}, X_{k_n+2}, \cdots)$ and, $\mathcal{B}_{k_n}$ and $\mathcal{C}_{k_n}$ being independent so are B_n and C_n. But this permutation leaves the distribution of $\overline{X}$ invariant while A, being exchangeable, remains the same and $A \triangle B_n$ is changed into $A \triangle C_n$ so that

$$P(A \triangle C_n) = P(A \triangle B_n) \to 0$$

hence $PC_n \to PA$; also

$$P(A \triangle B_nC_n) \leq P(A \triangle B_n) + P(A \triangle C_n) \to 0$$

hence $P(B_n C_n) \to PA$. Therefore, B_n and C_n being independent,

$$PA \leftarrow P(B_n C_n) = PB_n \cdot PC_n \to (PA)^2$$

so that $PA = 0$ or 1. The first assertion is proved and the second follows.

CONSEQUENCES.   1. $P[S_n \in A_n \text{ i.o.}] = 0$ or 1, liminf S_n and limsup S_n are degenerate.

2. THREE ALTERNATIVES.   For a (nondegenerate at 0) random walk $(S_1, S_2, \cdots)$ there are exactly three asymptotic alternatives:

   (i) $S_n \xrightarrow{a.s.} -\infty$ (drifts to $-\infty$)
   (ii) $S_n \xrightarrow{a.s.} +\infty$ (drifts to $+\infty$)
   (iii) $-\infty = \liminf S_n < \limsup S_n = +\infty$ a.s. (oscillates between $-\infty$ and $+\infty$).

Proof. Since liminf $S_n = c$ a.s. where the constant c may be finite or infinite, and $(S_2 - S_1, S_3 - S_1, \cdots)$ has the same distribution as $(S_1, S_2, \cdots)$, we have

$$\liminf (S_n - X_1) = \liminf S_n \text{ a.s.}$$

hence $c = X_1 + c$ a.s. The case $X_1 = 0$ a.s. being excluded (that is, is excluded the trivial alternative the random walk stays at 0 a.s.), we must have $c = +\infty$ or $c = -\infty$. Thus a.s.

$$\text{either liminf } S_n = -\infty \text{ or } \lim S_n = \liminf S_n = +\infty$$

and, changing X into $-X$, a.s.

$$\text{either limsup } S_n = +\infty \quad \text{or} \quad \lim S_n = \limsup S_n = -\infty.$$

The three alternatives assertion follows.

RANDOM TIMES.

Translations θ^n on $\overline{X} = (X_1, X_2, \cdots)$ *are defined by*

$$\theta^n \overline{X} = \theta^n(X_1, X_2, \cdots) = (X_{n+1}, X_{n+2}, \cdots), n = 1, 2, \cdots$$

so that
the terms of $\theta^n \overline{X}$ are iid with same common law $\mathfrak{L}(X)$ as the terms of $\overline{X}$.
Thus, $\theta^n \overline{X}$ has same distribution as $\overline{X}$ and therefore $\overline{X}$ is said to be *stationary* (see also 33.3). The random walks corresponding to $\overline{X}$ and to $\theta^n \overline{X}$ are, respectively, $(S_1, S_2, \cdots)$ and $(S_{n+1} - S_n, S_{n+2} - S_n, \cdots)$ with same distribution, and the σ-fields $\mathfrak{B}(S_1, \cdots, S_n) = \mathfrak{B}_n$ and $\mathfrak{B}(S_{n+1} - S_n, S_{n+2} - S_n \cdots) = \mathfrak{B}(X_{n+1}, X_{n+2}, \cdots) = \mathfrak{C}_n$ are independent.

These properties extend to "random times"—times $n(=1, 2, \cdots)$ becoming "degenerate" random times, as follows (see also 39.2 and 41.4 taking therein $T = (1, 2, \cdots)$ and $b = \infty$).

Given a nondecreasing sequence $(\mathfrak{B}_n)$ of sub σ-fields of events, a measurable function τ to $(1, 2, \cdots, \infty)$ is a $(\mathfrak{B}_n)$-*time* if $[\tau = n] \in \mathfrak{B}_n$, $n = 1, 2, \cdots$; if there is no confusion possible, we say that τ is a *random time*. Clearly, a random time τ is $\mathfrak{B}_\tau$-measurable with σ-field $\mathfrak{B}_\tau = \{$events $B: B[\tau = n] \in \mathfrak{B}_n, n = 1, 2, \cdots\}$. If $\tau < \infty$ a.s., we define $X_{\tau+k}(\omega)$ by $X_{\tau(\omega)+k}(\omega)$ so that the $X_{\tau+k}$ are r.v.'s, $k = 0, 1, \cdots$. Then the σ-field $\mathfrak{B}(X_{\tau+1}, X_{\tau+2}, \cdots)$ is denoted by $\mathfrak{C}_\tau$ and *translation by τ of $\overline{X}$* is defined by

$$\theta^\tau(X_1, X_2, \cdots) = (X_{\tau+1}, X_{\tau+2}, \cdots).$$

The above properties of translations by n remain valid as follows.

C. RANDOM TIMES TRANSLATIONS. *If a $(\mathfrak{B}_n)$-time $\tau < \infty$ a.s. then the σ-fields $\mathfrak{B}_\tau$ and $\mathfrak{C}_\tau$ are independent and the sequences $\overline{X} = (X_1, X_2, \cdots)$ and $\theta^\tau \overline{X} = (X_{\tau+1}, X_{\tau+2}, \cdots)$ have same distribution.*

Proof. The assertions mean that, for any pair of events $B_\tau \in \mathfrak{B}_\tau$ and $B \in \mathfrak{B}_\infty = \mathfrak{B}(X_1, X_2, \cdots)$,

(1) $$P(B_\tau[\theta^\tau \overline{X} \in B]) = PB_\tau P(\overline{X} \in B).$$

By definition, $\theta^\tau = \theta^n$ on $[\tau = n]$ hence

$$P(B_\tau[\theta^\tau X \in B]) = \sum_{n=1}^{\infty} P(B_\tau[\tau = n][\theta^n \overline{X} \in B]).$$

Since $B_\tau[\tau = n] \in \mathfrak{B}_n$ and $\theta^n \overline{X}$ is $\mathfrak{C}_n$-measurable, independence of $\mathfrak{B}_n$ and $\mathfrak{C}_n$ implies

$$P(B_\tau[\tau = n][\theta^n \overline{X} \in B]) = P(B_\tau[\tau = n]) \cdot P(\theta^n \overline{X} \in B).$$

Since $\sum_{n=1}^{\infty} P(\tau = n) = 1$, and $\theta^n \overline{X}$ has same distribution as $\overline{X}$, (1) becomes

$$P(B_\tau[\theta^\tau \overline{X} \in B]) = \sum_{n=1}^{\infty} P(B_\tau[\tau = n]) \cdot P(\overline{X} \in B) = PB_\tau \cdot P(\overline{X} \in B)$$

and the proof is terminated.

The above argument is characteristic of extensions of properties of times n to random times $\tau < \infty$ a.s.: use the definitions and the asserted property—valid on $[\tau = n]$, $n = 1, 2, \cdots$. For example, upon setting

$\tau_1 = \tau < \infty$ a.s. then defining τ_2 by $\tau_2 = \tau$ but on $\theta^{\tau_1}X$ in lieu of $\bar{X}$, and so on, it easily follows that

1. $S_{\tau_1}, S_{\tau_1+\tau_2} - S_{\tau_1}, \cdots$ *are iid r.v.'s.*

By the same procedure but with E in lieu of P, additivity of expectations, which for random walks becomes $ES_n = nEX$, extends to $\tau < \infty$ a.s. in lieu of n, upon using $E\tau = \sum_{n=1}^{\infty} nP(\tau = n) = \sum_{n=1}^{\infty} P(\tau \geq n)$ as follows.

2. WALD'S RELATION. $ES_\tau = E\tau \cdot EX$ *in the sense that if the right side exists so does the left one and then both are equal.*

Note that the right side exists when $E\tau < \infty$ and EX exists, or when $E\tau = \infty$ and EX is finite or $EX \geq 0$ or $EX \leq 0$.

Proof. Let $0 \leq EX \leq \infty$ and $E\tau \leq \infty$. Then

$$ES_\tau = \sum_{n=1}^{\infty} E(S_n[\tau = n]) = \sum_{n=1}^{\infty} \sum_{k=1}^{n} E(X_k[\tau = n])$$

$$= \sum_{k=1}^{\infty} \sum_{n=k}^{\infty} E(X_k[\tau = n]) = \sum_{k=1}^{\infty} E(X_k[\tau \geq k]$$

$$= \sum_{k=1}^{\infty} EX_k P[\tau \geq k] = EX \cdot E\tau.$$

The last but one equality is due to the fact that $[\tau < k]$ belongs to $\mathcal{B}_{k-1}$ hence so does its complement $[\tau \geq k]$, while X_k is $\mathcal{C}_{k-1}(=\mathcal{B}(X_k, X_{k+1}, \cdots))$-measurable, and $\mathcal{B}_{k-1}$ and $\mathcal{C}_{k-1}$ are independent. Changing X into $-X$ the same relation holds. The other cases follow from $EX = EX^+ - EX^-$ with EX^+ or EX^- finite.

We shall frequently encounter the *hitting* or *first visit* time τ_A of a linear Borel set A by a random walk $(S_1, S_2, \cdots)$:
$\tau_A(\omega) = \min\{n: S_n(\omega) \in A\}$ for $\omega \in \bigcup[S_n \in A]$ and $\tau_A(\omega) = \infty$ otherwise. Clearly τ_A is random walk time, since for every n,

$$[\tau_A = n] = [S_k \in A^c \text{ for } k < n, S_n \in A] \in \mathcal{B}(S_1, \cdots, S_n).$$

Similarly for other random times we shall encounter: In general, the fact that they are random walk times will be clear from their definitions.

ANDERSEN EQUIVALENCE.

"Finite exchangeability" alone suffices for a basic Andersen result for "finite fluctuations." We set $\bar{X}_n = (X_1, \cdots, X_n)$ and say that the random vector $\overline{X}_n$ is *exchangeable* or that its components $X_1, \cdots, X_n$ are *exchangeable* if the distribution of $\overline{X}_n$ is invariant under the $n!$ permutations of its components. We say that a measurable function $g(\overline{X}_n) = g(X_1, \cdots, X_n)$ is *exchangeable* if it is invariant under the $n!$ permutations of its arguments.

e. ANDERSEN EQUIVALENCE LEMMA. *Let $X_1, \cdots, X_n$ be exchangeable and let $S_0, \cdots, S_n$ be their partial sums $S_0 = 0$, $S_1 = X_1, \cdots, S_n = X_1 + \cdots + X_n$.*

If ν_n is the (random) number of positive terms in $(S_0, \cdots, S_n)$ and τ_n is the (random) time of occurrence of the first maximum of its terms, then ν_n and τ_n are identically distributed.

This result is an immediate consequence of a combinatorial lemma due to Feller whose elementary proof, modified by Joseph—as reported in Feller, follows.

f. COMBINATORIAL LEMMA. *To each permutation $(x_{k_1}, \cdots, x_{k_n})$ of $(x_1, \cdots, x_n)$ associate the sequence $0, x_{k_1}, \cdots, x_{k_1} + \cdots + x_{k_n}$ of its partial sums. Let $m = 0, 1, \cdots, n$.*

The number N_m of permutations with exactly m positive sums is the same as the number T_m of permutations in which the first maximum of partial sums occurs at time m.

Proof. Let N_{mk} and T_{mk} correspond to N_m and T_m when x_k is omitted in $(x_1, \cdots, x_n)$. We use induction: The assertion holds for $n = 1$ since, clearly, $x_1 \leqq 0$ implies $N_0 = T_0 = 1$ and $N_1 = T_1 = 0$ while $x_1 > 0$ implies $N_0 = T_0 = 0$ and $N_1 = T_1 = 1$. Suppose it holds for $n - 1 \geqq 1$, that is, $N_{mk} = T_{mk}$ for $k = 1, \cdots, n$ and $m = 0, \cdots, n - 1$; since trivially $N_{nk} = T_{nk} = 0$, it also holds for $m = n$.

We use the fact that by fixing x_k and permuting the $n - 1$ remaining x's then varying $k = 1, \cdots, n$, we obtain the $n!$ permutations of $x_1, \cdots, x_n)$.

If $s_n \leqq 0$ then N_m and T_m depend only on $x_1, \cdots, x_{n-1}$ hence, by induction hypothesis,

$$N_m = \sum_{k=1}^{n} N_{mk} = \sum_{k=1}^{n} T_{mk} = T_m.$$

If $s_n > 0$ then $N_m = \sum_{k=1}^{n} N_{m-1,k}$. As for T_m, consider all $(x_k, x_{k_1}, \cdots, x_{k_{n-1}})$ starting with x_k. Since $x_k + \cdots + x_{k_{n-1}} > 0$ the maximal terms of their partial sums cannot be s_0. Since the first maximum occurs for

$m (= 1, \cdots, n)$ if and only if the first maximum of partial sums of $(x_{k_1}, \cdots, x_{k_n-1})$ occurs for $m - 1$, we have

$$N_m = \sum_{k=1}^{n} N_{m-1,k} = \sum_{k=1}^{n} T_{m-1,k} = T_m.$$

By using an argument formulated by Spitzer, instead of proving **e** we can prove the more general.

D. EQUIVALENCE THEOREM. *Let $g(\overline{X}_n)$ be an integrable function of an exchangeable random vector $X_n = (\overline{X}_1, \cdots, X_n)$.*
If $g(\overline{X}_n)$ is exchangeable then, for $k = 0, 1, \cdots, n$,

$$E(g(\overline{X}_n)I_{[\nu_n=k]}) = E(g(\overline{X}_n)I_{[\tau_n=k]});$$

in particular,

$$E(e^{iuS_n}I_{[\nu_n=k]}) = E(e^{iuS_n}I_{[\tau_n=k]}), \; u \in R,$$

and

$$P[\nu_n = k] = P[\tau_n = k].$$

Proof. Let F_n be the d.f. of $\overline{X}_n$ and $\overline{X}_n = (x_1, \cdots, x_n)$.
Denote by Σ summations over the $n!$ permutations ω_n of $(1, \cdots, n)$. Since $g(\overline{X}_n)$ is exchangeable

$$E(g(\overline{X}_n)I_{[\nu_n=k]}) = \Sigma \frac{1}{n!} \int g(x_n)I_{[\nu_n=k]}(\omega_n\overline{x}_n)dF_n(\overline{x}_n)$$

and, by the combinatorial lemma,

$$\Sigma I_{[\nu_n=k]}(\omega_n x_n) = \Sigma I_{[\tau_n=k]}(\omega_n\overline{x}_n).$$

Thus the first sum equals the same sum but with τ_n in lieu of ν_n hence the expectation equals the one with τ_n in lieu of ν_n. The particular case with $g(\overline{X}_n) = e^{iuS_n}$ follows and then, setting $u = 0$, the last assertion—which is that of **e**, results.

By means of his equivalence, Andersen obtained his first limit theorem for finite fluctuations, namely

ARCSINE LAW. *Let $S_0(=0)$, $S_1, \cdots$ be partial sums of iid summands $X_1, X_2, \cdots$ with common law $\mathcal{L}(X)$.*
If $\mathcal{L}(X)$ is symmetric with $P(X = 0) = 0$ then

$$P(\nu_n/n < x) \to \frac{2}{\pi} \mathrm{Arcsin}\sqrt{x}, \; 0 \leqq x \leqq 1.$$

The Arcsine law was discovered by P. Levy in his study of Brownian motion, then obtained by Erdos and Kac as a limit theorem for sums of independent random variables with finite second moments and obeying Lindeberg condition (see also Chapter XII). Andersen's result which does not require second finite moments was unexpected and drew attention to his approach.

The proof is based upon the following considerations. The event $[\nu_n = k]$ consists in the occurrence of events $[S_k > S_0, \cdots, S_k > S_{k-1}]$ and $[S_{k+1} - S_k \leq 0, \cdots, S_n - S_k \leq 0]$. The first one belongs to $\mathcal{B}(X_1, \cdots, X_k)$ and the second one belongs to $\mathcal{B}(X_{k+1}, \cdots, X_n)$ and these two σ-fields are independent. Furthermore, $(X_{k+1}, \cdots, X_n)$ is distributed as $(X_1, \cdots, X_{n-k})$. It follows that

$$P(\nu_n = k) = P(\nu_k = k)P(\nu_{n-k} = 0)$$

and, by Andersen equivalence,

(1) $$P(\tau_n = k) = P(\tau_k = k)P(\tau_{n-k} = 0).$$

Let

$$p_n(k) = \frac{1}{2^{2n}} \frac{(2k)!}{k!k!} \frac{(2(n-k))!}{(n-k)!(n-k)!}, \quad k = 0, \cdots, n,$$

so that

$$p_n(k) = p_n(n-k), \quad \sum_{k=0}^{n} p_n(k) = 1.$$

We prove by induction that

(2) $$P(\nu_n = k) = p_n(k).$$

For $n = 1$, we have

$$P(\nu_1 = 0) = P(\nu_1 = 1) = \frac{1}{2} = p_1(0) = p_1(1).$$

If (2) holds for $n - 1$ hence, by (1), $P(\nu_n = k) = p_n(k)$ for $k = 1, \cdots, n - 1$, then

$$P(\nu_n = 0) + P(\nu_n = n)$$
$$= 1 - \sum_{k=1}^{n-1} P(\nu_n = k) = 1 - \sum_{k=1}^{n-1} p_n(k) = p_n(0) + p_n(n).$$

Since the hypothesis about $\mathcal{L}(X)$ implies easily that $P(\nu_n = 0) = P(\nu_n = n)$, it follows that

$$P(\nu_n = 0) = p_n(0) = P(\nu_n = n) = p_n(n).$$

Once (2) is proved, the Arcsine law follows by elementary computations using Stirling's formula (see Introductory Part, 7).

We shall use the foregoing basic implications without further comment.

26.2. Dichotomy: recurrence and transience. We recall that $x \in R$ is a *possible state* of a random walk $(S_1, S_2, \cdots)$ if for every neighborhood V_x, there is an $n = n(V_x)$ such that $P(S_n \in V_x) > 0$. We say that x is a *recurrent state* of the random walk, if, for every V_x, $P(S_n \in V_x \text{ i.o.}) = 1$; as usual "i.o." stands for "infinitely often," that is, for infinitely many n, and "f.o." for "finitely often" will stand for denial of "i.o.", that is, for "at most finitely many n." Thus, to say that x is recurrent is equivalent to $P(S_n \in V_x \text{ f.o.}) = 0$. Clearly, a recurrent state is possible and it suffices to consider neighborhoods V_x of the form $(x - \epsilon, x + \epsilon)$.

a. *If a random walk has a recurrent state x then all possible states are recurrent.*

Proof. If y is a possible state, that is, for every $\epsilon > 0$ there is a $k = k(\epsilon)$ such that $P(|S_k - y| < \epsilon) > 0$ then $x - y$ is recurrent: For then,

$$0 = P(|S_n - x| < 2\epsilon \text{ f.o.})$$
$$\geq P(|S_k - y| < \epsilon, |S_{n+k} - S_k - (x - y)| < \epsilon \text{ f.o.})$$
$$= P(|S_k - y| < \epsilon)P(|S_n - (x - y)| < \epsilon \quad \text{f.o.})$$

hence $P(|S_n - (x - y)| < \epsilon \text{ f.o.}) = 0$ and $x - y$ is recurrent. It follows that every possible state $y = x - (x - y)$ is recurrent and so is $x - x = 0$.

Thus we are led to a dichotomy: A random walk is *recurrent* if one hence all its possible states are recurrent, or it is transient if none of its possible states is recurrent.

As usual, $\mathcal{L}(X)$ denotes the common law of the iid random steps $X_1, X_2, \cdots$ which generate the random walk.

A. RECURRENCE THEOREM.   *Let X be L_d-distributed with $d \geq 0$.*
The random walk is recurrent if and only if one of its possible states is recurrent, and then L_d is the set of its states.

Proof. If the set $\mathcal{R}$ of recurrent states is not empty then, by **a,** the random walk is recurrent while the converse is trivially true.

Let the random walk be recurrent. Always by **a**, $\Re$ is closed under differences and $0 \in \Re$. It follows that $x \in \Re \Leftrightarrow -x = 0 - x\epsilon\Re$, and $\Re$ is an additive group. Furthermore, $\Re$ is topologically closed since, for any given V_x, if recurrent $x_n \to x$ then, from some n on, $x_n \in V_x$ hence $P(S_n \in V_x \text{ i.o.}) = 1$, and x is recurrent. Since the trivial case of random walks degenerate at 0 is excluded, $\Re \neq \{0\}$ and the only foregoing subgroups in R are of the form $\Re = L_{d'}$ with $d' \geq 0$. If $d' = 0$ then $d = 0$. If $d' > 0$ then $L_{d'} \subset L_d$ hence $d \leq d'$. Suppose $d < d'$ so that there is a possible state which is not recurrent. This contradicts the hypothesis that the random walk is recurrent. Thus $d = d'$, and the proof is terminated.

COROLLARY. *Let X be L_d-distributed with $d \geq 0$.*
Either $P(S_n \in V_x \text{ i.o.}) = 1$ for all bounded open sets V intersecting L_d, or $P(S_n \in V_x \text{ i.o.}) = 0$ for all such V.

B. DICHOTOMY CRITERION. *Let X be L_d-distributed with $d \geq 0$.*

(i) *If $\sum\limits_{n=1}^{\infty} P(S_n \in J) = \infty$ for some bounded open interval J, necessarily intersecting L_d, then the random walk is recurrent.*

(ii) *If $\sum\limits_{n=1}^{\infty} P(S_n \in J) < \infty$ for some bounded open interval J intersecting L_d, then the random walk is transient.*

Proof. By Borel–Cantelli lemma, the hypothesis in (ii) implies $P(S_n \in J \text{ i.o.}) = 0$ for some bounded open interval J intersecting L_d so that there is a possible state which is not recurrent hence, by **A**, no state is recurrent and the random walk is transient.

Let $\sum\limits_{n=1}^{\infty} P(S_n \in J) = \infty$ for some bounded open interval J with length $|J|$. Then, for every $\epsilon < |J|/2$ there is a $J_x = (x - \epsilon, x + \epsilon) \subset J$ such that $\sum\limits_{n=1}^{\infty} P(S_n \in J_x) = \infty$. Consider the time τ of the last visit by the random walk to J_x if any, and set $\tau = 0$ if none and $\tau = \infty$ if infinitely many. Thus, for $k = 1, 2, \cdots$

$$A_n = [\tau = n] = [S_n \in J_x, S_{n+k} \not\subset J_x \text{ for all } k], \quad n \geq 1,$$

and

$$A_0 = [\tau = 0] = P(S_n \not\subset J_x \text{ for all } n),$$

hence

$$P(\tau < \infty) = P(S_n \in J_x \text{ f.o.}) = \sum_{n=0}^{\infty} PA_k.$$

Since, for $n \geq 1$,

$$PA_n \geq P(S_n \in J_x, |S_{n+k} - S_n| \geq 2\epsilon \text{ for all } k)$$
$$= P(S_n \in J_x)P(|S_{n+k} - S_n| \geq 2\epsilon \text{ for all } k),$$

it follows that

$$1 \geq P(S_n \in J_n \text{ f.o.}) \geq P(|S_k| \geq 2\epsilon \text{ for all } k) \sum_{n=1}^{\infty} P(S_n \in J_x).$$

Thus, $\sum_{n=1}^{\infty} P(S_n \in J_x) = \infty$ implies that for every $\epsilon > 0$

(1) $$P(|S_k| \geq 2\epsilon \text{ for all } k) = 0.$$

This relation implies recurrence of 0 hence of the random walk, as follows.

Take $J_0 = (-\epsilon, +\epsilon)$, let $J_\delta = (-\delta, \delta)$ with $0 < \delta < \epsilon$, and define the corresponding A_n^0 as the A_n were defined but replacing x by 0. Note that, by (1), $PA_0^0 = P(|S_k| \geq \epsilon \text{ for all } k) = 0$. In fact, all $PA_n^0 = 0$ for $n \geq 1$: For, as $\delta \uparrow \epsilon$,

$$A_n^{0,\delta} = [S_n \in J_\delta, S_{n+k} \not\subset J_0 \text{ for all } k] \uparrow A_n^0$$

hence $PA_n^{0,\delta} \to PA_n^0$ and, by (1), $PA_n^0 = 0$ since

$$P(S_n \in J_\delta, S_{n+k} \not\subset J_0 \text{ for all } k)$$
$$\leq P(S_n \in J_\delta, |S_{n+k} - S_k| \geq \epsilon - \delta \text{ for all } k)$$
$$= P(S_n \in J_\delta)P(|S_n| \geq \epsilon - \delta \text{ for all } k) = 0.$$

Thus,

$$P(S_n \in J_0 \text{ f.o.}) = \sum_{n=0}^{\infty} PA_n^0 = 0$$

so that 0 is recurrent, and the proof is completed.

COROLLARY. *If for some bounded open interval* J *intersecting* L_d $\sum_{n=1}^{\infty} P(S_n \in J)$ *is either infinite or finite, then the same holds, respectively, for all such* J.

The elementary proofs of **A** and **B** are the original ones and are due to Feller, while the proof of **C** is due to Chung and Ornstein and that of **D** is due to Chung and Fuchs as modified by Feller.

The next proposition provides us with a dichotomy criterion in terms of *one* numerical characteristic of $\mathfrak{L}(X)$, namely in terms of *EX provided it exists.*

C. Expectation criterion.   *Let EX exist.   Then the random walk is recurrent if and only if EX = 0.   More precisely*

(i) *If EX = 0 then the random walk is recurrent and a.s.*

$$ -\infty = \liminf S_n < \limsup S_n = +\infty. $$

(ii) *If EX > 0 or EX < 0 then the random walk is transient and*

$S_n \xrightarrow{a.s.} +\infty$ *or* $S_n \xrightarrow{a.s.} -\infty$, *respectively.*

To prove this proposition we need the lemma below; we introduce $S_0 = 0$ and write $I(A)$ in lieu of I_A for any event A.

b. *For every $c > 0$ and every integer m*

$$ \frac{1}{2m} \sum_{m=0}^{\infty} P(|S_n| < mc) \leqq \sum_{n=0}^{\infty} P(|S_n| < c). $$

Proof.   Let the right side be finite; otherwise there is nothing to prove. Let **J** be an interval of length c and let $\nu = \sum_{n=1}^{\infty} I(S_n \in J)$ be the number of visits to **J** by the random walk $(S_1, S_2, \cdots)$ so that their expected number is $E\nu = \sum_{n=1}^{\infty} P(S_n \in J)$.   Set $\tau = \min\{n \geq 1 : S_n \in J\}$ when this set is not empty and $\tau = \infty$ when it is; τ is the time of the first visit to **J** and $E\nu = \sum_{n=1}^{\infty} E(\nu I(\tau = n))$.   On $[\tau = n]$, $I(S_k \in J) = 0$ for $k < n$ while $I(S_n \in J) = 1$ hence

$$ \nu I[\tau = n] = \sum_{k=1}^{\infty} I(S_k \in J) = 1 + \sum_{k=n+1}^{\infty} I(S_k \in J) $$

$$ = 1 + \sum_{k=n+1}^{\infty} I((S_k - S_n) + S_n \in J) $$

$$ \leqq 1 + \sum_{k=n+1}^{\infty} I(|S_k - S_n| < c) = 1 + \sum_{k=1}^{\infty} I(|S_k| < c) $$

$$ = \sum_{k=0}^{\infty} I(|S_k| < c). $$

Since $[\tau = n]$ and S_k are independent when $k > n$, it follows that

$$E\nu \leq \sum_{n=0}^{\infty} \left\{ P(\tau = n) \sum_{k=0}^{\infty} P(|S_k| < c) \right\} \leq \sum_{k=0}^{\infty} P(|S_n| < c).$$

Therefore,

$$\sum_{n=0}^{\infty} P(S_n \in J) \leq \sum_{n=0}^{\infty} P(|S_n| < c)$$

since $P(S_0 \in J) = 0$ unless $0 \in J$ when this inequality holds trivially, term by term. Upon replacing J by $J_j = [jc, (j+1)c)$ and summing over $j = -m, -m+1, \cdots, m-1$, the asserted inequality

$$\frac{1}{2m} \sum_{n=0}^{\infty} P(|S_n| < mc) \leq \sum_{n=0}^{\infty} P(|S_n| < c)$$

obtains.

Proof of C. By the SLLN, if $EX > 0$ then $S_n/n \xrightarrow{a.s.} EX > 0$ hence $S_n \xrightarrow{a.s.} +\infty$ and the random walk cannot be recurrent; similarly for $EX < 0$.

Let $EX = 0$ so that $S_n/n \xrightarrow{a.s.} 0$ and, a fortiori, $S_n/n \xrightarrow{P} 0$ hence, for given $\epsilon > 0$ and $n \geq n_\epsilon$ sufficiently large, $P(|S_n| < n\epsilon) < 1/2$. Therefore, for $m/\epsilon \geq n_\epsilon$,

$$\frac{1}{2m} \sum_{n=0}^{\infty} P(|S_n| < m) \geq \frac{1}{2}\left(\frac{m}{\epsilon} - n_\epsilon\right)/2m = 1/4\epsilon - n_\epsilon/4m$$

so that, by **b** with $c = 1$,

$$\sum_{n=0}^{\infty} P(|S_n| < 1) \geq \limsup_{m \to \infty}(1/4\epsilon - n_\epsilon/4m) = 1/4\epsilon \to \infty$$

as $\epsilon \to 0$, **B** applies and the random walk is recurrent. But then the (nondegenerate at 0) random walk cannot drift to $+\infty$ or to $-\infty$ and the only asymptotic alternative is a.s. $-\infty = \liminf S_n < \limsup S_n = +\infty$. The proof is terminated.

If a random walk obeys the infinite oscillations alternative it is not necessarily recurrent: Symmetric random walks, that is with $\mathcal{L}(X)$ symmetric, obey this alternative and we produce now such random walks which are transient.

Let $\mathcal{L}(X)$ be nondegenerate symmetric stable, that is, with $f(u) = \exp(-c|u|^\gamma)$, $c > 0, 0 < \gamma \leq 2$. According to end of 25.2, F' exists and is continuous and $0 \leq F'(x) \leq F'(0)$ with $F'(0) > 0$. Furthermore, $\mathcal{L}(X)$ being stable,

$\mathcal{L}(S_n/n^{1/\gamma}) = \mathcal{L}(X)$ for every $n = 1, 2, \cdots$. It follows that

$$P(|S_n| < 1) = P(|X| < 1/n^{1/\gamma})$$
$$= \int_{-n^{-1/\gamma}}^{n^{-1/\gamma}} F'(x) \, dx \sim 2F'(0)n^{-1/\gamma},$$

so that $\sum\limits_{n=1}^{\infty} P(|S_n| < 1)$ is finite or infinite, according as $\sum\limits_{n=1}^{\infty} n^{-1/\gamma}$ is finite or infinite hence, according as $0 < \gamma < 1$ or $1 \leq \gamma \leq 2$. Thus, by **B**, our *symmetric random walk is* transient for $0 < \gamma < 1$ and recurrent for $1 \leq \gamma \leq 2$; note that EX does not exist for $0 < \gamma \leq 1$.

Finally, we search for conditions for recurrence or transience in terms of the ch.f. f of $\mathcal{L}(X)$. (So far, they seem to provide the only approach for general random walks in euclidean spaces R^n, $n > 1$.) In what follows we use the immediate

Parseval relation: $\int f(u) \, dF_Y(u) = \int f_Y(X) \, dF_X(X)$ which obtains upon integrating $f_X(u) = \int e^{iux} \, dF(x)$ with respect to $F_Y(t)$, and two laws with

triangular pr. density:

$$F'(x) = \frac{1}{h}\left(1 - \frac{|x|}{h}\right) \vee 0, \, f(u) = 2\frac{1 - \cos hu}{h^2u^2}, \, h > 0,$$

triangular ch.f.:

$$f(u) = \left(1 - \frac{|u|}{h}\right) \vee 0, \, F'(u) = \frac{1}{\pi}\frac{1 - \cos hx}{hx^2}, \, h > 0.$$

D. Ch. f.'s and dichotomy. *Let f be the ch.f. of the common law $\mathcal{L}(X)$.*

(i) *The random walk is recurrent if there is a $\delta > 0$ with*

$$\limsup_{t \uparrow 1} \int_{-\delta}^{\delta} \frac{du}{1 - tf(u)} = \infty.$$

(ii) *The random walk is transient if there is a $\delta > 0$ with*

$$\sup_{0 < t < 1} \int_{-\delta}^{\delta} \frac{du}{1 - tf(u)} < \infty.$$

Proof. Let $S_0 = 0$.

1°. Parseval relation with triangular ch.f. and F_{S_n} yields

$$P(|S_n| < h) \geq \int_{-h}^{h} \left(1 - \frac{|x|}{h}\right) dF_{S_n}(x) = \frac{1}{\pi} \int \frac{1 - \cos hu}{hu^2} f^n(u) \, du.$$

Since $(1 - \cos hu)/hu^2 \geq ch$ for $|u| < 1/h$ and some $c > 0$ and

$$\Re \frac{1}{1 - tf(u)} \geq \frac{1 - t}{|1 - tf(u)|^2},$$

it follows that

$$\sum_{n=0}^{\infty} t^n P(|S_n| < h) \geq \frac{ch}{\pi} \int_{-1/h}^{1/h} \Re \frac{1}{1 - tf(u)} \, du = \frac{ch}{\pi} \int_{-1/h}^{1/h} \frac{du}{1 - tf(u)}.$$

Therefore, by hypothesis in (i), for $1/h < \delta$,

$$\sum_{n=0}^{\infty} P(|S_n| < h) \geq \frac{ch}{\pi} \limsup_{t \uparrow 1} \int_{-\delta}^{\delta} \frac{du}{1 - tf(u)} = \infty$$

and recurrence obtains by **B**.

2°. Parseval relation with triangular pr. density yields

$$\int \frac{1 - \cos hx}{h^2 x^2} dF_{S_n}(x) = \frac{1}{2h} \int_{-h}^{h} \left(1 - \frac{|u|}{h}\right) f^n(u) \, du$$

so that for $|x| < 2/h$ hence $(1 - \cos hx)/h^2 x^2 > 1/3$

$$\sum_{n=0}^{\infty} t^n P(|S_n| < 2/h) \leq \int \left(1 - \frac{|u|}{h}\right) \frac{du}{1 - tf(u)} \leq \frac{3}{2h} \int_{-h}^{h} \frac{du}{1 - tf(u)}.$$

Therefore, by hypothesis in (ii), for $h < \delta$,

$$\sum_{n=0}^{\infty} P(|S_n| < 2/h) \leq \frac{3}{2h} \sup_{0 < t < 1} \int_{-\delta}^{\delta} \frac{du}{1 - tf(u)} < \infty,$$

and transience obtains by **B**.

COROLLARY 1. *If, for some $\delta > 0$,*

$$\int_{-\delta}^{\delta} \frac{du}{1 - f(u)} = \infty,$$

then the random walk is recurrent.

CorollarY 2.   If $EX = 0$ *then the random walk is recurrent.*

Follows by elementary computations from the fact that, given $\epsilon > 0$, $EX = 0$ implies $0 \leq 1 - \Re ef(u) < \epsilon u$ for $|u| < \delta$ sufficiently small.

26.3. Fluctuations; exponential identities. We consider random variables defined on $(S_0, \cdots, S_n)$, say, the number of its positive terms or their maximum or times of occurrence of this maximum, etc. We shall find the explicit form of their laws in terms of "exponential identities." The method will be Fourier analytic. At its core lies a "Wiener-Hopf" factorization technique for the *generating characteristic* $1/(1 - tf)$ of the random walk $(S_0, S_1, \cdots)$.

In what follows, $0 < t < 1$, $u \in R$, A denotes a linear Borel set, and we set

$$f_A(u, t) = \exp\left\{\sum_{n=1}^{\infty} \frac{t^n}{n} \int_{[S_n \in A]} e^{iuS_n}\right\},$$

hence

$$f_{A^c}(u, t) = \exp\left\{\sum_{n=1}^{\infty} \frac{t^n}{h} \int_{[S_n \in A^c]} e^{iuS_n}\right\}.$$

a. Factorization lemma.

$$\frac{1}{1 - tf(u)} = f_A(u, t) f_{A^c}(u, t).$$

Results from

$$\frac{1}{1 - tf(u)} = \exp\left\{\log\left(\frac{1}{1 - tf(u)}\right)\right\} = \exp\left\{\sum_{n=1}^{\infty} \frac{t^n}{n} f^n(u)\right\}$$

by

$$f^n(u) = \int' e^{iuS_n} = \int_A e^{iuS_n} + \int_{A^c} e^{iuS_n}.$$

We shall be dealing with Fourier-Stieltjes transforms of functions of bounded variation on linear Borel sets, of the form

$$p(u) = \int_A e^{iux} \, dG(x)$$

with same affixes for p and G, if any. Exactly as for characteristic functions, the uniqueness theorem $p \leftrightarrow G$ (up to additive constants) is valid, their products pp' correspond to compositions $G*G'$ and, clearly, their sums and differences $p \pm p'$ are transforms of functions of bounded variation $G \pm G'$.

Let

$$P_A(u, t) = \sum_{n=0}^{\infty} P_n(u)t^n, \quad Q_{A^c}(u, t) = \sum_{n=0}^{\infty} q_n(u)t^n$$

where $P_0(u) = q_0(u) = 1$ and, for $n \geq 1$,

$$p_n(u) = \int_A e^{iux}\, dG_n(x), \quad q_n(u) = \int_{A^c} e^{iux}\, dG_n(x).$$

A. Unique factorization theorem. *If*

(i) $\dfrac{1}{1 - tf(u)} = P_A(u, t)Q_{A^c}(u, t)$ *or* (ii) $\dfrac{P_A(u, t)}{1 - tf(u)} = Q_{A^c}(u, t)$

then

$$P_A(u, t) = f_A(u, t) \text{ or } P_A(u, t) = f_A^{-1}(u, t)$$

and

$$Q_{A^c}(u, t) = f_{A^c}(u, t).$$

Proof. Because of **a**, it suffices to show that if the foregoing relations hold for $P_A'(u, t)$ and $Q_{A^c}'(u, t)$ then, for $n = 0, 1, \cdots, p_n(u) = p_n'(u)$ and $q_n(u) = q_n'(u)$, $u \in R$.

Upon identifying the coefficients of the t^n, (i) and (ii) then become, respectively,

(1) $$\sum_{k=0}^{n} p_k(u)q_{n-k}(u) = \sum_{k=0}^{n} p_k'(u)q_{n-k}'(u)$$

or

(2) $$\sum_{k=0}^{n} p_k(u)q_{n-k}'(u) = \sum_{k=0}^{n} p_k'(u)q_{n-k}'(u).$$

We proceed by induction: The assertion is trivially true for $n = 0$. If $p_k(u) = p_k'(u)$ and $q_k(u) = q_k'(u)$ for $k = 1, \cdots, n - 1$ then in (1) and (2) the first $n - 1$ terms in the left and right sums coincide so that

$$p_n(u) + q_n(u) = p_n'(u) + q_n'(u) \text{ or } p_n(u) + q_n'(u) = p_n'(u) + q_n(u).$$

Thus

$$p_n(u) - p_n'(u) = q_n'(u) - q_n(u) \text{ or } p_n(u) - p_n'(u) = q_n(u) - q_n'(u),$$

that is, for $u \in R$,

$$\int_A e^{iux} \, dH_n(x) = - \int_{A^c} e^{iux} \, dH_n(x) \text{ or } \int_A e^{iux} \, dH_n(x) = \int_{A^c} e^{iux} \, dH_n(x),$$

where the functions $H_n = G_n - G'$ are of bounded variation. Therefore, by the uniqueness property for Fourier-Stieltjes transforms, $I_A dH_n = \mp I_{A^c} dH_n$ so that both sides vanish. The assertion follows.

This proof as well as **B** are due to Baxter.

From now on, to simplify the writing, $\int_A e^{iuS_n} \equiv E(e^{iuS_n}I(A))$ will be denoted by $E(e^{iuS_n}: A)$ and, when A is of the form $[\cdots]$ we shall omit the square brackets. The first visit time of A by $(S_1, S_2, \cdots)$ will be called *hitting time* of A. When τ is a random time, for $\tau = \infty, t^{n+\tau} = 0$, $(0 < t < 1)$, $n = 0, 1, \cdots$; note that if τ is a time of $(S_1, S_2, \cdots)$ then $[\tau = 0] = \emptyset$.

B. RANDOM TIMES IDENTITIES. *Let τ be a time of $(S_1, S_2, \cdots)$.*

(i) *The following identities hold:*

$$E(t^\tau e^{iuS_\tau}) = \sum_{n=0}^{\infty} t^n E(e^{iuS_n}: \tau = n),$$

$$E\left(\sum_{n=0}^{\tau-1} t^n e^{iuS_n} \right) = \sum_{n=0}^{\infty} t^n E(e^{iuS_n}: \tau > n)$$

$$\frac{1 - E(t^\tau e^{iuS_\tau})}{1 - tf(u)} = E\left(\sum_{n=0}^{\tau-1} t^n e^{iuS_n} \right).$$

(ii) *When $\tau = \tau_A$ is hitting time of A then*

$$1 - E(t^{\tau_A} \exp iuS_{\tau_A}) = f_A^{-1}(u, t) = \exp\left\{ -\sum_{n=1}^{\infty} \frac{t^n}{n} E(e^{iuS_n}: S_n \in A) \right\}$$

$$E\left(\sum_{n=0}^{\tau_A-1} t^n e^{iuS_n} \right) = f_{A^c}(u, t) = \exp\left\{ \sum_{n=1}^{\infty} \frac{t^n}{n} E(e^{iuS_n}: S_n \in A^c) \right\}.$$

Proof. The first identity in (i) results at once from the definitions. The second one results from

$$E\left(\sum_{n=0}^{\tau-1} t^n e^{iuS_n}\right) = \sum_{n=1}^{\infty} E\left(\sum_{k=0}^{n-1} t^n e^{iuS_n}; \tau = k\right) = \sum_{n=0}^{\infty} E(t^n e^{iuS_n}; \tau > n).$$

Since τ is a time of the random walk $(S_1, S_2, \cdots)$,

$$E\left(\sum_{n=\tau}^{\infty} t^n e^{iuS_n}\right) = E\left(t^\tau e^{iuS_\tau} \sum_{n=0}^{\infty} e^{iu(S_{\tau+n}-S_\tau)}\right)$$

$$= E(t^\tau e^{iuS_\tau})E\left(\sum_{n=0}^{\infty} t^n e^{iuS_n}\right) = E(t^\tau e^{iuS_\tau})/(1 - tf(u))$$

and, replacing in

$$1/(1 - tf(u)) = E\left(\sum_{n=0}^{\tau-1} t^n e^{iuS_n}\right) + E\left(\sum_{n=\tau}^{\infty} t^n e^{iuS_n}\right),$$

the third identity obtains. By the unique factorization theorem, it implies the two identities in (ii).

Our main concern is with $A = (0, \infty)$ hence $A^c = (-\infty, 0]$, and we set $f_+ = f_{(0,\infty)}, f_- = f_{(-\infty, 0]}$ so that

$$f_+(u, t) = \exp\left\{\sum_{n=1}^{\infty} \frac{t^n}{n} E(e^{iuS_n}; S_n > 0)\right\},$$

$$f_-(u, t) = \exp\left\{\sum_{n=1}^{\infty} \frac{t^n}{n} E(e^{iuS_n}; S_n \leq 0)\right\}.$$

COROLLARY. *If $\tau = \tau_{(0, \infty)}$ then*

$$1 - E(t^\tau e^{iuS_\tau}) = f_+(u, t), \quad E\left(\sum_{u=0}^{\tau-1} t^n e^{iuS_n}\right) = f_-(u, t).$$

C. MAXIMA TIMES AND POSITIVE SUMS IDENTITIES.

(i) *If τ_n is the time of occurrence of the first maximum of $(S_0, S_1, \cdots, S_n)$ then*

$$E(e^{iuS_n}; \tau_n = k) = E(e^{iuS_n}; \tau_k = k) E(e^{iuS_{n-k}}; \tau_{n-k} = 0),$$

$$\sum_{n=0}^{\infty} t^n E(e^{iuS_n}; \tau_n = n) = f_+(u,t),$$

$$\sum_{u=0}^{\infty} t^n E(e^{iuS_n}: \tau_n = 0) = f_-(u,t),$$

$$\sum_{u=0}^{\infty} t^n E(s^{\tau_n}e^{iuS_n}) = f_+(u,st)f_-(u,t), \quad 0 < s \leqq 1.$$

(ii) *If v_n is the number of positive sums in $(S_0, S_1, \cdots, S_n)$ then the above identities remain valid when therein τ is replaced by v with same subscripts.*

(iii) *All above identities, including those in the above Corollary, remain valid when $(0, \infty)$ is replaced by $[0, \infty)$ provided τ_n is the time of the last maximum of $(S_0, S_1, \cdots, S_n)$ and v_n is the number of nonnegative sums in $(S_0, S_1, \cdots, S_n)$ while $S_n > 0$ and $S_n \leqq 0$ are replaced by $S_n \geqq 0$ and $S_n < 0$ in f_+ and in f_-, respectively.*

Proof. The identities in (i) are based upon a "sample space factorization": If $M_n = \max(S_0, \cdots, S_n)$ then the first time this maximum occurs is $\tau_n = \min\{0 \leqq k \leqq n: S_k = M_n\}$ and, by the very definition of $\tau_n = \tau_n(X_1, \cdots, X_n)$,

$$[\tau_n(X_1, \cdots, X_n) = k] = [\tau_k(X_1, \cdots, X_k) = k][\tau_{n-k}(X_{k+1}, \cdots, X_n) = 0].$$

Since the last two events are independent and so are S_k and $S_n - S_k$ while $S_n - S_k$ has the same distribution as S_{n-k}, it follows that

$$E(e^{iuS_n}: \tau_n = k) = E(e^{iuS_k}: \tau_k = k) \cdot E(e^{iuS_{n-k}}: \tau_{n-k} = 0).$$

Thus, upon multiplying by $s^k t^n$ and summing over $0 \leqq k \leqq n < \infty$,

$$\sum_{u=0}^{\infty} t^n E(s^{\tau_n}e^{iuS_n}) = P(u, st) Q(u, t)$$

where

$$P(u, t) = \sum_{n=0}^{\infty} t^n E(e^{iuS_n}: \tau_n = n),$$

$$Q(u, t) = \sum_{n=0}^{\infty} t^n E(e^{iuS_n}: \tau_n = 0).$$

For $s = 1$, the preceding relation becomes

$$\frac{1}{1 - tf(u)} = P(u, t) Q(u, t),$$

while $\tau_n = n$ implies $S_n > 0$ and $\tau_n = 0$ implies $S_n \leq 0$. The unique factorization theorem applies so that

$$P(u, t) = f_+(u, t), \quad Q(u, t) = f_-(u, t)$$

and the identities in (i) follow.

By Andersen equivalence, the sample space factorization for the times of first maxima is equivalent to the far from obvious sample space factorization for the numbers of positive sums:

$$[\nu_n(X_1, \cdots, X_n) = k] = [\nu_k(X_1, \cdots, X_k) = k][\nu_{n-k}(X_{k+1}, \cdots, X_n) = 0],$$

and (ii) for positive sums identities follows.

Finally, by using in the unique factorization theorem $[0, \infty)$ in lieu of $(0, \infty)$, (iii) results from the fact that all the foregoing arguments continue to apply to the corresponding τ_n and ν_n.

The following important identity, known in various guises and with various degrees of generality, has its origin in the basic Spitzer identity below (Pollaczec, Spitzer, Kemperman, Port, etc.).

D. MAXIMUM TIME AND VALUE IDENTITY. *If $M_n = \max(S_0, \cdots, S_n)$ and $\tau_n = \min\{0 \leq k \leq n : S_k = M_n\}$, then*

$$\sum_{n=0}^{\infty} t^n E(s^{\tau_n} e^{iuS_n + ivM_n}) = f_+(u + v, st) f_-(u, t), \quad 0 < s \leq 1.$$

Proof. Since $\tau_n = k \Leftrightarrow M_n = S_k$, by sample space factorization,

$$E(e^{iuS_n + ivM_n} : \tau_n = k) = E(e^{i(u+v)S_k + iu(S_n - S_k)} : \tau_n = k)$$
$$= E(e^{i(u+v)S_k} : \tau_k = k) \cdot E(e^{iuS_{n-k}} : \tau_{n-k} = 0).$$

Upon multiplying by $s^k t^n$ and summing for $0 \leq k \leq n < \infty$, it follows that

$$\sum_{n=0}^{\infty} t^n E(s^{\tau_n} e^{iuS_n}) = P(u + v, st) Q(u, t),$$

where P and Q are the functions introduced in the preceding proof and, as therein, the unique factorization theorem yields the asserted identity.

Particular cases. 1°. For $v = 0$ we obtain the last identity in **C**(i).

2°. For $s = 1$ and $u = 0$, changing v into u, we obtain the *Pollaczec-Spitzer identity*:

$$\sum_{n=0}^{\infty} t^u E(e^{iuM_n}) = \exp\left\{\sum_{n=1}^{\infty} \frac{t^n}{n} E(e^{iuS_n^+})\right\}.$$

It was first discovered by Pollaczec but remained unnoticed until rediscovered by Spitzer.

3°. For $s = 1$, interchanging u and v, we obtain

$$\sum_{n=0}^{\infty} t^n E(e^{iuM_n + ivS_n}) = f_+(u + v, t)f_-(v, t).$$

Upon setting $w = u + v$ then changing w into u, and v into $-v$, it becomes

$$\sum_{n=0}^{\infty} t^n E(e^{iuM_n + iv(M_n - S_n)}) = f_+(u, t)f_-(-v, t).$$

Finally, upon multiplying by

$$\frac{1}{1 - t} = \exp\left\{\sum_{n=1}^{\infty} \frac{t^n}{n} P(S_n > 0)\right\} \cdot \exp\left\{\sum_{n=1}^{\infty} \frac{t^n}{n} P(S_n \leq 0)\right\},$$

we obtain

$$\frac{1}{1 - t} \sum_{n=0}^{\infty} t^n E(e^{iuM_n + iv(M_n - S_n)}) = \exp\left\{\sum_{n=1}^{\infty} \frac{t^n}{n}(Ee^{iuS_n^+} + Ee^{ivS_n^-})\right\}$$

or

$$\sum_{u=0}^{\infty} t^n E(e^{iuM_n + iv(M_n - S_n)}) = \exp\left\{\sum_{n=1}^{\infty} \frac{t^n}{n}(Ee^{iuS_n^+} + Ee^{ivS_n^-} - 1)\right\},$$

and this is the *basic Spitzer identity* in its initial form.

EXTENSION. The basic exponential factors $f_+ (u, t)$ and $f_- (u, t)$ may still have meaning when $u \in R$ is replaced by complex z. In fact,

$$f_+(z, t) = \exp\left\{\sum_{n=1}^{\infty} \frac{t^n}{n} E(e^{izS_n} : S_n > 0)\right\}$$

is bounded and continuous for $\Im z \geq 0$ and regular for $\Im z > 0$,

$$f_-(z, t) = \exp\left\{\sum_{n=1}^{\infty} \frac{t^n}{n} E(e^{izS_n} : S_n \leq 0)\right\}$$

is bounded and continuous for $\Im z \leq 0$ and regular for $\Im z < 0$.
Thus the question arises whether the identities so far obtained remain valid for such z. The answer is in the affirmative for those identities in which figure only either f^+ or f^-; when both occur then, clearly, we must have $\Im z = 0$, that is, $z = u \in R$. These assertions result at once from the unicity lemma 15.2d, which yields (i) and (ii) below, while for (iii) we also use the fact that all the r.v.'s therein are nonnegative.

E. EXTENDED IDENTITIES. *The following identities are valid:*

(i) *For* $\Im z \geq 0$,

$$1 - E\left(t^\tau e^{iz S_\tau}\right) = f_+(z, t),$$

$$\sum_{n=0}^{\infty} t^n E(e^{iz S_n}: \tau_n = n) = f_+(z, t),$$

$$\sum_{n=0}^{\infty} t^n E(e^{iz S_n}: \nu_n = n) = f_+(z, t).$$

(ii) *For* $\Im z \leq 0$,

$$\sum_{n=0}^{\tau-1} t^n E(e^{iz S_n}) = f_-(z, t)$$

$$\sum_{n=0}^{\infty} t^n E(e^{iz S_n}: \tau_n = 0) = f_-(z, t)$$

$$\sum_{n=0}^{\infty} t^n E(e^{jz S_n}: \nu_n = 0) = f_-(z, t)$$

(iii) *For* $\Im z \geq 0$, $\Im z' \geq 0$,

$$\sum_{n=0}^{\infty} t^n E\left(e^{izM_n + iz'(M_n - S_n)}\right)$$

$$= \exp\left\{\sum_{n=1}^{\infty} \frac{t^n}{n}\left(E(e^{iz S_n^+}) + E(e^{iz' S_n^-})\right) - 1\right\}$$

and, in particular, for $\Im z \geq 0$,

$$\sum_{n=0}^{\infty} t^n E(e^{izM_n}) = \exp\left\{\sum_{n=1}^{\infty} \frac{t^n}{n} E\, e^{iz S_n^+}\right\}.$$

REMARK. In fact, the argument used for the unicity lemma 15.2d permits to prove simultaneously identities and a unique factorization theorem (Pollaczec, Ray, Kemperman). To fix the ideas, replace u by z in $P(u, t)$ and $\mathcal{Q}(u, t)$ used in the proof of **C**:

$$P(z, t) = \sum_{n=0}^{\infty} t^n E(e^{iz S_n}: \tau_n = n), \quad Q(z, t) = \sum_{n=0}^{\infty} t^n E(e^{iz S_n}: \tau_n = 0)$$

Note that $P(z, t)$ like $f_+(z, t)$ ($\mathcal{Q}(z, t)$ like $f_-(z, t)$) is bounded and continuous for $\Im z \geq 0$ ($\Im z \leq 0$) and regular for $\Im z > 0$ ($\Im z < 0$) while for $\Im z = 0$

$$f_+(z, t)f_-(z, t) = \frac{1}{1 - tf(z)} = P(z, t)Q(z, t).$$

Therefore, for $\Im z = 0$,

$$g(z) = \frac{P(z, t)}{f_+(z, t)} = \frac{f_-(z, t)}{Q(z, t)}$$

where the first (second) ratio is bounded and continuous for $\Im z \geq 0$ ($\Im z \leq 0$) and regular for $\Im z > 0$ ($\Im z < 0$). Thus, the two ratios are restrictions of a same bounded entire function $g(z)$ to $\Im z \geq 0$ and to $\Im z \leq 0$, respectively. By Liouville's theorem, $g(z)$ is a constant. But

$$\frac{P(z, t)}{f_+(z, t)} \to 1 \quad \text{as} \quad z \to +\infty$$

so that

$$P(z, t) = f_+(z, t) \text{ for } \Im z \geq 0, Q(z, t) = f_-(z, t) \text{ for } \Im z \leq 0.$$

This proves the corresponding extended identities together with unique factorization.

All preceding identities in f_+ and f_- which are in terms of exponentials, naturally, are called *exponential identities*. Their striking and unexpected feature is that the distributions of various fluctuation random variables are in terms of individual terms S_n of the random walk. The sample space factorizations

$$[\tau_n(X_1, \cdots, X_n) = k] = [\tau_k(X_1, \cdots, X_k) = k][\tau_{n-k}(X_{k+1}, \cdots, X_n) = 0]$$

and the equivalent one with τ replaced by ν are, naturally, called *extreme factorizations*. Their striking and unexpected feature is that the distributions of τ_n and of ν_n are determined by the pr.'s of their extreme values 0 and n.

26.4 Fluctuations; asymptotic behaviour. We relate now the asymptotic behaviour of the random walk to that of fluctuations r.v.'s τ_A, τ_n, ν_n, M_n; A denotes a linear Borel set.

a. Hitting time lemma. *If τ_A is hitting time of A then*

(i) $$1 - Et^{\tau_A} = \exp\left\{ -\sum_{n=1}^{\infty} \frac{t^n}{n} P(S_n \in A) \right\},$$

(ii) $$P(\tau_A = \infty) = \exp\left\{ -\sum_{n=1}^{\infty} P(S_n \in A)/n \right\},$$

(iii) $$E\tau_A = \exp\left\{ \sum_{n=1}^{\infty} P(S_n \in A^c)/n \right\} + \infty \cdot P(\tau = \infty)(\infty \cdot 0 = 0).$$

Proof. We use the elementary proposition: If the $a_n \geq 0$ and $\sum\limits_{n=0}^{\infty} a_n t^n$

converges for $0 < t < 1$, then $\sum\limits_{n=0}^{\infty} a_n t^n \to \sum\limits_{n=0}^{\infty} a_n \leq \infty$ as $t \uparrow 1$.

Set $\tau = \tau_A$. Identity (i) results from the first one in 26.3B(ii) with $u = 0$. Identity (ii) follows from (i) by letting $t \uparrow 1$ in

$$Et^{\tau} = \sum_{n=1}^{\infty} t^n P(\tau = n) \to \sum_{n=1}^{\infty} P(\tau = n) = P(\tau < \infty)$$

so that

$$P(\tau = \infty) \leftarrow 1 - Et^{\tau} \to \exp\left\{-\sum_{n=1}^{\infty} P(S_n \in A)/n\right\}.$$

Identity (iii) results from the second one in 26.3B(ii) with $u = 0$, by letting $t \uparrow 1$ so that

$$\exp\left\{\sum_{n=0}^{\infty} P(S_n \in A^c)/n\right\} \leftarrow E\sum_{n=0}^{\tau-1} t^n = \sum_{n=0}^{\infty} t^n P(\tau > n) \to \sum_{n=0}^{\infty} P(\tau > n)$$

and

$$E\tau = \sum_{n=0}^{\infty} P(\tau > n) + \infty \cdot P(\tau = \infty).$$

b. Finite interval lemma. *Let J be a finite interval and let τ be the hitting time of J^c. Then $E\tau^r < \infty$ for $r > 0$, and $ES_{\tau} = E\tau \cdot EX$ exists (and is finite) if and only if EX exists and is finite.*

The first assertion is Stein's lemma and the second one is Wald's relation, both obtained before general fluctuation theory.

Proof. The second relation was proved in 26.1 and it remains to prove the first one. To fix the ideas, let $J = [a, b]$. Since the only asymptotic alternatives are: a.s. $S_n \to -\infty$ or to $+\infty$ or $-\infty = \liminf S_n < \limsup S_n = +\infty$, there is an integer m such that $p(|S_m| \leq b - a) < 1$. But $[\tau > n + m]$ implies occurrence of independent events $[\tau > n]$ and $[|S_{n+m} - S_n| \leq b - a]$, where $S_{n+m} - S_n$ has the same distribution as S_m. Therefore,

$$P(\tau > n + m) \leq p P(\tau > n)$$

$$Ee^{iuM_\infty} = \exp\left\{\sum_{n=1}^{\infty} (Ee^{iuS_n^+} - 1)/n\right\},$$

$\nu_n \xrightarrow{a.s.} \nu_\infty,\ \tau_n \xrightarrow{a.s.} \tau_\infty$ *with common generating function*

$$Et^{\nu_\infty} = Et^{\tau_\infty} = \exp\left\{\sum_{n=1}^{\infty} (t^n - 1)P(S_n > 0)/n\right\}.$$

Note that the hypotheses in (i) and (ii) being contrary of each other, are equivalent to their conclusions.

Proof. By the above Corollary $\sum_{n=1}^{\infty} P(S_n > 0)/n = \infty$ is equivalent to

limsup $S_n = +\infty$ a.s. hence $M_\infty = \sup_n S_n^+ \geq$ limsup $S_n = +\infty$ a.s. It follows that

$$P(\nu_{n+1} = \nu_n + 1 \text{ i.o.}) = P(\tau_n = n \text{ i.o.}) = 1$$

so that $\nu_n \xrightarrow{a.s.} \infty$ and $\tau_n \xrightarrow{a.s.} \infty$. Assertions (i) are proved.

By the same Corollary, $\sum_{n=1}^{\infty} P(S_n > 0)/n < \infty$ is equivalent to limsup $S_n < \infty$ a.s., in fact, to lim $S_n = -\infty$ a.s. But, by definition of limsup, limsup $S_n < \infty$ a.s. implies $M_n \uparrow M_\infty < \infty$ a.s. and $P(\nu_{n+1} \neq \nu_n \text{ i.o.}) = P(\tau_{n+1} \neq \tau_n \text{ i.o.}) = 0$ hence $\nu_n \xrightarrow{a.s.} \nu_\infty < \infty$ and $\tau_n \xrightarrow{a.s.} \tau_\infty < \infty$.

We use now the classical Abel theorem: If the complex $a_n \to a$ finite then $(1 - t) \sum_{n=1}^{\infty} a_n t^n \to a$ as $t \uparrow 1$.

Since $M_n \uparrow M_\infty < \infty$ a.s., $Ee^{iuM_n} \to Ee^{iuM_\infty}$ hence, by Pollaczec-Spitzer identity, as $t \uparrow 1$,

$$Ee^{iuM_\infty} \leftarrow (1 - t) \sum_{n=1}^{\infty} t^n Ee^{iuM_n}$$

$$= \exp\left\{-\sum_{n=1}^{\infty} t^n/n\right\} \exp\left\{\sum_{n=1}^{\infty} t^n E(e^{iuS_n^+})/n\right\}$$

$$= \exp\left\{\sum_{n=1}^{\infty} t^n (Ee^{iuS_n^+} - 1)/n\right\} \to \exp\left\{\sum_{n=1}^{\infty} (Ee^{iuS_n^+} - 1)/n\right\}.$$

The last limit is an i.d. ch.f., since it is a product of i.d. ch.f.'s $e^{\psi_n(u)}$ with

$$\psi_n(u) = \frac{1}{n} \int (e^{iux} - 1) \, dF_{S_n}^{+}(x).$$

The first assertion in (ii) is proved.

Since $P(\nu_n = k) = P(\tau_n = k)$ for $k = 0, 1, \cdots, n$ and $n = 1, 2, \cdots$, $\nu_n \overset{a.s.}{\to} \nu$ and $\tau_n \overset{a.s.}{\to} \tau_\infty < \infty$ imply that $P(\nu_\infty = k) = P(\tau_\infty = k)$ for $k = 0, 1, \cdots$. Thus, to find the generating function of τ_∞ it suffices to find that

of ν_∞: $Et^{\nu_\infty} = \sum_{k=0}^{\infty} t^k P(\nu_\infty = k)$. Since $\nu_n \overset{a.s.}{\to} \nu_\infty < \infty$, it follows, by ex-

treme factorization, that

$$P(\nu_\infty = k) \leftarrow P(\nu_n = k) = P(\nu_k = k)P(\nu_{n-k} = 0) \to P(\nu_k = k)P(\nu_\infty = 0).$$

But, by **a**(ii),

$$P(\nu_\infty = 0) = P(\tau_{(0,\infty)} = \infty) = \exp\left\{-\sum_{n=1}^{\infty} P(S_n > 0)/n\right\}$$

while, by 26.3C(ii) and the second relation in (i) therein with $u = 0$,

$$\sum_{k=0}^{\infty} t^k P(\nu_k = k) = \exp\left\{\sum_{n=1}^{\infty} t^n P(S_n > 0)/n\right\}.$$

Therefore,

$$Et^{\nu_\infty} = \sum_{k=0}^{\infty} t^k P(\nu_k = k)P(\nu_\infty = 0) = \exp\left\{\sum_{n=1}^{\infty} (t^n - 1)P(S_n > 0)/n\right\},$$

and the proof is terminated.

This basic Spitzer theorem has the same striking and unexpected feature as the exponential identities: The limit distributions are in terms of individual sums S_n.

COMPLEMENTS AND DETAILS

As throughout this chapter, $X_1, X_2, \cdots$ are iid summands with common non-degenerate law $\mathcal{L}(X)$, d.f. F, ch.f. f, and $S_0 = 0$, $S_n = X_1 + \cdots + X_n$. Slowly varying functions will be denoted by $h(x)$ with or without affixes.

1. Let F_k, $k = 1, 2$, be d.f.'s and let $x \to \infty$.

If $1 - F_k(x) \sim x^{-\alpha} h_k(x)$ then $1 - (F_1 * F_2)(x) \sim x^{-\alpha}(h_1(x) + h_2(x))$.

If $1 - F(x) \sim x^{-\alpha} h(x)$ then $1 - F_{S_n}(x) \sim n x^{-\alpha} h(x)$. Deduce similar propositions for $F_k(-x)$, $F(-x)$.

2. Extrema. Let $Y_n = \max\limits_{1 \le k \le n} X_k$.

If $P(X < c) = 1$ for some constant c, then $\mathfrak{L}(Y_n) \to \mathfrak{L}(c)$.

If $P(X < x) < 1$ for every $x \in R$ then there exist scale factors $b_n > 0$ such that $\mathfrak{L}(Y_n/b_n) \to \mathfrak{L}(Y)$ nondegenerate if and only if $1 - F(x)$ varies regularly with exponent $a < 0$, and then $F_Y(x) = 0$ or e^{-cx^a} with $c > 0$, according as $x < 0$ or $x > 0$. What about $Z_n = \min\limits_{1 \le k \le n} X_k$?

3. Let S be a (λ, f)-compound Poisson: $f_S = e^{\lambda(f-1)}$. Let $x \to \infty$.

If $1 - F(x) \sim x^{-a}h(x)$ then $1 - F_s(x) \sim \lambda x^{-a}h(x)$. Is there a similar proposition about $F(-x)$ and $F_S(-x)$?

4. Let F be an i.d. d.f. with $f = e^\psi$, $\psi = (\alpha, \beta^2, L)$. Let $x \to \infty$.

If $L(x) = x^{-a}h(x)$ then $1 - F(x) \sim L(x)$. Is there a similar proposition about $L(-x)$ and $F(-x)$?

5. Norming. Let $\mathfrak{L}(X)$ be attracted by a nondegenerate stable $\mathfrak{L}_\gamma$, $0 < \gamma \le 2$, that is, $\mathfrak{L}(S_n/b_n - a_n) \to \mathfrak{L}_\gamma$ for suitable $b_n > 0$ and a_n.

(a) Let $\mu_2(t) = \int_{-t}^{+t} x^2 dF(x)$ and $q(t) = 1 - F(t) + F(-t)$. Let $t \to \infty$ and use 25.1.**D**.

If $r < \gamma$ then $\dfrac{t^{2-r}}{\mu_2(t)} \int_{|x| < t} |x|^r dF(x) \to \dfrac{2-\gamma}{\gamma - r}$.

If $r > \gamma$ when $\gamma < 2$ then $\int_{|x| < t} |x|^r dF(x) \sim \dfrac{\gamma}{r - \gamma} t^r q(t)$. Deduce that $E|X|^r < \infty$ for $r < \gamma$ and $E|X|^r = \infty$ for $r > \gamma$ when $\gamma < 2$.

(b) *Centering constants.* If $0 < \gamma < 1$ we can take $a_n = 0$. If $1 < \gamma < 2$, we can take $a_n = EX$: Use (a).

(c) *Scale factors.* All suitable scale factors b_n are of the form $b_n = n^{1/\gamma}h(n)$: Use $|f^n(u/b_n)| = e^{-c|u|/\gamma}(1 + o(1))$, replace n by nk then $1/b_{nk}$ by $(b_n/b_{nk})/b_n$, note that $o(1) \to 0$ uniformly in every given finite interval, show that if the sequence (b_n/b_{nk}) is not bounded then $e^{-ck} = 1$—impossible, and finally $b_{nk}/b_n \to k^{1/\gamma}$.

6. Standard domains of attraction. We say that $\mathfrak{L}(X)$ belongs to the *standard* domain of attraction of a nondegenerate stable $\mathfrak{L}_\gamma$ if $b_n = bn^{1/\gamma} > 0$ are suitable scale factors. (The usual but confusing term is "normal" not "standard.")

$\mathfrak{L}(X)$ belongs to the standard domain of attraction of a nondegenerate stable $\mathfrak{L}_\gamma$ with $0 < \gamma < 2$, if and only if, as $x \to \infty$, $x^\gamma(1 - F(x)) \to b^\gamma cp$ and $x^\gamma F(-x) \to b^\gamma cq$, $c > 0$, $p, q \ge 0$.

$\mathfrak{L}(X)$ belongs to the standard domain of attraction of $\mathfrak{N}(0,1)$ if and only if $EX^2 < \infty$, and then $b_n = \sigma n^{1/2}$ with $\sigma = \sigma X$.

7. Estimates for $E|S_n|$. Let $\mathfrak{L}(X)$ with $EX = 0$ belong to the standard domain of attraction of $\mathfrak{L}_\gamma(Y)$ with $1 < \gamma < 2$.

(a) $\mathfrak{L}(S_n/n^{1/\gamma}) \to \mathfrak{L}_\gamma(Y)$, $F(-x) \le cx^{-\gamma}$ and $1 - F(x) \le cx^{-\gamma}$ for some constant $c > 0$.

(b) There is a positive a independent of n such that for $x \ge x_0$ independent of n, $P(|S_n|/n^{1/\gamma} > x) \le a/x^2$.

(c) For $0 \le r < \gamma$ there is a positive $b = b(r)$ independent of n such that $E(|S_n/n^{1/\gamma}|^r) \le b$.

(d) $E(S_n/n^{1/\gamma}) \to EY$ and $E(|S_n/n^{1/\gamma}|^r \to E|Y|^r$ for $0 \leq r < \gamma$.

8. *Partial attraction.* $\mathcal{L}(X)$ is said to belong to the domain of *partial attraction* of a nondegenerate $\mathcal{L}(Y)$ if there is a subsequence (k_n) of integers such that, for suitable $b_n > 0$ and a_n, $\mathcal{L}(S_{k_n}/b_n - a_n) \to \mathcal{L}(Y)$. It is a property of types of laws. Discuss the propositions below in whichever order is preferred.

(a) Every $\mathcal{L}(X)$ belongs to the domain of partial attraction of either no type or of one type or of an uncountable family of types.

(b) If $\mathcal{L}(X)$ belongs to the domain of partial attraction of only one type, then this type is stable.

(c) A symmetric distribution with slowly varying two-sided tail belongs to no domain of partial attraction.

(d) If f belongs to the domain of attraction of an i.d. e^{ψ} so does the i.d. e^{f-1}. An i.d. law need not belong to its own domain of partial attraction: Use the first statement and (c).

(e) If $\mathcal{L}(X)$ is partially attracted by $\mathcal{L}(Y)$ which is partially attracted by $\mathcal{L}(Z)$ then $\mathcal{L}(X)$ is partially attracted by $\mathcal{L}(Z)$. The domain of partial attraction of a stable law is strictly larger than its domain of attraction.

(f) Let i.d. $f_n = e^{\psi_n}$ have bounded ψ_n. Set $\phi(u) = \sum_{n=1}^{\infty} \psi_n(b_n u)/k_n$. There are $b_n > 0$ and integers $k_n \to \infty$ such that $k_n\phi(u/b_n) - \psi_n(u) \to 0$, $u \in R$.

(g) If f is partially attracted by i.d. $e^{\psi_n} \to e^{\psi}$ then it is partially attracted by e^{ψ}. Is i.d. property of the e^{ψ_n}, e^{ψ} needed?

(h) Every i.d. $f = e^{\psi}$ has a nonempty partial domain of attraction: Note that there are compound Poisson $e^{\psi_n} \to f$, and use $\lim_n e^{k_n\phi(u/a_n)} = \lim e^{\psi_n(u)} = f$.

(i) Lévy example: $f = e^{\psi}$ with $\psi(u) = 2\sum_{k=-\infty}^{\infty} 2^{-k}(\cos 2^k u - 1)$ is i.d. Find its Lévy function. Show that $f^{2^n}(u) = f(2^n u)$; f is not stable but partially attracts itself.

(j) Every sequence of i.d. laws has an i.d. law belonging to the domain of partial attraction of each of its terms.

(k) *Doblin universal laws.* There are i.d. laws belonging to the domain of partial attraction of every i.d. law. Consider the countably many i.d. laws—ordered into a sequence $e^{\psi_1}, e^{\psi_2}, \cdots$, whose Lévy functions are purely discontinuous with only rational discontinuities and only rational jumps, every i.d. e^{ψ} is limit of a subsequence of (e^{ψ_n}), and use (j).

9. Consider random walk on lattices with, to simplify, span 1.

(a) Such a random walk forms a constant Markov chain with a countable number of states. What is its transition matrix?

(b) Interpret the concepts and results in the Introductory Part III in terms of those in §26.

(c) Discuss the Introductory Part CDIII in terms of §26 and complete it.

10. (a) A truly two-dimensional random walk with zero expectations and finite variances is recurrent.

(b) A truly three-dimensional random walk is always transient. What about m-dimensional random walks with $m > 3$?

For (a) and (b) use ch.f.'s analogously to the one-dimensional case in 26.2.

11. ES_r. Let $EX = 0$ and $\sigma^2 = \sigma^2 X (\leqq \infty)$.

(a) $E|S_n|/n^{1/2} \geqq a$ for some constant a and all n. In fact, $E|S_n|/n^{1/2} = 2ES_n^+/n^{1/2} \to \sigma\sqrt{2/\pi}$.

(b) Let $A = (-\infty, 0]$ or $(0, \infty)$.

$\sigma^2 < \infty = ES_{r_A}$ and $ES_{r_{A^c}}$ are both finite, and then

$$ES_{r_A} = \frac{\sigma}{\sqrt{2\pi}} \exp\left\{\frac{1}{2} - P(S_n \in A)\right\}.$$

12. Arcsine law. (a) Complete the computations in the proof of Arcsine law in 26.1.

(b) Let $c = \sum_{n=1}^{\infty}\left(\frac{1}{2} - P(S_n > 0)\right)$ be finite. Then

$$P(\nu_n = 0) \sim e^c/\sqrt{2\pi n}, \ P(\nu_n = n) = e^{-c}\sqrt{2\pi n}$$

and the Arcsine law holds.

(c) *Andersen and Spitzer generalizations.* Let $a_n = P(S_n < 0)$.

$$(a_1 + \cdots + a_n)/n \to a \Leftrightarrow \mathcal{L}(1 - \nu_n/n) \to \mathcal{L}(Y)$$

with $\mathcal{L}(Y) = \mathcal{L}(1)$ for $a = 0$, $\mathcal{L}(Y) = \mathcal{L}(0)$ for $a = 1$ and, for $0 < a < 1$,

$$P(Y < y) = \frac{\sin\pi a}{a}\int_0^y x^{-a}(1 - x)^{a-1}dx;$$

if $(a_1 + \cdots + a_n)/n$ does not converge then $\mathcal{L}(1 - \nu_n/n)$ does not converge. (Andersen case: $a_n \to a$.) If $a = 1/2$, $\mathcal{L}(Y)$ is Arcsine law.

Use Kemperman's recurrence relation: Let $b_n(k) = E(n - \nu_n)^k$; $b_n(0) = 1$, $b_n(1) = n - (a_1 + \cdots + a_n)$, $b_0(k) = 0$ for $k = 1, 2, \cdots$. Then

$$b_n(k + 1) = nb_n(k) - \sum_{m=0}^{n-1} a_{n-m}b_m(k).$$

When $(a_1 + \cdots + a_n)/n \to a$ then $(1 - \nu_n/n) \to (Y)$ with $EY^k = (1 - a)(1 - a/2) \cdots (1 - a/k)$; apply Ch. IV,CD*10* (Spitzer).

13. Identities and limit distributions. Let $\nu_n, \nu'_n, \bar{\nu}_n, \bar{\nu}'_n$ be respectively the number of positive nonnegative, negative, nonpositive sums in $(S_0, \cdots, S_n)$. Let $\tau_n, \tau'_n, \bar{\tau}_n, \bar{\tau}'_n$ be respectively the time of occurrence of the first maximum M_n, the last maximum M_n, the first minimum $\bar{M}_n$, the last minimum $\bar{M}_n$ of $(S_0, \cdots, S_n)$.

(a) The equivalence relation $P(\nu_n = k) = P(\tau_n = k)$ remains valid if same affixes above are added simultaneously to ν and to τ; similarly for $E(e^{iuS_n}: \nu_n = k) = E^{(iuS_n}: \tau_n = k)$ and, more generally, for $E(f_n: \nu_n = k) = E(f_n: \tau_n = k)$ in 26.1. What about extreme factorizations?

(b) Which exponential identities in 26.3 and results in 26.4 remain valid or have to be modified accordingly when the same affixes are added?

14. Ranked sums (order statistics). Order the sums as follows: $S_i(\omega)$ precedes $S_j(\omega)$ if $S_i(\omega) < S_j(\omega)$ or $S_i(\omega) = S_j(\omega)$ but $i < j$. For every $k = 0, \cdots, n$, let $R_{nk}(\omega)$ be the kth from the bottom of $S_0(\omega), \cdots, S_n(\omega)$ according to this ordering. Let $\tau_{nk}(\omega)$ be the index of corresponding $S_j(\omega)$, that is, $R_{nk}(\omega) = S_j(\omega) \Leftrightarrow \tau_{nk}(\omega) = S_j(\omega)$. Note that $R_{n0} \leq \cdots \leq R_{nn}$, $R_{n0} = \overline{M}_n$ is the first minimum occurring at time $\overline{\tau}_{n0} = \tau_n$ and $R_{nn} = M_n$ is the last maximum occurring at time $\nu_{nn} = \nu'_n$.

Discuss the following Wendel identities:

$$Es^{\tau_{nk}} = Es^{\tau_{kk}} \cdot Es^{\tau_{n-k,0}},$$

$$Ee^{ivR_{nk}} = Ee^{ivM_k} \cdot Ee^{ivM_{n-k}},$$

$$E(e^{iuS_n + ivR_{nk}}) = E(e^{iuS_k + ivM_k}) \cdot E(e^{iuS_{n-k} + ivM_{n-k}}).$$

BIBLIOGRAPHY

Titles of articles the results of which are included in cited books by the same author are omitted from the bibliography. Roman numerals designate books, and arabic numbers designate articles.

The references below may pertain to more than one part or chapter of this book.

INTRODUCTORY PART: ELEMENTARY PROBABILITY THEORY

I. Bernoulli, J. *Ars conjectandi* (1713).
II. Bernstein, S. *Theory of probabilities*—in Russian (1946).
III. Borel, E. *Principes et formules classiques* (1925).
IV. Chung, K. L. *Notes on Markov chains*—mimeographed, Columbia University (1951).
V. Delteil, R. *Probabilités géometriques* (1926).
VI. Feller, W. *An Introduction to probability theory and its applications* (1950).
VII. Fréchet, M. *Méthode des fonctions arbitraires. Théorie des événements en chaine dans le cas d'un nombre fini d'états possibles* (1938).
VIII. Fréchet, M. *Les probabilités associees à un systeme d'événements compatibles et dependants* I (1940), II (1943).
IX. Gnedenko, B. *Course in theory of probabilities*—in Russian (1950).
X. Laplace. *Traité des probabilités* (1801).
XI. Loève, M. *Probability methods in physics*—mimeographed, University of California (1948).
XII. Markov. *Wahrscheinlichkeitsrechnung*—translated from Russian (1912).
XIII. Parzen, E., *Modern probability theory and its applications* (1960).
XIV. Perrin, F. *Mécanique statistique quantique* (1939).
XV. Neyman, J. *First course on probability and statistics* (1950).
XVI. Uspensky, J. *Introduction to mathematical probability*.

1. Kac, M. Random walk in the presence of absorbing barriers. *Ann. Math. Stat.* 16 (1945).
2. Kolmogorov, A. Markov chains with a countable number of possible states—in Russian. *Bull. Math. Univ. Moscow* 1 (1937).
3. Loève, M. Sur les systèmes d'événements. *Ann. Univ. Lyon* (1942).
4. Polyà, G. Sur quelques points de la théorie des probabilités. *Ann. Inst. H. Poincaré* 1 (1931).

PART ONE: NOTIONS OF MEASURE THEORY

I. Banach, S. *Théorie des opérations lineaires* (1932).
II. Bourbaki, N. *Eléments de mathématique* (1939-).

  III. Glivenko, J. *Stieltjes integral*—in Russian (1936).
  IV. Fréchet, M. *Espaces abstraits* (1928).
   V. Hahn and Rosenthal. *Set functions* (1948).
  VI. Halmos, P. *Measure theory* (1950).
 VII. Hausdorff, F. *Mengenlehre* (1914).
VIII. Lebesgue, H. *Leçons sur l'intégration et la recherche des fonctions primitives* (1928).
  IX. Riesz, F. and Sz-Nagy, B. *Leçons d'analyse fonctionnelle* (1952).
   X. Saks, S. *Theory of the integral* (1937).

CHAPTER I.   SETS, SPACES, AND MEASURES

1. Andersen, E. S. and Jessen, B. On the introduction of measures in infinite product sets. *Danske Vid. Selsk, Mat-Fys. Medd.* **22** (1946).
2. Daniell, P. J. Functions of limited variation in an infinite number of dimensions. *Ann. Math.* **21** (1919–1920).
3. Doubrosky, V. On some properties of completely additive set-functions and passage to the limit under the integral sign. *Izv. Ak. Nauk SSSR* **9** (1945).
4. Kelley, J. L. Convergence in topology. *Duke Math. J.* (1952).

CHAPTER II.   MEASURABLE FUNCTIONS AND INTEGRATION

5. Bochner, S. Integration von Funktionen, deren Werte die Elemente eines Vektorraumes sind. *Fund. Math.* **20** (1933).
6. Dunford, N. Integration in general analysis. *Trans. Amer. Math. Soc.* **37** (1942).
7. Kolmogorov, A. Untersuchungen über den Integralbegriff. *Math. Ann.* **103** (1930).
8. Nikodym, O. Sur une généralisation des integrales de Radon. *Fund. Math.* **15** (1930).
9. Robbins, H. Convergence of distributions. *Ann. Math. Stat.* **19** (1948).
10. Tulcea, I. Mesures dans les espaces produits. *Atti Accad. Naz. Lincei Rend.* **7** (1949, 1950).
11. Scheffé, H. A useful convergence theorem for probability distributions. *Ann. Math. Stat.* **18** (1947).

PART TWO: GENERAL CONCEPTS AND TOOLS OF PROBABILITY THEORY

   I. Bochner, S. *Fourierische integrale* (1932).
  II. Cramér, H. *Mathematical methods of statistics* (1946).
 III. Fréchet, M. *Généralités sur les probabilités. Eléments aléatoires* (1937).
 IV. Kolmogorov, A. *Grundbegriffe der Wahrscheinlichkeitsrechnung* (1933).
  V. Lévy, P. *Calcul des probabilités* (1925).
 VI. Lukacz, E. *Characteristic functions* (1960).

CHAPTER III.   PROBABILITY CONCEPTS

1. Bienaymé. Considérations à l'appui de la découverte de Laplace sur la loi des probabilités dans la méthode des moindres carrés. *C. R. Acad. Sci. Paris* **37** (1853).

2. Cantelli, F. P. Una teoria astratta del calcolo delle probabilitá. *Ist. Ital. Attuari* **3** (1932).

3. Lomnicki, A. Nouveaux fondements du calcul des probabilités. *Fund. Math.* **4** (1923).

4. Steinhaus, H. Les probabilités denombrables et leur rapport à la théorie de la mesure. *Fund. Math.* **4** (1923).

5. Tchebichev. Des valeurs moyennes. *J. de Math.* **12** (1867).

CHAPTER IV. DISTRIBUTION FUNCTIONS AND CHARACTERISTIC FUNCTIONS

6. Alexandrov, A. D. Additive set functions in abstract spaces. *Matem. Sbornik* **8, 9, 13** (1940, 1941, 1943).

7. Billingsley, P. Weak convergence of measures: applications in probability. *Soc. Indust. and Appl. Math.* (1971).

8. Bochner, S. Monotone Funktionen, Stieltjes Integrale, und harmonische Analyse. *Math. Ann.* **108** (1933).

9. Bray, H. E. Elementary properties of the Stieltjes Integral. *Ann. Math.* **20** (1919).

10. Dugué, D. Analyticité et convexité des fonctions caractéristiques. *Ann. Inst. H. Poincaré* **XXII** (1952).

11. Fortet, R. Calcul des moments d'une fonction de repartition à partir de sa characteristic. *Bull. Sc. Math.* **88** (1944).

12. Fréchet, M. and Shohat, J. A proof of the generalized second limit theorem in the theory of probability. *Trans. Am. Math. Soc.* **33** (1931).

13. Helly, E. Ueber lineare Funktionaloperationen. *Sitz. Nat. Kais. Akad. Wiss.* **121** (1949).

14. Kawata, T. and Udagawa, M. On infinite convolutions. *Kadai Math. Sem.* **3** (1949).

15. Le Cam. Convergence in distribution of stochastic processes. *Univ. Cal. Publ. Stat.* **2**, no 4 (1957).

16. Marcienkiewicz. Sur les fonctions indépendantes. *Fund. Math.* **31** (1939).

17. Marcienkiewicz. Sur une propriété de la loi de Gauss. *Math. Zeit.* **44** (1939).

18. Parzen, E. On uniform convergence of families of sequences of random variables. *Univ. Calif. Publ. Stat.* **2** (1954).

19. Polya, G. Remark on characteristic functions. *Proc. First Berkeley Symp. on Stat. and Prob.* (1949).

20. Prohorov, U. V. Convergence of random processes and limit theorems of probability theory in Russia. *Teoria Veroyatnostey,* **1** (1956).

21. Zygmund, A. A remark on characteristic functions. *Ann. Math. Stat.* **18** (1947).

PART THREE: INDEPENDENCE

I. Breiman, L. *Probability* (1968).

II. Chung, K. L. *A course in probability theory,* second edition (1974).

III. Cramer, H. *Random variables and probability distributions* (1937).

IV. Feller, *An introduction to probability theory and its applications,* Vol. II, second edition (1971).

V. Gnedenko, B. and Kolmogorov, A. *Limit distributions for sums of independent random variables* (1954).

VI. Kac, M. *Statistical independence in probability, analysis and number theory* (1959).

VII. Kemperman, J. H. B. *The passage problem for a stationary Markov chain* (1961).

VIII. Khintchine, A. *Asymptotische Gesetze der Wahrscheinlichkeitsrechnung* (1933).

IX. Khintchine, A. *Limit laws for sums of independent random variables*—in Russian (1938).

X. Lévy, P. *Theorie de l'addition des variables aleatoires*, second edition (1954).

XI. Petrov, V. V. *Sums of independent random variables*—in Russian (1972).

XII. Spitzer, F. *Principles of random walk*, second edition (1976).

XIII. Tucker, H. G. *A graduate course in probability* (1967).

CHAPTER V. SUMS OF INDEPENDENT RANDOM VARIABLES

1. Birnbaum, Z. and Zuckerman, H. An inequality due to Hornich. *Ann. Math. Stat.* **15** (1944).
2. Borel, E. Les probabilités dénombrables et leurs applications arithmétiques. *Rend. Circ. Mat. Palermo* **27** (1909).
3. Brunk, H. D. The strong law of large numbers. *Duke Math. J.* **15** (1948).
4. Cantelli, F. P. Su due applicazioni di un teorema di G. Boole. *Rend. Accad. Naz. Lincei* **26** (1917).
5. Chung, K. L. The strong law of large numbers. *Proc. Second Berkeley Symp. on Stat. and Prob.* (1951).
6. Daly, J. On the use of the sample range in an analogue of Student's test. *Ann. Math. Stat.* **17** (1946).
7. Feller, W. The general form of the so-called law of the iterated logarithm. *Trans. Am. Math. Soc.* **54** (1943).
8. Kolmogorov, A. Ueber die Summen durch den Zufall bestimmter unabhängiger Grössen. *Math. Ann.* **99** (1928), **102** (1929).
9. Prokhorov, U. V. On the strong law of large numbers—in Russian. *Dokl. Akad. Nauk USSR* **69** (1949).
10. Robbins, H. On the equidistribution of sums of independent random variables. *Proc. Am. Math. Soc.* **4** (1953).
11. Tukey, J. An inequality for deviations from medians. *Ann. Math. Stat.* **16** (1945).

CHAPTER VI. CENTRAL LIMIT PROBLEM

12. Bawly, G. Über einige Verallgemeinerungen der Grenzwertsätze der Wahrscheinlichkeitsrechnung. *Mat. Sbornik* **43** (1936).
13. Berry, A. The accuracy of the Gaussian approximation to the sum of independent variates. *Trans. Am. Math. Soc.* **49** (1941).
14. Doblin, W. Sur les sommes d'un grand nombre de variables aléatoires indépendants. *Bull. Sc. Math.* **63** (1939).
15. Doblin, W. Sur l'ensemble de puissances d'une loi de probabilité. *Ann. Ec. Norm. Sup.* (1947).
16. Esseen, G. Fourier Analysis of distribution functions. A mathematical study of the Laplace-Gaussian law. *Acta Math.* (1944).

17. Feller, W. Ueber den Zentralen Grenzwertsatz der Wahrscheinlichkeits-rechnung. *Math. Zeit.* **40** (1935), **42** (1937).
18. Feller, W. Ueber das Gesetz der grossen Zahlen. *Acta Univ. Szeged* **8** (1937).
19. Feller, W. On the Kolmogoroff-P. Lévy formula for infinitely divisible distribution functions. *Proc. Yougoslav Acad. Sc.* **82** (1937).
20. de Finetti, B. Sulle funzioni a incremento aleatorio. *Rend. Accad. Naz. Lincei* **10** (1929).
21. Liapounov, A. Nouvelle forme du theorème sur la limite de la probabi-lité. *Mem. Acad. St-Petersbourg* **11** (1900), **12** (1901).
22. Lindeberg, J. W. Ueber das Exponentialgesetz in der Wahrscheinlich-keitsrechnung. *Math. Zeit.* **15** (1922).
23. Loève, M. Nouvelles classes de lois limites. *C. R. Ac. Sc.* (1941), *Bull. Soc. Math.* (1945).
24. Loève, M. Ranking limit problem. *Second Berkeley Symp.* (1956).
25. Loève, M. A l'intérieur du problème limite central. *Ann. Inst. Stat. Paris* (1958).
26. Raikov, D. A. On the decomposition of Gauss and Poisson laws—in Russian. *Izv. Akad. Nauk USSR* **2** (1938).

CHAPTER VII. INDEPENDENT IDENTICALLY DISTRIBUTED SUMMANDS

27. Andersen, Sparre. On sums of symmetrically dependent random vari-ables. *Skand. Akturaetid.* **36** (1953).
28. Andersen, Sparre. On the fluctuations of sums of random variables. *Math. Scand.* **1** (1953), **2** (1954).
29. Baxter, G. An analytic approach to finite fluctuation problems in prob-ability. *J. d'Anal. Math.* **9** (1961).
30. Blackwell, D. Extension of a renewal theorem. *Pac. J. Math.* **3** (1953).
31. Chung, K. L. and Fuchs, W. H. J. On the distribution of values of sums of random variables. *Mem. Am. Math. Soc.* **6** (1951).
32. Chung, K. L. and Ornstein, D. On the recurrence of sums of random variables. *Bull. Am. Math. Soc.* **68** (1962).
33. Doblin, W. Sur l'ensemble de puissances d'une loi de probabilité. *Studia Math.* **9** (1940).
34. Gnedenko, B. V. Some theorems on the powers of distribution functions. *Uchenye Zapiski Moskow Univ. Mat.* **45** (1940).
35. Hewitt, E. and Savage, E. L. Symmetric measures on Cartesian products. *Trans. Am. Math. Soc.* (1955).
36. Katz, M. A note on the weak law of large numbers. *Ann. Math. Statist.* **39** (1968).
37. Owen, W. L. An estimate for $E|S_n|$ for variables in the domain of normal attraction of a stable law of index α, $1 < \alpha < 2$. *Ann. of Prob.* **1** (1973).
38. Pollaczec, F. Uber eine Aufgabe der Wahrscheilichkeitstheorie I, II. *Math. Zeitsch.* **32** (1930).
39. Pollaczec, F. Fonctions characteristiques de certaines répartitions dé-finies au moyen de la notion d'ordre. Applications à la theorie des at-tentes. *Comptes Rendus* **234** (1952).

40. Polyà, G.   Uber eine Aufgabe der Wahrscheinlichkeitsrechnung betreffend
    die Irrfahrt in Strassennetz.   *Math. Ann.* **84** (1921).
41. Port, S. C.   An elementary probability approach to fluctiation theorie.
    J. *Math. Anal. and Appl.* **6** (1963).
42. Ray, D.   Stable processes with an absorbing barrier.   *Trans. Am. Math.
    Soc.* **89** (1958).
43. Spitzer, F.   A combinatorial lemma and its application to probability
    theory.   *Trans. Am. Math. Soc.* **82** (1956).
44. Wendel, J. G.   Order statistics of partial sums.   *Ann. Math. Statist.* **31**
    (1960).

INDEX

Graduate Texts in Mathematics

Soft and hard cover editions are available for each volume up to Vol. 14, hard cover only from Vol. 15